The Palgrave Handbook of Critical Physical Geography

Rebecca Lave • Christine Biermann
Stuart N. Lane
Editors

The Palgrave Handbook of Critical Physical Geography

palgrave
macmillan

Editors
Rebecca Lave
Indiana University Bloomington
Bloomington, IN, USA

Christine Biermann
University of Washington
Seattle, WA, USA

Stuart N. Lane
Université de Lausanne
Gryon, Switzerland

ISBN 978-3-319-71460-8   ISBN 978-3-319-71461-5   (eBook)
https://doi.org/10.1007/978-3-319-71461-5

Library of Congress Control Number: 2017960189

Cover illustration: Getty/Image Source

Printed on acid-free paper

This Palgrave Macmillan imprint is published by the registered company Springer International Publishing AG part of Springer Nature
The registered company address is: Gewerbestrasse 11, 6330 Cham, Switzerland

Preface

One could imagine the origins of Critical Physical Geography (CPG) as a moment of forehead-smacking revelation while staring at an obviously eco-social landscape: the Tijuana Estuary, bifurcated by a particularly formidable segment of the US/Mexico border wall, or an expanse of sugarcane being grown to produce ethanol on former rainforest land in the Brazilian Amazon. But in fact the origins of CPG lie in a windowless conference room in Milwaukee, Wisconsin, where, in 2010, Mona Domosh gave it to Rebecca Lave as a belated birthday present at a conference panel on Geography and Science and Technology Studies (co-organized with Matthew Wilson). During a great discussion among the panelists and audience, someone asked if there were any physical geographers who engaged with science and technology studies (STS). Mona Domosh replied from the audience, "You mean a critical physical geographer?" And then she and several other people pointed at Rebecca. It was one of those rare moments in life when something big happens and you recognize it at the time: instead of being the odd duck (even within the remarkably expansive field of geography) who was trying to cobble together fluvial geomorphology, political ecology, and STS, Rebecca was a *critical physical geographer*. Amazing, the power of a name!

Talking it over after the session, the name was immensely appealing but also a bit provocative. Whilst some loved the implicit call for a more Critical Physical Geography, and a more physical Critical Human Geography, others were concerned that it would appear as either an invitation to critique Physical Geography, or to turn Critical Human Geography positivist. In the end, CPG stuck because despite the potential for misunderstanding, it had the strength of immediately raising questions about the kinds of research, pedagogy, and

v

political practices needed in order to actually address our profoundly eco-social world.

The next step, for Rebecca, was to build the field, and to encourage others to begin the kind of integrative research that would fall under the CPG umbrella. Following in the footsteps of generations of scholars who have made important contributions to integrative thinking (including Doreen Massey, Bruce Rhoads, Keith Richards, Louise Bracken, Elizabeth Oughton, and many others), Rebecca wrote a call for papers that spanned Physical and Critical Human Geography. This was revelatory, as in starting to write what was meant to be a clarion call to integrated research, a large number of people seemed to be *already* doing work that Rebecca recognized as CPG. As the list of "but see" citations grew from an initial handful to more than twenty, it finally dawned on her that what was needed was not the construction of CPG from scratch, but the introduction all the various people who were already doing such research to each other, and then the announcement of the field as already arrived. A series of conference sessions, journal articles, special issues, and workshops followed, bringing visibility to an already existing and rapidly growing body of work that combines, in our collective initial definition, critical attention to relations of social power with deep knowledge of biophysical science in the service of social and environmental transformation. This Handbook was initiated through Rebecca's commitment to grow the field.

One of those people already attempting to do CPG research was Christine Biermann, who was at the time a graduate student struggling to build a framework within which to combine training in biogeography and dendrochronology with an interest in the politics of science and knowledge production. Christine joined the CPG intellectual project at the 2012 Association of American Geographers conference, where a series of panels ultimately culminated in a team-authored paper reviewing existing CPG work and arguing for its practical and intellectual relevance (Lave et al., 2014). In CPG she found that it might indeed be possible to reconcile a critique of the quest for universal Truth with the practice of natural science, but the nuts and bolts of how to most effectively *do* this type of work continued to elude her. When approached by Rebecca about working together on this CPG Handbook, she accepted, hoping that it would provide an opportunity to think through her lingering questions.

It was to a CPG workshop linked to the 2015 AAG that Rebecca invited Stuart Lane to speak. Bemused by both the notion that Physical Geography needed to be more critical (when critique is the hallmark for him of being a scientist) and that someone thought that he might have something to say about CPG, he accepted the invitation. Uninspired by some of the many attempts to "cross the divide" that have circulated over the last couple of decades, he was intrigued to find in CPG a wealth of creative scholarship founded upon both; the material interest that follows from a scientific interest in "things"; and a commitment to explanation of those things that was not constrained by their material nature, and the scientific method that typically follows. Struggling to understand the relationship between CPG and his own scientific journey, and faced with a much more fissiparous tendency for Geography in Europe, he took the bait, and agreed to support Rebecca in the production of this Handbook.

In writing this handbook, we three have come together to produce a volume that we hope introduces CPG: its epistemology, methodologies, genealogy, and core tenets. Perhaps because we are closet empiricists (defined in the broadest sense), the core of the collection is a set of papers where we seek to demonstrate the explanatory power of CPG research through examples from across the spectrum of subjects that might typically be treated by physical geographers or human geographers in isolation. However, we also bracket these papers with the first full attempt to develop some basic tenets of CPG, ones that distinguish it from other attempts to cross the divide as well as from other sub-disciplines like political ecology.

We hope that whatever your field of environmental study, you will find chapters here that cause you to re-examine your research questions, field methods, pedagogy, and political practice, as well as the deep inter-relations among them.

We would like to thank those who have helped with the publication of the Handbook, the reviewers of each chapter, our universities for supporting workshops in financial and other ways and Palgrave Macmillan for committing to and supporting this project, particularly our editor, Rachel Ballard.

Bloomington, IN   Rebecca Lave
Seattle, WA   Christine Biermann
Gryon, Switzerland   Stuart N. Lane
September 2017

Contents

Notes on Editors and Contributors

Notes on Editors

Christine Biermann is an Assistant Professor of Geography at the University of Washington, Seattle. A human-environment geographer, her work focuses on socio-ecological forest dynamics, the science and politics of biodiversity conservation, and the use of molecular technologies for species restoration and conservation.

Stuart Lane is a Professor of Geomorphology at the University of Lausanne, Switzerland, having held positions previously at the Universities of Cambridge, Leeds, and Durham in the UK. His work focuses on rivers, including flooding, sediment transport, aquatic ecology, river restoration, and river response to rapid climate change. He has a particular interest in the ways in which the practice of science might become a more democratic process.

Rebecca Lave is an Associate Professor in the Geography Department at Indiana University Bloomington. Her research focuses on the contradictory relations among markets, science, and the state in attempts to manage and restore the physical landscape, particularly fluvial systems, the neoliberalization of environmental science, and the construction of scientific expertise outside the academy.

Notes on Authors

Javier Arce-Nazario is an Associate Professor of Geography at the University of North Carolina at Chapel Hill and an Adjunct Researcher at the Institute of Interdisciplinary Research at the University of Puerto Rico at Cayey.

Xavier Arnauld de Sartre is a geographer and senior researcher (directeur de recherche) at the French National Center for Scientific Research (Centre National de la Recherche Scientifique). After having worked with local communities in the Amazon and argentine Pampas on the relationships between agriculture and environment, he currently works on environmental governance. He coordinated a research project entitled "Political ecology of ecosystem services," which aimed to understand the origins, uses, and possibilities of the ecosystem service concept.

Peter Ashmore is a Professor in the Department of Geography, University of Western Ontario in London, Canada. His research is primarily in fluvial geomorphology processes and morphodynamics of gravel bed rivers. Research on urban rivers and river restoration over the past 15 years, and co-teaching philosophical issues in Geography, led him to think outside the usual geomorphological box.

Joel Baker is a Master's student in Geography and Environmental Systems at University of Maryland-Baltimore County.

Elizabeth Barron is an environmental geographer whose work focuses on conservation, knowledge, value, and, most recently, on place. She is an Assistant Professor in the Department of Geography & Urban Planning and the Environmental Studies Program at the University of Wisconsin Oshkosh, and currently serves as the Associate Director for the newly formed Sustainability Institute for Regional Transformations at UWO.

Wiebke Bebermeier joined the Institute of Geographical Sciences at the Freie Universität Berlin in 2008 as a postdoctoral fellow; she has been a Junior Professor for Physical Geography with a concentration on landscape archaeology since 2011. Her research interests include (pre)historic human-environmental interactions, present and ancient watershed management, and landscape archaeology.

Dawn Biehler is an Associate Professor of Geography and Environmental Studies at the University of Maryland, Baltimore County. She is author of *Pests in the City: Flies, Bedbugs, Cockroaches, and Rats* (University of Washington, 2013).

Yinka Bode-George holds a Master's degree from the Bloomberg School of Public Health at Johns Hopkins University.

Danielle Bodner holds a Master's degree from the Department of Environmental Science and Technology at the University of Maryland, College Park.

Bilal Butt is an Associate Professor in the School for Environment and Sustainability at the University of Michigan. His research is focused on African drylands and examines the interactions between pastoralists, livestock, and wildlife. His recent publications have appeared in *Journal of Applied Ecology* (2017), *Humanity: An International Journal of Human Rights, Humanitarianism, and Development* (2016), and *Human Ecology* (2015).

Monica Castro was trained in Biology at the University of Los Andes, Colombia. She has a Master's degree in Ethnoecology (National Natural History Museum, France) and a PhD in Geography (Ecole des Hautes Etudes en Sciences Sociales). From a political ecology perspective, her research focuses on international environmental policies effects on rural livelihoods.

Katherine Clifford is a PhD Candidate in Geography at the University of Colorado Boulder. Her research explores the consequences of how we study, measure, and manage environmental change, with a specific focus on how dust escapes science and regulation in the American West.

Diana K. Davis is a geographer and veterinarian and a Professor of History and Geography at the University of California, Davis. She is the author of the award-winning *Resurrecting the Granary of Rome: Environmental History and French Colonial Expansion in North Africa* (Ohio), *The Arid Lands: History, Power, Knowledge* (MIT), and many articles and book chapters.

Martin Doyle is a Professor at Duke University's Nicholas School of the Environment. His research focuses on hydrology and ecology of river systems, along with policy and finance of infrastructure and conservation projects.

Simon Dufour obtained a BSc and PhD in Geography from the University of Lyon, France. He is currently an Assistant Professor of Geography at Rennes 2 University (LETG, CNRS), where he specializes in research on fluvial landscapes, especially on spatial patterns of riparian buffers at large scales, interactions between vegetation and hydro-geomorphic processes, fluvial landscape evolution, river and floodplain management and restoration, remote sensing uses, and ecosystem services.

Chris Duvall is an Associate Professor in the Department of Geography and Environmental Studies at the University of New Mexico. His research focuses on people-plant interactions in Western Africa and the African Atlantic Diaspora. His recent publications include the books *Cannabis* (2015) and *Mariamba: African Roots of Marijuana* (2018).

Salvatore Engel-Di Mauro is an Associate Professor at the Geography Department of SUNY New Paltz. He teaches courses in physical and people-environments geography and studies and publishes on soil degradation processes, society-environment relations, urban soils and cultivation, and ideologies about soils. He is author of *Ecology, Soils, and the Left*, and he is chief editor of *Capitalism Nature Socialism*.

Nicole Gillett has a BA in Environmental Science from Colorado College and completed her MS in Geography at the University of Massachusetts Amherst working on the RiverSmart Communities project. She recently took a position with Tucson Audubon Society where she continues to work on the nexus of people and their environments.

Mara J. Goldman is an Associate Professor in the Department of Geography and a Faculty Associate in the Institute for Behavioral Sciences at the University of Colorado Boulder. She received her PhD from the University of Wisconsin-Madison and was a postdoctoral fellow at the International Livestock Research Institute (ILRI), in Nairobi. Goldman's research is situated in human-environment geography and can best be described as political ecology strongly influenced by science and technology studies with a focus on wildlife conservation, climate change, and pastoral development in East Africa.

Heather Goodman is an Associate Research Specialist and Lab Manager at Cary Institute of Ecosystem Studies.

Michel Grimaldi is Doctor in Agronomy, area Soil Science (University of Rennes 1, 1981) and was until 2016 Research Director at the Institute of Research for Development (IRD, France). He worked at the Institute of Ecology and Environmental Sciences of Paris (iEES Paris) and coordinated the work on soil ecosystem services in the AMAZ project.

Henri D. Grissino-Mayer is the James R. Cox Professor of Geography at the University of Tennessee, Knoxville, and Director of the Laboratory of Tree-Ring Science. He studies ecosystem disturbance processes and uses dendrochronology, the science of tree rings, to learn how environments have changed over time. His research concentrates on using tree-ring data to analyze the history of wildfires, the history of past climate, and the dating of historic structures and objects.

Christine Hatch is an Extension Associate Professor in Geosciences and Research-Extension Liaison at the University of Massachusetts Amherst and was Co-Lead Investigator on RiverSmart Communities project. Her research focuses on rivers and surface water-groundwater interactions. She strives to understand these systems and aims to educate others about how they function so that together we may effectively preserve and protect our most basic and precious natural resource: water.

Rebecca Jordan is Director of the Program in Science Learning in the School of Environmental and Biological Sciences Professor at Rutgers University.

Lisa Kelley is a graduate of UC Berkeley's Environmental Science, Policy, and Management PhD program and an Assistant Professor of Geography at the University of Hawaii at Manoa. Her work integrates remote sensing, political ecology, and critical agrarian studies to understand changing agrarian land uses and livelihoods in Southeast Asia.

Leonora King is a PhD Candidate in Geography at the University of British Columbia, where she studies surface meltwater channels and catchments on the Greenland Ice Sheet. Her primary methods are remote sensing and spatial analysis. More broadly her interests include the role of ethics and reflexivity in geoscience.

Daniel Knitter is a postdoc within the CRC1266 Scales of Transformation, where he investigates different modeling approaches to the dynamics of human-environmental relationships. Between 2013 and 2016 he worked in the Topoi Lab of Research Area A. In 2013 he obtained his doctorate in Berlin on Central Places and the Environment—Investigations of an Interdependent Relationship. His research interests include human-environmental interactions, (pre)historic landscape development under human influence, and theoretical geography.

Jan Krause joined the Institute of Geographical Sciences, Freie Universität Berlin, as a research assistant in 2003 and has served as project coordinator of Research Area A in the Excellence Cluster 264 Topoi since 2008. His research interests include GIS, paleohydrology, past and present morphodynamics, drylands, and landscape archaeology.

Christian Kull researches the political ecology of resource management issues like fire, introduced plants, peasant agriculture, and forest management in places like Madagascar, Vietnam, and around the Indian Ocean rim. Trained in the United States, he has taught in Canada, in Australia, and since 2015 at the *Institut de géographie et durabilité* in the University of Lausanne, Switzerland.

Shannon LaDeau is Associate Scientist at the Cary Institute of Ecosystem Studies.

Chris Larsen is an Associate Professor in the Department of Geography, University at Buffalo, the State University of New York. He is a biogeographer who studies the impacts of natural and anthropogenic disturbances on forest structure and composition using dendrochronological and historical ecological methods.

Justine Law is an Assistant Professor of Ecology and Environment Studies in the Hutchins School of Liberal Studies at Sonoma State University. She is a human-environment geographer with interests in natural resource governance, political ecology, terrestrial ecology, and rural livelihoods. To date, most of her research has focused on forest management in North America.

Solen Le Clec'h has a PhD in Geography from Rennes 2 University, France. She is currently a postdoctoral fellow in the Agricultural Economics and Policy (AECP) group, ETH Zürich, Switzerland. She focuses her researches on human-nature interactions and environmental management tools through the spatial and quantitative dimensions of ecosystem services supply and demand in agricultural areas.

Devin Lea is a PhD student at the University of Oregon interested in applying a Critical Physical Geography lens to flood science and policy in the United States. More broadly, his academic interests include environmental hazards, geomorphology, and political ecology.

Paul T. Leisnham is an Associate Professor in the Department of Environmental Science and Technology at the University of Maryland, College Park.

Alexander Liebman is a researcher in Political Ecology and Plant-Soil Agroecology, completing an MSc in agronomy at the University of Minnesota in December 2017. He has conducted experiments in plant decomposition and soil ecology, as well as projects exploring the politics of knowledge production in development agronomy institutions in Colombia and dynamics of historical and contemporary US agrarian social movements.

Melanie Malone is an interdisciplinary scientist whose education began with physical and metaphorical roots in soils and geology. Her research interests involve explaining how environmental issues arise from the combination of biophysical, institutional, political, and cultural dynamics. Melanie received her PhD in Earth, Environment, and Society at Portland State University in June 2017, and she currently works as an Assistant Professor at The Oregon Extension.

Greta Marchesi is a postdoctoral fellow at Dartmouth College. Her research combines environmental history, political economy, and critical soil science.

Marissa Matsler is an interdisciplinarian working at the intersection between ecological, technological, and social systems, focusing on interactions between these domains within urban infrastructures. Her research has focused on urban ecosystem services, restoration ecology, ecological sanitation, and green infrastructure. She recently received her PhD in Urban Studies from Portland State University, and is currently a postdoctoral research associate at the Cary Institute of Ecosystem Studies working on the NSF-funded Urban Resilience to Extremes Sustainability Research Network (UREx SRN).

Abigail Neely is an Assistant Professor in the Department of Geography at Dartmouth College. In her research, she seeks to explain relationships between the material world (microbes, crops, and economies) and the way people understand that world (as mitigated through institutions, culture, and experience). Her recent publications have appeared in *Health & Place* (2017), *Annals of the American Association of Geographers* (2015), and *Handbook of Political Ecology* (2015).

Johan Oszwald is an Assistant Professor of Geography at Rennes 2 University (Littoral - Environnement - Télédétection - Géomatique, Centre Nationnal de la Recherche Scientifique). His research is on the endurance capacity of tropical forest ecosystems and conservation issues, with a particular focus on the complexity of studying the interactions between the environment and biodiversity when human activities induce changes in environmental conditions at several spatial and temporal scales.

John-Henry Pitas is a PhD student in Geography and Environmental Systems at University of Maryland, Baltimore County.

Emily Reisman is a PhD candidate in Environmental Studies at the University of California, Santa Cruz, where she applies more-than-human geographic theory to the political ecology of agri-food systems. Her current work examines the deep roots of divergent almond production paradigms in California and Spain.

David Robertson is an Associate Professor in the Department of Geography at the State University of New York College at Geneseo. He is a cultural and historical geographer whose research focuses on landscape change, place, and identity.

Morgan Robertson is an Associate Professor in the Department of Geography at the University of Wisconsin-Madison. His research focuses on resource governance and market-based environmental policy, with a special focus on ecosystem and habitat credit markets.

Megan Saunders is a PhD student in Ecosystem Health and Natural Resources Management at the University of Maryland, College Park.

Nathan Sayre is a Professor and Chair of Geography at the University of California, Berkeley. He specializes in the history and politics of rangeland conservation and management. His books include *Working Wilderness: The Malpai Borderlands Group and the Future of the Western Range*; *Ranching, Endangered Species, and Urbanization in the Southwest: Species of Capital*; and *The Politics of Scale: A History of Rangeland Science*.

Brigitta Schütt is a Professor since 2002, at the Institute of Geographical Sciences, Freie Universität Berlin, and since 2010 Vice President for Research at the Freie Universität Berlin. Her research interests include past and present soil erosion, Late Quaternary paleoenvironments, paleohydrology, past and present morphodynamics, drylands, and watershed management.

Gregory Simon is an Associate Professor in the Department of Geography and Environmental Sciences at the University of Colorado Denver. He has recently held research appointments at Stanford University, UCLA, and the University of Colorado Boulder. Among other positions, he has been Chair of the Cultural and Political Ecology (CAPE) Specialty Group of the Association of American Geographers and a Core Advisor to the United Nations Foundation.

Jai Singh is an eco-hydrologist with cbec eco engineering, a water resources engineering and environmental restoration consulting firm based in Northern California. He works on a broad range of process-based ecosystem restoration projects for rivers, wetlands, and estuaries.

Noah Slovin has a BA in Science of Earth Systems from Cornell University and completed his MS in Geosciences at the University of Massachusetts Amherst working on the RiverSmart Communities project. Noah is now an Environmental Scientist at Milone & MacBroom, Inc., where he develops hazard mitigation and coastal resilience planning documents.

Amanda E. Sorensen is a postdoctoral researcher at the University of Nebraska-Lincoln School of Natural Resources.

Marc Tadaki is a PhD candidate in the Department of Geography at the University of British Columbia. His PhD research examines the interrelations between freshwater science, policy, and political economy in New Zealand. His other interests include environmental valuation, the politics of scientific knowledge, and disciplinary debates in Geography.

Steve Tulowiecki is an Assistant Professor in the Department of Geography at the State University of New York College at Geneseo. As a GIScientist and biogeographer, he studies forested ecosystems, with a focus on the application of quantitative modeling for understanding forest conditions prior to European settlement in the Northeastern United States. Dr. Tulowiecki's research examines the factors that shaped past geographic distributions of tree species, as well as methodological issues surrounding this area of inquiry.

Matthew Turner is a Professor of Geography at the University of Wisconsin-Madison. His work concerns social and environmental change in the Sahelian region of West Africa with particular emphases on the political ecology of environmental governance and the critical analysis of environmental scientific practice.

Michael Urban is an Associate Professor and Chair of Geography at the University of Missouri, Columbia, whose research specialty is geomorphology, water resources, and environmental management. The focus of much of his work has been on how river systems and water resources have changed in response to patterns of climate, human behavior, and human impacts on the environment over the past century.

Eve Vogel is an Associate Professor of Political and Environmental Geography at the University of Massachusetts Amherst and was Co-Lead Investigator on RiverSmart Communities project. Her research investigates the human-environmental dynamics and histories of rivers. She focuses in particular on river governance institutions and policy and their interaction with wide ecological and social processes and needs.

Sacoby Wilson is an Associate Professor at the Maryland Institute for Applied Environmental Health at the University of Maryland.

List of Figures

List of Tables

Part I

Introduction

1

Introducing Critical Physical Geography

Rebecca Lave, Christine Biermann, and Stuart N. Lane

Critical Physical Geography (CPG) is an emerging body of work that brings together social and natural science in the service of eco-social transformation, combining attention to power relations and their material impacts with deep knowledge of particular biophysical systems (Lave et al. 2014). By studying material landscapes, social dynamics, and knowledge politics together, CPG answers the periodic calls for integrating geographic research (e.g. Thornes 1981; Goudie 1986; Massey 1999; Clifford 2002; Harrison et al. 2004, 2006, 2008 special issue of *Geoforum;* Bracken and Oughton's 2009a special issue of *Area*). This mission is particularly timely given the explosion of interest in 'the Anthropocene' (Fig. 1.1) and the widespread understanding that the material world is now shaped by deeply intermingled social and biophysical processes. If the biophysical world that surrounds us is now an eco-social hybrid, our research must be, too.

Yet CPG differs in significant ways from other calls for integration in light of the Anthropocene, challenging a dominant discourse that reduces eco-social relations to the unidirectional influence of humans on the environment

R. Lave (✉)
Department of Geography, Indiana University, Bloomington, IN, USA

C. Biermann
Department of Geography, University of Washington, Seattle, WA, USA

S. N. Lane
Institute of Earth Surface Dynamics, Université de Lausanne,
Lausanne, Switzerland

© The Author(s) 2018
R. Lave et al. (eds.), *The Palgrave Handbook of Critical Physical Geography,*
https://doi.org/10.1007/978-3-319-71461-5_1

3

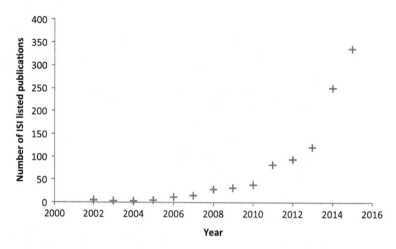

Fig. 1.1 ISI listed papers using the term 'Anthropocene'. No reference was made in the ISI to the term before 2002

and often precludes a deeper understanding of complex power relations that shape and are shaped by the biophysical world. The current conversation leaves a number of fundamental issues at the margins, including those actions that the Anthropocene is being used to legitimate, the presumptions that underpin environmental science and decision-making (e.g. a preoccupation with GDP as a goal), and the diverse suite of eco-social relations that comprise the Anthropocene. Methodologically, research on the Anthropocene has tended toward global-scale modeling and highly simplified understandings of human actions, failing to consider the material realities of day-to-day life that might give rise to very different definitions of what is important in the Anthropocene. Aspiring to a richer and more open consideration of the Anthropocene, CPG not only rethinks and breaks down the divides between conventional disciplines but also engages with fundamental questions about the conditions within which we find ourselves as a society and the role of scientific inquiry in shaping those conditions.

In this Handbook, we advocate and demonstrate careful integrative work that addresses crucial geoscientific questions while taking seriously the power relations, economic systems, and socio-cultural and philosophical presumptions upon which modern society has been built. This body of work showcases what Castree (2014, p. 244) calls 'engaged analysis', where researchers 'get their hands dirty in the places … scientists operate' while simultaneously 'questioning scientific representations of the world' and recognizing that scientific knowledge profoundly affects the systems it purports to know. We term this emerging field 'Critical Physical Geography', pointing to the integration of insights, methods, and theories from both critical Human Geography and Physical Geography.

While CPG includes a wide range of environmental topics, research methods, and epistemological commitments, it is centered on three core intellectual tenets. First, most landscapes are now deeply shaped by human actions and structural inequalities around race, gender, and class. These power relations are not social drivers, external to nature and shaping it from the outside. Rather, structural power relations incorporate and draw on the materiality of nature, creating inextricably eco-social systems. Thus, it no longer makes sense (if it ever did) to concentrate natural science research on pristine systems or to separate research on the environment into the natural sciences and the social sciences (Urban, this volume). Second, the same power relations that shape the landscapes we study also shape who studies them and how we study them. Both natural and social science are inextricably imbricated in social, cultural, and political-economic relations that affect the questions we ask (or ignore), the way we conduct our research, and even our findings (King and Tadaki, this volume). Finally, the knowledge we produce has deep impacts on the people and landscapes we study. The myth of the ivory tower is just that: a myth. Our research has unavoidably political consequences; our choice is thus not between being political or apolitical but among different possible political commitments (Law, this volume).

Taking these three core tenets seriously requires us to ask different questions or to add layers to the questions we already ask. For example, while a soil scientist might start and end their study of lead concentrations in urban soils in Oakland, California, with measurements of soil chemistry and spatial analysis, a critical physical geographer of soils, such as Nathan McClintock (2015), would add additional layers of inquiry (Table 1.1).

Table 1.1 Questions raised by a CPG approach to soil science

- What are the concentrations of Pb in soils across Oakland, CA?
- How do political-economic factors, past and present, shape the uneven spatial distribution of Pb?
- What impacts do they have on human health and well-being?
- How are studies of urban soils shaped by particular intellectual commitments of soil scientists (e.g. soil classification systems with little capacity to engage the range of human impacts)?
- How do soil scientists' aversion to engaging issues of social and environmental justice reinforce existing inequalities in Oakland?
- How is past and current research on soil contamination being taken up in the political debate, and how does that research thus in turn shape Oakland's landscape?

Table 1.2 Questions raised by a CPG approach to desertification

- What arguments are mustered in support of the desertification hypothesis in francophone North Africa, and how have those arguments persisted or changed over time?
- What political-economic interests are at stake in these debates (e.g. colonial and state attempts to control resources and nomadic populations)?
- How do archival sources, including travellers' accounts, support or disprove desertification in North Africa?
- What physical evidence is there for or against desertification from pollen analysis, climate data, and so on?
- How have these historical and biophysical data been shaped by social, cultural, and political-economic priorities?
- What are the material impacts of anti-desertification environmental policies on the people and landscapes of francophone North Africa?

Similarly, while a sociologist might begin and end a study of desertification with analysis of the rhetoric used in environmental policy debates, a critical physical geographer, such as Diana Davis (2007), would move from discourse into a range of material concerns (Table 1.2).

Tables 1.1 and 1.2 are just two examples. We could chart a similarly expanded set of questions for any of the chapters in this Handbook and for the existing body of CPG research (e.g. Wilcock et. al. 2013; Engel-Di Mauro 2014; Lave and Lutz 2014; Barron et al. 2015; Doyle et al. 2015; Hatvany et al. 2015; Sayre 2015; Van Dyke 2015; Blue and Brierley 2016; Cullum et al. 2016; Penny et al. 2016; Simon 2016; Ashmore and Dodson 2017; Holifield and Day 2017; Lane 2017; Laris et al. 2017; Sarmiento et al. 2017; Zimmerer et al. 2017). The point is that CPG *allows us to investigate material landscapes, social dynamics, and knowledge politics together, as they co-constitute each other.* CPG is thus an intellectually and politically robust response to the implications of 'the Anthropocene'.

We hope that the examples above begin to shed light on the name 'CPG'. For physical geographers, we argue, a more Critical Physical Geography means paying attention to: (1) how knowledge is constructed in Physical Geography, through the myriad ways in which we frame what it is we wish to research and how we actually go about researching it and (2) the historical origins of the particular ways we have come to conceptualize the subject of physical geographical enquiry (see Sherman 1996). We use the word 'Critical' not to claim that physical geographers are inherently uncritical but to argue that Physical Geography might benefit from a parallel version of the transition Human Geography went through in the 1970s, highlighting both a more reflexive attention to knowledge production and a consideration of the social inequalities and power relations that are implicitly bound up with what we study and which may be invoked inadvertently when such relations are overlooked. Similarly, our insertion of the word 'Physical' into Critical Geography is an argument that critical human geographers need to engage far more deeply with natural science. The social and environmental injustices on which critical human geographers focus are profoundly material, and we cannot understand their co-constitutive relations without studying biophysical and social processes together.

Barriers to Interdisciplinary[1] Research

Is interdisciplinary research actually a good idea in practice? Why would we go through the extra effort needed to conduct integrative research rather than staying within the comforting confines of a particular field? 'Interdisciplinarity' now seems to be considered an obvious good in much of the academic world. There have been dozens of articles and books advocating integrative research (e.g. Wear 1999; Ramadier 2004; Bracken and Oughton 2009b; Hall et al. 2012; Barry and Born 2013), but the continued advocacy of the need to be interdisciplinary suggests that response remains slow.

It is easy to hypothesize why calls for integrated geographical research might go unheeded, as there are formidable barriers to such work. Sometimes the barriers are physical: in many European universities, physical and human geographers are increasingly based in different administrative units and sometimes even housed in separate buildings, preventing the casual interactions and intellectual familiarity on which collaborations are often built. For other disciplines this physical separation is even more pronounced: Anthropology and Chemistry rarely share a building, much less a department.

There are also logistical barriers. It has until recently been quite difficult to get funding for integrative research, with a tendency for such projects to be supported through programs directed to applied, pre-defined questions rather than more open-ended research. In many countries, there are separate grant agencies for natural science and social science, making it impossible to fund integrated research. Even when the same agency funds a wide range of research, finding reviewers qualified to review interdisciplinary proposals can be challenging. Similarly, the vast majority of journals publish either natural or social science but not both; journals that publish across the divide struggle, like funding agencies, to find qualified reviewers. There is some hope for substantive change on this front, however, as the rise of the Anthropocene concept and the increasing insistence that research demonstrate practical impact have catalyzed integrative funding calls and journals.

Another barrier is a lack of cross-training that renders even basic research methods unfamiliar across the physical-social divide. Most natural science programs do not require cross-training in the social sciences and vice versa. While both Physical and Human Geography courses used to be a staple of graduate programs in Geography, many departments have reduced or even eliminated these requirements, diminishing our ability to understand the importance of our colleagues' research questions and the strengths and weaknesses of their methods. This mutual ignorance inhibits collaboration, as it prevents us from evaluating the rigor and understanding the intellectual value of our colleagues' research, surely both prerequisites for working together.

Mutual disrespect is also a formidable barrier to integrated research. Spurred in part by the lack of cross-training mentioned above, natural scientists and social scientists are sometimes quietly dismissive of each others' approaches, in other cases openly hostile. For example, one of us received an accidentally forwarded mass email to river scientists praising her work that began, 'I know social scientists are navel-gazing idiots, but this woman has something to say that you actually want to hear!' Similar disrespect flows from those social scientists who view natural scientists as 'naïve positivists'. This mutual disregard is a very serious obstacle in the way of interdisciplinary collaboration. It is perhaps most commonly seen when the word 'jargon' is assigned to a particular person or approach. Labeling someone's work 'jargon' is as much an opportunity missed to learn something new as it is a failure to agree to a common terminology.

A final barrier is the potential career risk from pursuing an unconventional research program (Lane 2017). In many academic fields, the boundaries of acceptable inquiry are far more narrowly drawn than in Geography. Entrenched power structures protect disciplinary norms as to what constitutes appropriate publication outlets, research questions, and even course topics. Within such fields, taking up an integrative research program is highly risky, particularly for graduate students and those without stable, tenured employment. Even within Geography, CPG approaches pose some risk. Physical geographers have put considerable effort into establishing their field as a serious natural science (Thrift 2002); embracing social science, a less authoritative form of knowledge, risks loss of perceived status. Critical human geographers who embrace natural science risk ejection from their field, which defines itself in opposition to realist research approaches. One impetus for the development of CPG as a field is to provide institutional shelter from at least some of these risks. Ironically, doing so will need to invoke some of the same processes of boundary creation and maintenance that make CPG research risky in the first place.

Doing CPG Research: Structure and Methods

The barriers we have just outlined are substantial, but they are not impassable. The growing body of CPG research demonstrates the feasibility and intellectual strength of integrative research on the environment. What enables CPG to transcend the obstacles just described?

One aspect is simply the flipside of barriers described above: mutual respect (for interdisciplinary teams) and sufficient cross-training either to carry out a project solo or to function smoothly as a team. Equally important is a set of research questions that *requires* both biophysical and social analysis to answer. Without integrated questions, it is very easy to slip into a multi-disciplinary framework in which results from different parts of a study are simply juxtaposed at the end or in which ties between the different parts disintegrate altogether rather than informing each other in any way. This points to another central characteristic of CPG work: iterative analysis, in which researchers work back and forth between their biophysical and social findings, modifying their research plans in one area in response to new data or questions in another. In this sense, CPG reflects a call for science to return to being more scientific, through the ways in which the empirical (in the broadest sense) can be allowed to 'speak back', to sow seeds of doubt about what it is we think we know and slowly engender new questions about the world around us (Stengers 2013). Finally, collaborative writing up and presentation of results deepen integration as researchers hone their findings. The chapters in this Handbook present many variations on these key qualities of successful CPG research.

While there is clearly a shared structure, integrated and iterative, to research that sails under the CPG flag, there is no standard suite of research methods. Because CPG researchers address contingent problems across a broad range of environmental topics, they have to be able to choose methods best suited to the problem at hand. But while CPG cannot be delimited by a pre-defined methodological toolkit, it can be characterized by an emphatically mixed-methods approach. Figure 1.2 presents a heuristic for thinking about CPG research methodology. There are two distinctions at work in this figure: natural versus social science and quantitative versus qualitative research; it is important not to conflate them. While it is easy to assume that there is a one-to-one match between natural science and quantitative methods and social science and qualitative methods, actual research practices are far more varied. There is a long and distinguished tradition of descriptive, qualitative natural science research, such as Charles Darwin's *The Origin of Species*, which continues today as an important complement to quantitative research in practices of classification, analysis of aerial photographs, and so on. Similarly, there is a

	Quantitative Methods	Qualitative Methods
Natural Science	frequency/magnitude curves geospatial analysis hydraulic modeling soil chemistry	descriptions of species and ecosystems soil classification aerial photograph analysis
Social Science	surveys social network analysis Q-method econometrics	ethnography/participant observation interviews document analysis archival

Fig. 1.2   Methods four-square

long and distinguished tradition of quantitative research across the social sciences in surveys and econometric approaches, among many others. What distinguishes CPG research methodologically is thus not use of a particular suite of methods but a reach across traditional ideas of what are admissible methods, whether in the natural or social sciences, and often a reach across at least three of the four squares in Fig. 1.2 (e.g. see Fig. 1.3). This vastly increases the explanatory power of CPG research by allowing triangulation among many different data sources and forms of analysis. It is worth noting that while triangulation may increase explanatory power, a mixed-methods approach can also yield contradictory data; such contradictions are important results themselves, particularly given CPG's explicit recognition that research findings are inextricably imbricated in social, cultural, and political-economic relations.

Epistemology

As with topics and methodological toolkits, there is no single epistemological position that defines CPG research. Figure 1.4 lays out the range of epistemological positions and the ways in which scholars along that spectrum adjudicate

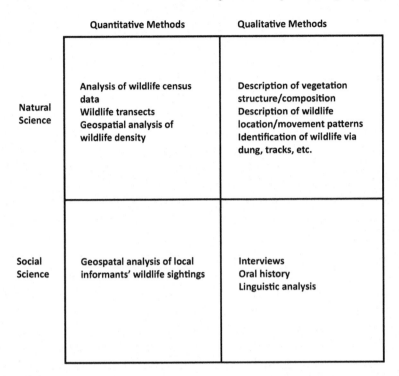

	Quantitative Methods	Qualitative Methods
Natural Science	Analysis of wildlife census data Wildlife transects Geospatial analysis of wildlife density	Description of vegetation structure/composition Description of wildlife location/movement patterns Identification of wildlife via dung, tracks, etc.
Social Science	Geospatal analysis of local informants' wildlife sightings	Interviews Oral history Linguistic analysis

Fig. 1.3   Example of CPG methods: circulating wildlife (see Goldman, this volume)

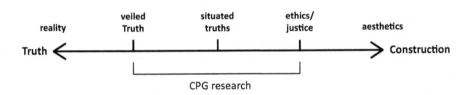

Fig. 1.4   The epistemological spectrum

among rival knowledge claims. At the far left end is capital 'T' Truth: science as the mirror of nature. At this end of the spectrum, the core epistemological assumption is that scientists have unmediated, entirely objective, and neutral access to the world. A knowledge claim is either correct or incorrect, and the test of that is entirely empirical. Moving right, we come to scholars who still argue for capital 'T' Truth but who argue that that Truth is veiled or obscured by social relations which shape the questions we ask and our understanding of the world around us. This is the classical critical realist position (e.g. Bhaskar 1975, 1979). Knowledge claims are still adjudicated with reference to material reality but with the assumption that obvious commonsense explanations are themselves objects of study, not arbiters of correctness.

The next position to the right claims no single reality. Instead of capital 'T' Truth, what we have are myriad little 't' truths that are situated in the lived experience of those who claim them. In this thinking, the lived reality of a heterosexual, white, homeless male is importantly different than the lived reality of a homosexual, dark-skinned, upper-class female. Neither of their truths is more correct; what is important is to ground truth claims in the power relations that shape them. Moving further to the right end of the spectrum, we reach strong constructivist positions in which there is still an external world, but it does not pre-exist humans: it is entirely co-produced and deeply shaped by our actions and intentions. Here a knowledge claim is not true or false but better or worse depending on its ethical implications; there are no longer correct or incorrect truth claims, even if they are only true for particular situated bodies, as in the center of the epistemological spectrum. Finally, at the far right end is capital 'C' Constructivism, which argues that there is no material reality at all, only a collectively or solipsistically constructed world to which we have no verifiable access. Here, a knowledge claim is superior to another only on aesthetic grounds.

Given CPG's core tenets, it is unsurprising that no CPG scholarship occupies either end of the spectrum. Instead, as the chapters in the body of this volume demonstrate, the field's core commitments to reflexively examining the production of knowledge, to careful analysis of the biophysical landscape, and to social and environmental justice direct scholars into the middle of the spectrum.

This may strike casual observers of natural science as strange. Would not, perhaps should not, natural science fall on the far left end of the epistemological spectrum? Put differently, can any work to the right of the arrowhead in Fig. 1.4 still be considered natural science? It is important to realize that the environmental sciences are a long way from the certainties of Newtonian physics. Most natural scientists acknowledge that what they study and how they study it have unavoidably social and political constraints in terms of priorities for research funding, institutional politics, intellectual property concerns, and a laundry list of other factors that shape scientists' day-to-day research practices. This messiness extends outward from academia into the field. Environmental scientists (natural and social) study complex, particular, deeply interconnected systems and their knowledge claims are correspondingly specific and partial. Fluvial geomorphologists, for example, are far better at explaining why particular systems behave the way they do than at generalizing their findings into rules that predict how other systems will behave (Phillips 2007). In many cases, even arriving at broadly accepted explanations can be difficult because the best available techniques are imprecise (as in sediment measurement) or because scientists do not agree about which methods are best. One notorious example among river scientists is the Water Division I court case, in which opposing teams of researchers led by two of the most respected figures in twentieth-century geomorphology (Luna Leopold and Stan Schumm) were sent out by the judge to gather basic data on the same stretch of river and came back with different numbers (Gordon 1995). There is tremendous uncertainty in environmental science and broad acknowledgment that scientists have not reached the standard of replicability or falsification expected of lab-based sciences. Thus there is far more compatibility between the epistemological positions of critical environmental social and natural scientists than is immediately apparent, and environmental science can indeed be found to the right of capital 'T' Truth on the epistemological spectrum.

Relations to Cognate Fields

Lane et al. (this volume) trace in detail the genealogy of CPG, including its relationship to the history of and debates regarding integration in Geography. Here we briefly address the cognate fields to which CPG relates, including political ecology, science and technology studies (STS), and land use/land cover (LU/LC) change research. First, CPG has deep roots in political ecology, particularly the initial formulation of the field in the 1980s. In this early work, scholars such as Piers Blaikie, Susanna Hecht, and Michael Watts brought together agrarian political economy and climate science, ecology, and pedology in a powerful and intellectually robust critique of the core claims of development practice, such as Malthusian and Tragedy of the Commons arguments (Blaikie 1985; Blaikie and Brookfield 1987; Hecht 1985; Watts 1985). Political ecology has for the most part moved away from this integrative approach, however, and even in its early days few political ecologists conducted their own natural science research. Political ecologists today typically give little attention to natural science; the landscape has become a backdrop to political research rather than an important object of analysis (Walker 2005; but see Turner 2015). By contrast, CPG's first core tenet is the importance of employing natural and social science approaches together to better understand the co-produced landscapes we inhabit today (see Urban, this volume).

Second, STS research findings have deeply influenced CPG, grounding its focus in the inextricably social character of scientific knowledge production. CPG also draws on the STS emphasis on following the sites of knowledge production as they are revealed rather than reducing research to the orchestration of pre-defined research plans. CPG's second core tenet (King and Tadaki, this volume) is a direct response to decades of STS research. Yet, the two fields differ importantly in their object of study. STS is a social science: natural science is a primary analytical object, not a central aspect of STS scholars' intellectual practice. CPG thus differs markedly from STS in its methodological emphasis on combining social and natural science research.

Third, LU/LC research has also been deeply influential on CPG research, in part by demonstrating the practical and intellectual value of integrative environmental research. However, the two fields have quite different methodological approaches. CPG embraces any research method appropriate for the topic at hand, while LU/LC's emphasis not just on explanation, but on prediction, leads to a strong preference for quantitative and spatial analysis and modeling. CPG's emphasis on the inextricably social character of scientific knowledge production is also quite different from the positivist commitments of most LU/LC research (but see Munroe et al. 2014, the authors of which have contributed to the development of CPG and are encouraging the LU/LC community to move in similar directions). LU/LC and CPG are thus distinct but complementary endeavors.

In summary, we wish to emphasize that while CPG is different from political ecology, STS, and LU/LC research, CPG research is both enriched by and very much in conversation with these fields. Our intention is to build a complementary body of research, not to replace them.

Structure of this Handbook

This Handbook is organized into three sections. The first section introduces CPG as a field. This introduction and a chapter on CPG's genealogy by the editors define the scope of CPG and explore its intellectual roots, situating it in relation to the history of integrative science in Geography. Three additional chapters then provide detailed treatments of each of the three core tenets of the field. Michael Urban explains the focus on 'crappy' rather than pristine landscapes. Leonora King and Marc Tadaki lay out the knowledge politics that shape not only the practice of science but also its findings. Section One ends with a chapter by Justine Law that explores the physical, social, and environmental justice impacts of scientific research and knowledge claims.

The second section of the Handbook makes the case for CPG research empirically by demonstrating the intellectual and political utility of CPG approaches for a range of environmental topics. This section is subdivided into five parts by topic (in the print version only): landscapes, plants, animals, soil, and water. In the first of these sections, Chris Duvall, Bilal Butt, and Abigail Neely reveal the ambiguous and sometimes troubling history of 'savanna' landscapes, and of environmental classifications more broadly, particularly in the colonial context of Africa. This is followed by Diana Davis' critique of centuries of Eurocentric views of the semi-arid and arid landscapes of the Mediterranean region as ruined, deforested, and desertified. Gregory Simon then demonstrates how the actual causes of fire in the American West, including the political economy of US housing markets, are down-played and de-politicized. And finally, Daniel Knitter, Wiebke Bebermeier, Jan Krause, and Brigitta Schütt examine the challenges of conducting integrative research in landscape archaeology.

The next set of chapters showcases CPG research on plant species. Christine Biermann and Henri Grissino-Mayer explore the potential for integrative, reflexive, and engaged scholarship in dendroclimatology. David Robertson, Chris Larsen, and Steven Tulowiecki present the results of a meta-analysis of the scientific literature on forest land-use legacies, showing that while this cognate field shares some CPG characteristics, it could benefit from stronger engagement with CPG's core tenets. Christian Kull then calls for integrating CPG into the study of invasive species to create a critical invasion science that questions the terminology, spatial and biological scale, social implications, and privileging of scientific authority characteristic of invasion science today. Simon Dufour, Xavier Arnauld de Sartre, Monica Castro, Michel Grimaldi, Solen Le Clec'h, and Johan Oszwald close this sub-section by demonstrating the perils of overly simplified mapping of ecosystem services in the Brazilian Amazon.

Animals are the focus of the subsequent set of chapters which present a range of CPG approaches to mosquitos, wildlife, and livestock. Dawn Biehler, Joel Baker, John-Henry Pitas, Yinka Bode-George, Rebecca Jordan, Amanda E. Sorensen, Sacoby Wilson, Heather Goodman, Megan Saunders, Danielle Bodner, Paul T. Leisnham, and Shannon LaDeau analyze the intellectual and political transformation of their urban ecological study of mosquitos in a mostly black neighborhood in Baltimore, Maryland. Switching continents, but paying similar attention to the tensions between local and academic knowledge claims, Mara J. Goldman analyzes wildlife conservation in Tanzania. Nathan Sayre concludes this section by revealing the capitalist and racist assumptions that underpinned the foundational principles of range science in the US West.

The next set of chapters illustrate CPG approaches to soil, from erosion and acidification to nutrient cycling and fungi. Greta Marchesi examines populist programs in Columbia in the 1920s–1940s that worked to prevent soil

erosion and degradation in small-scale coffee farming through holistic attention to soil health. Elizabeth Barron then demonstrates how fungal conservation's poor fit with traditional conservation efforts opens up space for
reconsidering how we value biodiversity more broadly. Salvatore Engel-
DiMauro reveals the linked biophysical and social relations that lead to soil
acidification in the Northern Drava basin in Hungary. Finally, Matthew
Turner analyzes the ideological roots and political implications of nutrient
budgets, a common tool for evaluating the sustainability of African
agriculture.

The final group of chapters in Section Two focuses on water. Rebecca Lave,
Martin Doyle, Morgan Robertson, and Jai Singh explore the biophysical
impacts of market-based environmental management of streams in North
Carolina. Javier Arce Nazario combines water chemistry and political ecology
to argue that water-quality regulations intended to promote environmental
justice can in fact undermine it, based on a case study of community water
systems in Puerto Rico. Peter Ashmore concludes Section Two by demonstrating that it is only possible to understand the evolution of fluvial systems
through a socio-geomorphological approach that attends both to the biophysical dynamics of rivers and to changing social priorities around flooding and
conservation.

Section Three steps back from the case studies that make up the bulk of this
volume to address the importance of pedagogy in enabling CPG research in
two chapters with graduate students as lead authors. First Nicole Gillett, Eve
Vogel, Noah Slovin, and Christine Hatch address the challenges and opportunities of CPG pedagogy during the course of a single research project: the
RiverSmart Communities project. Then Lisa Kelley, Katherine Clifford,
Emily Reisman, Devin Lea, Marissa Mattsler, Alex Liebman and Melanie
Malone explain how to successfully navigate the challenges of conducting
CPG research at different stages of graduate school, drawing on their diverse
experiences in a wide range of graduate programs. The volume closes with the
editors' critical reflections on the distinctiveness, risks, and benefits of CPG
research.

Conclusion

As we argue explicitly in Chap. 2, and implicitly throughout this Handbook,
a CPG approach enables researchers to take up the gauntlet thrown down by
the Anthropocene concept: if the world we inhabit is widely understood to be
shaped by social and biophysical processes, it is unreasonable to assume that

we as scholars can investigate either in isolation. Yet arguments for CPG both pre-date and stretch beyond debates over the Anthropocene and are inspired both philosophically by calls to undermine nature-culture dualisms and practically by the deeply co-constituted world we see at our field sites. Accepting that our biophysical systems are profoundly social (and vice versa) is not the ultimate objective of CPG but rather the starting point.

Why is Geography the field in which this critical, deeply integrated natural and social science research has emerged? One factor is clearly Geography's intellectual diversity. It is the original interdiscipline: many Geography departments span the full breadth of the university from natural science to social science to the humanities. Physical geographers regularly hear about social science research in colloquia and more casual conversation with colleagues and visitors, just as human geographers are routinely exposed to natural science research. Within many Geography departments, there is a broad methodological toolbox in use and a wide range of respected publication outlets, topical foci, and pedagogical approaches. This produces tremendous intellectual freedom: geographers can pursue a strikingly broad range of research questions while still remaining comfortably within disciplinary bounds. Another important factor is Geography's origin in place-based research. The long-standing tradition of 'muddy boots' in Geography has meant that generations of researchers delved deeply into the specificities of particular locations. This focused attention to a particular place makes eco-social relations more visible than they would be from the vantage point of the laboratory or the library, building on a tradition of research on human-environment relations that extends back to the early nineteenth century (Turner 2015). We also see within Geography a more normative take on the eco-social worlds we inhabit than in the other geosciences, a perspective which challenges the often technocratic nature of the integration imperative that has come to dominate calls for interdisciplinary problem-led science.

That said, even within Geography, CPG's deeply integrative approach can be challenging and even a serious risk, as it requires scholars to move beyond familiar intellectual comfort zones, to work across long-established disciplinary boundaries, and to seek relevance and legibility among academic communities with differing norms, expectations, and disciplinary practices (Lane 2017). It is certainly worth asking whether such research is indeed worth the effort. How does CPG advance our intellectual and political agendas?

Our advocacy of CPG is part of a broader agenda to attend more directly to the practical and political consequences of our research. A CPG approach recognizes that scholarship is unavoidably political and that the knowledge we produce has deep impacts on the people and landscapes we study. As such, we are motivated by Feyerabend's (1978) observation that there is a need to challenge the socialization and enculturation that produce natural scientists who are unable (or at least only partly able) to think freely despite being exceptionally able, through their claims to knowledge authority, to place limits on what others can think. Put differently, we cannot escape David Harvey's (1972, p. 114) question: 'who is going to control whom, in whose interest is the controlling going to be, and if control is exercised in the interest of all, who is going to take it upon themselves to define the public interest?' These questions apply not only to how we do our work but also to the eco-social relations we study. Explanation that does not combine attention to power relations and their material impacts with deep knowledge of particular biophysical systems (Lave et al. 2014) will produce knowledge that is incomplete at best, and incorrect and unjust at worst.

Notes

1. Perhaps symptomatic of the increasing number of calls for interdisciplinary research, there are a number of different terms for such work, including interdisciplinary, transdisciplinary, post-normal, triple helix, and Mode II research (Gibbons et al. 1994). Here, our starting point is interdisciplinary research, but we argue for a particular kind of interdisciplinarity, one that provides a much stronger attention to the nature of the things we study and their capacity to make us redefine how we study them. Our use of the term integrative is designed to capture the disciplinarily interwoven character of CPG inquiry.

References

Ashmore, Peter, and Belinda Dodson. 2017. Urbanizing physical geography. *The Canadian Geographer* 61 (1): 102–106.

Barron, E.S., C. Sthultz, D. Hurley, and A. Pringle. 2015. Names matter: Interdisciplinary research on taxonomy and nomenclature for ecosystem management. *Progress in Physical Geography* 39 (5): 640–660.

Barry, A., and G. Born. 2013. *Interdisciplinarity: Reconfiguration of the social and natural sciences*. London: Routledge.

Bhaskar, R. 1975. *A realist theory of science*. London: Verso.

———. 1979. *The possibility of naturalism*. Atlantic Highlands, NJ: Humanities Press.

Blaikie, P. 1985. *The political economy of soil erosion in developing countries*. New York: John Wiley & Sons Inc.

Blaikie, P., and H. Brookfield. 1987. *Land degradation and society*. London: Methuen.

Blue, B., and G. Brierley. 2016. 'But what do you measure?' Prospects for a constructive Critical Physical Geography. *Area* 48: 190–197.

Bracken, L.J., and E. Oughton, eds. 2009a. Special issue: Interdisciplinarity within and beyond geography. *Area* 41 (4): 371–481.

———. 2009b. Interdisciplinary research: Framing and reframing. *Area* 41 (4): 385–394.

Castree, N. 2014. The Anthropocene and the environmental humanities: Extending the conversation. *Environmental Humanities* 5: 233–260.

Clifford, N.J. 2002. The future of geography: When the whole is less than the sum of its parts. *Geoforum* 33 (4): 431–436.

Cullum, C., K.H. Rogers, G. Brierley, and E.T. Witkowski. 2016. Ecological classification and mapping for landscape management and science: Foundations for the description of patterns and processes. *Progress in Physical Geography* 40 (1): 38–65.

Davis, D.K. 2007. *Resurrecting the granary of Rome: Environmental history and French colonial expansion in North Africa*. Athens, OH: Ohio University Press.

Doyle, M.W., J. Singh, R. Lave, and M.M. Robertson. 2015. The morphology of streams restored for market and nonmarket purposes: Insights from a mixed natural-social science approach. *Water Resources Research* 51 (7): 5603–5622.

Engel-Di Mauro, S. 2014. *Ecology, soils, and the left: An ecosocial approach*. New York: Palgrave Macmillan.

Feyerabend, P. 1978. *Science in a free society*. London: Routledge.

Gibbons, Michael, Camille Limoges, Helga Nowotny, Simon Schwartzman, Peter Scott, and Martin Trow. 1994. *The new production of knowledge: The dynamics of science and research in contemporary societies*. London: Sage.

Gordon, N. 1995. Summary of technical testimony in the Colorado water division 1 trial. Overview; January–December 1990. Fort Collins, CO: Rocky Mountain Forest and Range Experiment Station.

Goudie, A.S. 1986. The integration of human and physical geography. *Transactions of the Institute of British Geographers* 11 (4): 454–458.

Hall, K.L., A.L. Vogel, B.A. Stipelman, D. Stokols, G. Morgan, and S. Gehlert. 2012. A four-phase model of transdisciplinary team-based research: Goals, team processes, and strategies. *Behavioral Translational Medicine* 2 (4): 415–430.

Harrison, S., D. Massey, K. Richards, F. Magilligan, N. Thrift, and B. Bender. 2004. Thinking across the divide: Perspectives on the conversations between Physical and Human Geography. *Area* 36 (4): 435–442.

Harrison, S., D. Massey, and K. Richards. 2006. Complexity and emergence (another conversation). *Area* 38 (4): 465–471.

———, eds. 2008. Special issue: Conversations across the divide. *Geoforum* 39 (2): 549–686.

Harvey, D. 1972. Revolutionary and counter revolutionary theory in Geography and the problem of ghetto formation. *Antipode* 4 (2): 110–125.

Hatvany, M., D. Cayer, and A. Parent. 2015. Interpreting salt marsh dynamics: Challenging scientific paradigms. *Annals of the Association of American Geographers* 105 (5): 1041–1060.

Hecht, S. 1985. Environment, development and politics: Capital accumulation and the livestock sector in Eastern Amazonia. *World Development* 13 (6): 663–684.

Holifield, R., and M. Day. 2017. A framework for a Critical Physical Geography of 'sacrifice zones': Physical landscapes and discursive spaces of frac sand mining in western Wisconsin. *Geoforum* 85: 269–279.

Lane, S.N. 2017. Slow science, the geographical expedition and Critical Physical Geography. *The Canadian Geographer* 61: 84–101.

Laris, P., M. Koné, S. Dadashi, and F. Dembele. 2017. The early/late fire dichotomy: Time for a reassessment of Aubréville's savanna fire experiments. *Progress in Physical Geography* 41 (1): 68–94.

Lave, R., and B. Lutz. 2014. Hydraulic fracturing: A Critical Physical Geography review. *Geography Compass* 8 (10): 739–754.

Lave, R., M.W. Wilson, E. Barron, C. Biermann, M. Carey, C. Duvall, L. Johnson, et al. 2014. Critical Physical Geography. *The Canadian Geographer* 58 (1): 1–10.

Massey, D. 1999. Space-time, 'science' and the relationship between physical geography and human geography. *Transactions of the Institute of British Geographers* NS 24 (3): 261–276.

McClintock, N. 2015. A Critical Physical Geography of urban soil contamination. *Geoforum* 65: 69–85.

Munroe, D., K. McSweeney, J.L. Olson, and B. Mansfield. 2014. Using economic geography to reinvigorate land-change science. *Geoforum* 52 (1): 12–21.

Penny, D., G. Williams, J. Gillespie, and R. Khem. 2016. 'Here be dragons': Integrating scientific data and place-based observation for environmental management. *Applied Geography* 73: 38–46.

Phillips, J.D. 2007. The perfect landscape. *Geomorphology* 84 (3–4): 159–169.

Ramadier, T. 2004. Transdisciplinarity and its challenges: The case of urban studies. *Futures* 36: 423–439.

Sarmiento, F.O., J.T. Ibarra, A. Barreau, J.C. Pizarro, R. Rozzi, J.A. González, and L.M. Frolich. 2017. Applied montology using critical biogeography in the Andes. *Annals of the American Association of Geographers* 107 (2): 416–428.

Sayre, N.F. 2015. The Coyote-Proof Pasture Experiment: How fences replaced predators and labor on US rangelands. *Progress in Physical Geography* 39 (5): 576–593.

Sherman, D. 1996. Fashion in geomorphology. In *The scientific nature of geomorphology*, ed. C.E. Thorn and B.L. Rhoads, 87–114. New York City: Wiley.

Simon, G.L. 2016. *Flame and fortune in the American West: Urban development, environmental change, and the great Oakland hills fire.* Berkeley, CA: University of California Press.

Stengers, I. 2013. *Une autre science est possible.* Paris: La Découverte.

Thornes, J.E. 1981. A paradigmatic shift in atmospheric studies? *Progress in Physical Geography* 5 (3): 429–440.

Thrift, N. 2002. The future of geography. *Geoforum* 33 (3): 291–298.

Turner, M. 2015. Political ecology II: Engagements with ecology. *Progress in Human Geography* 40 (3): 413–421.

Van Dyke, C. 2015. Boxing daze–using state-and-transition models to explore the evolution of socio-biophysical landscapes. *Progress in Physical Geography* 39 (5): 594–621.

Walker, P. 2005. Political ecology: Where is the ecology? *Progress in Human Geography* 29 (1): 73–82.

Watts, M.J. 1985. Social theory and environmental degradation: The case of Sudano-Sahelian West Africa. In *Desert development: Man and technology in sparselands*, ed. Y. Gradus, 14–32. Dordrecht: Reidel.

Wear, D.N. 1999. Challenges to interdisciplinary discourse. *Ecosystems* 2: 299–301.

Wilcock, Deirdre, G.J. Brierley, and Richard Howitt. 2013. Ethnogeomorphology. *Progress in Physical Geography* 37 (5): 573–600.

Zimmerer, K.S., H. Córdova-Aguilar, R. Mata Olmo, Y. Jiménez Olivencia, and S.J. Vanek. 2017. Mountain ecology, remoteness, and the rise of agrobiodiversity: Tracing the geographic spaces of human–environment knowledge. *Annals of the American Association of Geographers* 107 (2): 441–455.

2

Towards a Genealogy of Critical Physical Geography

Stuart N. Lane, Christine Biermann, and Rebecca Lave

Introduction

The chapters featured in this volume draw on and are informed by a variety of geographic (and non-geographic) subfields, each with their own particular histories, conversations, and tensions. Because of this diversity, then, it is difficult to develop a genealogy of Critical Physical Geography (CPG) that adequately reflects the range of influences on the field. Yet here we endeavour to do just that: to identify some of the various bits and pieces of geographic scholarship that have preceded, prefigured, and informed the work featured in this volume. Of course, our genealogy is a 'whiggish' history in that it is written from the present looking back through the past (Livingstone 1992) and with a particular goal of showing why we think CPG is an opportunity for Geography as a discipline. Following Livingstone (1992, 4–5), this means that our genealogy needs to be appreciated from three critical perspectives: (1) our view of the past is from a particular perspective, that is strongly framed through a set of tenets that we have ourselves defined; (2) in this chapter, we

S. N. Lane (✉)
Institute of Earth Surface Dynamics, Université de Lausanne,
Lausanne, Switzerland

C. Biermann
Department of Geography, University of Washington, Seattle, WA, USA

R. Lave
Department of Geography, Indiana University, Bloomington, IN, USA

© The Author(s) 2018
R. Lave et al. (eds.), *The Palgrave Handbook of Critical Physical Geography*,
https://doi.org/10.1007/978-3-319-71461-5_2

have selected material that we think helps to understand why CPG has come to be what it is; and (3) as a result of (1) and (2), the reader should be careful of how we provide a set of claims that might appear to speak for themselves as we make them and accompany them with a kind of narrative that passes from the dark ages to the golden ages of a truer, purer, and, above all, better kind of (critical) Physical or (more materialized) Human Geography. Rather, what we present is selected and interpreted, partly to justify our own means. Acknowledgement of this history is needed to understand the key tenets of CPG that we outlined in the introductory chapter and which are addressed in more depth in the four chapters that follow. In particular, we need to attend to four main themes, which provide the structure for this chapter.

First, CPG advocates integration, so we need to understand the origins of the apparent fission that justifies such advocacy, within a discipline that was, in the early part of its modern history, a strongly integrative project (Livingstone 1992). We begin with early academic Geography which we argue, following others (e.g. Castree 2011), sets up the bipolarity in Geography that frames subsequent debate regarding integration.

Second, and as others have argued (e.g. Goudie 2017), this focus on schism may overlook the ongoing presence of integration within the discipline. At least two kinds of such integration have been described, the study of human impacts on environmental processes (including recent concerns with the Anthropocene) and systems science (including its most recent manifestation as Earth System Science [ESS]). We present and critique both of these intellectual projects to show that the kind of integration CPG envisages is of a different and deeper kind than envisaged by either human impact or systems science, even if it retains a focus upon the pervasiveness of human impacts on the environment (Urban, this volume).

Third, as we outline in the Introduction, the notion of being critical is central to the kind of integration we advocate in CPG. Science, broadly defined, is a discipline that makes progress through being critical, in extremis, by seeking to demonstrate it is wrong. However, the 'Critical' that we envisage in CPG has parallels with a more specific form of being critical, one that follows from similar ideas initially introduced into Human Geography in the 1970s. Thus, we think through the history of the engagement of Physical Geography with, and in, more critical approaches to the discipline. Following from this, we argue that being critical in CPG is about being more than just analytical. It is also about being normative in ways that are commonly excluded in scientific practice. CPG is concerned with a type of Physical Geography that is about more than just describing what the world is: it is also concerned with what the world ought to be. In turn, this raises questions regarding the traditional emphasis of Physical Geography upon 'value-free' scientific investigation, meriting a more politically aware Physical Geography (King and Tadaki, this volume). It also explains the interest of CPG in a more situated or contingent science, one which recognizes the pervasive impacts of the knowledge that we produce (Law, this volume), that is more open about the relationship between science and scientist, and which profits from new thinking regarding participation in science.

Much of the chapter is written from the perspective of Physical Geography because the object of enquiry in CPG is commonly the environment, and so Physical Geography, we believe, should be profoundly concerned by some of the issues that we raise. However, we also argue that Human Geography itself has a need for 'rematerialization' and we illustrate this through a discussion of the branch of Critical Human Geography that touches the environment most directly—political ecology. Some of what are now seen as the classic texts in political ecology (e.g. Watts 1983; Blaikie 1985) began by challenging the simplistic ways in which human activities were being conceptualized as a cause of environmental risk or degradation. To strengthen this challenge, early political ecologists drew on both physical and social data, though they only rarely collected the former themselves (Walker 2005). This distinctly integrative approach fell out of fashion in the 1990s. In this sense we argue that in genealogical terms, CPG is not only about bringing back political ecology's attention to the materiality of environmental issues but also deepening that engagement very substantively to provide more balanced physical and social inquiry in the field.

Early Academic Geography: From Integration to Schism

The notion that there is a need for an academic project that is integrative, at the interface of the natural sciences and social sciences in the context of the surface of the earth, has a long history. Livingstone (1992, 177) describes how from the second half of the nineteenth century, in turning from "natural theology to evolution theory, the founders of professional Geography embarked on, … the geographical experiment—an experiment in keeping nature and culture under the one conceptual umbrella". It was Darwin who laid the foundations for biology to be incorporated into social theory, but this initial interest in Darwin itself evolved to a multitude of related but competing alternatives, most important of which in relation to the geographical experiment was a Neo-Lamarckian movement (Livingstone 1992). Following Livingstone, this movement was key because of two imperatives: (1) a focus on time (the transfer of life experience between generations) and (2) the role of the environment and environmental change in influencing organisms and, eventually, social processes. Livingstone (1992) argues that it is this emphasis upon the environment that provided the opportunity for the development of a new academic Geography, centred on bridging the gaps between the natural and human sciences that integrated society and environment (Mackinder 1887). According to Mackinder (1887, 143), Geography should be "the science whose main function is to trace the interaction of man in society and so much of his environment as varies locally", in which the varying environment, taken to be the focus of Physical Geography, is used as the building block for explanation in Political Geography. Implicit is a kind of aerial differentiation that may explain Rhoads' (1999) observation of 'space' and 'spatial patterns' as a unifying entity. Although Mackinder's account focuses on how the environment influences society, he recognized (1887, 157) that, in turn, society can shape its environment: "Man alters his environment, and the action of that environment on his posterity is changed in consequence". Livingstone (1992) notes that this crossing of the divide was not new, rather it was being crossed in evolutionary (i.e. scientific) rather than theological terms.

To attribute the idea that society and environment are implicitly indivisible to Mackinder alone would not be correct. Livingstone (1992) traces an environmental determinism to the German geographer Friedrich Ratzel as well as US geologist Nathaniel Shaler. Although the latter preferred to see Geography as the geology of the present, it was William Morris Davis, a student of Shaler, who from the end of the nineteenth century sought to develop a US form of the geographical experiment in integration (Leighly 1955) but one which was clearly geographical rather than geological (Livingstone 1992). Although Davis is more widely known within Physical Geography for his geomorphological contribution, he set the course for an early-twentieth-century US Geography focused on "social plasticity, environmental causation, and cultural inheritance" (Livingstone 1992, 212).

The key point in the above is that early in the days of academic Geography, there was the notion that its raison d'être was the service that it could provide by being integrative: "One of the greatest of all gaps lies between the natural sciences and the study of humanity. It is the duty of the geographer to build one bridge over an abyss which in the opinion of many is upsetting the equilibrium of our culture. Lop off either limb of Geography and you maim it in its noblest part" (Mackinder 1887, 145). Interestingly, this was to be a normative project, with Mackinder (1887) arguing, "The mind which has vividly grasped in their true relations the factors of the environment is likely to be fertile in the suggestion of new relations between the environment" (144). However, as Castree (2011) argues, the rhetorical appeal of this approach itself gave way to the kind of schism that is commonly bemoaned today. Castree observes that from the 1920s, and albeit with some exceptions such as in the work of Carl Sauer and in the well-known edited volume of Thomas (1956), "Man's Role in Changing the Face of the Earth", the successors to these early academic geographers retreated into distinctly physical and human approaches to Geography and often into specific sub-disciplines within those approaches (e.g. geomorphology, economic Geography). Explanations grounded in areal differentiation, whether in terms of spatial explanation and ways to map it, or area studies, became dominant over the imperative of integration (Pattison 1964). Nature-society interactions of the kind pioneered by George Perkins Marsh (1864, and others; see Urban, this volume) in the middle of the nineteenth century became confused with and constrained by evermore crude forms of environmental determinism (Pattison 1964) which, as illustrated in Barrows' (1923) human ecology, reduced Physical Geography to the description of the earth's surface necessary to explain the spatial distribution of human activities (Leighly 1955). Those works that sought to retain a less extreme vision of how the environment influenced society tended to revert to straightforward descriptions of the physical environment with only passing attention given to its impacts upon humans (Leighly 1955). In 1955, Leighly himself called for a Physical Geography that was allowed to exist with a focus on processes, the effects of the laws of nature on the earth. Thus, by the middle of the twentieth century, emerging schisms between natural science and social science, Physical Geography and Human Geography, were becoming increasingly institutionalized (Castree 2011).

It is this apparent institutionalization that merits comment as the question of whether and how to integrate Human and Physical Geography has been raised repeatedly over recent decades. On the one hand, there are those who argue that we need more 'conversations across the divide' between nature and society/culture/people (e.g. Harrison et al. 2004, 2006, 2008), to deal with geographical questions that are becoming ever more interdisciplinary. On the other hand (e.g. Goudie 2017), there are those who argue that whilst this divide may exist, there is plenty of evidence of boundary crossings, enough to guarantee the kinds of integration sought in the early origins of modern academic Geography. In the next two sections, we consider two of the kinds of integration that CPG is not: (1) a simple account of human-environment relations, as reflected most recently in some geographical approaches to the Anthropocene and (2) a restatement of the long-established idea of Geography as the integrated study of systems.

Man (Sic) as a 'Unit Process'

The relationship between humans and their environment is not simply a geographical concern (see Rasmussen and Arler 2010 for a review) even if it remains one of primary interest to the discipline (Zimmerer 2010). The idea that somehow the environment constrains human activity, albeit in a less crude form than expressed in notions of environmental determinism, remains important (Goudie 1986). Equally, physical geographers are actively involved in describing and quantifying human impacts upon the earth's surface in areas such as geomorphology (e.g. Syvitski et al. 2005; Chin 2006; Hooke 2006; Haff 2010; Lewin 2013; Harden 2014; Tarolli and Sofia 2016; Brown et al. 2017), land-use change science (e.g. Serra et al. 2008; Ellis 2011), and biogeography (e.g. Clement and Horn 2001; Dyer 2010; Francis et al. 2012). On this basis, there appears to be a rich and growing exchange between Human Geography and Physical Geography (Goudie 2017). However, others have argued that we are more in a state of outward valence than inward cohesion (Clifford 2002; see also Demeritt 2009; Whatmore 2013) where our research (and possibly also our teaching) commitment is more to other disciplines than it is to Geography (Maddrell 2010).

The dichotomy in these views is well-explained by Johnston (1986), who suggests that in advocating integration, we have confused two very different connotations of Geography: the vernacular, relating to the subjects that geographers typically address, and the academic, relating to what geographers do (such as where physical geographers study those processes that produce the physical environment). He argues that physical geographers, as academics, have argued for the integration of Human Geography into their studies, but they have tended to interpret Human Geography in a vernacular rather than an academic sense, stripping humans of the very complexity that is the focus of academic Human Geography (see also Cooke 1992 and Gregory 2000) and tending to bolt social science on to (physical) geographical research projects (Demeritt 2009). Where humans are allowed to be present they are treated as uncomplicated agents that change the rate of operation of earth surface processes (Fischer-Kowalski and Weisz 1999), in common with the ways in which early models in Human Geography reduced people to rational, economic, 'men'.

Our argument here is that, in Johnston's (1986) terms, CPG is concerned with an integration that goes beyond that which is implied when man (sic) is reduced to a unit process capable of changing some element of the physical environment (see also Rhoads 1999). Johnston (2006, 8) has recently bemoaned the fact that "nobody has ever convincingly shown what such an 'integrated Geography' would be" other than "a tendency to "bang on about its necessity-aided by the almost compulsory Venn diagrams". In Johnston's (1986) terms, we situate CPG as combining a double academic reading of particular vernacular issues, double in the sense that CPG should embrace the complexity that comes from both a human geographical and a physical geographical interpretation of the world around us.

Systems and Integration

The second often advocated means of integration is more epistemological. Notwithstanding the schisms within academic Geography present in the 1950s, the latter part of the quantitative revolution that was happening in Human Geography quite clearly and Physical Geography to a certain extent (Chorley and Kates 1969) led to a new kind of integrative Geography, based around systems. By the 1960s, arguments emerged (e.g. Ackermann 1963; Eyre 1964; Haggett 1965; Chisholm 1967; Harvey 1969; Stoddart 1965) that without some kind of dialogue between physical and human geographers, Geography would cease to exist as a discipline capable of holistic analysis. There was a call for a new generation of physical geographers willing and able to face up to the needs of the whole subject and concentrate on those parts of the physical environment that are most relevant to a human-oriented Geography (Chorley and Kates 1969). Chorley and Kates saw this arising through: (1) development of shared techniques that could be applied equally to human and physical systems (e.g. Bennett and Chorley 1978; Haggett and Chorley 1969; Woldenberg and Berry 1967) and (2) 'resource' Geography, with a methodological combination of those elements of the biophysical and biochemical world that tie natural and social sciences together (e.g. water). Systems were advocated as a means of achieving the "coming together of natural potential and of human need and aspiration" that "provides a unique focus for geographic study" (Chorley and Kates 1969, 4). As systems thinking evolved, the focus became more and more on developing a unified methodology based on systems analysis (Stoddart 1965). Nowhere was this clearer, or more extreme, than in Bennett and Chorley's *Environmental Systems: Philosophy, Analysis and Control* (1978), a book which as the title suggests saw systems not only as a unifying analytical framework but one which should become central to understanding how to make interventions in a complex world.

What followed the 1960s calls for systems thinking as a geographical approach has been well-rehearsed: many human geographers progressively abandoned it (Malanson 2014); physical geographers found it hard to handle the large-scale time-dependent coupled systems with which they were working. Even recent attempts to revisit systems analysis through complexity science (e.g. Harrison et al. 2006) have tended to focus on the description of the emergent properties of systems phenomena (the rank-size rule that describes city size being a good example) in ways that are largely descriptive. Associated simulation models (e.g. Bithell et al. 2008; Macmillan and Huang 2008) have highly simplified representations of people, even if there are 'agents' or 'processes' who are somehow sensitive to the milieu in which they are simulated (Clifford 2008). In their most extreme case, systems approaches have been accused of being more concerned with the mathematical tools that interested their proponents than they were with the feasibility of these tools in genuinely integrating humans and their environment (Kennedy 1979). That said, systems as an integrating concept has been maintained and also seen a partial resurgence more recently through two related dimensions.

The first area of resurgence is the progressive treatment of natural systems and social systems as integrated socio-ecological systems. Here, the interactions between the nature and society are conceived in terms of connections that drive physical exchanges (e.g. of materials or energy) and integrated into a metabolic system (Fischer-Kowalski and Haberl 2002). This witnessed substantial interest during the 1970s, following from growing concerns regarding human impacts on the environment. Socio-ecological systems research was instrumental in revealing cultural differences in the ways in which environmental resources subsidized human activities (e.g. with respect to agriculture, Bayliss-Smith 1982). More recently, the analysis of socio-ecological systems in terms of physical exchange has been argued to be a key means of quantifying the dependencies that exist between cultural and biophysical processes that lead to social change and so a valuable framework for linking environmental change to the human world (Fischer-Kowalski and Weisz 1999). However, socio-ecological systems analysis has been subject to a number of critiques including insufficient attention being given to capital flows (Gandy 2002), political economy (Swyngedouw and Heynen 2003), social systems (Brand 2016), and the material detail of the social processes (Demaria and Schindler 2016) that produce environmental or ecological crises.

The second area of resurgence of systems thinking is in relation to ESS, its focus on human-induced change within the Earth system (Pitman 2005), and concerns that there is a growing complexity central to contemporary environmental problems (Richards and Clifford 2008). ESS was argued to be an opportunity for Geography because of the latter's traditional interest in more holistic accounts of the world and its willingness, even if not always pursued in practice, to consider humans as important components of explanation (Pitman 2005). However, the treatment of humans in ESS remains very much simplified, and the critique of the representation of the human world directed at physical geographers (op. cit., Johnston 1986) has been also directed at ESS (Richards and Clifford 2008, 1324). More directly, ESS has been described as an approach where human geographers bring "a few extra terms to build into climate-change equations" (Johnston 2006, 9). Indeed, ESS raises questions regarding the ways in which its missionary zeal (Richards and Clifford 2008) prioritizes certain research questions to the exclusion of others.

In summary, the questions that concern CPG are different from the kinds of integration both made explicit and left implicit in systems thinking. CPG seeks to go beyond the descriptive and the analytical focus of systems analysis to a more normative position concerned with how things ought to be. In the next section, we argue that this emphasis on a more CPG comes at least in part from the ways in which some physical geographers have developed a wider interest in philosophical questions.

Physical Geography and Engagements with Philosophy

Critical Human Geography grew out of a central concern that our research involves concepts that are constructed by researchers and, as such, are at least partially contingent, reflecting both the world as it is but also how we have been conditioned and have responded to that world. As such, our concepts merit continuous examination and criticism (Billinge et al. 1983). Human geographers realized the potential of philosophical discourse for developing alternative approaches to the discipline in the 1970s and early 1980s. Physical Geography remained largely anodyne to such debates, content with the traditional scientific belief that there should be a clear separation between the researcher and the researched. However, a small minority of physical geographers have made significant contributions to thinking about the nature of Physical Geography (Thornes 1981; Richards 1990; Rhoads and Thorn 1994, 1996; Richards 1996; Rhoads 1999; Thornes and McGregor 2003). Writing about atmospheric sciences, Thornes (1981) raised the possibility of alternative approaches to meteorology and climatology that were not unnecessarily restricted by the dominant 'positivist paradigm', which "gives us no direction as to how to apply the knowledge that its methods produce" (p. 429). Stressing the relationship between science and society, Thornes (1981) emphasizes that novel approaches might allow geographers to apply their knowledge more readily and meaningfully and also provide one of the earliest post-quantitative revolution calls for a more normative Physical Geography.

There followed a series of re-engagements by physical geographers in questions regarding the philosophy of some areas of the discipline (e.g. Richards 1990; Rhoads and Thorn 1994) including a major treatise on philosophy, method, and geomorphology (Rhoads and Thorn 1996). Richards (1990) drew explicitly on ideas drawn from realist approaches to the social sciences that recognized a difference in status between 'reality' and our measures of that reality. Rhoads and Thorn (1994) urged geomorphologists to engage with philosophies of science. Such philosophical introspection, they argue, allows physical scientists to address questions about what constitutes scientific validity, how methods align with theory, and how physical geographic knowledge shapes and is shaped by social, cultural, and political conditions. It is a means of encouraging the kind of examination and criticism that is commonplace in Human Geography. In a later appeal to physical geographers to explore ontological and epistemological issues, Rhoads (1999) noted obstacles preventing the integration of philosophical inquiry into physical geographic work. Besides the general issue of lack of training in philosophy, he also suggested that philosophical work may have difficulty contributing to the often very specific cutting-edge questions that are driving particular subfields at any given point in time. Perhaps because of this difficulty, these calls for philosophical engagement in Physical Geography appear to have gone relatively unheeded, with only a few exceptions (e.g. Urban 2002; Harrison and Dunham 1998).

In parallel, and outside the discipline, research in the 1970s showed that scientific practice needs some kind of critical interrogation, illuminating science as a social practice (Latour and Woolgar 1979). Scientific knowledge is in some senses constructed (Collins and Evans 2002), even co-constructed. As Sheila Jasanoff wrote in 2004 (33), "the realities of human experience emerge as the joint achievements of scientific, technical, and social enterprise: science and society, in a word, are co-produced, each underwriting the other's existence". However, such philosophical discussions, whilst actively pursued in Human Geography, have all too often been relegated to the status of 'anti-science' by physical geographers who, even as early as the mid-1990s, were being accused of carrying "on more or less as always, largely oblivious to the more arcane debates of their colleagues in Human Geography" (Demeritt 1996, 485) (for an earlier observation in the same vein, see Thornes 1981).

This is notwithstanding two points. The first point is the implicit contradiction in the claim made by certain advocates of the scientific method that, if pursued in a certain way, science is ethically neutral and ideology free, allowing greater credence to be given to knowledge produced through a 'scientific' method. However, such a claim is itself an ideological one because of the assumption that it makes about what kind of knowledge should count and a political one because of whose knowledge is given primacy in accounts of what constitutes the world. The statement that values should be excluded from research in Physical Geography is itself a value statement and emphasizes Harvey's (1974) observation that the principles of scientific method should be seen as normative (Harvey 1974). The second point is that research has shown the relevance of seeing as socially constructed the kinds of research areas that are of interest to Physical Geography including soil science (e.g. Wynne 1992; Latour 1999), hydrology (e.g. Lane 2014), hydraulics (e.g. Bijker 2007; Wesselink et al. 2009), geomorphology (e.g. Ashmore 2015), and climate science (e.g. Darier et al. 1999; Demeritt 2001, 2006; Lahsen 2005; Sundberg 2009).

If what physical geographers do, how they do, and even what they research (Jasanoff 2004) is at least in some part socially constructed, then *what* should be integrated, *how* it is being done, and *by whom* become key questions (King and Tadaki, this volume). Thus, explanation in Physical Geography that does not combine attention to power relations and their material impacts with deep knowledge of particular biophysical systems (Lave et al. 2014) will produce knowledge that is incomplete at best, incorrect at worst. If research explanation is being sought to serve its own disciplinary ends, then this may not matter. But, in the neo-liberal academy, such as through the 'impact' agenda (see Stengers 2013), we are increasingly being asked to make contributions to socio-environmental transformation. The *what*, the *how*, and the *by whom* in these contributions become more important as they may shape directly the kind of transformation that results. After Mercer (1983), then, the reason we advocate the word critical in CPG reflects not simply a desire to negotiate balance between apparently different views of what constitutes the world. Nor is it simply to explain in some more integrated kind of way what might constitute that world. Rather, it is to advocate an interrogation of the world as we think that we see it, both biophysically and socially, such that we end up with both new ways of knowing the world and an understanding of the ways in which our own worldviews, and those of others, have made the subjective realities (Mercer 1983) that concern us (see also Thornes 1981). This is where CPG shares parallels with the advocates of a more Critical Human Geography almost five decades earlier (see Gregory 1978).

A Genealogy of a More Critical Physical Geography

We argue that there are three core tenets that follow from a 'critical turn' in Physical Geography and a 'physical turn' in Critical Human Geography, addressed consecutively in the chapters that follow. The first relates to the sites of CPG research (Urban, this volume). In emulating classical methods of scientific abstraction, we argue that Physical Geography has often (but not always—see e.g. Petts et al. 2008; Pollard et al. 2008; Wainwright 2008; Francis 2014; Ashmore and Dodson 2016) sought to abstract the natural from the social/cultural setting within which it is found, achieved through a focus on either largely pristine environments or where such environments can externalize human impacts to being some kind of forcing or driving variable (e.g. human-driven climate change impacts on ice sheet response). It is perhaps justified by the combination of an academic approach to Physical Geography with a vernacular approach to Human Geography, of the kind described by Johnston (1986) and which allows the treatment of humans as some kind of unit process. Of course, current focus on the Anthropocene points to the fact that finding pristine landscapes is going to become progressively more difficult. Excluding 'crappy landscapes' (Urban, this volume) simultaneously excludes those landscapes where the majority of people live. A more critical approach considers why certain sites have been privileged over others (e.g. why not those sites where large numbers of people live and where the acute effects of rapid environmental change are already being felt) and what this might mean for the kinds of questions that result. Who is determining this privilege and how the results of such research become used are necessary parts of a more critical investigation (Stengers 2013; Lane 2017). In advocating other sites for research, there is also a recognition that a simple treatment of the social/cultural can no longer be justified, traditional methods of scientific abstraction and generalization need to be challenged, and a focus on the more contingent processes operating in particular places becomes necessary (Rhoads et al. 1999; Phillips 2001; Clifford and Richards 2005; Richards and Clifford 2008; Görg 2007; Ashmore and Dodson 2016). This ties CPG directly into some of the more critical readings of the way the Anthropocene is being pursued within global science agendas (Castree 2015, 2016).

The second tenet (King and Tadaki, this volume), implicit to the first, recognizes the way in which our research practices are structured, at scales from the personal (what we know, who we know), through the communities within which we work (what is an admissible research question or admissible knowledge) through to broader social and cultural forces on the academy and its practices (e.g. Pain 2014; Mountz et al. 2015). Power is not distributed evenly between those able to influence our research practices, making the practice of Geography (Physical and Human) also amenable to political analysis. Such analysis shows that scientists have the agency to make choices in the way that they practise their science and these choices can be politically shaped. Thus, the second tenet is not simply concerned with analysis of the ways in which the practice of research is politically motivated. It goes further to recognize the more normative interpretation of being critical that is described above: the need, through the practice of research, to challenge what research is being done, how and by whom, so making space for news kinds of research questions.

The third tenet addresses the critical dimension in CPG through recognizing that our research has an impact (Law, this volume), whether directly through the ever greater pressure that we face to demonstrate the economic or social utility of the research we do or more simply because, as King and Tadaki (this volume) address, knowledge can be a means of acquiring power in decision-making, such that those who produce knowledge may have greater capacity to instigate change than others. As Law (this volume) argues, even the production of knowledge in places where there are no people, or where people are explicitly excluded from scientific investigation (as can be the case in classical enquiry in natural science), may have impacts because of the consequences, intended or otherwise, of the knowledge produced. The notion that research has an impact itself has a genealogy, in relation to advocates of the need to change the relationship between the researcher and the researched. In Geography, it has had traction more generally in Human Geography and this can be traced back to the late 1960s and early 1970s (e.g. Bunge 1973). In reflecting on the 'Detroit Geographical Expedition', Bunge (1979) describes how it was a personal and material (i.e. physical) displacement of working in a community in Detroit that had forced him to dissociate from the axioms of theoretical Geography that dominated the 1960s academy, that is to escape a particular structuring of his own academic practice. Through a material engagement with that community, Bunge's conventional reading was slowed down, his accumulated wisdom unsettled, and he came to think about his research in a markedly different way. As Bunge reflected, he had to *sight* rather than *cite* (Bunge 1979, 172) and use this *sight* to find a new way of consulting with what he was researching, one where the subjects of his research were given a much stronger *right of reply*. Bunge's personal assessment is important because it challenges the traditional tenet of scientific enquiry that the researcher must retain some distance from the researched in order to have some kind of 'objective' position. The greater this distance, the more 'sight' is lost, to the point at which certain kinds of knowledge (e.g. aesthetics) go unnoticed at best, are excluded at worst (see Tuan 1989; also Kennedy 1979). Such separation may not only be an ideal that is rarely achieved (as Science Technology Studies research has shown) but also it is likely to produce particular kinds of explanation, reinforcing the notion of a vernacular treatment of humans in Geography. The detachment of knowledge from meaning, as Jasanoff (2010) has argued happened around climate science and policy, does not serve either humans or the crappy landscapes of which they are an active part. As Law (this volume) shows, our growing concern with the Anthropocene reinforces the need for physical geographers, as well as human geographers, to give traction to rethinking the relationship between the researcher and the researched.

Towards a More Physical Critical Human Geography

If much of the above had advocated a more 'critical' turn in Physical Geography, then there is a second critical perspective of equal importance: the need for a more physical turn in Critical Human Geography. Since the 1970s, not only has Human Geography distanced itself from the kind of scientific method advocated by Harvey (1969), even the use of evidence from the natural sciences has become increasingly rare in Human Geography. The expansion of human-environment Geography seemed like it might reverse this trend but, as we note above (Johnston 1983, 1986), research that truly combines Physical and Human Geography is uncommon. This is particularly the case for research that integrates physical science with the theory-laden critiques of Critical Human Geography. The field of political ecology illustrates this well.

Political ecology was built on a combination of critical social and physical science evidence. Early work in the field (Watts 1983; Blaikie 1985; Hecht 1985) paired strong critiques of environmental injustice with more sober presentations of physical evidence and built quantitative social science data into clear exposés of the politics of environmental science. Social science was clearly dominant, physical science providing simply the material setting or the template upon which politics was played out. For instance, of political ecology's pioneers, only Hecht conducted physical science research. Given how peripheral natural science (and even quantitative social science) are today in political ecology, it is striking how integral environmental data was in the formative work in the field. Early political ecology was deeply focused on disproving dominant explanatory frameworks for environmental degradation, particularly Malthusian, Tragedy of the Commons, and ignorant peasant explanations, all of which blame land managers for their own predicaments. By contrast, political ecology's pioneers argued that, "… environmental problems in the Third World, …, are less a problem of poor management, overpopulation, or ignorance, as of social action and political economic constraints, … [Analysis should thus concentrate on] market integration, commercialization, and the dislocation of customary forms of resource management." (Peet and Watts 1996, 4–5).

Exposing the regressive politics underlying these supposedly neutral explanations required robust explanatory frameworks. Watts, for example, challenged prevailing explanations of famine through a combination of quantitative social and physical data that included historical and contemporary quantitative data on demographics, food availability, and climate; ethnographic data on farmers' sophisticated ability to respond to substantial variation in rainfall quantity, geography, and timing; oral histories of farming and food shortage; and quantitative social science data from household surveys. There are similarly broad evidence bases in the early work of Blaikie (1985) and Hecht (1985), though the empirical focus was soil rather than famine. Early political ecologists challenged dominant explanatory frameworks by not only undermining the physical science 'data' on which they were based by exposing their colonialist, racist, classist biases but also by replacing them with new kinds of data, exposing new kinds of understanding.

In the 1990s, political ecology underwent a 'post-structural turn', which refocused the field empirically and epistemologically. Instead of focusing on the materiality of nature, political ecologists shifted their attention to representations of nature. Thus, whilst there are a few notable exceptions (some of whom have contributed chapters to this volume), political ecology's engagement with natural science today is characterized best not by critical ecology (or pedology or hydrology) but instead with unpacking the notion of what constitutes nature, as well as how nature and related concepts are represented (Peet and Watts 1996; Walker 2005; Turner 2015). The result is that political ecology research is no longer characterized by its fusion of critical social and physical science evidence, and it has never been typified by the integrated physical and social analysis CPG advocates. While CPG has some overlap with the initial formulation of political ecology, its emphasis on integrating physical and critical social science analysis is also quite different.

We have chosen political ecology as an example of the 'dematerialization' of investigation in one area within Critical Human Geography. Yet, at least since the 1990s (Jackson 2000), there has been a renewed focus within Human Geography on the material practices surrounding day-to-day life. This rematerialization was not really conceived as a re-engagement with 'things physical' but more cultural, although there are examples of a rematerialization of social scientists' interest in environmental questions. The latter reflects the observation that if the environment is co-produced or co-constituted through science, technology, and the social (Jasanoff 2004), then material concerns cannot be overlooked. For instance, within political ecology there has been a renewed interest in a more integrative approach such as that illustrated in the 'new' non-equilibrium ecology for integrative understanding of environmental change (Zimmerer 1994, 2000).

Conclusions

In this chapter, we have presented the genealogy of CPG as a tension between the fissiparous tendencies of a discipline, Geography, that straddles the social and natural sciences and the integrating tendencies that can be traced through method (e.g. the kind of integration envisaged in systems analysis), substance (e.g. the perception that environmental processes may need to be explained by human activity), and philosophy.

There are three main threads to this genealogy. The first is work relating to the philosophy of Physical Geography, and scholars like Bruce Rhoads and Keith Richards, that showed that the kinds of debates that have characterized major upheavals in the recent history of Human Geography might merit at least discussion in Physical Geography. This philosophical introspection is important because it raises debate around the traditional scientific method that remains the backbone of research in Physical Geography and points to the possibility that the methodological debates surrounding the nature and practice of Human Geography may not be as arcane as imagined to Physical Geography.

The second genealogical thread is the commitment to integrative study. Following Johnston (1986), we noted the tendency of both Physical Geography and Human Geography to construct integrative collaborations that overly simplified the nature of the other. The modern history of the discipline is replete with debates and calls for more (or less) integration. Early academic Geography had clear elements of an integrative project but one that had clear schisms by the mid-twentieth century. The quantitative revolution and the subsequent systems analysis brought Physical and Human Geography together again, only for a marked divergence to begin during the 1970s. The interest in the idea that there is scope for a shared, integrated project remains, and there have been attempts to find what that project might be conceptually (e.g. in complexity thinking). But, the genealogy shows that the kind of integrative work that CPG advocates is only partially related to some kind of attempt to build disciplinary integrity. Rather, integration is argued as needed because the kinds of things that geographers study necessitate research grounded in both the natural sciences and the social sciences. The pervasive nature of 'crappy' landscapes (Urban, this volume), the idea that our understandings of such landscapes are shaped by socio-political processes (King and Tadaki, this volume), the recognition of the pervasive impacts of geographical enquiry (Law, this volume), and the recognition that people are more than just unit processes reinforce the importance of a more-than-vernacular integrative project. CPG advocates that geographers are uniquely placed to embark upon such a project precisely because of their long-standing ambition for academic integration. The need to do so is becoming more acute with the growth of pressure on academics to realize impacts beyond the academy.

The third thread through the genealogy is 'critique'. Here we use analogies with the development of Critical Human Geography to advocate for an approach that it is implicitly normative. This is partly inspired by the engagement of physical geographers with questions philosophical, but its need is sustained by the legacy of work that has shown that science is partly, if not entirely, a social process, work that has been equally a focus of human geographers. If what we do and how we do it is at least in some part socially informed, then we should be thinking through what should be integrated, how it is being done, and by whom. That is, we cannot escape a normative element in our research and, rather than trying to exclude it, we should learn to work with it. We, ourselves, work within communities that are constrained by others (e.g. the growing social and cultural constraints on the academy), as well as being self-constraining (e.g. systems of peer review). Such constraints may restrict our ability to be sensitive to the kinds of questions being asked of us by the world around us (material and non-material). We need to find a means of being critical about the established norms associated with what we do and to find ways of placing ourselves in situations that allow us to understand the world in different ways (Stengers 2013; Lane 2017). Thus, the displacement that CPG envisages is not just away from those pristine environments readily subject to more conventional scientific experimentation and towards those environments strongly impacted by humans. It is also an epistemological displacement in terms of what and who is allowed to be engaged in the science that we do.

Doing the kind of work that CPG advocates is becoming more socially acceptable. Physical geographers have become more sophisticated in their attempts to engage with the roles of human agency, perception, and culture. This includes research in geomorphological systems (Gregory 2006; Jones and MacDonald 2007; Urban 2002; Wilcock et al. 2013). In hydrology, new kinds of participation with local communities have been the basis of challenges to the dominant focus of both flood risk science and flood risk management (Lane et al. 2011; Lane 2014). In biogeography, debates about the degree to which the vegetation and disturbance regimes of the Americas were impacted by Native Americans have remained active for over two decades (Denevan 1992, 2011; Clement and Horn 2001; Tulowiecki and Larsen 2015; Vale 2002) and have cross-fertilized with broader discussions in ecology about the value of studying peopled landscapes and the role of humans in shaping ecological patterns. And finally, in atmospheric studies, the relationship between climate, climate science, and society has provided new lines of inquiry, with Thornes and McGregor (2003) introducing a 'cultural climatology', which highlights the "impacts of climate on culture and culture on climate" (p. 190), and Hulme (2008) leading a re-examination of climate change "with contributions from the interpretive humanities and social sciences, married to a critical reading of the natural sciences, and informed by a spatially contingent view of knowledge" (p. 5). On the surface, these developments involve a relatively small number of researchers and have little overlap with dialogues occurring in other subfields. They are not necessarily motivated by CPG, but we argue that they illustrate the continued salience of calls within Geography to integrate the human and the physical and the increasing need for a subdisciplinary home within which to pursue this integration.

Acknowledgements The basis of this chapter was a consultation as to key articles in the history of Geography that have influenced the authors of other chapters in this Handbook and we acknowledge the responses received from Louise Bracken, Simon Dufour, Chris Duvall, Salvatore Engel-Dimauro, Daniel Knitter, Javier Arce Nazario, Nathan Sayre, and Marc Tadaki.

References

Ackermann, E.A. 1963. Where is a research frontier? *Annals of the Association of American Geographers* 53: 429–440.

Ashmore, P.E. 2015. Towards a sociogeomorphology of rivers. *Geomorphology* 251: 149–156.

Ashmore, P., and B. Dodson. 2016. Urbanizing physical geography. *The Canadian Geographer* 61: 102–106.

Barrow, H. 1923. Geography as human ecology. *Annals of the Association of American Geographers* 13: 1–14.

Bayliss-Smith, T.P. 1982. *The ecology of agricultural systems*, 112. Cambridge: Cambridge University Press.

Bennett, R.J., and R.J. Chorley. 1978. *Environmental systems: Philosophy, analysis and control*, 624. Princeton: Princeton University Press.

Bijker, W.E. 2007. Dikes and dams, thick with politics. *ISIS* 98: 109–123.

Billinge, M., K. Gregory, and R. Martin, eds. 1983. *Recollections of a revolution: Geography as spatial science*, 235. London: Macmillan.

Bithell, M., J. Brasington, and K. Richards. 2008. Discrete-element, individual-based and agent-based models: Tools for interdisciplinary enquiry in geography? *Geoforum* 39: 625–642.

Blaikie, P. 1985. *The political economy of soil erosion in developing countries*. London and New York: Longman.

Brand, U. 2016. How to get out of the multiple crisis? Contours of a critical theory of social-ecological transformation. *Environmental Values* 25: 503–525.

Brown, A.G., S. Tooth, J.E. Bullard, D.S.G. Thomas, R.C. Chiverrell, A.J. Plater, J. Murton, et al. 2017. The geomorphology of the Anthropocene: Emergence, status and implications. *Earth Surface Processes and Landforms* 42: 71–90.

Bunge, W.W. 1973. The geography. *The Professional Geographer* 25: 331–337.

———. 1979. Perspectives on theoretical geography. *Annals of the Association of American Geographers* 69: 169–174.

Castree, N. 2011. Nature and society. In *The Sage handbook of geographical knowledge*, 287–299. London: Sage.

———. 2015. Geography and global change science: Relationships necessary, absent and possible. *Geographical Research* 53: 1–15.

———. 2016. Geography and the new social contract for global change research. *Transactions of the Institute of British Geographers* 41: 328–347.

Chin, A. 2006. Urban transformation of river landscapes in a global context. *Geomorphology* 79: 460–487.

Chisholm, M. 1967. General systems theory. *Transactions of the Institute of British Geographers* 42: 45–52.

Chorley, R.J., and R.W. Kates. 1969. Introduction. In *Water, earth and man*, 1–7. London: Methuen.

Clement, R.M., and S.P. Horn. 2001. Pre-Columbian land-use history in Costa Rica: A 3000-year record of forest clearance, agriculture and fires from Laguna Zoncho. *The Holocene* 11: 419–426.

Clifford, N.J. 2002. The future of geography: When the whole is less than the sum of its parts. *Geoforum* 33: 431–436.

————. 2008. Models in geography. *Geoforum* 39: 675–686.

Clifford, N., and K. Richards. 2005. Earth system science: An oxymoron? *Earth Surface Processes and Landforms* 30: 379–383.

Collins, H.M., and R.J. Evans. 2002. The third wave of science studies: Studies of expertise and experience. *Social Studies of Science* 32: 235–296.

Cooke, R.U. 1992. Common ground, shared inheritance: Research imperatives for environmental geography. *Transactions of the Institute of British Geographers* 17: 131–151.

Darier, E., S. Shackley, and B. Wynne. 1999. Towards a 'folk integrated assessment' of climate change? *International Journal of Environment and Pollution* 11: 351–372.

Demaria, F., and S. Schindler. 2016. Contesting urban metabolism: Struggles over waste-to-energy in Delhi, India. *Antipode* 48: 293–313.

Demeritt, D. 1996. Social theory and the reconstruction of science and geography. *Transactions of the Institute of British Geographers* 21: 484–503.

————. 2001. The construction of global warming and the politics of science. *Annals of the Association of American Geographers* 91: 307–337.

————. 2006. Science studies, climate change and the prospects for constructivist critique. *Economy and Society* 35: 453–479.

————. 2009. From externality to inputs and interference: Framing environmental research in geography. *Transactions of the Institute of British Geographers* 34: 3–11.

Denevan, W.M. 1992. The pristine myth: The landscape of the Americas in 1492. *Annals of the Association of American Geographers* 82: 369–385.

————. 2011. The "Pristine Myth" revisited. *Geographical Review* 101: 576–591.

Dyer, J.M. 2010. Land-use legacies in a central Appalachian forest: Differential response of trees and herbs to historic agricultural practices. *Applied Vegetation Science* 13: 195–206.

Ellis, E.C. 2011. Anthropogenic transformation of the terrestrial biosphere. *Philosophical Transactions of the Royal Society of London A: Mathematical, Physical and Engineering Sciences* 369: 1010–1035.

Eyre, S.R. 1964. Determinism and the ecological approach to Geography. *Geography* 49: 369–376.

Fischer-Kowalski, M., and H. Haberl. 2002. Sustainable development: Socio-economic metabolism and colonization of nature. *International Social Science Journal* 50: 573–587.

Fischer-Kowalski, M., and H. Weisz. 1999. Society as hybrid between material and symbolic realms: Toward a theoretical framework of society-nature interrelation. *Advances in Human Ecology* 8: 215–251.

Francis, R.A. 2014. Urban rivers: Novel ecosystems, new challenges. *WIREs Water* 1: 19–29.

Francis, R.A., J. Lorimer, and M. Raco. 2012. Urban ecosystems as 'natural' homes for biogeographical boundary crossings. *Transactions of the Institute of British Geographers* 37: 183–190.

Gandy, M. 2002. *Concrete and clay: Reworking nature in New York City*. Cambridge: MIT Press.

Görg, C. 2007. Landscape governance: The "politics of scale" and the "natural" conditions of places. *Geoforum* 38: 954–966.

Goudie, A.S. 1986. The integration of human and physical geography. *Transactions of the Institute of British Geography* NS11: 454–458.

———. 2017. The integration of human and physical geography revisited. *The Canadian Geographer* 61: 19–27.

Gregory, D.J. 1978. *Ideology, science and human geography*. London: Hutcherson.

Gregory, K.J. 2000. *The changing nature of physical geography*, 368. London: Arnold.

———. 2006. The human role in changing river channels. *Geomorphology* 79: 172–191.

Haff, P.K. 2010. Hillslopes, rivers, plows, and trucks: Mass transport on Earth's surface by natural and technological processes. *Earth Surface Processes and Landforms* 35: 1157–1166.

Haggett, P. 1965. *Locational analysis in human geography*. New York: St. Martin's Press.

Haggett, P., and R.J. Chorley. 1969. *Network models in geography*. London: Arnold.

Harden, C.P. 2014. The human-landscape system: Challenges for geomorphologists. *Physical Geography* 35: 76–89.

Harrison, S., and P. Dunham. 1998. Decoherence, quantum theory and their implications for the philosophy of geomorphology. *Transactions of the Institute of British Geographers* 23: 501–514.

Harrison, S., D. Massey, K. Richards, F.I. Magilligan, N. Thrift, and B. Bender. 2004. Thinking across the divide: Perspectives on the conversations between physical and Human Geography. *Area* 36: 435–442.

Harrison, S., D. Massey, and K. Richards. 2006. Complexity and emergence (another conservation). *Area* 38: 465–471.

———. 2008. Conservations across the divide. *Geoforum* 39: 549–551.

Harvey, D. 1969. *Explanation in geography*. New York: St. Martin's Press.

———. 1974. Population, resources, and the ideology of science. *Economic Geography* 50: 256–277.

Hecht, S. 1985. Environment, development and politics: Capital accumulation and the livestock sector in Eastern Amazonia. *World Development* 13: 663–684.

Hooke, J.M. 2006. Human impacts on fluvial systems in the Mediterranean region. *Geomorphology* 79: 311–335.

Hulme, M. 2008. Geographical work at the boundaries of climate change. *Transactions of the Institute of British Geographers* 33: 5–11.

Jackson, P. 2000. Rematerializing social and cultural geography. *Social and Cultural Geography* 1: 9–14.

Jasanoff, S. 2004. *Ordering knowledge, ordering society. Chapter 2 in States of knowledge: The Co-production of science and the social order*, 25–98. London: Routledge.

———. 2010. A new climate for society. *Theory, Culture and Society* 27: 233–253.

Johnston, R.J. 1983. Resource analysis, resource management and the integration of human and physical geography. *Progress in Physical Geography* 7: 127–146.

———. 1986. Fixations and the quest for unity in geography. *Transactions of the Institute of British Geographers* 11: 449–453.

———. 2006. Geography (or geographers) and earth system science. *Geoforum* 37: 7–11.

Jones, P., and N. Macdonald. 2007. Getting it wrong first time: Building an interdisciplinary research relationship. *Area* 39: 490–498.

Kennedy, B.A. 1979. A naughty world. *Transactions of the Institute of British Geographers* 4: 550–558.

King, Leonora, and Marc Tadaki. this volume. A framework for understanding the politics of science (Core Tenet #2).

Lahsen, M. 2005. Seductive simulations? Uncertainty distribution around climate models. *Social Studies of Science* 35: 895–922.

Lane, S.N., N. Odoni, C. Landström, S.J. Whatmore, N. Ward, and S. Bradley. 2011. Doing flood risk science differently: An experiment in radical scientific method. *Transactions of the Institute of British Geographers* 36 (1): 15–36.

Lane, S.N. 2014. Acting, predicting and intervening in a socio-hydrological world. *Hydrology and Earth System Sciences* 18 (3): 927–952.

Lane, S.N. 2017. Slow science, the geographical expedition, and Critical Physical Geography. *The Canadian Geographer* 61: 84–101.

Latour, B. 1999. *Pandora's hope: Essays on the reality of science studies*, 336. Harvard: Harvard University Press.

Latour, B., and S. Woolgar. 1979. *Laboratory life: The construction of scientific facts*. Princeton: Princeton University Press.

Lave, R., M.W. Wilson, E.S. Barron, C. Biermann, M.A. Carey, C.S. Duvall, L. Johnson, et al. 2014. Intervention: Critical Physical Geography. *The Canadian Geographer* 58: 1–10.

Law, Justine. this volume. The impacts of doing environmental research (Core Tenet #3).

Leighly, J.B. 1955. What has happened to physical geography? *Annals of the Association of American Geographers* 45: 309–318.

Lewin, J. 2013. Enlightenment and the GM floodplain. *Earth Surface Processes and Landforms* 38: 17–29.

Livingstone, D. 1992. *The nature of geography*, 434. Oxford: Blackwell.

Mackinder, H.J. 1887. On the scope and methods of geography. *Proceedings of the Royal Geographical Society and Monthly Record of Geography* 9: 141–160.

Macmillan, W., and H.Q. Huang. 2008. An agent-based simulation model of a primitive agricultural society. *Geoforum* 39: 643–658.

Maddrell, A. 2010. Academic Geography as terra incognita: Lessons from the 'expedition debate' and another border to cross. *Transactions of the Institute of British Geographers* 35: 149–153.

Malanson, G.P. 2014. Biosphere-human feedbacks: A physical geography perspective. *Physical Geography* 35: 50–75.

Marsh, G.P. 1864. *Man and nature, C*, 560. New York: Scribner.

Mercer, D. 1983. Unmasking technocratic geography. In *Recollections of a revolution: Geography as spatial science*, ed. M. Billinge, K. Gregory, and R. Martin, 153–199. London: Macmillan.

Mountz, A., A. Bonds, B. Mansfield, J.M. Lloyd, J. Hyndman, M. Walton-Roberts, R. Basu, et al. 2015. For slow scholarship: A feminist politics of resistance through collective action in the neoliberal university. *ACME: An International E-Journal for Critical Geographies* 14: 1235–1259.

Pain, R. 2014. Impact: Striking a blow or working together? *ACME: An International E-Journal for Critical Geographies* 13: 19–23.

Pattison, W.D. 1964. The four traditions of geography. *Journal of Geography* 63: 211–216.

Peet, R., and M. Watts, eds. 1996. *Liberation ecologies: Environment, development, social movements*. London: Routledge.

Petts, J., S. Owens, and H. Bulkeley. 2008. Crossing boundaries: Interdisciplinarity in the context of urban environments. *Geoforum* 39: 593–601.

Phillips, J.D. 2001. Human impacts on the environment: Unpredictability and the primacy of place. *Physical Geography* 32: 321–332.

Pitman, A.J. 2005. On the role of geography in earth system science. *Geoforum* 36: 137–148.

Pollard, J.S., J. Oldfield, S. Randalls, and J.E. Thornes. 2008. Firm finances, weather derivatives and geography. *Geoforum* 39: 616–624.

Rasmussen, K., and F. Arler. 2010. Interdisciplinarity at the human-environment interface. *Geografisk Tidsskrift-Danish Journal of Geography* 110: 37–45.

Rhoads, B.L. 1999. Beyond pragmatism: The value of philosophical discourse for physical geography. *Annals of the Association of American Geographers* 89: 760–771.

Rhoads, B.L., and C.E. Thorn. 1994. Contemporary philosophical perspectives on physical geography with emphasis on geomorphology. *Geographical Review* 84: 90–101.

———, eds. 1996. *The scientific nature of geomorphology*, 484. Chichester: Wiley.

Rhoads, B.L., D. Wilson, M. Urban, and E.E. Herricks. 1999. Interaction between scientists and nonscientists in community-based watershed management: Emergence of the concept of stream naturalization. *Environmental Management* 24: 297–308.

Richards, K.S. 1990. Real' geomorphology. *Earth Surface Processes and Landforms* 15: 195–197.

Richards, K. 1996. Samples and cases: Generalisation and explanation in geomorphology. In *The scientific nature of geomorphology*, ed. B.L. Rhoads and C.E. Thorn, 171–190. Chichester: Wiley.

Richards, K., and N. Clifford. 2008. Science, systems and geomorphologies: Why LESS may be more. *Earth Surface Processes and Landforms* 33: 1323–1340.

Serra, P., X. Pons, and D. Saurí. 2008. Land-cover and land-use change in a Mediterranean landscape: A spatial analysis of driving forces integrating biophysical and human factors. *Applied Geography* 28: 189–209.

Stengers, I. 2013. *Une autre science est possible! Manifeste pour un ralentissement des sciences*, 216. Paris: Les Empêcheurs de penser en rond.

Stoddart, D. 1965. Geography and the ecological approach. The ecosystem as a geographical principle and method. *Geography* 50: 242–251.

Sundberg, M. 2009. The everyday world of simulation modeling: The development of parameterizations in meteorology. *Science, Technology and Human Values* 34: 162–181.

Swyngedouw, E., and N.C. Heynen. 2003. Urban political ecology, justice, and the politics of scale. *Antipode* 35: 898–918.

Syvitski, J.P.M., C.J. Vörösmart, A.J. Kettner, and P. Green. 2005. Impact of humans on the flux of terrestrial sediment to the global coastal ocean. *Science* 308: 376–380.

Tarolli, P., and G. Sofia. 2016. Human topographic signatures and derived geomorphic processes across landscapes. *Geomorphology* 255: 140–161.

Thomas, W.L. 1956. *Man's role in changing the face of the earth*, 1193. Chicago: University of Chicago Press.

Thornes, J.E. 1981. A paradigmatic shift in atmospheric studies? *Progress in Physical Geography* 5: 429–440.

Thornes, J.E., and G.R. McGregor. 2003. Cultural climatology. In *Contemporary meanings in physical geography*, ed. S. Trudgill and A. Roy, 73–197. London: Springer.

Tuan, Y.F. 1989. Surface phenomena and aesthetic experience. *Annals of the Association of American Geographers* 79: 233–241.

Tulowiecki, S.J., and C.P. Larsen. 2015. Native American impact on past forest composition inferred from species distribution models, Chautauqua County, New York. *Ecological Monographs* 85: 557–581.

Turner, M. 2015. Political ecology II: Engagements with ecology. *Progress in Human Geography* 40: 413–421.

Urban, M.A. 2002. Conceptualizing anthropogenic change in fluvial systems: Drainage development on the upper Embarras River, Illinois. *The Professional Geographer* 54: 204–217.

Urban, Michael A. this volume. In defence of crappy landscapes (Core Tenet #1).

Vale, T.R. 2002. *The pre-European landscape of the United States: Pristine or humanized*. Washington, DC: Island Press.

Wainwright, J. 2008. Can modelling enable us to understand the rôle of humans in landscape evolution? *Geoforum* 39: 659–674.

Walker, P.A. 2005. Political ecology: Where is the ecology? *Progress in Human Geography* 29: 73–82.

Watts, M. 1983. *Silent violence: Food, famine, and peasantry in Northern Nigeria*. Berkeley: University of California Press.

Wesselink, A.J., H.J.d. Vriend, H.J. Barneveld, M.S. Krol, and W.E. Bijker. 2009. Hydrology and hydraulics expertise in participatory processes for climate change adaptation in the Dutch Meuse Water. *Water Science and Technology* 60: 583–595.

Whatmore, S.J. 2013. Where natural and social science meet? Reflections on an experiment in geographical practice. In *Interdisciplinarity: Reconfigurations of the social and natural sciences*, ed. A. Barry and G. Born, 161–177. London: Routledge.

Wilcock, D., G. Brierley, and R. Howitt. 2013. Ethnogeomorphology. *Progress in Physical Geography* 37: 573–600.

Woldenberg, M.J., and B.J.L. Berry. 1967. Rivers and central places: Analogous systems? *Journal of Regional Science* 7: 129–139.

Wynne, B. 1992. Uncertainty and environmental learning—reconceiving science and policy in the preventive paradigm. *Global and Environmental Change—Human and Policy Dimensions* 2: 111–127.

Zimmerer, K.S. 1994. Human geography and the "new ecology": The prospect and promise of integration. *Annals of the Association of American Geographers* 84: 108–125.

———. 2000. The reworking of conservation geographies: Nonequilibrium landscapes and nature-society hybrids. *Annals of the Association of American Geographers* 90: 356–369.

———. 2010. Retrospective on nature–society geography: Tracing trajectories (1911–2010) and reflecting on translations. *Annals of the Association of American Geographers* 100: 1076–1094.

3

In Defense of Crappy Landscapes
(Core Tenet #1)

Michael A. Urban

Introduction

> When I do the pebble count, what should I do about things like shopping carts, bed springs, and old tires? Should I measure the B axes and record them as bedload? (Graduate student as quoted in Harden 2013, 35)

Students of Physical Geography at the University of Königsberg in the eighteenth century would have experienced a curious combination of instruction focusing on both societies and the environment. The Professor, the Philosopher Immanuel Kant, coupled his Physical Geography with anthropology in order that students would have a practical moral guide for how to behave and interact with the world (Eldon 2011). Consideration of the human role in not only living in but shaping the physical environment is even more pertinent in the twenty-first century than the eighteenth-century Germany of Kant. In our own time, the rapid rate of environmental change coupled with the sheer magnitude of human drivers have propelled geographers to increasingly examine landscapes that are as much a byproduct of social policy or individual action as the laws of physics, chemistry, or genetics. The question of how best to ethically behave in the world has implications not just for our well-being as moral individuals but the integrity and functioning of biophysical systems. Despite the fact that few sciences or disciplinary traditions interweave the social complexities of human existence with the

M. A. Urban (✉)
Department of Geography, University of Missouri, Columbia, MO, USA

© The Author(s) 2018
R. Lave et al. (eds.), *The Palgrave Handbook of Critical Physical Geography*,
https://doi.org/10.1007/978-3-319-71461-5_3

biophysical complexities of environmental systems, geographers are well equipped to take on this challenge. After all, the historical drive to maintain physical "nature" and human "culture" under one conceptual umbrella delineates the fundamental characteristic of the geographical experiment that stems back to Kant himself (Livingstone 1992).

This chapter builds on much of the recent work that has extended the conceptual territory of Physical Geography to explicitly include the role of people in intentionally and unintentionally shaping modern landscapes. In geomorphology, this territory has been defined alternatively as anthropogeomorphology (Urban 2002), cultural geomorphology (Gregory 2006), ethnogeomorphology (Wilcock et al. 2013), and socio-geomorphology (Ashmore 2015). The commonality in all these approaches is that the importance of human influence is not simply relegated to a matter of classifying the genetic origin of landscape features but rather, social action is included in biophysical systems as a process or physical force. Extending the boundaries of the science of Physical Geography to explicitly include people requires that we examine human perception, individual behavior, public policy, social and economic structures, and ethical considerations as feedback loops altering biophysical processes themselves (Chin et al. 2014; Harden 2014). Echoing Kant, Critical Physical Geography (CPG) expands the scope of our study with the explicit intent of creating a cosmopolitan knowledge that can guide our behavior and inform our ethics.

In these pages, I highlight the idea that many of these composite landscapes heavily influenced by people are so widely viewed as ordinary, unappealing, or despoiled that they are often neglected in our investigations. Calling such landscapes "crappy" is, admittedly, a provocation. The goal, however, is to describe how such a provocation can be converted into a celebration of ordinary, vernacular, or degraded landscapes. In the same way that J.B. Jackson's cultural study of roads and highway landscapes in the 1950s was initially seen as a useless exercise in examining the tawdry and tasteless landscapes of blight, reclaiming crappy landscapes as fascinating and critically important consequences of human-environment interaction allows us to continue to expand the territory of Physical Geography in the time of the Anthropocene (Jackson 1984; Davis 2003).

The Age of Us

The Anthropocene has become an important organizing principle that is increasingly informing how many physical sciences conceptualize environmental change. The rapid proliferation of scientific journals, workshops, and conferences on the Anthropocene, as well as the highly publicized search for the golden spike that would allow us to quantify the exact moment in time where humans become significant geological agents, all highlight the notion that the separation of the social and physical sciences can no longer be as rigidly fixed as disciplinary boundaries have traditionally delineated. In a broader context, the concept has sparked the public imagination in recent years and has become a touchstone around which environmental issues of all sorts are being examined. In an initial iteration of the Anthropocene, Crutzen (2002) proposed a new and distinct geological epoch that elevates human impacts on environmental systems to a level of such significance that humanity must now be considered a geologic force. As evidence, people have fundamentally altered the terrestrial carbon cycle and other biogeochemical cycles, modified and appropriated water cycles, and pushed species into extinction (Steffen et al. 2011). The spikes in human population growth, landscape transformation, appropriation of primary productivity, and fossil fuel usage have had widespread environmental impacts distinctly different from Charles Lyell's original conception of the Holocene as the most recent interglacial period (Lyell 1830; Crutzen and Stoermer 2000; Haberl et al. 2007). While the exact designation of when such a period would have begun is debated (e.g. Crutzen and Stoermer 2000; Ruddiman 2003; Zalasiewicz et al. 2011), there is a general acceptance that the rate and scope of human intervention has accelerated since the middle of the twentieth century (Steffen et al. 2011).

In some ways, the core assertions of the Anthropocene are not really all that new or novel. The discipline of Geography has long been concerned with the ways by which people impact and interact with the biophysical environment (Lane et al. this volume). Marsh (1864) was, if not the first, certainly the most influential of these early voices pointing to the magnitude of human impacts on biophysical systems. Stoppani (1873) proposed early on the creation of an Anthropozoic Era saying "the creation of man [sic] constitutes the introduction into nature of a new element with a strength by no means known to ancient worlds." While the Anthropozoic itself did not catch on in the nineteenth century, other attempts to subsequently formalize these ideas continued: LeConte (1877) offered up the Psychozoic Era, Sherlock (1922) focused on "man as a geologic agent," and Vernadsky (1926) proposed the concept of the noösphere to accommodate the world of human thoughts. In the 1950s, luminaries such as Carl Sauer, Clarence Glacken, Paul Sears, and Lewis Mumford revisited the works of G.P. Marsh to expand on his ideas that people are not passively impacted by the environment but take an active role in shaping it (Thomas 1956). Indeed, theories trying to resolve the human-environment relation have never truly disappeared in Geography.

Today, it is almost inconceivable to think of any landscape as completely untouched by direct or indirect human action. In locations that have supported significant human populations, the physical environment is commonly degraded, damaged, simplified, or in some way modified to satisfy our needs or appetites. Indirect ecologic, hydrologic, or climatic effects extend our physical imprint far beyond obviously settled landscapes. Various areas of science have tried to resolve the ways in which humans are impacting the biophysical environment. The concept of the human footprint defines the extent to which infrastructure is continuing to expand the physical imprint of societies into an ever greater percentage of land surfaces and ecosystems around the world (Sanderson et al. 2002; Magnani et al. 2007; Venter et al. 2016). International reports such as the Millennium Ecosystem Assessment (2005) have helped to redirect conversation within the scientific community toward the extent and rate in which ecosystems and the geochemical cycles integrated within those environments have been transformed by human action. Similarly, the notion of ecosystem services has highlighted this social seizure of natural capital and transformation of land cover (Foley et al. 2005; Rockström et al. 2009; Foley et al. 2011; Hansen et al. 2013). The widespread substitution of natural ecosystems with agricultural cultivation or human appropriation of net primary productivity (HANPP) has been shown to be accelerating in recent decades as populations rise and technology expands (Haberl et al. 2001, 2007; Krausmann et al. 2013).

Despite all the attention paid to human impacts over the years, the conceptual focus within Geography has always maintained a certain distance between humans and biophysical systems. Physical geographers looking to the hard sciences for exemplars of rigor pushed the human to the margins of investigations (Lane et al. this volume). In this view, humans were seen to impact biophysical systems but not to be a part of them. Forms or features in nature may be ephemeral but they are generated by environmental forces that are perpetual, knowable, and bound by the laws of physics and chemistry. The natural system can only be known when the "distorting effects of human projection" and subjectivity are removed from scientific inquiry (Evernden 1992, 58). Trimble (1992) suggests that within Geography this is the result of disciplinary specialties radiating out from the traditional core of human-environment interaction, leading to an unresolved tension with increased distance from the center. Yet the unrelenting transformation of environments into landscapes where the biophysical is diminished or subsumed by human agency complicates scientific assessment by introducing the unpredictability and chaotic nature of human behavior and perception as a causal factor in landscape development. People introduce a level of systemic complexity which makes it difficult to isolate the biological, physical, or chemical variables at the heart of any environmental system, which in and of itself is another reason why such landscapes are viewed as undesirable as the focus of scientific investigations.

In many ways, environmental problems are as much social phenomena as they are physical or biological (Urban 2002). Yet classifying humans as a distinct set of influences implicitly outside the scope of Physical Geography limits our ability to conceptualize systemic function in situations where people are playing a critical role *within* the system. A more explicit focus by physical geographers on environmental questions associated with the increasing imprint of human agency could doubtless augment our ability to investigate problems and management challenges that are deemed socially relevant. But perhaps a more significant rationale for such a focus is that crappy landscapes which are the composite of human agency and biophysical function are a more accurate representation of how many systems actually function in the Anthropocene.

In the same way that the concept of *nature* became the focus of a tremendous amount of scholarship decades ago (e.g. Evernden 1992; Soper 1995; Castree 2005), the Anthropocene has become a vehicle for bridging the human-physical divide (Castree 2014). On the surface, there is significant resonance between the aforementioned traditions of examining human-environment interaction and the newer concept of the Anthropocene. But recent discussions extend the conversation further than previous geographical traditions by positing human influence as a high-level variable. The Anthropocene is a radical break from preceding concepts for two main reasons: the magnitude of human impact is not limited to local disturbance but has also been seen at the scale of global operation of environmental processes, and these impacts are best understood as internal to biophysical systems in that they are contingent on feedback mechanisms. In a humanized Earth, people are a local disturbance acting on biophysical systems from without. In the Anthropocene, cascading outcomes of human behavior transcend local impacts and have the potential to alter biophysical systems on much broader spatial and temporal scales, and human agency is mediated by systemic feedback and intentionality. In this sense, the Anthropocene is "a fundamental rupture from that which preceded it" (Hamilton and Grinevald 2015). Because we are now altering the ways in which nutrient cycling, water cycling, land cover change, and environmental fragmentation occur, humans are less an external disturbance and more an internal force.

For physical geographers, this is the true utility of the idea of the Anthropocene. It not only allows for the formal inclusion of people within our investigations of environmental systems but rather reframes our conception of these systems. Humans are not simply viewed as the source of external disturbance to systems. Rather, the scale and magnitude of accumulated impacts are generating change that alters the very configuration and function of the system itself (e.g. Urban and Rhoads 2003; Urban 2005). In environments where humans are a dominant force acting to shape the landscape and subvert the "natural" function of biophysical process, a threshold has been passed. In cities, intensive agricultural regions, or locations with rapidly shifting microclimates, significant changes in biota, erosion rates, and other markers of the physical landscape are subsuming legacy environments. Within the Anthropocene, humans must be internalized within environmental systems precisely because of the magnitude and persistence of these systemic effects over time. Interactions or impacts are mediated by feedback from the system itself in the form of perception, experience, local knowledge, and scientific investigation and can no longer be truly defined as external (Rhoads et al. 1999). For all practical purposes, every contemporary landscape subject to geographical investigation already contains the stigma of human activity. While the degree of contagion varies greatly all are affected. Because of this, the tacit separation of social phenomenon from biophysical process is conceptually incongruous and no longer makes any sense.

Disturbance

…man is everywhere a disturbing agent. Wherever he plants his foot, the harmonies of nature are turned towards discords. (Marsh 1874, 34)

Lake Poopó, the second largest lake in Bolivia, disappeared in December, 2015. In its place lies a dry, salty expanse occasionally punctuated by decaying fish. In the past, the lake had experienced both periods when it shrank and conversely when it flooded and spilled over into the Coipasa Saltpan to the southeast. What distinguishes these past events from the disappearance of Poopó in 2015 is not the fact that it dried up but rather the processes that led to this outcome. Though the proximal cause of the lake's disappearance is easy to identify—a lack of water—Poopó's desiccation can be traced back directly to systemic human drivers. There is a lack of water precisely because people have reduced the availability of the water flowing into the basin.

On the South American Altiplano, water scarcity defines critical factors such as ecosystem stability, erosional forces, and the economic stability of local settlements. Surface hydrology is highly sensitive to cyclical drought patterns; indeed, there are long periods in the historical record where the region was abandoned due to drought (Núñez et al. 2002; Bush et al. 2010). Despite this aridity, the plateau is home to Bolivia's two largest lakes, Titicaca in the north and Poopó to the south, and irrigated agriculture has been practiced for thousands of years dating back to the early Tiwanaku Civilization (AP 2016).

Across the plateau, precipitation rates vary seasonally and are significantly impacted by regional atmospheric anomalies such as El Niño and La Niña. The southern Altiplano where Poopó is located is extremely dry, relying on a complex routing of precipitation and glacial meltwater flowing from Lake Titicaca into the Desaguadero River and eventually draining into Poopó (Canedo et al. 2016). In 2015, the area around Lake Poopó was largely abandoned by people and animals as the lake completely evaporated. The fishery crashed, migratory birds left, and many of the local people relocated.

Climatic trends, cyclical atmospheric anomalies, surface hydrology, and topography have always driven irregular rhythms of water flow on the Altiplano (Zolá and Bengtsson 2006; Calizaya et al. 2010). Yet, these forcing mechanisms have been further complicated by increased human intervention in the hydrologic system. International water treaties have limited the amount of water flowing out of Lake Titicaca toward the Desaguadero River, climate change has reduced regional runoff coming from the highlands, increased irrigation driven in part by the global market price of quinoa has increased the volume of water being taken from regional streams, and water diversions for the booming mining industry have led to heavy metal accumulations and accelerated rates of sedimentation. As a result, Lake Poopó is a degraded landscape acutely impacted by both intentional and indirect human agency.

Over 8000 km away, the Niger Delta in West Africa provides a very different type of case study than the high-altitude Altiplano in Bolivia. Where the Altiplano is defined by aridity, the Niger Delta has emerged from water. Spanning a distance of over 500 km, the coastline stretches from the Benin River in the west to the Imo in the east. The Delta is home to the largest system of wetlands in Africa and the third largest in the world (Ebeku 2004). Consisting of a vast mosaic of coastal barrier islands, mangrove forests, freshwater swamps, and lowland rainforests, the region is one of the most important and biodiverse wetland ecosystems in the world (Uluocha and Okeke 2004; Ebeku 2004; Ayanlade and Proske 2015).

Globally, wetlands are under more pressure and are being lost at a faster rate than any other major ecosystem type (Agardy and Alder 2005). It has been estimated that over half of the total wetland area that existed throughout Europe and the United States has been drained or severely degraded (Finlayson et al. 1999). Though many of these same drivers of change are operating within the Niger Delta, it remains difficult to assess definitively the exact extent of wetland loss because it has never been comprehensively examined (Okonkwo et al. 2015). The Delta is spread over nine different states in Nigeria and is home to an estimated 41.5 million people (Okonkwo et al. 2015). Many of the physical changes occurring within the Delta are the direct result of increasing human encroachment on sensitive ecosystems. The primary productivity generated by mangrove forests is increasingly being replaced by rice and sugarcane production or dramatically reduced altogether by logging for timber and fuel (Haberl et al. 2007; Adeloke and Mitchell 2011). When confronted with the social dilemma of what to do about high rates of unemployment and poverty compounded by rising population, the response has typically been "drain the swampy areas" (Oyatomi and Umoru 2009). As a result, wetland "reclamation" continues to be aggressively pursued by the Nigerian government, despite the area being named a global biodiversity hotspot by the FAO in 1997 (FAO 1997). Upstream dams on the Niger River have interrupted sediment supply to the Delta and combined with reclamation efforts have led to significant changes in water chemistry, flow, turbidity patterns, and habitat fragmentation (Okonkwo et al. 2015). As momentous as all these changes are to the environmental dynamics of the Niger Delta, many observers view all of this as secondary to the net effects of oil and natural gas production. The first oil wells began pumping in the Delta in 1956 and the industry has since come to dominate the Nigerian economy, currently accounting for around 90% of the nation's total export (Atakpo and Ayolabi 2009).

Oil and gas exploration continue to directly disrupt hydrological processes throughout the Niger Delta. Construction of transportation routes and physical infrastructure have degraded ecosystems and exacerbated the acute pressures for housing and industrial development (Uluocha and Okeke 2004). Since the 1990s, the juxtaposition of economic wealth associated with oil and natural gas exports and the lack of opportunities available to the local population has transformed the Delta itself into the backdrop for periodic uprisings, the rise of militant groups such as the recently formed Niger Delta Avengers, and catastrophic environmental damage. Much of the social unrest is driven by perceived injustices related to foreign companies and government officials extracting and profiting from oil reserves while Delta residents bear the burden of pollution, loss of biodiversity, and environmental degradation (Adeloke and Mitchell 2011). It has been estimated that 9–13 million barrels of oil have spilled into the Delta since extraction began. In recent years, this has accelerated because locals and militant groups alike have begun to illegally tap into pipelines to gain a greater share of the oil wealth at the expense of the government and foreign oil companies. These assaults on pipeline infrastructure crisscrossing the Delta have led to widespread spills and periodic fires, especially when locals attempt to process the crude oil themselves.

Trying to understand the social dimensions of political unrest and instability in isolation from a comprehensive knowledge of the human drivers underwriting fundamental transformations of the physical environment and how these changes in the landscape in turn trigger social unrest and instability is insufficient and fragmentary. Segregating the social and physical from one another provides only fragmentary and incomplete insight into coupled human-environment systems. Within the discipline of Geography, we have long cultivated a myth of the field being part biophysical science, part humanities, part social science. Yet these differences rarely coalesce into coherent work that combines them all. Human geographers are busy with people, physical geographers are concerned with the biophysical, and there is a much smaller group nibbling at the middle, mostly in applied studies or modular research that combines disparate elements like Lego pieces without ever altering or impacting them. The promise of CPG lies in the potential for this approach to not only provide a conceptual frame allowing us to bring values and their impact on how we approach research as scientists to the foreground but also by improving the practice of science by explicitly allowing us to explain and to investigate the various ways in which social elements of perception, valuation, power, politics, and scale become biophysical forcing mechanisms (and vice versa).

The overall human imprint on Lake Poopó and the Niger Delta has fundamentally altered functions of these environmental systems yet, in many ways, these locations are exemplars of the types of landscapes emerging in the *Anthropocene*. While Poopó and the Niger Delta are illustrative of these changes, they are not exceptional. From the flooding of the Salton Sea and the desiccation of Owens Lake in California to the agricultural appropriation of tall grass prairie in Illinois or wetland glades in Florida, landscapes have been degraded by human actions. The recent creation of an artificial sea on the Yangtze River through construction of the Three Gorges Dam and the drainage of the Mesopotamian marshes of southern Iraq represent hydro- and ecosystems that are catastrophically altered. Though they represent incredibly diverse climates, surficial dynamics, hydrology, and ecosystems, all of these landscapes inexorably respond to the increasing magnitude of human agency. But we need not look so far. Physical landscapes are being impacted all around us. Everyday places such as urban brownfields, abandoned lots, weedy fields, and trash-filled gullies echo these faraway exemplars. If we expand the scope of our science to explicitly include the physical investigation of environments whose human imprint is significant, systemic, or even catastrophic, every mundane landscape becomes a potential field site that can contribute to our understanding of how these complex amalgams of physical, biological, and human impulses operate. For many, these landscapes are instead dismissed as broken, diminished, and "crappy."

Crappy Landscapes

Our discipline as a whole has been plagued by numerous dichotomies, none of which is more troublesome than the split between the physical and the human sides of the discipline. (Rhoads 1999, 767)

Crappy is a coarse word that may make some readers uncomfortable. It is a vulgar term used to denote banal elements around us defined by their own crudeness. Dismissing a landscape as crappy indicates it is worthless, second rate, extremely poor in quality. It is a direct commentary on how (little) people value these environments. In any research enterprise, both values and the scientific method are critical to the ways in which we generate rigorous and reliable information. But they are distinct from one another. Fundamentally, values manifest themselves as heuristics or cognitive shortcuts allowing the researcher to make broad conceptual leaps without having to constantly renegotiate these pathways. Values in this sense serve less as logical or analytical constructs and more as an aesthetic. Because it is largely an aesthetic judgment used to deride or to dismiss the referent, calling an object or landscape *crappy* highlights what we value as much as what we think is worthless.

Heavily managed, degraded, or colonized landscapes to one degree or another are often seen as less pleasant, lower quality, or worthless than pristine, untouched natural spaces (Cronon 1995). Domesticated landscapes typified by agriculture, reforestation or secondary growth, scrublands, or even urban brownfields, all contain a complex legacy of people and biophysical processes mutually determining environmental configuration (Urban 2002). Such landscapes are a shadow or a ghost of what they once were. These qualities coincide directly with the magnitude of human control or intervention in biophysical systems, especially when there is a considerable aesthetic variation from our expectations of how the natural environment should be configured. Crappy landscapes not only involve degradation, they also involve some level of devaluation. Some environments or landscapes are quite clearly seen as better than others (although perspective matters: one person's crappy landscape may be another's valuable food source).

We should be clear here: acknowledging that some landscapes are perceived as crappy is not intended to dilute Physical Geography by shifting the primary focus of investigation to philosophical concerns or biases held by scientific researchers. There is a danger in that such a dilution can devolve quickly into the well-worn territory of trying to define and distinguish what is natural from what is human (e.g. Collingwood 1945; Glacken 1967; Soper 1995; Braun and Castree 1998; Castree 2005; Urban and Rhoads 2003). While these debates have productively illustrated the range of philosophical assumptions grounding our science and illuminate the ways by which imagination can limit or shape how we practice Geography, they can also distract from the goal of elucidating the mechanisms by which physical landscapes are altered over time and space.

Rather, the tacit, or even the explicit, acknowledgment that some landscapes are understood to be less interesting because they have been significantly impacted by human agency allows Physical Geography to maintain its topical integrity and foundation as a scientific enterprise while still incorporating humans. Studies of human-environment interaction have a long history within Physical Geography. In practice, they have been most often accomplished by considering human behavior and influence as a type of *externality* or *disturbance* of the natural (Johnston 1986). Thus, the logic behind how this external influence operates is beyond the focus of the problem. Investigations begin at the point where the physical system is "disturbed" or altered or such disturbance is noted and physically manifests itself as a changed environmental state, flux, or parameter. This approach, treating human influence as an externality or a disturbance, has been very productive in allowing biophysical scientists to address socially relevant environmental problems— natural hazards, environmental management, disturbance response, and human modification of the environment—but in a very constrained and modular conceptual framework. The totality of humans is never considered in this model.

Those landscapes bereft of charm or so influenced by people that we begin to minimize their importance as proper subjects for physical geographical research require us to approach them differently. First we must recognize the potential these environments have as focal points for scientific inquiry. We cannot just think of such landscapes as tainted or fake (Elliot 1982). Crappy landscapes are worth investigating in part because they provide unique challenges for us to understand and to explain the manifold landscape dynamics at play when biophysical processes are not isolated from people. In this way, the conceptual scope of our discipline can be explicitly expanded to include the driving force behind human behavior and the multifarious sociocultural influences behind these physical manifestations. Physical changes then drive feedback loops which alter human perception, behavior, and policy (Urban and Rhoads 2003). At a moment in time when biophysical systems are becoming increasingly complex precisely because of the magnitude and ubiquity of human influence, we would do well to resist the increasing pressure that exists throughout academia for scientific reductionism and specialization (Harden 2013).

As few landscapes can truly be considered as isolated from some measure of direct or indirect human influence, this is one way physical geographers can reclaim a domesticated Earth. The pervasiveness of human influence in environmental systems today also points to the social relevance of our work as scientists. Environmental problems can themselves be defined as deleterious, misguided, or unanticipated ways in which human behavior has created landscapes we subsequently perceive as crappy. Practicing Physical Geography in the Anthropocene requires us to rethink how we are defining biophysical systems and landscapes of importance. Through investigation of crappy landscapes, we have the potential to determine how and why such places are seen as losing value or utility.

A second way in which we need to approach the investigation of crappy landscapes differently from traditional practice is by fully incorporating the confounding variable, in this case people, within the scope of our systems themselves. The consideration of human behavior and the reaction of biophysical systems to this behavior is the beginning. People should not be considered an externality or a disturbance only perturbing biophysical processes during discrete events (Urban 2002). Recalling the distinctions made above in our definition of the Anthropocene, the practice of Physical Geography needs to reflect the notion that we are no longer simply trying to understand and explain a humanized Earth but rather define landscape dynamics where human agency is mediated by systemic feedback and intentionality.

Critical Theory and Crappy Landscapes: From Science to Intervention

> Though men now possess the power to dominate and exploit every corner of the natural world, nothing in that fact implies that they have the right or the need to do so. (Abbey 1991)

Critical theory has, in recent decades, become a powerful explanatory device in social sciences such as Geography, yet the "critical" in critical theory is not a fixed referent and has over time meant different things to different practitioners. Because it is our intent to apply critical frameworks to physical systems, it is especially imperative to be very clear about what and how critical theory can provide more robust explanations for pressing questions in the field and point our new questions we had not previously considered. How can critical theory help inform the ways in which we practice Physical Geography? The model of critical theory as utilized in Human Geography offers some clarity but should be viewed with a certain amount of caution. Critical theory was embraced by human geographers at a time when the foundations of radical geography were being called into question, but while *radical* and *critical* concepts are sometimes used synonymously, the two traditions have different origins and presuppositions.

One of Horkheimer's (1972) original distinctions between "traditional" theory and "critical" thought was the inclusion of a normative impulse to fundamentally alter the phenomena represented or reflexively analyzed *for an explicit purpose* which ultimately benefits society through emancipation, liberation, or the mitigation of caustic power relations. Though this impulse manifests itself with slight variations in different fields, the core components of critical theory always seem to come back to these implicit normative goals. In Human Geography, critical practice can be seen as moving beyond critical reflection into an activist and transformative space. Similarly, the promise of incorporating the critical into Physical Geography lies not merely in challenging current practice or acting as a provocation but rather in the possibility of expanding the scope of our science to explicitly incorporate the human element and using such cybernetic[1] feedback to guide how people should best alter physical systems to achieve ethical aims or goals. In this way, crappy landscapes are central to geographical inquiries investigating the mechanistic ways by which humans are driving environmental change as well as wider considerations of the ethics, morality, and propriety of individual behavior, management policy, or societal imprint.

Through the lens of critical theory, CPG can query and investigate the various ways in which social elements of perception, valuation, power, politics, and scale become forcing mechanisms for the biophysical (and vice versa). Of course, we can extend this reasoning to the physical landscape only if there is an explicit linkage between these social elements and the ways in which the biophysical environment is shaped and modified over time. In this way, we extend the social, cultural, and political into the biophysical. CPG has the potential to explode our vision of how things work, why environmental systems function the way they do, and how we, as geographers, can become more critically engaged with influencing or changing these interactions. People and their unintentional or intentional impacts on ordinary landscapes do indeed matter.

Conclusion

CPG pushes physical geographers toward relevance by critically engaging not just with science but also public policy and decision-making. While this enterprise may be understood simply as ways of making our work more relevant, it is perhaps more accurate to recognize the potential CPG has to bring us closer to the core of the traditional geographic enterprise. In acknowledging the interplay between human agency and biophysical forces, CPG reorients our science away from the pristine and centralizes the tainted. Viewing the environment through the lens of the Anthropocene, landscapes influenced by agriculture, forest and range management, urbanization, or degradation are not anomalous. Such domesticated or vulgar places typify the contemporary biophysical forces shaping landscapes, ecosystem, and climate alike. People are powerful agents of change. To understand, explain, and mitigate this, people must be re-conceptualized into the Geography of physical systems. CPG, as described within this handbook, delineates a path forward. It reasserts the practical and philosophical importance of crappy landscapes to the science of Physical Geography as well as the relevance of our work to environmental management and solving real social problems.

Notes

1. In Strahler's (1980) description of systems analysis in Physical Geography, he delineates five distinct types of approaches delimited by their level of complexity. The fifth and most complex approach is defined by intelligent feedback being used to intentionally control process-response variables within systems. This he defined as cybernetic feedback related primarily to human perception and decision-making.

References

Abbey, E. 1991. *A voice crying in the wilderness =: Vox clamantis in deserto: Notes from a secret journal.* New York: St. Martin's Press.

Adeloke, O., and G. Mitchell. 2011. The Niger Delta wetlands: Threats to ecosystem services, their importance to dependent communities and possible management measures. *International Journal of Biodiversity Science, Ecosystem Services & Management* 7 (1): 50–68.

Agardy, T., and J. Alder. 2005. Coastal systems. In *Millennium ecosystem assessment. Vol. 1, Ecosystems and human wellbeing: Current state and trends. Findings of the conditions and trends working group.* Washington, DC: Island Press.

AP. 2016, January 21. Bolivia's second-largest lake dries up and may be gone forever, lost to climate change. *Guardian.* https://www.theguardian.com/world/2016/jan/22/bolivias-second-largest-lake-dries-up-and-may-be-gone-forever-lost-to-climate-change

Ashmore, P. 2015. Towards a sociogeomorphology of rivers. *Geomorphology* 251: 149–156.

Atakpo, E.A., and E.A. Ayolabi. 2009. Evaluation of aquifer vulnerability and the protective capacity in some oil producing communities of western Niger Delta. *Environmentalist* 29 (3): 310–317.

Ayanlade, A., and U. Proske. 2015. Assessing wetland degradation and loss of ecosystem services in the Niger Delta, Nigeria. *Marine and Freshwater Research* 67 (6): 828–836.

Braun, B., and N. Castree. 1998. *Remaking reality: Nature at the millennium.* Philosophy and Geography, 1. New York: Routledge.

Bush, M.B., J.A. Hanselman, and W.D. Gosling. 2010. Nonlinear climate change and Andean feedbacks: An imminent turning point? *Global Change Biology* 16 (12): 3223–3232.

Calizaya, A., O. Meixner, L. Bengtsson, and R. Berndtsson. 2010. Multi-criteria decision analysis (MCDA) for integrated water resources management (IWRM) in the Lake Poopo basin, Bolivia. *Water Resources Management* 24 (10): 2267–2289.

Canedo, C., R.P. Zolá, and R. Berndtsson. 2016. Role of hydrological studies for the development of the TDPS system. *Water (Switzerland)* 8 (4): 144.

Castree, N. 2005. *Nature,* 312. Hoboken: Routledge.

———. 2014. Geography and the anthropocene II: Current contributions. *Geography Compass* 8 (7): 450–463.

Chin, A., J.L. Florsheim, E. Wohl, and B.D. Collins. 2014. Feedbacks in human-landscape systems. *Environmental Management* 53: 28–41.

Collingwood, R.G. 1945. *The idea of nature.* New York: Oxford University Press.

Cronon, W. 1995. The trouble with wilderness; or, getting back to the wrong nature. In *Uncommon ground: Rethinking the human place in nature,* ed. W. Cronon, 69–90. New York: W. W. Norton & Co.

Crutzen, P.J. 2002. Geology of mankind. *Nature* 415 (6867): 23.

Crutzen, P.J., and E.F. Stoermer. 2000. The "anthropocene". *IGBP Newsletter* 41: 17–18.

Davis, T. 2003. Looking down the road: J.B. Jackson and the American highway landscape. In *Everyday America: Cultural landscape studies after J.B. Jackson,* ed. C. Wilson and P. Groth. Berkeley, CA: University of California Press.

Ebeku, K.S.A. 2004. Biodiversity conservation in Nigeria: An appraisal of the legal regime in relation to the Niger Delta area of the country. *Journal of Environmental Law* 16 (3): 361–375.

Eldon, S. 2011. Reintroducing Kant's geography. In *Reading Kant's geography*, ed. S. Eldon and E. Mendieta. Albany, NY: SUNY Press.

Elliot, R. 1982. Faking nature. *Inquiry (United Kingdom)* 25 (1): 81–93.

Evernden, N. 1992. *The social creation of nature*, 200. Baltimore: Johns Hopkins University Press.

Finlayson, C.M., N.C. Davidson, A.G. Spiers, and N.J. Stevenson. 1999. Global wetland inventory—Current status and future priorities. *Marine and Freshwater Research* 50 (8): 717–727.

Foley, J.A., R. DeFries, G.P. Asner, C. Barford, G. Bonan, S.R. Carpenter, F.S. Chapin, et al. 2005. Global consequences of land use. *Science* 309 (5734): 570–574.

Foley, J.A., N. Ramankutty, K.A. Brauman, E.S. Cassidy, J.S. Gerber, M. Johnston, N.D. Mueller, et al. 2011. Solutions for a cultivated planet. *Nature* 478 (7369): 337–342.

Food and Agriculture Organization of the United Nations (FAO). 1997. *Review of the state of the world fishery resources: Marine resources*, 173. FAO fisheries circular no. 920 Firm/C920, Rome.

Glacken, C.J. 1967. *Traces on the Rhodian Shore: Nature and culture in western thought from ancient times to the end of the eighteenth century.* Berkeley, CA: University of California Press.

Gregory, K.J. 2006. The human role in changing river channels. *Geomorphology* 79: 172–191.

Haberl, H., K.-H. Erb, and F. Krausmann. 2001. How to calculate and interpret ecological footprints for long periods of time: The case of Austria 1926–1995. *Ecological Economics* 38 (1): 25–45.

Haberl, H., K.H. Erb, F. Krausmann, V. Gaube, A. Bondeau, C. Plutzar, S. Gingrich, W. Lucht, and M. Fischer-Kowalski. 2007. Quantifying and mapping the human appropriation of net primary production in earth's terrestrial ecosystems. *Proceedings of the National Academy of Sciences of the United States of America* 104 (31): 12942–12947.

Hamilton, C., and J. Grinevald. 2015. Was the anthropocene anticipated? *Anthropocene Review* 2 (1): 59–72.

Hansen, J., P. Kharecha, M. Sato, V. Masson-Delmotte, F. Ackerman, D.J. Beerling, P.J. Hearty, et al. 2013. Assessing "dangerous climate change": Required reduction of carbon emissions to protect young people, future generations and nature. *PLoS One* 8 (12): e81648.

Harden, C.P. 2013. Geomorphology in context: Dispatches from the field. *Geomorphology* 200: 34–41.

———. 2014. The human-landscape system: Challenges for geomorphologists. *Physical Geography* 35 (1): 76–89.

Horkheimer, M. 1972. *Critical theory*. New York: Seabury Press. Reprinted Continuum: New York, 1982.

Jackson, J.B. 1984. *Discovering the vernacular landscape*. New Haven, CT: Yale University Press.

Johnston, R.J. 1986. Fixations and the quest for unity in geography. *Transactions of the Institute of British Geographers* 11: 449–453.

Krausmann, F., K.-H. Erb, S. Gingrich, H. Haberl, A. Bondeau, V. Gaube, C. Lauk, C. Plutzar, and T.D. Searchinger. 2013. Global human appropriation of net primary production doubled in the 20th century. *Proceedings of the National Academy of Sciences of the United States of America* 110 (25): 10324–10329.

LeConte, J. 1877. 'Psychozoic era'. In: On critical periods in the history of the earth, and their relation to evolution; on the quaternary as such a period. *American Naturalist* 11 (9): 540–557.

Livingstone, D. 1992. *The geographical tradition: Episodes in the history of a contested enterprise*. Cambridge, MA: Blackwell.

Lyell, C. 1830. *Principles of geology, being an attempt to explain the former changes of the earth's surface, by reference to causes now in operation*. London: John Murray, 3 Vols. (Reprint with a new Introduction by Rudwick MJS, Chicago, IL: University of Chicago Press, 1990, 3 vols).

Magnani, F., M. Mencuccini, M. Borghetti, P. Berbigier, F. Berninger, S. Delzon, A. Grelle, et al. 2007. The human footprint in the carbon cycle of temperate and boreal forests. *Nature* 447 (7146): 848–850.

Marsh, G.P. 1864. *Man and nature, or, physical geography as modified by human action*. New York: Scribner, Armstrong and Company.

———. 1874. *The earth as modified by human action*. New York: Scribner, Armstrong and Company.

Millennium Ecosystem Assessment. 2005. *Ecosystems and human well-being: Synthesis report*, 160. Washington, DC: Island Press.

Núñez, L., M. Grosjean, and I. Cartajena. 2002. Human occupations and climate change in the Puna de Atacama, Chile. *Science* 298 (5594): 821–824.

Okonkwo, C.N.P., L. Kumar, and S. Taylor. 2015. The Niger Delta wetland ecosystem: What threatens it and why should we protect it? *African Journal of Environmental Science and Technology* 9 (5): 451–463.

Oyatomi, K., and H. Umoru. 2009. Don't blame Yar'Adua if he is slow. *IBB Nigerian Guardian*. Lagos, Nigeria.

Rhoads, B.L. 1999. Beyond pragmatism: The value of philosophical discourse in physical geography. *Annals of the Association of American Geographers* 89: 760–771.

Rhoads, B.L., D. Wilson, M. Urban, and E.E. Herricks. 1999. Interaction between scientists and nonscientists in community-based watershed management: Emergence of the concept of stream naturalization. *Environmental Management* 24: 297–308.

Rockström, J., W. Steffen, K. Noone, Å. Persson, F.S. Chapin, E.F. Lambin, T.M. Lenton, et al. 2009. A safe operating space for humanity. *Nature* 461 (7263): 472–475.

Ruddiman, W. 2003. The anthropogenic greenhouse era began thousands of years ago. *Climatic Change* 61 (3): 261–293.

Sanderson, E.W., M. Jaiteh, M.A. Levy, K.H. Redford, A.V. Wannebo, and G. Woolmer. 2002. The human footprint and the last of the wild. *BioScience* 52 (10): 891–904.

Sherlock, R.L. 1922. *Man as a geological agent: An account of his action on inanimate nature*. London: H.F. & G. Witherby.

Soper, K. 1995. *What is nature?: Culture, politics and the non-human*, 304. Hoboken: Wiley-Blackwell.

Steffen, W., J. Grinevald, P. Crutzen, and J. Mcneill. 2011. The anthropocene: Conceptual and historical perspectives. *Philosophical Transactions of the Royal Society A: Mathematical, Physical and Engineering Sciences* 369 (1938): 842–867.

Stoppani, A. 1873. *Corso di Geologia*, 868. Milan: Bernardoni & Brigola.

Strahler, A. 1980. Systems theory in physical geography. *Physical Geography* 1 (1): 1–27.

Thomas, W.L., Jr. 1956. *Man's role in changing the face of the earth*, 1236. Chicago: University of Chicago Press.

Trimble, S.W. 1992. Preface. In *The American environment: Interpretations of past geographies*, ed. L.M. Dilsaver and C.E. Colten. Tontowa, NJ: Rowman and Littlefield.

Uluocha, N.O., and I.C. Okeke. 2004. Implications of wetlands degradation for water resources management: Lessons from Nigeria. *GeoJournal* 61 (2): 151–154.

Urban, M.A. 2002. Conceptualizing anthropogenic change in fluvial systems: Drainage development on the Upper Embarras River, Illinois. *The Professional Geographer* 54 (2): 204–217.

———. 2005. An uninhabited waste: Transforming the Grand Prairie in nineteenth century Illinois, USA. *Journal of Historical Geography* 31 (4): 647–665.

Urban, M.A., and B.L. Rhoads. 2003. Catastrophic human-induced change in stream channel planform and geometry in an agricultural watershed, Illinois, USA. *Annals of the Association of American Geographers* 93 (4): 783–796.

Venter, O., E.W. Sanderson, A. Magrach, J.R. Allan, J. Beher, K.R. Jones, H.P. Possingham, et al. 2016. Sixteen years of change in the global terrestrial human footprint and implications for biodiversity conservation. *Nature Communications* 7: 125580.

Vernadsky, V.I. 1926. *The biosphere*, 1998 ed, 178. Trans. D.B. Langmuir. Göttingen: Copernicus.

Wilcock, D., G.J. Brierley, and R. Howitt. 2013. Ethnogeomorphology. *Progress in Physical Geography* 37 (5): 573–600.

Zalasiewicz, J., M. Williams, A. Haywood, and M. Ellis. 2011. The anthropocene: A new epoch of geological time? *Philosophical Transactions of the Royal Society A: Mathematical, Physical and Engineering Sciences* 369 (1938): 835–841.

Zolá, R.P., and L. Bengtsson. 2006. Long-term and extreme water level variations of the shallow Lake Poopó, Bolivia. *Hydrological Sciences Journal* 51 (1): 98–114.

4

A Framework for Understanding the Politics of Science (Core Tenet #2)

Leonora King and Marc Tadaki

Introduction

[S]ocio-biophysical landscapes are as much the product of unequal power relations, histories of colonialism, and racial and gender disparities as they are of hydrology, ecology, and climate change. (Lave et al. 2014: 3)

Whereas the first core tenet of Critical Physical Geography recognizes that environments are shaped in a biophysical sense by human action, a second tenet—one that we elaborate here—recognizes the human shaping of environmental *science*. Recent calls for ethical reflexivity in environmental science have addressed diverse aspects of scientific practice, recognizing that how we conduct environmental science and the conclusions we draw about the natural world reflect both the biophysical world and social values and institutions (Salas-Zapata et al. 2013; Forsyth 2015; Fernández 2016; Salmond et al. 2017). Work in this vein ranges from cultivating awareness about the framing effects of scientific representations such as invasive species or sustainable development (e.g. Larson 2011) to debates about the validity of specific methods and forms of environmental management (e.g. ecological classification and modelling, see Cullum et al. 2016). In this chapter, we assemble these diverse concerns into a coherent and practical framework for thinking through—and taking responsibility for—the practices and outputs of environmental scientists.

L. King (✉) • M. Tadaki
Department of Geography, University of British Columbia, Vancouver, BC, Canada

© The Author(s) 2018
R. Lave et al. (eds.), *The Palgrave Handbook of Critical Physical Geography*,
https://doi.org/10.1007/978-3-319-71461-5_4

We develop a framework for environmental scientists to systematically examine and connect different aspects of scientific practice to their societal meanings. We focus on the *choices* scientists make that require consideration beyond their scientific intent and on the societal and environmental *consequences* of those choices. The chapter explores how scientists exercise agency in the choices they make throughout the scientific process and how various internal and external drivers act to influence the types of choices made. We identify five moments of choice in a generic scientific process and consider their specific rationalities and consequences through illustration by examples from glaciology. Glaciology proves a particularly illustrative example as it is often considered a 'pure' science dealing with the physical modelling of (often) remotely sensed environments that are distal to human activities (e.g. see O'Reilly 2017).

We begin in "Science, Facts, and Big-P Politics" with the conventional wisdom that the scientific process aspires to uncover natural 'truths' through an objective process of hypothesis testing, using appropriate and reproducible methods. In this conventional view of science, the scientific process takes place inside an insulated box, becoming politicized only after the fact, when scientific findings move into the public sphere and are used to support the agendas of various parties. We term this view of politicized science Big-P Politics.

However, Big-P Politics reflects only one set of choices that scientists can make about how their work interacts with the world. Throughout the ordinary practices of science, scientists must make small-p political choices about theories, data collection, methods, and how and for whom to apply their research. They also make choices about how to conduct themselves in various roles such as supervisors, peer reviewers, lecturers, and administrators. All of these choices involve value-judgements that are not simply about representing the biophysical environment as it is, and such choices are also structured by our worldviews, social relationships, and institutional settings. Thus, in contrast to *explicit* Big-P Politics, section "Scientific Choices and Their Consequences: A Small-p Politics of Science" looks at how, inside the 'black box' of science, the scientific method is *inherently* political. We explore three small-p political choice-contexts relating to theory, methodology, and data used in the environmental sciences. We consider how choices in these contexts are not value—neutral, as well as why they demand ethical debate and justification. Section "Application-Driven Environmental Science: Why and for Whom?" considers how application-driven environmental science might be thought of as uniquely political in the way it connects scientists to specific human communities and their environments. In section "Institutions:

Structuring Our Scientific Choices", we introduce a concept of *institutions* as norms and expectations that shape scientific decision-making in a variety of settings. By thinking more explicitly about the nature of institutions shaping our choices, we can be more reflexive in making and justifying our choices and actions as scientists.

In the spirit of a Critical Physical Geography that seeks to empower (and not just critique) environmental scientists (see Tadaki 2017), we have attempted to make this chapter accessible and relevant to environmental scientists both in and beyond physical geography. Overall, we hope readers will emerge with a tangible sense of how we might see and take responsibility for the ways in which environmental science can reinforce and rework power relations and socio-political agendas.

Science, Facts, and Big-P Politics

In conventional wisdom, science refers to the organized generation of reproducible knowledge of the natural world. By testing scientific theories against observations and consistencies of logic, it is widely understood that science 'progresses' towards providing more accurate understandings of the natural world over time (Sismondo 2010: 6). In this way of thinking, scientists are understood to be authoritative representatives of the natural world, with a responsibility to produce and to test scientific claims rigorously against the collective reasoning of the scientific community as well as reproducible forms of evidence from the biophysical world. This positioning of scientists and scientific truth claims outside the political fray is often viewed as one of the key sources of scientific authority and one of the defining features of scientific practice (Merton 1973). Fundamentally, this understanding of science as solely accountable to the material reality of the natural world conveys a vision of science as separate (or black boxed) from the political environment (Fig. 4.1). Once claims are produced and validated through science, *then* other actors (including some scientists) may opt to interpret scientific claims for circulation in wider public and political spheres (see Callon 1999).

Claims that reach the level of common sense become social 'facts' (see Latour 1987), shaping public understanding of the natural world and calls for action associated with that understanding. We refer to this championing of scientific facts into social facts as Big-P Politics. Through a range of Big-P Political activities, scientists engage directly and *intentionally* with contested public values to champion selected scientific claims and mobilize support for specific public decisions. For complex and high-stakes public decisions such

Fig. 4.1 Big-P Politics involves scientists producing and validating claims within the 'black box' of science and then stepping outside of the scientific community and into the political environment, where they argue for particular forms of action based on specific facts and values

as the formation of laws or regulations, Politics is about directly altering formal decisions through offering a correct understanding of 'the facts'.

There are many Big-P mechanisms through which scientists may explicitly promote particular scientific claims into positions of prominence, authority, and influence in public consciousness (Callon 1999). Through actions such as protests (e.g. 350.org), expert testimonies, scientific advisory panels, and publishing public blog posts and opinion editorials in mass media, scientists champion certain claims as 'facts' within the public realm. Some scientists deliberately eschew and even explicitly warn against such efforts at explicit politicization, based on the assumption that making political claims undermines the credibility of science and scientists (Delborne 2008). However, across the board, scientists are increasingly being encouraged to produce and evidence societal 'impact' resulting from their science (Castree 2016; Lane 2017). Through including societal impact criteria in publication metrics, funding applications, and professional development evaluations, these pressures are driving many scientists to undertake various forms of Big-P Politics.

In this emerging milieu of societally impactful environmental science, what intellectual and ethical norms should be used? There are no rigorous standards for attempting to influence public common sense, and there is no single true or correct way to translate scientific claims into arenas of contested public values (Sarewitz 2004; Delborne 2008). Different public actions may result if even the same scientific facts are linked with different societal values (Castree 2016).

We can usefully illustrate these ideas with examples from glaciology. In glaciology, public and private interest in the effects of climate change necessitate engagement from scientists in expert panels and assessments. The Intergovernmental Panel on Climate Change (IPCC) was created as an expert international advisory group to issue authoritative 'state of knowledge' reports on climate change (Spencer and Lane 2017). The IPCC assembles a range of glaciological observations into social facts that are meaningful to the IPCC's constituent governments (Vaughan et al. 2013). The desire for science-driven policy often requires informal processes of filling in gaps in observations and system knowledge, as well as the prioritization of particular observations to draw conclusions from incomplete or sometimes conflicting scientific understandings (O'Reilly 2017). These synthesized observations become scientific facts through the process of consensual aggregation, and they become social facts when they are connected to and stabilized with particular societal meanings and policy mandates (e.g. focussing on global-scale observations rather than local effects or modelling, see Spencer and Lane 2017). By assembling observations of glaciers into summative assessments and producing scientific consensus statements, scientists create social facts that can be used in the struggle to influence national and international climate change policies (Forsyth 2015).

Big-P Politics offers one way of understanding the common-sense notion that science is politicized after-the-fact in the pursuit of particular agendas. Although different human actors may draw from a shared pool of scientific claims, the specific values and political goals of actors can lead to different strategies for claim politicization. The moment of choice for scientists engaged in Big-P Politics lies in selecting which environmental claims produced by (a black-boxed) science are championed into public consciousness and how they are translated through particular societal meanings and linked to particular actions (see Castree 2016). Big-P Politics leaves the black box of science itself unquestioned. While understanding Political initiatives is both important and necessary, this high-level conception only helps us to understand part of the politics of science. Scientists who do not see themselves as participating in Big-P projects to politicize science (such as the IPCC or public communication initiatives) may feel exempt from discussion about politics because it is understood as something external to the scientist and the traditional practices of science, as something extra or optional that one can opt into or not.

Scientific Choices and Their Consequences: A Small-p Politics of Science

For scientists wondering about the ethical content of their theories, traditions, and communities, thinking about Big-P Politics provides a limited set of tools for self-reflection and action. To take us beyond Politics, we need an expansive concept of politics that considers both the external (Political) as well as internal (e.g. scientific norms and communities) activities of scientists (Tadaki 2017). In this section, we look inside the black box of science to consider the choices of scientists and how they produce small-p outcomes in the world (Fig. 4.2a). We define the small-p politics of science as the ways in which scientists make (intentional or unintentional) value-laden choices within the scientific realm that produce distinct consequences (social meanings, inequalities, power relations) for real people and environments. This concept of politics allows us to follow the practices of scientists and to trace their meanings into and beyond their intended communities. We identify three moments of choice within the black box of science (theory, methodology, and data) (Fig. 4.2b), and we discuss how these moments (1) involve value-laden *choices* about the environment and society, that (2) produce material *effects* on social and environmental life (i.e. they are political choices).

Theory

The definition of theory in physical geography and the environmental sciences is widely debated (see Rhoads and Thorn 2011). We do not engage with these

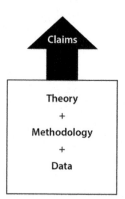

Fig. 4.2a Inside the 'black box' of science, scientists make choices about theory, methodology, and data. These choices are involved in the production of scientific claims

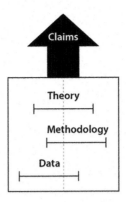

Fig. 4.2b Choices about theory, methodology, and data are made from a wide (though perhaps not infinite) spectrum of potential practices. The vertical dotted line indicates the combination of specific choices made from these spectra and links them to the claims resulting from the process

debates, but rather we use the idea of theory simply to help understand some choices that are made within the scientific process. We broadly define theory as the question of 'what to study': what is the research question, what elements of the system are important, and where and why should they be studied? In an ideal world, we might study everything—every system and process would be given equal consideration, across the full range of time and space and systems. In reality, however, science is both constrained and directed—it is not practical, efficient, or even useful to study everything, and so we make choices about what elements of a system to study. This moment of choice can be referred to as 'closure', as it involves specifying the components and boundaries of the system under investigation (Lane 2001; Blue and Brierley 2016).

We make these choices as individuals within the context of a pre-existing scientific field, whose research focus and institutional arrangements have been structured according to the priorities and interests of lab groups, scientific associations, research communities, universities, and funding organizations over long periods of time (e.g. Kuhn 1996; Sismondo 2010). While these choices are related to biophysical environments, observations, and technologies, they are also a result of deliberation with particular social groups. The utility and popularity of particular theoretical paradigms can reflect a range of social and contextual relationships, including: the popularity and eminence of the originator (Sherman 1996), the prestige of the institutions that support training in and proliferation of ideas (Kennedy 2006), the alignment of a theory with corporate and government interests (and thus funding) (Sismondo 2010), and public interest and enthusiasm (e.g. interest in climate change research and glacier retreat (Davenport et al. 2015)).

The development of glaciological theory illustrates many of these points. Glaciologists have historically been associated with a romantic enthusiasm for nature, motivated by an interest in landforms and features of the physical environment (Clarke 1987; Carey et al. 2016; O'Reilly 2017). Knight (1999: 220) suggests a 'clear and distinct route from childhood fascination with the great outdoors, mountains, and wilderness into the study of the most quintessential of wilderness phenomena'. So intertwined were science and outdoor pursuits in early debates about glacier motion that early proponents attempted to validate their particular theories through prominent discussion of their alpinistic triumphs, which captured the imaginations of the scientific and public community (Carey et al. 2016). Many glaciologists continue to justify their work to the public, funding agencies, and other scientists through public accounts of mountaineering heroics (e.g. Davenport et al. 2015; see also Carey et al. 2016). However, since the adoption of glaciology by members of the physics community during the quantitative revolution of the 1950s, this tradition of fieldwork-based glaciology gave way to a new tradition with a strong emphasis on mathematical representations of glaciers (Clarke 1987). Knight (1999) suggests that for this second dominant group of glaciologists, 'the ice sheet represents one of the most simple and elegant expressions of physics amenable to mathematical treatment that the surface of this planet offers'. Whether field scientists or mathematical modellers, the glaciological community is composed of self-selecting groups of researchers who share particular personality traits, scientific worldviews, and social class positions (e.g. see Carey et al. 2016). They iconize glaciers for different and sometimes opposing reasons, with different implications for our understanding of glaciological science. Empirical and often fieldwork-oriented 'holistic' representations of spatially heterogeneous glacier systems are juxtaposed against the interests of modellers intent on reduction, simplification, and generalization (O'Reilly 2017).

This has ongoing consequences for the ways in which we understand glaciers. An increasingly dominant perspective of the physics-oriented scientists has de-emphasized glacial processes that are not (or were not thought to be) immediately necessary for understanding glacier movement and dynamics. For example, in the 1970s and the 1980s, meltwater-formed channels on glaciers were the subject of much curiosity-driven, empirical field research by fluvial geomorphologists who were interested in comparing channels carved in ice with channels carved into bedrock and sediment (e.g. Dozier 1976). In contrast, for the modelling-oriented glaciological community, hydrological questions were of little interest until it became clear that sub-glacial (under the ice) hydrology influences glacier movement (e.g. Iken 1981). Perhaps owing to major public and scientific interest in climate change modelling, englacial (within the ice) and sub-glacial hydrology are receiving increasing attention from physics-based glaciological modellers, yet surface hydrology and meltwater channels remain little studied. When surface channels are studied, it is rarely because of an interest in their intrinsic properties; it is almost always in reference to their role in shaping glacier responses to climate change (e.g. Smith et al. 2015).

Thus, what gets studied and what does not get studied in glacial systems is not purely a scientific choice. Glacial surface hydrology has remained understudied not because it has less inherent scientific merit but because it is of less scientific interest under the prevailing popular theories on glaciers. Surface channels, lacking in sediment and fish, are of minor interest to fluvial geomorphologists and are of insufficient interest to glaciologists (although interest has grown significantly in recent years). Although simplified representations of surface hydrology may be 'adequate' for models of glacier sliding, for example, the very notion of 'adequate' representation is a value judgement on a representation of a system, and choosing one representation over another is a value-laden choice. Value is assigned to theory based on our perceptions of what is important; glacial movement and ice dynamics have been important because of a more generalized push for quantification, closer ties with physics and mathematics, fewer ties with traditional geographical disciplines, and the popularity of particular ways of thinking, all of which reflect broader societal meanings and social relations.

Not only are choices about theory shaped by external influences, but they also have external effects beyond the scientific community. The choices we make about 'what to include' in our theoretical specification of an environmental system creates exclusions which are of political *consequence*. When a glacier system is specified in one paradigm and not another, it includes (and thus privileges and takes note of) certain environmental processes and outcomes and excludes others (see Biehler et al. and King and Tadaki, this volume). Theoretical specifications of glacier runoff in a changing climate often frame the problem as a glacier, water, and climate change issue system, which is an appealing justification to funding agencies and the public. However, such a framing conceals from view the many inequitable socio-economic relationships driving water scarcity, and this sanitized view of glacier science reinforces notions of environmental determinism by presenting downstream communities as vulnerable and static in particular ways (Carey et al. 2017).

Methodology and Data Availability

Just as environmental theories shape the world politically, *how* scientists study the environment involves choices which both shape and are shaped by social relations. Acting upon a research question involves choices about:

- How the world will be represented; for example, will scientists observe the physical world, or will they simulate reality?
- What kinds of data will be collected, and what types of data will be ignored?
- How will this data be used to summarize the world and support our conclusions? How will this data be analysed and interpreted?

Often, choices about method will be shaped by the choices of theory, but not always. In deciding how to analyse the environment under study, a methodological decision may be to utilize the physical environment, including observational or experimental field studies or the use of remotely sensed imagery. Alternatively, a researcher may choose to abstract their research object from the complexities and scale of the real world through laboratory experimentation or numerical models (Church 2011). The choice between these two generic approaches will often explicitly include considerations other than the purely scientific appropriateness of the method to the question, such as logistics and cost. Each of these choices is made within a practising scientist's context (e.g. available technology, community norms), and the substance of these choices will in turn shape the meaning of research 'outcomes' (see Lane 2017). Therefore, we must ask, in what ways do our findings reflect the sum of the choices and values we have imbued in our methods?

Choices made about methodologies are always value-laden and involve privileging particular forms of knowledge and evidence relative to others (Church 2011). Methodologies rise in popularity and prominence, much as theories do, and the choice of a particular methodology (e.g. field observation vs modelling) may reflect the popular appeal of that method as much as its utility for the particular research question (e.g. Sherman 1996; Bauer et al. 1999). Part of this process involves considering what kinds of methods are valued beyond the scientific communities: which methods are granted prestige and authority?

Answers to this question can be sought from outside the scientific community as much as within it. When government agencies collaborate with scientific groups to promote the use of remote sensing, for example, we can observe a strong push towards utilizing particular types of data sets and particular methods of analysis. Stemming from recent collaborations between NASA, the National Geospatial-Intelligence Agency, the National Science Foundation, and (privately owned) DigitalGlobe, there has been a proliferation of high-resolution digital elevation models of the Arctic which are rapidly becoming go-to research tools and are concentrated within a few specific institutions (e.g. Noh and Howat 2015; Shean et al. 2016). In this way, the choice of method incorporates what interested and powerful social groups want from environmental research. Consequently, the methodological choices scientists make effectively narrow the range of what is considered legitimate knowledge or an acceptable perspective on an issue, thus reinforcing particular knowledge-power dynamics relating to the environment (Goldman et al. 2011). Seldom do geoscientists, for example, make efforts to promote or legitimize alternative forms of knowledge about a system, such as indigenous or traditional knowledge (Cruikshank 2005; Carey et al. 2016). Rather, such scientists often position their evidence as superior to local and lay forms of environmental knowledge (Forsyth 2003).

Particular methodologies may perpetuate this marginalization. Remote sensing and modelling, for instance, abstract researchers and research from the material and historical context of the landscapes they are studying (O'Reilly 2017). It is possible to empirically describe or model the Greenland ice sheet numerically without ever interacting with the Inuit people who live and have traditionally lived on its periphery. Field work, similarly, creates privileges and exclusions. Field work is costly, requires unique access to the outdoors, and often relies on pre-existing infrastructure such as research stations, which in turn narrow the spatial diversity of field studies (Carey et al. 2016). Glaciological field work is particularly rife with these limitations. Glaciers can be dangerous and are far from many Western universities today, making field work costly and difficult. The expense, logistics, and culture of glaciological field research embed and reinforce hierarchies of particular groups of researchers who are able to secure access and make authoritative claims about glaciers from the field (O'Reilly 2017).

In addition to the motivations shaping the choice of methodology and theory, the availability of data also affects how environments are framed, studied, and represented. Again, such choices often have pragmatic elements: data sources may be variable and access might be patchy (especially as satellite and other data are privately produced), and requisite analytical capabilities (e.g. computational power) are unevenly distributed (see Salmond et al. 2017). However, in addition to these concerns, the geopolitical and administrative nature of data also matters. The particular types/locations/timings of data *production* lead to geographical path dependencies whereby data *analysis* is controlled by the gatekeepers of data (e.g. Thatcher et al. 2016). In glaciology, this often takes the form of what Knight (1999: 221) calls 'siege glaciology', whereby groups of people develop long-term research programmes on particular glaciers, concentrating and controlling access to glaciological installations and data, involvement in projects, and even the heredity of research programmes. We derive large quantities of what we know about glaciers from these particular besieged glaciers, and the bulk of published material on these glaciers can be attributed to the dominant 'siege team', often with links to just one or two institutions (Knight 1999).

In glaciology, data availability is often inversely related to the distribution of human populations that are impacted by glacial dynamics. Rather than being concentrated where glacial dynamics have immediate consequences (e.g. flooding, water resources), data access is often controlled by dominant scientific agencies and institutions in the first world (see also Spencer and Lane 2017). There have been significant resources and efforts invested into producing high-resolution digital topographic maps of the Arctic, as well as the Greenland and Antarctic ice sheets (e.g. Noh and Howat 2015). However, there is, to our knowledge, little or no free, high-resolution topographic data for Himalayan or Andean glaciers, despite these glaciers having very direct consequences for downstream nearby communities. People who are the least (directly) affected by glacier change control the majority of the production of glacier data and access to it. This disconnect between data availability and glacier-dependent communities enables the abstraction of glaciological research from the communities that directly experience glacial change.

Application-Driven Environmental Science: Why and for Whom?

If the Big-P narrative of environmental science involves championing scientific facts to public and private actors beyond the scientific community in a general sense, it is important to consider how this relates to—and differs from—conventional ways of doing applied science for human decision-making (Clark et al. 2016; Castree 2016). For us, application-driven environmental science represents a style of doing science that is explicitly and intentionally conducted to serve the interests of particular social groups or interests. As such, application constitutes another key moment where value-laden choices are made that affect the meanings and consequences of the scientific process (Fig. 4.3).

For definitional purposes, we consider application-driven environmental science to be investigation which largely accepts its research question from

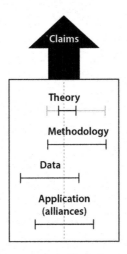

Fig. 4.3 Application-driven environmental science can be added to our framework through two operations. First, by definition, clients will constrain the range of scientific choices (e.g. theory) available to the analyst. We illustrate this with a narrowed range of possible theoretical choices, but this could be applied to methodology or data also. Second, a new moment of choice confronts the scientist in relation to the spectrum of possible clients and alliances

outside of the scientific community, usually from public agencies but sometimes from private actors (Church 2009). In contrast to Big-P Politics, which seeks to champion particular scientific claims after they have been produced through 'pure' (black-boxed) science, application-driven environmental science generates claims within explicitly pre-specified constraints. This has often been referred to in the literature as Mode 2 science, which is problem—driven rather than investigator—driven (e.g. see Castree 2016). The key question then becomes: who defines the problem, and based on whose interests?

The politics of application-driven environmental science relate to—but are distinct from—the politics of theory, methodology, and data. Environmental applications still involve choices about theories, methods, analysis, data sources, and so on, but (1) some of these choices (such as theory) are made by the client instead of the scientist and (2) the effects of these choices come to have direct consequences for specific people and places. In a consultancy, for example, scientists might be hired to conduct a cost-benefit analysis of glacier melt for a geographic area in the context of crop irrigation, using existing climate datasets. In turn, the production of such an analysis affects local conversations about prospective local land use futures in the context of climate change (Carey et al. 2017). The scientist is confronted more directly with the prospect of intervening in questions about the distributive implications of environmental policies and investments. Perhaps the most important questions for applied scientists are: whose version of the problem will be given credibility through the rigor of our work and with what consequences?

Through scientists' choices around social alliances, their scientific projects intervene into historical and place-based struggles over the framing and use of the environment and natural resources (Forsyth 2003; Goldman et al. 2011). Applied science, by definition, involves a human subject (an individual or group) who frames the 'problem' to be answered and generally stands to gain by framing the problem in a way that is advantageous to them, especially in situations of conflict. As Church (2009) notes, the problem-framing power of the client effectively turns researchers into advocates for particular solutions.

By producing an economic analysis of a glacier as a skiing asset for a tourist operator (e.g. Olefs and Fischer 2008), applied science grants authority to claims that skiing is a legitimate and even desirable social aspiration. Consider a counter-case: if scientists had worked with indigenous groups to produce a model of the glacier as 'ancestor' (e.g. Cruikshank 2005), such a move might have instead granted legitimacy to a different set of social actors and claims about what the environment is and how it should be governed (see also Goldman et al. 2011). Legitimizing particular notions of glaciers has important implications in a world where glaciers represent for some cultural assets, for others mining or tourism opportunities, water resources for irrigation, and so on (see Carey et al. 2017). Applied environmental science connects with specific storylines about environmental change, embedding assumptions, and/or preferences about: (1) what environments and processes are meaningful, (2) whose understandings of environmental change are validated, contested, or marginalized, (3) which environmental metaphors are promoted and sustained in a particular place, and (4) which actors and interests are elevated and benefitted (see Forsyth 2015). This is not to say that scientists should refrain from applied environmental science. Rather, by accepting that applied science frames the world in a particular way and favours particular actors and claims (Fig. 4.3), scientists can become more aware of and responsible for the material impacts of applied science.

Institutions: Structuring Our Scientific Choices

Throughout this chapter, we have considered the ways in which choices in the scientific process have consequences beyond their scientific intent. We have identified moments of choice relating to: (1) Big-P Political activities, (2) theory, (3) methodology, (4) data, and (5) applications, and for these moments we see how such choices produce material consequences in the world. While it may seem that any choice is possible, we have alluded to constraints and incentives that affect choices. Put another way, there are guiding norms and values that make certain choices (of theory, method, etc.) more likely or desirable than others.

To understand how scientific choices are guided, we use a basic concept of institutions as *enduring patterned social relationships and practices, organized in pursuit of collective aims.* Institutions are those codes of normal, expected, or desired ways of doing things, based on specific societal aspirations. For example, in facing choices about where to publish scientific results, scientists are often guided towards publishing (as often as they can) in 'high-impact' journals which value particular types of research over others (Schekman 2013). There are many institutions involved in making this aspiration appear normal. Promotion reviews, researcher evaluations, and funding applications all privilege and value high-impact journal publications, which means that while scientists technically have a choice as to where and how to publish their results, their choices are influenced by this network of norms. Consequently, these institutions direct and incentivize particular patterns of scientific choices: not all choices are equally likely.

If we expand our purview beyond publishing, we might also consider institutions (norms, incentives, habits) shaping scientific practices relating to funding, data access, expectations of social impact, collaboration across disciplines and organizations, or the valuing of teaching and mentorship, among others. The institutions involved in guiding these practices cut across government, industry, academic and research institutions such as universities or scholarly societies, journals, the popular media, (increasingly) social media, all the way down to the scale of interpersonal, collegial relationships. Corporations, for example, might affect scientific norms by incentivizing a range of scientific behaviours (such as journal publishing guidelines), by getting involved in the regulation of science and technology or through creating demand for a scientific workforce with specific skill sets. Media organizations affect what is seen as authoritative, rigorous, and relevant science, which in turn affects political investments into scientific projects and practices.

All of these actors, roles, and relationships contribute towards shaping what is thought to be a normal or desirable choice of theory, method, and so on. These institutions influence choices about what, why, and how we study environments by creating push and pull factors to select for particular intersections of theory, method, data, analysis, and application. Figure 4.4 illustrates this idea by adding a distribution curve to each set of choices about theory, methodology, and so on. For each of these domains, institutional incentives, constraints, and norms contribute towards making some types of choices more likely or desirable than others.

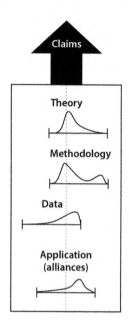

Fig. 4.4 Not all possibilities relating to theory, method, data, and application are equally likely. Institutions affect the relative likelihood of each choice, and this can be represented as a probability distribution mapped onto each moment of choice. The peak of the curve represents the most popular, prestigious, or feasible choice for a researcher in a given context

Table 4.1 summarizes a selection of prominent trends in the field of glaciology and some of the emerging norms shaping these. Climate change is increasingly a motivational driver for glaciological theory and research, incentivizing scientists to explicitly imbue their work with practical meaning and champion its public importance. As public concern about climate change has continued to increase, scientific communities, national governments, and international bodies (e.g. the United Nations) have collectively directed funding towards studies that support prediction of climate change impacts, particularly global impacts such as sea level rise. Similarly, private sector interests (such as insurance companies) have also funnelled money into climate change research, for purposes ranging from assessing impacts relevant to their operations (O'Reilly 2017) through to debunking climate change claims (Forsyth 2015). As a part of this, there is a new 'prestige' for climate change research within the scientific community that amplifies the reward structure and incentives for scientists to make particular decisions about theory. The International Glaciological Society hosts symposia and conferences around climate change themes, privileging and strengthening particular scientific values, theories, and choices.

Table 4.1 A selection of prominent trends in the field of glaciology and some of their institutional drivers

Aspect	Trend	Institutional drivers
Theory	Anthropogenic climate change as a motivational framework	Public and government interest; enthusiasm and prestige within scientific community; funding availability
Method	Numerical modelling and prediction	Continued prestige in scientific community; growing opportunities for training and publishing, desire for prediction
Data	Remote sensing data	Interest in global claims; power centres of data production, industry data providers
Applications	Empirical description and analysis of particular glaciers	Government need for climate change planning for adaptation/livelihoods; industry interest in economic costs/ opportunities of climate change

High-impact-factor journals prioritize 'big' claims about climate change processes and pathways, often detracting from locally grounded studies. Collectively, these institutions mainstream climate change as a framework for theory, which may lead to fewer opportunities for glaciological research that cannot be justified in this framework. This narrows our representations and interpretations of glaciers.

Trends in methods, data, and applications mirror these trends in theory. Methods that support global-scale prediction and integration with climate change models grow in popularity (Spencer and Lane 2017). Improvements in computing facilitate the propagation of complex glaciological theories that can be tested numerically (Knight 1999). These computing technologies provide a material motivation for utilizing data from remote sensing. However, the drivers of change in methods and analysis are not simply about practicalities; governments and scientific organizations increasingly value 'big data' applications, global-scale claims, prediction, and complex numerical theory in and of themselves.

Through funding, training, publishing opportunities, and other social norms, these institutions (patterned practices) incentivize the proliferation of research that is already abstracted from material reality, requiring numerically complex and yet reductionist views of glaciers and ice sheets. Although such trends provide new opportunities for large-scale insights, they decrease the apparent need for and efforts to create intensive field-based measurements. Furthermore, they support a shift towards data provision by private industry (e.g. from U.S. government-owned Landsat satellites to private sector remote sensing companies such as DigitalGlobe) and create power centres in data providers (e.g. NASA) (Shean et al. 2016). On top of this, growing interest in the social and economic impacts of climate change is increasing demand for application-driven climate change research; however, certain types of approaches (narrow, reductionist) and impacts (mainly economic) are receiving the lion's share of investment (e.g. see Castree 2016; Carey et al. 2017).

These examples are merely illustrative and are not intended as a comprehensive account of dominant trends in glaciological research, and the trend categories we have chosen could be split or repackaged to be more precise and/or accurate along particular dimensions. With regard to our broader conceptual framework, we have used simple categories such as theory, methodology, data, and application as starting points for general illustration, rather than complete representation. For example, one could easily identify multiple choices about theory that a scientist must face. There is not just one axis of measurement for theory or any other type of choice in science. One might consider the gender composition of different communities of theory, for example, or the extent to which different glaciological theories align with funding opportunities. This would suggest two different moments of choice relating to theory, and in our model this could suggest that theory represents not just one but at least two types of choices that can be plotted. Our categories were intended only as heuristic starting points, and as such we encourage scientists to identify important moments of choice within their own research process, as a way to compare and evaluate the consequences of different choices.

Conclusion

Scientists make many choices in environmental research. These choices are shaped by relationships and values that are internal and external to the scientific investigator (see Merton 1973). This chapter has constructed a framework for understanding the politics of environmental science in order to identify and interrogate these dynamics, in order to encourage scientists to take deeper responsibility for the societal outcomes of their work. By opening up the black box of science to consider how science itself is small-p political (involving value-laden choices and consequences), we have made the case that environmental science itself is a socio-natural hybrid (e.g. see Ashmore 2015). Rather than sitting outside of social relationships and representing a singular biophysical 'truth' about environmental processes, environmental science is enmeshed in social relations, from questions about theory, method, and data to application contexts and activist scientists. Across all of these domains, scientists are making consequential choices about their practices, and while these choices might be pushed in certain directions by various norms and institutions, scientists themselves still have agency in deciding what to do and why.

We began by offering Big-P Politics as a conventional way of thinking about the relationships between scientists, values, and society. Activism in this vein takes the claims of science and seeks to relate these claims to particular societal values as well as other scientific claims. While this model of Politics is a valuable starting point, it is not the only way in which scientists can choose to affect society. We proceeded to identify some important moments of choice *within* the black box of science, and we explored how these choices are both shaped by—and in turn shape—the socio-political contexts and meanings of science. To recognize that science is political does not make it untrue or unhelpful (e.g. see Sarewitz 2004). Rather, recognizing the political nature of environmental science helps us realize how even the most rigorous claims produced through science reflect value-laden choices and thus only represent a particular cross section of reality (Forsyth 2015). That cross section is justified, designed, and mediated by social relationships, and if we as scientists can understand and take responsibility for these relationships, then perhaps we can choose to conduct a different kind of work to produce a different kind of society.

References

Ashmore, P. 2015. Towards a sociogeomorphology of rivers. *Geomorphology* 251: 149–156.

Bauer, B.O., T.T. Veblen, and J.A. Winkler. 1999. Old methodological sneakers: Fashion and function in a cross-training era. *Annals of the American Association of Geographers* 89 (4): 679–687.

Blue, B., and G. Brierley. 2016. "But what do you measure?" Prospects for a constructive Critical Physical Geography. *Area* 48 (2): 190–197.

Callon, M. 1999. The role of lay people in the production and dissemination of scientific knowledge. *Science, Technology and Society* 4 (1): 81–94.

Carey, M., et al. 2016. Glaciers, gender, and science: A feminist glaciology framework for global environmental change research. *Progress in Human Geography* 40 (6): 1–24.

———. 2017. Impacts of glacier recession and declining meltwater on mountain societies. *Annals of the American Association of Geographers* 107 (2): 350–359.

Castree, N. 2016. Geography and the new social contract for global change research. *Transactions of the Institute of British Geographers* 41 (3): 328–347.

Church, M. 2009. Relevance: The application of physical geographic knowledge. In *Key concepts in geography*, ed. N.J. Clifford et al. London: SAGE.

———. 2011. Observation and experiments. In *The SAGE handbook of geomorphology*, ed. K. Gregory and A.S. Goudie, 121–142. London: SAGE.

Clark, W.C., et al. 2016. Crafting usable knowledge for sustainable development. *Proceedings of the National Academy of Sciences* 113 (17): 4570–4578.

Clarke, G.K.C. 1987. A short history of scientific investigations on glaciers. *Journal of Glaciology* (Special Issue): 4–24.

Cruikshank, J. 2005. *Do glaciers listen? Local knowledge, colonial encounters, and social imagination*. Vancouver: UBC Press.

Cullum, C., et al. 2016. Ecological classification and mapping for landscape management and science. *Progress in Physical Geography* 40 (1): 38–65.

Davenport, C., et al. 2015. Greenland is melting away. *The New York Times*. http://www.nytimes.com/interactive/2015/10/27/world/greenland-is-melting-away.html.

Delborne, J.A. 2008. Transgenes and transgressions: Scientific dissent as heterogeneous practice. *Social Studies of Science* 38 (4): 509–541.

Dozier, J. 1976. An examination of the variance minimization tendencies of a supraglacial stream. *Journal of Hydrology* 31: 359–380.

Fernández, R.J. 2016. How to be a more effective environmental scientist in management and policy contexts. *Environmental Science and Policy* 64: 171–176.

Forsyth, T. 2003. *Critical political ecology: The politics of environmental science*. Oxon: Routlledge.

———. 2015. Integrating science and politics in political ecology. In *The international handbook of political ecology*, ed. R.L. Bryant, 103–116. Cheltenham, UK: Edward Elgar.

Goldman, M.J., P. Nadasdy, and M.D. Turnder, eds. 2011. *Knowing nature: Conversations at the intersection of political ecology and science studies*. Chicago: University of Chicago Press.

Iken, A. 1981. The effect of the subglacial water pressure on the sliding velocity of a glacier in an idealized numerical model. *Journal of Glaciology* 27 (97): 407–421.

Kennedy, B.A. 2006. *Inverting the earth: Ideas on landscape development since 1740*. Malden, MA: Blackwell.

Knight, P.G. 1999. *Glaciers*. Cheltenham, UK: Stanley Thornes (Publishers) Ltd.

Kuhn, T.S. 1996. *The structure of scientific revolutions*. 3rd ed. Chicago: University of Chicago Press.

Lane, S.N. 2001. Constructive comments on D Massey's 'Space-time, "science" and the relationship between physical geography and human geography'. *Transactions of the Institute of British Geographies* 26 (2): 243–256.

———. 2017. Slow science, the geographical expedition, and Critical Physical Geography. *The Canadian Geographer* 61 (1): 84–101.

Larson, B. 2011. *Metaphors for environmental sustainability: Redefining our relationship with nature*. New Haven: Yale University Press.

Latour, B. 1987. *Science in action: How to follow scientists and engineers through society*. Cambridge, MA: Harvard University Press.

Lave, R., et al. 2014. Intervention: Critical Physical Geography. *The Canadian Geographer* 58: 1–10.

Merton, R.K., 1973. *The sociology of science: Theoretical and empirical investigations.* Ed. N.W. Storer. Chicago: University of Chicago Press.

Noh, M.-J., and I.M. Howat. 2015. Automated stereo-photogrammetric DEM generation at high latitudes: Surface Extraction with TIN-based Search-space Minimization (SETSM) validation and demonstration over glaciated regions. *GIScience & Remote Sensing* 52 (2): 1–20.

O'Reilly, J. 2017. *The technocratic Antarctic: An ethnography of scientific expertise and environmental governance.* Ithaca and London: Cornell University Press.

Olefs, M., and A. Fischer. 2008. Comparative study of technical measures to reduce snow and ice ablation in Alpine glacier ski resorts. *Cold Regions Science and Technology* 52 (3): 371–384.

Rhoads, B.L., and C.E. Thorn. 2011. The role and character of theory in geomorphology. In *The SAGE handbook of geomorphology*, ed. K.J. Gregory and A.S. Goudie, 59–77. London: SAGE.

Salas-Zapata, W.A., L.A. Rios-Osorio, and A.L. Trouchon-Osorio. 2013. Typology of scientific reflections needed for sustainability science development. *Sustainability Science* 8: 607–612.

Salmond, J.A., M. Tadaki, and M. Dickson. 2017. Can big data tame a "naughty" world? *The Canadian Geographer* 61 (1): 52–63.

Sarewitz, D. 2004. How science makes environmental controversies worse. *Environmental Science and Policy* 7: 385–403.

Schekman, R. 2013. How journals like nature, cell and science and damaging science. *The Guardian.* https://www.theguardian.com/commentisfree/2013/dec/09/how-journals-nature-science-cell-damage-science.

Shean, D.E., et al. 2016. An automated, open-source pipeline for mass production of digital elevation models (DEMs) from very-high-resolution commercial stereo satellite imagery. *ISPRS Journal of Photogrammetry and Remote Sensing* 116: 101–117.

Sherman, D.I. 1996. Fashion in geomorphology. In *The scientific nature of geomorphology*, ed. B.L. Rhoads and C.E. Thorn, 87–114. Chichester, UK: John Wiley & Sons Ltd.

Sismondo, S. 2010. *Introduction to science and technology studies.* Chichester, UK: John Wiley & Sons Ltd.

Smith, L.C., et al. 2015. Efficient meltwater drainage through supraglacial streams and rivers on the southwest Greenland ice sheet. *Proceedings of the National Academy of Sciences of the United States of America* 112 (4): 1001–1006.

Spencer, T., and S.N. Lane. 2017. Reflections on the IPCC and global change science: Time for a more (physical) geographical tradition. *The Canadian Geographer* 61 (1): 124–135.

Tadaki, M. 2017. Rethinking the role of critique in physical geography. *The Canadian Geographer* 61 (1): 73–83.

Thatcher, J., D. O'Sullivan, and D. Mahmoudi. 2016. Data colonialism through accumulation by dispossession: New metaphors for daily data. *Environment and Planning D: Society and Space* 34 (6): 990–1006.

Vaughan, D.G., J.C. Comison, et al. 2013. Observations: Cryosphere. In *Climate change 2013: The physical science basis. Contribution of working group I to the Fifth Assessment Report of the Intergovernmental Panel on Climate Change*, ed. T.F. Stocker et al. Cambridge, UK and New York: Cambridge University Press.

5

The Impacts of Doing Environmental Research (Core Tenet #3)

Justine Law

Introduction

In the previous chapter, King and Tadaki addressed the politics of scientific inquiry, as well as the politics around who gets to be a scientist. They demonstrated that environmental scientists—who, like all of us, are enmeshed in social relations—make value-laden choices about the theories, methods, and analytical tools they use. This chapter expands on King and Tadaki's exploration of the framing of scientific questions to consider the socioecological impacts of asking questions and producing answers. I am interested, in other words, in the impacts of doing research.

Researchers are not passive observers of phenomena, hidden away behind rocks, bushes, or buildings. Rather, through our practices of research and our production of knowledge, we become agents of change. We create new research landscapes, wherein the social and biophysical features of the landscape are altered *by our study of them*. A number of these changes happen via fieldwork, since fieldwork (unlike laboratory experiments, secondary data analysis, or computer modeling) takes place in preexisting, "open air," and often peopled spaces that researchers cannot control. Examples of fieldwork impacts include removing samples, modifying social relations, or recalibrating social imaginaries in the research site. Other socioecological changes occur when our research is published and/or incorporated into environmental policy and practice, such as the forcible removal of "out-of-place" people and species, as determined by

J. Law (✉)
Hutchins School of Liberal Studies, Sonoma State University,
Rohnert Park, CA, USA

© The Author(s) 2018
R. Lave et al. (eds.), *The Palgrave Handbook of Critical Physical Geography*,
https://doi.org/10.1007/978-3-319-71461-5_5

academic classification systems and discourses, from conservation spaces (e.g. Davis and Kull, both this volume). And this knowledge diffusion inevitably aligns our research with particular applications and/or agendas and therefore particular politics—a process that is only accelerating given current pressures to produce scholarship with explicit social relevance. I consider both of these areas of impact in this chapter.

We cannot be blasé about these research impacts, especially in the Anthropocene, an era in which we are engaged in global-scale experimentation (see Latour 2011; Callon et al. 2009). In this new era, many scientists, philosophers, and practitioners are christening our planet an "experimental society" (Krohn and Weyer 1994), a "laboratory Earth" (Grossman 2016) filled with "climate change experiments" (Bulkeley and Broto 2013). But it is critical to recognize that a "laboratory Earth" or "experimental society" is not globally homogenous. Research does not drape itself across the landscape like a blanket. Rather, it produces uneven contours as it differentially enrolls and affects humans and nonhumans. Some places may experience a heavy research hand, while others may escape examination and experimentation. This unevenness, and the winners and losers resulting from it, provide a wide opening for the insights of a Critical Physical Geography fluent in the knowledge politics of environmental science—both its production and application.

I do not intend for this chapter to be an all-inclusive list of research impacts. Instead, I have assembled impacts that I believe to be particularly salient to Critical Physical Geography. As noted above, I divide research impacts into two broad categories: fieldwork impacts and impacts of our research once published or released. I begin with perhaps the more straightforward of the two: fieldwork impacts on biophysical environments. In the following sections, I consider fieldwork impacts on communities and society. Then, I briefly outline some of the political implications of mapping, data collection, and data classification; in essence, I discuss why making research spaces legible might matter to the humans and nonhumans in them. The final, and longest, body section delves into the impacts of producing knowledge in the Anthropocene. I conclude with suggestions for managing these research impacts, as well as a few directions for future research.

Fieldwork Impacts on Biophysical Environments

This first point is almost self-evident: the study of biophysical environments may lead to minor or major modifications of those environments. To illustrate the ubiquity of these modifications, I offer the example of the forest reserve I took my students to multiple times last semester. In that 300+ acre reserve, researchers have extracted tree cores, planted trees, harvested trees, trapped small mammals and amphibians, installed deer fencing, taken soil samples, mowed clearings, and sprayed glyphosate on nonnative plant species. Each of these field methods changed the structure and function of the reserve's ecosystem. Some of these changes may be small, and some may be transformative; however, regardless of the scale of impact, the above list demonstrates how forest research often can be a form of forest management.

Of course, forestry and forest science are not the only fields that affect biophysical environments through fieldwork. Geologists remove rock samples and fossils. Geomorphologists, glaciologists, and hydrogeologists drill boreholes. Hydrologists install weirs and gauges to measure rainfall and river flow. Fire scientists set controlled burns to test the effects of different fire frequencies and intensities. Wildlife biologists manipulate animal behaviors and habitats or even reintroduce extirpated species into ecosystems. Soil scientists use portable rainfall simulators to produce measurable soil erosion in remote areas. And, even when we collect evidence remotely via satellite, internet databases, and aerial photography, we still can change biophysical environments by changing terminology, data sets, classification systems, and management regimes. Indeed, when you consider how many sites researchers have used over the last few centuries, it is likely that we have altered millions of acres of Earth's surface. This point, again, is a simple one, but it is worth stating since so many physical geographers engage in fieldwork practices that change environments—and therefore human and nonhuman uses and perceptions of those environments.

Fieldwork Impacts on Social Relations

Whether we study biophysical or social phenomena, we tend to interact with people in and around our research sites. For researchers who use ethnographic methods, these interactions *are* the research. For those who use biophysical methods, these interactions include hiring guides, interpreters, and research assistants or getting advice from local practitioners or community members about the best field sites. The methodology, however, is immaterial. The point I want to make is that researchers cannot interview everyone, hire everyone, or associate with everyone in a community, and these differences in treatment help create uneven research landscapes in which some community members benefit (or are harmed) more than others. In other words, our research practices inevitably reshape social relations in our field sites.

This is particularly true in regions where researchers have much more wealth and power than the research subjects and/or local community members. For example, in my own research on wood energy in Vermont and Michigan (Law 2017), I interviewed foresters, loggers, business owners, and power plant managers who had more wealth and power than I did as a graduate student—or, frankly, than I do now as a faculty member—and they had little to gain from their interactions with me. If anything, it is more likely that their willingness to participate in the research roused the suspicion of their coworkers and subtly affected their social networks in that way. In contrast, in regions where a guide or research assistant's salary can become a meaningful component of a family's livelihood, or a conversation with a university-appointed scholar can assign prominence to a community member, the consequences of our fieldwork decisions can be immense.

Paige West documents the power of our research practices in her book *Conservation is Our Government Now: The Politics of Ecology in Papua New Guinea* (2006), an account of the tensions between a conservation project and a local village in Papua New Guinea. This conservation project, the creation of the Crater Mountain Wildlife Management Area (CMWMA) by the Research and Conservation Foundation of Papua New Guinea (RCF), has brought in researchers from all over the world to study the island's biodiversity. A second component of the project's mission is to give the local village "development" in exchange for their participation in, and land for, "conservation" (West 2006). This partnership between the village and the CMWMA was fairly novel for the 1990s, a time when most conservation projects still sought to exclude local communities from protected areas (Neumann 2004).

A large portion of the "development" the community received came in the form of wages for carrying researchers' bags, acting as guides, and assisting the researchers in a variety of other ways. Another portion came in the form of funding for community projects (e.g. teachers, water wells) in exchange for the long-term monitoring of various research projects (e.g. on harpy eagle populations). However, as suggested earlier, some community members "derived status and benefits from [their] association with the project[s]" and some did not (West 2006, p. 194). Women, in particular, became further marginalized as men shifted their labor to "conservation" (West 2006).

The village, as West shows, recognized how the CMWMA altered the social relations between its inhabitants. For example, community members complained when the RCF paid one person to do a job but not another or paid the school fees of one resident's child but not another's (West 2006). Women also became increasingly frustrated by their lack of enrollment in the CMWMA's research and began to protest these inequalities at conservation management committee meetings. Perhaps the most compelling evidence of the impact of research on the village, though, is a knife fight over who is "in charge of conservation" (2006, p. 16).

Overall, the example of the CMWMA in Papua New Guinea is a striking demonstration of the new social landscapes that research can create, and it is important to note that West does not exempt herself from this burden. Her own research practices, no matter how reflexive or attentive to local power dynamics, still provoked fights over who should receive her payments and still prioritized some villagers over others (e.g. her "parents" in the village, the beneficiaries of money for doctor's visits). None of us are exempt from exerting these kinds of influences on our field sites.

Fieldwork Impacts on Socioecological Imaginaries

The previous two sections outlined types of impacts that researchers should consider before and during their fieldwork. Many grant applications, for example, include a section on potential impacts, and all researchers working with human subjects need to complete Institutional Review Board (IRB) training. Typically, however, we are not forced to consider how the sheer act of asking questions, of doing research, might disturb the spaces in which we conduct research. To be sure, plenty has been written about the "observer effect," which suggests that researchers always influence the phenomena they study (Monahan and Fisher 2010), but this is not quite what I mean. Here I am arguing that, simply by studying particular phenomena, we give weight to them, and this added weight may cause people in our study regions to view these phenomena differently and/or behave differently.

Consider an example: physical geographer Jeff La Frenierre conducts research on glacial change and water availability in the Ecuadorian Andes. Although his primary sources are climate records and hydrologic data, La Frenierre also interviews residents in the villages around Chimborazo, the tallest mountain in Ecuador, about their perceptions of the environmental changes occurring around them (La Frenierre and Mark 2017). When talking about his work, La Frenierre describes how the local residents he interviews are all very aware of the glacial retreat in the region but, in these interviews, tend to blame local resource use and land management practices for the retreat. The interviews, therefore, often introduce the notion that this glacial change is the result of global climate changes, which may alter residents' perspectives about glacial change. And La Frenierre is particularly careful when talking about water. He recognizes that his work (i.e. his questions about the potential loss of glacial melt and corresponding increase in water scarcity) could incite water conflict. The Andean communities he studies are already feeling the effects of glacier change on water availability, and his research suggests that water scarcity will only get worse, and there is very little they can do about it (La Frenierre and Mark 2017). As such, La Frenierre worries about causing panic in the communities in which he works.

In my own fieldwork, I have noticed how particular lines of questioning can cause research participants to examine an issue from a new perspective, to modify their statements partway through an interview, or to view me and my research collaborators with (heightened) suspicion. In one instance (Law and McSweeney 2013), in fact, a landowner I interviewed ended up changing components of his forest management plan after his participation in the research project caused him to discover a new interest in non-timber forest products. In sum, even without removing samples, altering the biophysical environment of our field sites, or embroiling some community members in our research, we can change the landscapes and communities in which we work. Simply being there, and asking questions, is enough to accomplish this.

The Impacts of Our Research Results

Once conducted, our research results can have similarly consequential socio-ecological effects by making legible that which was not legible before and/or by influencing management policy and practice. For example, even the simple act of placing a camera on a drone changes our understanding of glacial retreat in mountain landscapes (Wigmore and Mark 2016), of immigrant movements along the United States-Mexico border (Nixon 2016), and of the decline of forest cover in southeastern Asia (Koh and Wich 2012)—and consequently how we and others respond to these phenomena.

In many cases, this increased legibility may promote positive changes in local communities. To go back to the Andes, if a community is better able to respond to water scarcities caused by the loss of glacier melt, then the research landscape has been changed for the better. But increased legibility, whether by remote methods or field methods, may bring about an increase in the primary accumulation of resources, in the potential for outsider occupation, and in the marginalization of vulnerable populations as well. Others have written extensively about the ethics of this with regard to, for example, the Bowman Expeditions' first project, *México Indígena* (see Grossman 2012). The stated goal of *México Indígena*, which was led by University of Kansas

Professor Peter Herlihy, was to use participatory mapping and GIS to investigate how a state land certification program altered land use and land tenure in Mexican indigenous communities (Herlihy 2010). The project was funded, in part, by the Foreign Military Studies Office, and controversy arose when some community members insisted that Herlihy's research team never disclosed its funding sources or the eventual recipients of its findings, one of which is a military intelligence and weapons contractor for the US Department of Defense (Grossman 2012). In short, the community members were concerned that the research findings might be used against them and/or used for applications to which they did not consent.

A separate set of concerns arises when we classify the spaces we have made legible (e.g. as a glacier vs. a moraine, as a high-traffic vs. low-traffic immigration route, or as a deciduous vs. coniferous forest). Such classification systems, which reflect both on-the-ground realities *and* our perceptions of these realities, tend to prescribe management regimes—whether we intend for them to or not. For example, forest categories may be used to advocate for local forest ownership (Nightingale 2003) or to rubber stamp nonnative shrub plantations that have little value to local farmers (Robbins 2001). Similarly, classifying an ecosystem as "arid," which is a category that we often view as less structurally or functionally desirable than "temperate," may result in attempts to "fix" that ecosystem (Davis, this volume). And such "fixes," of course, will have impacts on the landscape and the people in it.

Moreover, the increasing emphasis on producing "relevant" research in many academic settings both obscures the myriad ways in which research already had such impacts *and* funnels those impacts into a potentially narrower set of channels. As many observers have noted, there has been a clear trend since the 1990s to encourage—and increasingly to require—academics to produce knowledge that meets social and/or economic needs, as defined by funding agencies and administrators, not by communities themselves (e.g. Canaan and Shumar 2008; Gibbons et al. 1994; Lave et al. 2010). This has the virtue of emphasizing the transformative potential of research, even to those academics who had not previously considered their work in this light. At the same time, the emphasis on impactful research, paradoxically, has made it more difficult to produce research with positive impacts. And, notably, as we shift our work away from basic problems and into "specific programmes funded by external agencies for defined purposes" and are compelled to jump from topic to topic in search of funding (Gibbons et al. 1994, p. 78), our ability to address complex environmental issues seems far more likely to decrease than to improve.

Research in the Anthropocene

In this new geologic era, the Anthropocene, many scholars and popular writers are talking about the "global experiment" in which we are now engaged (e.g. Latour 2011; Powell 2007; Rojas 2015; Krohn and Weyer 1994; Ackerman 2015; Yong 2017). The Earth has become, the argument goes, a laboratory within which we are testing a whole range of techniques for dealing with environmental change: designing and managing novel eco-systems (Marris 2011), manipulating species' genes for purity (Biermann 2014), growing biomass plantations for energy (Ragauskas et al. 2006), geoengineering our climate (Keith 2000), creating global carbon markets (Callon 2009), paying villages for ecosystem services (Jayachandran et al. 2017), introducing engineered mosquitoes into tropical areas to prevent the spread of disease (Harris et al. 2011), and devoting half of Earth's land to preservation (Wilson 2016), just to name a few compelling examples.

Such global, "open-air" experiments alter research norms in important ways. First, global experiments often require powerful partners. As such, research in the Anthropocene demands more collaboration with NGOs, government agencies, and private sector firms. These collaborators help scientists set the terms of the research, design the research questions, develop the methodologies, and execute the project. The Large-Scale Biosphere-Atmosphere Experiment in Amazonia (LBA) that Rojas writes about, for instance, was a collaboration between NASA, the Brazilian Ministry of Science and Technology, and dozens of universities, firms, institutes, and government agencies (2015). Second, these global experiments require researchers to navigate the public realm more visibly (and carefully) than they ever had to before. As a result, many fields are having conversations about the "scholar/activist divide" that researchers must straddle (e.g. Castree 2016; Roston 2017; Epstein 1996). Third, the research itself becomes higher stakes. These global experiments are seen, as Karvonen and Van Heur explain, as "collective learning processes with contingent boundaries" (2014, p. 387). They continue: "real-world experimentation is founded on the idea that one is compelled to act despite uncertainties and gaps in knowledge" (Karvonen and Heur 2014, p. 387). Latour, likewise, claims:

> Far from waiting for absolute certainty before moving the little finger, we know we have to experiment and distribute equally audacity and what in German is called so beautifully *Sorge* and what we call in French *le souci*. Care and caution go together with risk-taking. (2011, p. 13)

For some, the transfer of experimentation out of the laboratory and into society is a positive step. In the same text as cited above, Latour argues that "we have all been made (most of the time unwillingly) co-researchers and we are all led to formulate research problems" (2011, p. 14). His point is that putting research out in the world, where its questions, goals, and methods are negotiated by non-scientists, may make research more democratic. Callon makes a similar point in his investigation of carbon markets (2009). He sees carbon markets as big, messy experiments that we need to embrace. We need, he says, to get scientists and other stakeholders at the economists' table if we want to be able to "civilize" these markets (Callon 2009). And, importantly, if these experiments work, we may be able to civilize other markets as well. In sum, there are reasons to view the global experiments of the Anthropocene with hope.

Others are much less hopeful. The critiques of collaborative global experimentation tend to center on one set of related questions: who, exactly, has a seat at the table in these collaborations? Do only the wealthy and powerful have a voice—or, at least, the loudest voice—during negotiations? And, if this is the case, will research become a tool for legitimizing capitalist agendas or neoliberal governance strategies? Rojas, for example, argues that "celebratory conclusions regarding open-air experimentation are misguided" in his analysis of scientists who were researching soil, forest, and agricultural systems in the Brazilian Amazon (2015, p. 136). Although these scientists were interested in nonhuman responses to human-driven environmental change, and although they tried to demonstrate that these responses were occurring "*because* of ecological disruptions driven by capitalist operations and lifestyles" (Rojas 2015, p. 141), their research was subordinate to the goals of the agro-industrial actors in the region. In Rojas's words, they:

> Did not command but rather fiddled within capitalist networks working alongside non-human entities under the premise that agro-industrial intensification could not be halted … their combination of experimental forest management strategies and capitalist agro-industrial experiments would create new worlds indeed, but scientists expected these worlds to be less than hospitable. (2015, p. 142)

Similarly, when discussing a rewilding project in the Netherlands, Lorimer and Driessen worry that, because novel ecology is a bit wishy-washy in its prescriptions and/or abilities to diagnose ecological failures, "rewilding could offer a convenient gloss for cutting expensive subsidies, waiving restrictive conservation legislation and even the accelerated implementation of markets in ecosystem services" (2014, p. 179). Evans likewise suggests that scientific theories and methods can frame our governance options too narrowly, and she offers examples of sustainable urban design projects wherein "truth becomes synonymous with success" (2011, p. 232). In other words, if a project succeeds economically for urban developers, it has produced a "truth" about urban design. The success of small businesses, the success of mixed-income housing projects, or the success of pollinators are not considered, demonstrating the (explicit or implicit) political commitments of the researchers. Thus, when research extends its scope into the world—into the "open air" to deal with the challenges of the Anthropocene—the resulting landscape may be less sustainable, less equitable than the one we started with.

In many ways, the challenges of "open-air" and "real world" experiments in the Anthropocene are nothing new. Researchers always have had to bring their work out of the laboratory to demonstrate its value and applicability (Latour 1983). Additionally, as Gross points out using the example of Jane Addam's Hull House, we have been using the real world as our laboratory for a long time (2009). And research has certainly furthered both desirable agendas, such as polio vaccination, and undesirable agendas, such as colonialism, before. Still, the stakes feel a bit higher in the Anthropocene. This is partly because of the pace of environmental change, partly because of the scale of the research that is being done (e.g. geoengineering), and partly because this research is more widely reported—more in your face, so to speak. But perhaps the greatest difference between research in the Anthropocene and research in previous eras is that the Anthropocene has been cast as *a time and place for experimentation* (Yang 2017; Hawken 2017; Schmitz 2016; Purdy 2015; Brondizio et al. 2016). In other words, open-air, and sometimes global, experiments have become normalized, even encouraged. As such, this new era of research demands more critical examination. If we are to experiment with the world, we need to craft these experiments reflexively, thinking critically about what political commitments shape our research agendas and how our findings might impact the landscapes and communities under study.

All of this means that in the Anthropocene, research will rapidly—and perhaps dramatically—rearrange socio-biophysical landscapes. To refer back to the list I enumerated at the beginning of this section: geoengineering may alter the amount of carbon on Earth and in its atmosphere and/or change the amount of sunlight hitting Earth's surface; conservation projects may change the genetics of nonhuman species; biodiverse forests may become plantations of *Miscanthus* or poplar grown to support "green" energy production; and novel, designed coral reefs may cover our shallow oceans. These changes will matter greatly to humans and nonhumans, particularly because they will produce winners and losers (Ogden et al. 2013). For instance, an artificial coral reef may support macro-algal species but not corrallivorous fish. It also might support an island's tourist industry but not its fishermen and women. Meanwhile, an energy economy dependent on woody biomass plantations might benefit deer, birds, and large landowners but not lichen, forest carnivores, and small landowners. The big questions, then, are: what are the ethics of this real-world experimentation? Who gets to determine the winners and losers in each of these experiments? And do we even understand biophysical or social systems well enough to make such determinations in a time of rapid environmental change? These are difficult questions, but, with its emphases on theorizing power *and* possessing a deep knowledge of biophysical systems, Critical Physical Geography may be well positioned to answer them (Lave et al. 2014).

Discussion and Future Directions

In sum, the fieldwork we do and the knowledge we produce as researchers deeply impact the spaces and communities in which we work—even if not in the distinguished and/or auspicious ways we might hope. Given this, I would like to suggest three directions forward for critical physical geographers. First, we need to be more reflexive about the impacts of our fieldwork practices and the fieldwork practices of other environmental scientists. We need to consider how we alter the biophysical landscape, how we decide who to talk to and who to enroll in our research, how we discuss our work, and how our presence, in and of itself, might disturb local politics, environmental management practices, and livelihoods. That does not mean that simply by being reflexive we become a benevolent force in our field sites; however, we can attempt to aim for more socially just, environmentally innocuous outcomes.

Second, to build upon a point introduced by King and Tadaki (this volume) we need more research *on* environmental research. How do scientists operate in the field? What are they attendant to? What impacts of their fieldwork do they overlook? How do they select collaborators, engage in responsible decision-making in "open-air" experimentation, and fit into the tangled research networks many scholars now work within? How do they view the politics of their work? And how does their research get selected for, and ultimately used in, real-world applications? Rebecca Lave's *Field and Streams* is an excellent example of this research *on* environmental research (2012). In the text, Lave shows how stream restoration science morphed into a form that fit seamlessly within its neoliberal context—a change that has altered stream ecosystems (for better or worse) through the United States (2012). Irus

Braverman's *Wild Life: The Institution of Nature* is another text that delves into the messy, unsettled convictions of conservation scientists (2015). We need much more work like this if we truly want to understand how our work impacts landscapes and communities around the world. Moreover, we need this work to speak to the community it is critiquing. A critique of forest ecology, for instance, the biophysical field I am most familiar with, achieves very little *in* forest ecology if it cannot speak a forest ecologist's language. And this is one place where critical physical geographers have an upper hand over others working in critical theory; because critical physical geographers are able to speak the language of forest ecology or fluvial geomorphology or environmental engineering, their arguments (hopefully) gain more traction in those fields.

Third, we must recognize that the "scholar/activist divide" is, at best, a permeable divide. In some cases, scholars intentionally tear down this divide to engage the public and inspire change. The public outreach efforts of climate change researchers like James Hansen (Hansen and Kivlehan 2017) and Michael Mann (Mann and Toles 2016) are clear examples of such activism. In other cases, scholars promote their work's "broader impacts" on society—in fact, this is now a requirement for successful National Science Foundation grant proposals. But even if a scholar does not intend to embrace activism or improve society, the knowledge she produces is political. A quantitative analysis of the availability of geothermal energy, for example, still presupposes that geothermal energy could be a desirable option for a region (Gross 2016), and it therefore can be just as political as the more normative (and sometimes explicitly activist) research of a political ecologist. Or to consider a related example: a viability analysis for an experimental wood biofuel plant may include the valuable insights of engineers, county commissions, and energy executives but not the community members who will be impacted by the project (Law 2017). As such, critical physical geographers need to pay particular attention to the politics of our work, as well as the work of other environmental scientists, particularly in our "experimental society" (Krohn and Weyer 1994). We need to be thoughtful about who our collaborators are (and who is not at the table—see Law, this volume); who is calling the shots with regard to research design (and what their vision is for human-nonhuman relations in the Anthropocene); who is funding the research (and what do they want); what real-world projects our work will bolster (and which it will make invisible); and who will benefit from the knowledge we produce (and who will suffer). If we can use this awareness to be as inclusive and as intellectually rigorous as possible, we may be able to achieve some of the benefits of a more democratic science wherein we have "sovereignty over our research agendas" in the Anthropocene (Latour 2011).

References

Ackerman, D. 2015. *The human age: The world shaped by us.* New York: W.W. Norton & Company.

Biermann, C. 2014. Biodiveristy, purity, and death: Conservation biology as biopolitics. *Environment and Planning D: Space and Society* 32: 257–273.

Braverman, I. 2015. *Wild life: The institution of nature.* Stanford, CA: Stanford University Press.

Brondizio, E., K. O'Brien, X. Bai, F. Biermann, W. Steffen, F. Berkhout, C. Cudennec, et al. 2016. Re-conceptualizing the anthropocene: A call for collaboration. *Global Environmental Change* 39: 318–327.

Bulkeley, H., and V. Broto. 2013. Government by experiment? Global cities and the governing of climate change. *Transactions of the Institute of British Geographers* 38: 361–375.

Callon, M. 2009. Civilizing markets: Carbon trading between in vitro and in vivo experiments. *Accounting, Organizations and Society* 34: 535–548.

Callon, M., P. Lascoumes, and Y. Barthe. 2009. *Acting in an uncertain world: An essay on technical democracy.* Cambridge, MA: MIT Press.

Canaan, J., and W. Shumar. 2008. *Structure and agency in the Neoliberal University.* New York: Routledge.

Castree, N. 2016. Geography and the new social contract for global change research. *Transactions of the Institute of British Geographers* 41: 328–347.

Epstein, S. 1996. *Impure science: AIDS, activism, and the politics of knowledge.* Berkeley, CA: University of California Press.

Evans, J. 2011. Resilience, ecology and adaptation in the experimental city. *Transactions of the Institute of British Geographers* 36: 223–237.

Gibbons, M., C. Limoges, H. Nowotny, S. Schwartzman, P. Scott, and M. Trow. 1994. *The new production of knowledge: The dynamics of science and research in contemporary societies.* London: SAGE.

Gross, M. 2009. Collaborative experiments: Jane Addams, Hull House and experimental social work. *Social Science Information* 48: 81–95.

———. 2016. Give me an experiment and I will raise a laboratory. *Science, Technology, & Human Values* 41: 613–634.

Grossman, Z. 2012. Geographic controversy over the Bowman Expeditions/Mexico Indigena. https://academic.evergreen.edu/g/grossmaz/bowman.html

Grossman, D. 2016. Laboratory earth. Pulitzer Center. http://pulitzercenter.org/projects/laboratory-earth

Hansen, J., and S. Kilvehan. 2017. Ok, US government, see you in court. *The Boston Globe.* August 14.

Harris, A., D. Nimmo, A. McKemey, N. Kelly, S. Scaife, C. Donnelly, C. Beech, W. Petrie, and L. Alphey. 2011. Field performance of engineered male mosquitoes. *Nature Biotechnology* 29: 1034–1037.

Hawken, P. 2017. *Drawdown: The most comprehensive plan ever proposed to reverse global warming.* New York: Penguin Books.

Herlihy, P. 2010. Self-appointed gatekeepers attack the American Geographical Society's first Bowman Expedition. *Political Geography* 29: 417–419.

Jayachandran, S., J. de Laat, E. Lambin, C. Stanton, R. Audy, and N. Thomas. 2017. Cash for carbon: A randomized trial of payments for ecosystem services to reduce deforestation. *Science* 357: 267–273.

Karvonen, A., and B. Heur. 2014. Urban laboratories: Experiments in reworking cities. *International Journal of Urban and Regional Research* 38: 379–392.

Keith, D. 2000. Geoengineering the climate: History and prospect. *Annual Review of Energy and the Environment* 25: 245–284.

Koh, L., and S. Wich. 2012. Dawn of drone ecology: Low-cost autonomous aerial vehicles for conservation. *Tropical Conservation Science* 5: 121–132.

Krohn, W., and J. Weyer. 1994. Society as a laboratory: The social risks of experimental research. *Science and Public Policy* 21: 173–183.

La Frenierre, J., and B. Mark. 2017. Detecting patterns of climate change at Volcán Chimborazo, Ecuador, by integrating instrumental data, public observations, and glacier change analysis. *Annals of the Association of American Geographers* 107: 979–997.

Latour, B. 1983. Give me a laboratory and I will raise the world. In *Science observed: Perspectives on the social study of science*, ed. K. Knorr-Cetina and M. Mulkay. London: SAGE Publications.

———. 2011. From multiculturalism to multinaturalism: What rules of method for the new socio-scientific experiments? *Nature and Culture* 6: 1–17.

Lave, R. 2012. *Fields and streams: Stream restoration, neoliberalism, and the future of environmental science*. Athens, GA: University of Georgia Press.

Lave, R., P. Mirowski, and S. Randalls. 2010. Introduction: STS and neoliberal science. *Social Studies of Science* 40: 659–675.

Lave, R., M. Wilson, E. Barron, C. Biermann, M. Carey, C. Duvall, L. Johnson, et al. 2014. Intervention: Critical Physical Geography. *Canadian Geographer-Geographe Canadien* 58: 1–10.

Law, J. 2017. The other questions we need to be asking about wood bioenergy. *Journal of Forestry* 115: 128–133.

Law, J., and K. McSweeney. 2013. Looking under the canopy: Rural smallholders and forest recovery in Appalachian Ohio. *Geoforum* 44: 182–192.

Lorimer, J., and C. Driessen. 2014. Wild experiments at the Oostvaardersplassen: Rethinking environmentalism in the Anthropocene. *Transactions of the Institute of British Geographers* 39: 169–181.

Mann, M., and T. Toles. 2016. *The madhouse effect: How climate change denial is threatening our planet, destroying our politics, and driving us crazy*. New York: Columbia University Press.

Marris, E. 2011. *Rambunctious garden: Saving nature in a post-wild world*. New York: Bloomsbury USA.

Monahan, T., and J. Fisher. 2010. Benefits of 'observer effects': Lessons from the field. *Qualitative Research* 10: 357–376.

Neumann, R. 2004. Nature-state-territory: Toward a critical theorization of conservation enclosures. In *Liberation ecologies*, ed. R. Peet and M. Watts. London: Routledge.

Nightingale, A. 2003. A feminist in the forest: Situated knowledges and mixing methods in natural resource management. *ACME* 2: 77–90.

Nixon, R. 2016. Drones, so useful in war, may be too costly in border duty. *The New York Times*. November 2.

Ogden, L., N. Heynen, U. Oslender, P. West, K. Kassam, and P. Robbins. 2013. Global assemblages, resilience, and Earth Stewardship in the Anthropocene. *Frontiers in Ecology and the Environment* 11: 341–347.

Powell, R. 2007. Geographies of experiment. *Environment and Planning A* 39: 1790–1793.

Purdy, J. 2015. *After nature: A politics for the anthropocene*. Cambridge, MA: Harvard University Press.

Ragauskas, A., C. Williams, B. Davison, G. Britovsek, J. Cairney, C. Eckert, W.J. Frederick Jr., et al. 2006. The path forward for biofuels and biomaterials. *Science* 311: 484–489.

Robbins, P. 2001. Fixed categories in a portable landscape: The causes and consequences of land-cover categorization. *Environment and Planning A* 33: 161–179.

Rojas, D. 2015. Environmental management and open-air experiments in Brazilian Amazonia. *Geoforum* 66: 136–145.

Roston, M. 2017. The March for Science: Why some are going, and some will sit out. *The New York Times*. April 17.

Schmitz, O. 2016. *The new ecology: Rethinking a science for the anthropocene*. Princeton, NJ: Princeton University Press.

West, P. 2006. *Conservation is our government now: The politics of ecology in Papua New Guinea*. Durham, NC: Duke University Press.

Wigmore, O., and B. Mark. 2016. UAV mapping of debris covered glacier change, Llaca Glacier, Cordillera Blanca, Peru. In *Proceedings of the 73rd Annual Meeting of the Eastern Snow Conference*. Byrd Polar and Climate Research Center, Columbus, Ohio.

Wilson, E.O. 2016. *Half-earth: Our planet's fight for life*. New York: Liveright Publishing.

Yang, W. 2017. Is the 'Anthropocene' epoch a condemnation of human interference, or a call for more? *The New York Times Magazine*. February 14.

Yong, E. 2017. Artificial intelligence: The park rangers of the Anthropocene. *The Atlantic*. March 24.

Part II

CPG in Practice

6

The Trouble with *Savanna* and Other Environmental Categories, Especially in Africa

Chris S. Duvall, Bilal Butt, and Abigail Neely

The concept of environment is useful for understanding geography as the integration of biophysical and sociocultural, objective and subjective, material and nonmaterial processes (Gregory 2017; Radcliffe 2009). A formal definition of environment is the biophysical conditions, aesthetic values, and human activities that exist in a given location (Mayhew 2009: 159). Other terms can be approximately synonymous, including land cover, biome, region, and landscape. By categorizing environmental conditions, geographers can characterize and compare locations. Yet environmental categories pose inescapable problems of geographic philosophy—in short, how to identify and delimit meaningful features. These problems are resolved only through subjective sociocultural consensus rather than objective reasoning.

In this chapter, we argue that practitioners of Critical Physical Geography (CPG) must query the philosophical and sociocultural significance of environmental categories as much as their biophysical meaning. Environmental categories are simplifications meant to enable geographic generalization, which is necessary to develop predictive knowledge of biophysical reality.

C. S. Duvall (✉)
Department of Geography and Environmental Studies, University of New Mexico, Albuquerque, NM, USA

B. Butt
School for Environment and Sustainability, University of Michigan, Ann Arbor, MI, USA

A. Neely
Department of Geography, Dartmouth College, Hanover, NH, USA

© The Author(s) 2018
R. Lave et al. (eds.), *The Palgrave Handbook of Critical Physical Geography*,
https://doi.org/10.1007/978-3-319-71461-5_6

107

However, the environmental categories used in physical geography also provided bases for generalizing about human conditions. For instance, environmental determinism—the idea that environment controls human characteristics and behaviors—is founded upon ideas about biophysical conditions. In terms of transformative physical geographic scholarship (Tadaki et al. 2015), environmental categorization is an important point of engagement because it is: a social process, explicitly centered on simplification and generalization, and significant broadly across science and society (Gregory 2017; Simon 2016).

The remainder of this chapter comprises four sections. First, we outline geographic theory regarding the social nature of environmental categories. Second, we describe the history of the category of 'savanna' in Western geography. Third, we sketch the social meaning that savanna has had, particularly in Africa. In these sections, we analyze historical works from the perspective of postcolonial ecocriticism, which argues that environmental knowledge arises across whole societies and not separately in scientific and aesthetic contexts (Huggan and Tiffin 2010). Finally, we conclude that physical geographers must conscientiously avoid overgeneralizing the environmental context.

Geographic Philosophy and Society

Western environmental thought has long emphasized the relative abundance of trees versus grasses, and privileged trees over other flora (Dove 2004; Davis, this volume). Within this frame of reference, savanna represents a real but irresolvable biophysical condition. Globally, tree-grass abundance exhibits clinal variation associated primarily with water availability. There are locations where trees dominate, with humid climates and/or high soil moisture. At the other extreme, there are locations where vegetation is sparse and ephemeral because conditions are too dry for most plants. For economic reasons—particularly livestock husbandry—Europeans have long noticed the middle condition of tree-grass co-abundance. The extremes and the middle are the most distinct positions when viewed as independent samples along the continuum of tree-grass co-abundance. Nonetheless, "[a]ny definition of the limits of savanna on this continuum is unavoidably arbitrary" (Scholes 1997: 258).

The trouble with environmental categories begins with geographic ontology, or the fundamental properties of and relationships between geographic concepts (Smith 2001; Smith and Mark 2003). Features such as mountain, city, or lake seem unquestionably to exist; they are, indeed, basic concepts in the informal knowledge people use to navigate reality. Yet the existence of most physical geographic features is difficult to demonstrate formally. Few features have bona fide (natural) boundaries that correspond to objective discontinuities on the Earth's surface, such as oceans, rivers, cliffs, or highways. Most features are some part of a field of continuous variation, such as surface elevation or plant density, so they have fiat (socially constructed) boundaries because they are not discrete objects, even if conceptualized as such (Smith 2001; Smith and Mark 2003; Smith and Varzi 2000; Varzi 2001). Any boundaries defined for geographic features within continuous fields of variation are inherently arbitrary (Bennett 2001).

Defined geographic features are thus merely one among many possibilities for differentiating portions of these continuous fields. As such, concepts like desert, moraine, or peneplain have historical ontologies: they emerged at particular moments and have changed within broader sociocultural change (Hacking 2002). Surprisingly, few physical geographic features are natural kinds, or entities recognized pan-culturally (Smith and Mark 2001, 2003). Rather, different sociocultural groups—including academic geographers—have unique geographic ontologies, featuring seemingly self-evident categorizations that are in fact specific to each group (Burenhult 2008; Mark et al. 2011; Blaut 1979).

Social consensus is prerequisite for constructing shared geographic knowledge, including environmental categories. Thus proposed geographic features, such as savannas or deserts, gain existence alongside shared theories of knowledge (that is, epistemologies) that make the underlying concepts observable in the real world (Inkpen 2005; Blaut 1979; Simon 2010, 2016). For instance, where does forest end and something else, such as savanna, begin? The range of answers to this question demonstrates that the physical arrangement of trees is secondary to social context in answering this question (Helms 2002; Körner 2007: 317).

A deep-seated epistemological debate in Western geographic thought is of generalization versus specificity (or nomothetic versus idiographic approaches) (Cresswell 2013: 84–88). Scholars debate how broadly or narrowly to define environmental categories, but all categories simplify reality. Placing multiple locations in a single category means that differences between locations must be overlooked to greater or lesser degrees, depending upon how generalizing or specifying the category's definition might be.

As social constructs, physical geographic features relate to ideas shared widely in society. The concept of *tropical rainforest*, for example, arose within nineteenth-century European botany but reflected contemporaneous belief in social Darwinism, imperatives for colonial expansion, and imagined geographies of 'the tropics' (Stott 1999). These broadly shared ideas helped enable European powers to gain and exert political-economic authority worldwide, to the detriment of many non-European peoples. Environmental categories have had especially important social roles as bases for environmental determinism, which is an idea that has sustained racist explanations of human difference (Correia 2013; Raleigh et al. 2014), even if it is less controversial in explanations of non-human biogeographies (Rosindell et al. 2012). Environmental determinism subtly shapes resource management in many locations (Robbins 2001; Simon 2010, 2016). Most relevant for this chapter are examples from colonial and post-colonial contexts in the Global South (Bassett and Crummey 2003; Cinnamon 2003; Dove 1992, 1997, 2004; Ickowitz 2006). People and activities that are categorized as belonging to one environment are considered inappropriate, inauthentic, or incapable when in locations that are categorized differently. Socially constructed environmental mismatches have enabled authorities to claim rights or obligations to intervene in resource management. For instance, in colonial and post-colonial Guinea, ethnic groups categorized as "savanna people" were considered inimical to forest, and thus the became the targets of repressive policies intended to conserve forest vegetation (Fairhead and Leach 1996: 34).

It is important in CPG to recognize that environmental categories are useful in studying biophysical reality but also inescapably social constructions that enable thought about how external conditions affect biotic and abiotic objects. All categories are historically and geographically contingent concepts.

Historical Ontology of Savanna

By some accounts, savanna is deeply rooted in human experience (Domínguez-Rodrigo 2014). The savanna hypothesis in evolutionary psychology is that our East African origins make us prefer environments with "high resource-providing potential", "distant views and low, grassy ground cover" and scattered trees (Orians and Heerwagen 1992: 559). Regardless of human paleontology, the category of savanna is neither timeless nor natural but a historically traceable concept.

Innumerable works claim that savanna simply *is*, presenting it as a self-evident natural feature and thus ignoring the social context that allows us to delineate part of a continuum as a discrete category. For example, "The savanna biome comprises a mix of trees and grasses. The trees do not form a continuous canopy, and lack of shade allows grass to grow. Savanna covers large areas of Africa, South Asia, South America, and Australia" (Rubenstein 2008: 27). Many works present savanna as a primary environmental category for African and tropical geography more broadly, composing with forest and desert the basic environmental structure for these expansive locations (Murray 1990: 14).

Ostensibly, savanna is identifiable through observations of features like trees and grasses, but scholars have long struggled to define the category in a way that consistently aligns with the range of conditions that various people have considered characteristic of savanna. By 1965, for instance, scholars had identified savanna in terms of "climate, soil, pedology, landforms, planation, hydrology, environment, landscape, fauna, agriculture, culture, economy, people, crops, [and] cattle", not to mention vegetation (Hills 1965: 216). Repeatedly, since the 1920s, some have argued that no proposed criteria are sufficient to define savanna, whether globally or in particular locations (Cole 1963; Eiten 1986; Lawesson 1994; Pratt et al. 1966; Shantz and Marbut 1923: 7–8). Predictably, maps of savanna do not correspond (Fig. 6.1).

Despite the challenges of defining the category, savanna has been widely accepted because there is a spatial middle condition between global extremes of tree-grass abundance, as well as a conceptual middle between forest and desert. The absence of bona fide boundaries means that savanna is hardly constrained spatially. Similarly, its conceptual amorphousness—evident in the lack of a reliable definition—allows highly malleable fiat boundaries. Savanna might be any low-latitude, seasonally dry location; any place with lots of grass and some trees; areas where people exhibit certain behaviors; or places otherwise somehow between forest and desert.

In English, savanna dates to a 1555 translation of Peter Martyr's Latin representation of Spanish borrowings of Native American words encountered before 1530 (Oxford English Dictionary 2006). Not surprisingly, the loanword's meaning and origin are uncertain. Martyr identified savanna in two locations. In eastern Panama, he described an indigenous dominion that included "a playne of twelue leages in breadth and veary frutefull. This playne, they caule *Zauana* [Savanna]" (Arber 1885: 148). The relevant language was perhaps Kuna, but Martyr's text is so imprecise that *Zauana* could refer to topography, political unit, or something else. The other location was Hispaniola, where the now-extinct Taino language was spoken. Hispaniola's *Zauana* was a "lordshyp", which encompassed forested mountains and "plaines [...] withowt trees, whether the earth be with grasse or withowt" (Arber 1885: 169, 173, 212).

The concept taken into English came from Martyr's table of contents, which described the Panamanian place as "the large and frutefull playne of

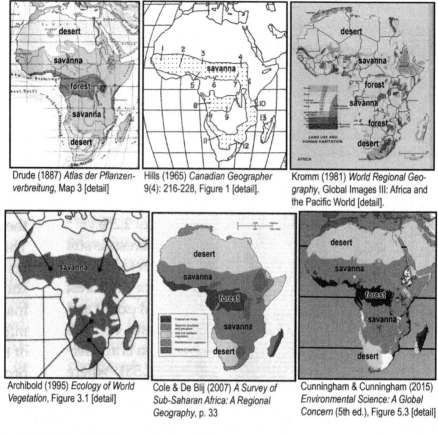

Drude (1887) *Atlas der Pflanzen-verbreitung*, Map 3 [detail]

Hills (1965) *Canadian Geographer* 9(4): 216-228, Figure 1 [detail].

Kromm (1981) *World Regional Geography*, Global Images III: Africa and the Pacific World [detail].

Archibold (1995) *Ecology of World Vegetation*, Figure 3.1 [detail]

Cole & De Blij (2007) *A Survey of Sub-Saharan Africa: A Regional Geography*, p. 33

Cunningham & Cunningham (2015) *Environmental Science: A Global Concern* (5th ed.), Figure 5.3 [detail]

Fig. 6.1 Maps of African environmental geography. Hills and Archibold portray only savanna. Drude, Cole & de Blij, and Cunningham & Cunningham include scattered highland zones, Mediterranean areas at the northernmost and southernmost latitudes, and various other small patches, particularly temperate grassland (primarily South Africa and Ethiopia). Kromm includes tundra (sic; Kenya-Tanzania border) and various patches related to population density and agricultural productivity.

Zauana" (Arber 1885: 395). In the following centuries, the concept of savanna appeared widely in the New World and subsequently elsewhere (Fig. 6.2).

The pre-scientific category of savanna was not formally defined, but characteristically had grassy vegetation and planar topography, which enabled sweeping views and facilitated travel; its plants, soil, and livestock or game seemingly promised high productivity. Early uses compared savanna to pasture and meadow, economically significant concepts that by the fourteenth century were locations used for grazing livestock (Oxford English Dictionary 2006). Savanna began appearing in truly imagined contexts by 1700, showing that the concept was sufficiently generalized to become placeless. Most

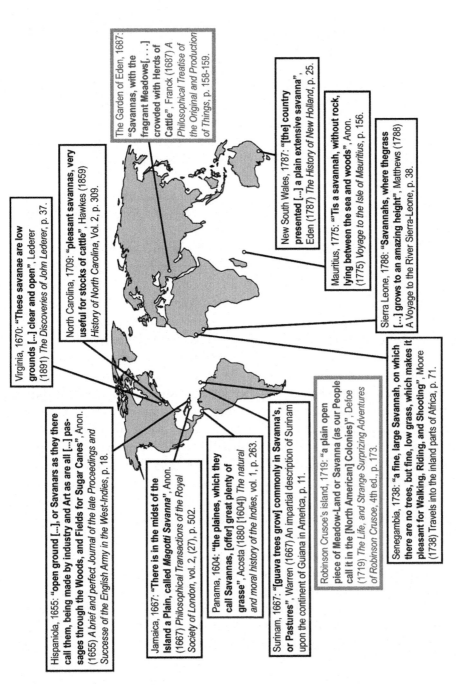

The Garden of Eden, 1687: "Savannas, with the fragrant Meadows[. . . .] crowded with Herds of Cattle", Franck (1687) *A Philosophical Treatise of the Original and Production of Things*, p. 158–159.

New South Wales, 1787: "[the] country presented [...] a plain extensive savanna", Eden (1787) *The History of New Holland*, p. 25.

Mauritius, 1775: "'Tis a savannah, without rock, lying between the sea and woods", Anon. (1775) *Voyage to the Isle of Mauritius*, p. 156.

Sierra Leone, 1788: "Savannahs, where thegrass [...] grows to an amazing height", Matthews (1788) *A Voyage to the River Sierra-Leone*, p. 38.

Virginia, 1670: "These savanae are low grounds [...] clear and open", Lederer (1891) *The Discoveries of John Lederer*, p. 37.

North Carolina, 1709: "pleasant savannas, very useful for stocks of cattle", Hawkes (1859) *History of North Carolina*, Vol. 2, p. 309.

Hispaniola, 1655: "open ground [...], or Savanars as they there call them, being made by industry and Art as are all [...] passages through the Woods, and Fields for Sugar Canes", Anon. (1655) *A brief and perfect Journal of the late Proceedings and Successe of the English Army in the West-Indies*, p. 18.

Jamaica, 1667: "There is in the midst of the Island a Plain, called *Magotti Savanna*", Anon. (1667) *Philosophical Transactions of the Royal Society of London*, vol. 2, (27). p. 502.

Panama, 1604: "the plaines, which they call Savannas, [offer] great plenty of grasse", Acosta (1880 [1604]) *The natural and moral history of the Indies*, vol. 1, p. 263.

Surinam, 1667: "[guava trees grow] commonly in Savanna's, or Pastures", Warren (1667) *An impartial description of Surinam upon the continent of Guiana in America*, p. 11.

Robinson Crusoe's Island, 1719: "a plain open piece of Meadow-Land or Savanna (as our People call it in the [North American] Colonies)", Defoe (1719) *The Life, and Strange Surpizing Adventures of Robinson Crusoe*, 4th ed., p. 173.

Senegambia, 1738: "a fine, large Savannah, on which there are no trees, but fine, low grass, which makes it pleasant for Walking, Riding, and Shooting", Moore (1738) *Travels into the inland parts of Africa*, p. 71.

Fig. 6.2 Local appearances of savanna in English, 1600–1800. Non-fictional uses in black boxes, fictional uses in gray

importantly, in the 1810s, *The Swiss Family Robinson*'s South Pacific island had three main environments: forests, a desert, and a grassy plain abounding with game and "interspersed at agreeable distances with little woods" (Wyss 2007 [1816]: 281). In the first English edition (1816) this area was unnamed; in the second edition (1848) it was "savanna" (Locke 1848: 198).

Savanna circulated via incompletely known social contexts (Domínguez-Rodrigo 2014). In English, into the 1800s it was recognized as a Spanish loanword, and often italicized and defined, suggesting novelty. Adoption varied spatially and between European language communities. For instance, in North America in 1799, English speakers widely said savanna while French speakers said *prairie* (Winterbotham 1799: 487), even though French *savane* was written decades before in Central America, the Caribbean, and Senegambia (Adanson 1757; Anonymous 1684: 40, 360, 417; Raveneau de Lussan 1690: 254).

About 1700, the concept of savanna began to enter scientific discourse. Its placelessness made savanna useful but imprecise, as English philosopher John Locke observed: savanna, like woodland, mountain, and plain, enabled "loose Description" of a country, but did not offer "the more useful Observations of the Soil, Plants, Animals and Inhabitants, with their several Sorts of Properties" (Locke 1706: 78). The generalizing power of savanna was attractive during the Age of Sail, when many locations outside Europe were seen briefly, with limited knowledge. Yet naturalists also sought to characterize place-specific conditions. They struggled to do so without having formally considered the epistemology of environmental categories, that is, how they might be defined and thus identified in the real world. Savanna posed particular problems, because, as a popular geography of North America recognized, "the dubious boundaries of the savannas, ris[e] imperceptibly toward the forests" (Pinkerton 1802: 584). Generalizing proved easier than specifying place-specific conditions. In Jamaica, for instance, Hans Sloane inconsistently defined savanna through comparisons and contrasts with other categories: he variously wrote "Savanna or Meadow", "low Land Woods and Savanna's", "Savanna's or Plains", "Savanna Woods", "woody parts of the *Savanna's*, or Low-Lands", "Low-Land or Savanna Woods", "woody Savanna's", and "Savanna's and Woods" (Sloane 1725: ix, 19, 24, 25, 26, 39, 131, 173, and elsewhere).

Alexander von Humboldt initiated formal discussion of environmental ontology and epistemology (Humboldt 1819: 148–154). His concern arose with regard to environments with few trees: "Europe has *bruyères* [moors], Asia *steppes*, Africa *deserts*, America *savannas*; but this distinction establishes contrasts that are founded neither in the nature of things, nor in the genius of language" (Humboldt 1819: 148). He considered vegetation the most meaningful criterion for environmental categories and divided environments with few trees into deserts ("bare lands, without a trace of plants") and "savannas or steppes" ("lands covered with grasses") (Humboldt 1819: 149). He also contrasted savanna and forest, at least in South America, where he consistently reduced environmental variation to a binary opposition as embodied by "savanna peoples" versus "forest peoples" in Brazil (Humboldt 1819: 609). Despite the implication that his categories were obvious and self-evident, Humboldt did not construct a consistent epistemology. He took local concepts as equivalent to global categories, challenged the applicability of general concepts in specific instances, and inserted criteria beyond his stated preference for vegetation. For example, he observed that "the *puszta* of Hungary are veritable savannas" but "the savannas of America, especially those in the temperate zone, [are often called] *prairies*; but this word seems to me hardly applicable to pasturage [that is] often very dry. Instead, environments like] the *Llanos* and *Pampas* of southern America are veritable [savannas or steppes]" (Humboldt 1819: 149). Humboldt's savanna was not explicitly tropical but within his New World "equinoctial regions", and he consistently contrasted low-latitude and mid-latitude environments with few trees.

Humboldt's forest-savanna-desert model undergirded subsequent scholarship, though some preferred narrower categories. In particular, French geographer Conrad Malte-Brun argued that Humboldt's broad, vegetation-based categories must be subdivided by soil, hydrology, and topography to be meaningful (Malte-Brun 1819: 215–220). Nonetheless, subsequent writers maintained savanna as a concept loosely defined only through juxtaposition with other categories. By 1860, science writers began writing explicitly of "tropical savanna" (Drude 1887; Grisebach 1872; Hartwig and Guernsey 1876; Mangin 1872; Müller 1857), despite its persistence in the southeastern United States. Economic and aesthetic values together shaped understandings of savanna, which commonly seemed to signify presumed productive potential as much as biophysical conditions. In 1889, for instance, forest occupied windward slopes in Fiji, while drier leeward slopes "offer[ed] only savannas [...] where colonists find the most favorable lands, already suited for agriculture or raising livestock" (Reclus 1889: 874). Nineteenth-century scholars explicitly took aesthetic notions as a basis for environmental categorization (Warming 1909: 137), mixing beliefs about imagined geographic regions with Western preferences for trees over grasses, and verdant over senescent vegetation (Dove 2004; Stott 1999).

Vegetation structure was privileged in environmental geography because it was considered dependent upon climate (Grisebach 1872; Herbertson 1905; Köppen 1900; Schimper 1903).

The presence or absence of water in the Tropical World exerts an influence upon all forms of animal and vegetable life [...]. Wherever water is absolutely wanting, the country is given over to barrenness. Wherever water is perpetual and abundant, the soil is clothed with lofty forests and a profusion of lush vegetation. Midway between these extremes are vast tracts dry at one season and wet at another. These regions [...] we may call savannas [...]. (Hartwig and Guernsey 1876: 499)

This epistemology—that vegetation indicated meaningful climatic differences—reflected early twentieth-century interest in environmental determinism. Tree-grass co-abundance became diagnostic of Köppen's savanna climate zone. Nonetheless, named categories never corresponded precisely. Warming's "savannah-vegetation" (1909) differed from Schimper's "savanna forest" (1903) and Grisebach's "tropical savanna" (1872). Differences reflected contrasting beliefs about vegetation ecology, and thus different interests to promote in scientific discourse (Shantz and Marbut 1923; Tansley 1920).

Generalizing concepts like savanna masked variation among environmental conditions in particular locations, which scholars tried to indicate with the plural form savannas or qualified forms like humid savanna or shrub savanna. Debates arose on the appropriateness of qualified terms versus alternatives like steppe, prairie, and grassland (Walter 1971: 238–239). Qualified and alternative forms represent competing ecological beliefs, as well as different social contexts (Domínguez-Rodrigo 2014; Eiten 1986; Gleave and White 1969; Hills 1965; Lawesson 1994). As the savanna lexicon expanded, prominent scholars unsuccessfully tried to impose specific epistemologies via standardized definitions (Forsberg 1967; Trochain 1957; UNESCO 1973; Walter 1971).

In the 1980s, leading scholars of savanna promulgated scale- and resolution-specific criteria for knowing savanna. This epistemology remains dominant, and makes savanna "those forms of vegetation that occur between the equatorial rain forests and the mid-latitude deserts and have a continuous grass stratum that is either treeless or [with] trees and shrubs [at] variable [...] densit[ies]" (Cole 1986: 6). Thus, savanna is meaningful only for "general" vegetation description (White 1983: 18) because it refers to "a continuum of physiognomic [vegetation] types" (Menaut 1983: 110). This epistemology posits that ultimately (or at least theoretically) climate produces environmental geography; savanna represents precipitation seasonality due to the annual north-south shift of the subtropical high-pressure belt and inter-tropical convergence zone (Nix 1983).

This epistemology allows simultaneous generalization and specificity. Savanna is globally real, yet locally unobservable. For instance, one study states, "[we] call this collection of ecosystems 'savannas,' except when we need to distinguish particular habitats. We note that savannas range from well-wooded areas [...], to open habitats with few trees" (Loarie et al. 2009: 3100). This epistemology acknowledges fine-scale environmental variability, yet con-tradictorily allows scholars to describe particular conditions as deviations from idealized conditions, rather than environments requiring particularized descriptions. Thus, one study described a site as including "tree savanna", "wooded savanna", "open shrub savanna", "dense shrub savanna", "disturbed savanna", "grass savanna", "savanna grassland", "humid savanna grassland", "humid savanna", "dry savanna grassland", and also "open forest", without further differentiating these categories (Bassett and Koli Bi 2000).

The currently dominant epistemology does not resolve any objective geo-graphic feature, but justifies socially grounded belief that savanna is real. "Although it may be difficult to define the term 'savanna' precisely," one paper acknowledges, "the general concept of a tropical or subtropical mixed tree (or shrub)-grass community is widely accepted" (Jeltsch et al. 2000: 161). Savanna "is widely accepted"—has socially determined validity—because of its history in Western thought, and its social roles outside biophysical science.

Social Roles of African Savanna

The savanna concept enables generalizations about biophysical conditions but also sustains facile characterizations of human geography that are sustained broadly across society.

The concept of savanna first appeared in Africa—in Senegambia—in 1738 (Moore 1738), but many subsequent European travelers did not see it. Mungo Park (West Africa, 1790s–1800s) and David Livingstone (Southern Africa, 1850s–1860s) saw pasture, prairie, and meadow, while Richard Burton (Central and East Africa, 1860s–1870s) and Henry Morton Stanley (East, Southern, and Central Africa, 1860s–1870s) sometimes encountered savanna. The Portuguese generally saw *campina*, Germans *Steppe*, Afrikaaners *veld*, and the French *savane*, if not *prairie*. European scholars increasingly described African environments as savanna, beginning with Carl Ritter's 1822 descrip-tion of what is now Benin (Ritter 1822: 297). By 1880, savanna covered 21.3% of Africa, 37% by 1923, and 65% by 1995 (White 1892: 53; Archibold 1995: 61; Shantz and Marbut 1923: 57).

Humboldt's forest-savanna-desert model, described above, has been the backbone for environmental geographies of Africa, garnished with other, marginal categories (Fig. 6.1).

By 1900, the forest-savanna-desert ontology supported deterministic portrayals of Africans (Richards 1996; Salazar 2009; Tilley 2011). Environmental stereotypes varied across the continent. Most importantly, in northern Africa, the Hamitic hypothesis flourished. This idea was that any vestiges of civilization were the heritage of Caucasian Hamites—descendants of Noah's son Ham—who entered Africa from the north. This hypothesis became environmental because of the north-to-south progression perceived for the desert-savanna-forest categories. Thus, desert pastoralists surpassed savanna agriculturalists, who surpassed more southerly forest peoples (Mangin 1872: 181).

Savanna was, at least theoretically, observable in bodies: "Peoples of the Savannas are taller and, though often darker, are more mixed in [racial] origin than the negroes of the [forested] south" (Harrison Church et al. 1964: 213). Supposed linkages between environment, livelihood, and race generated corporeal consequences, particularly because the three environments were commonly portrayed as in conflict. In Rwanda and Burundi, for instance, the Tutsi identity arose under Belgian colonial rule, and linked pastoralism, savanna, and arrival from elsewhere (that is, a Hamitic past); this simplistic history was variously taken to show the benign superiority or domineering usurpation of Tutsi over Hutu, an identity linked to farming, indigeneity, and forest, which was considered to have declined as savanna expanded (Eltringham 2006; Hintjens 1999; Jefremovas 1997). Repeated conflicts since 1959, including genocide in 1972 and 1994, centered on these identities.

Environmental generalizations about people have contributed to conflicts elsewhere (Gruley and Duvall 2012; Richards 2001). Popular media disseminate such generalizations, including an educational website that tells us that: "The people of Africa's vast savanna are united by their strong identity with the sprawling plains that surround them. [Amongst] these pastoral groups […] the Maasai have held the most tenaciously to their wanderlust. These tall, dark skinned herdsmen [now] share the plains with the Kikuyu, traditionally a nation of farmers […]" (PBS.org 2001).

Colonial ecological science portrayed savanna as if in existential struggle against forest and desert. Forest was nature's aboriginal, tropical lushness; desert was barren nature (Stott 1999; Verstraete 1986). The potential bounty of savanna could be lost through "desertification" (see Davis and Sayre, this volume), although savanna itself could represent degradation—the "savannization" of forest (most notably: Aubréville 1949, 1962). Savanna people were believed to cause both. This narrative was especially prominent in West Africa, where it justified colonial and neo-colonial interventions (Bassett and Crummey 2003; Cinnamon 2003; Fairhead and Leach 1996; Ickowitz 2006). Ecologists widely found savanna productivity below their expectations, interpreting this as evidence that so-labeled savanna livelihood practices were maladaptive. Savanna located near forest was qualified as "derived", to indicate belief in past (but unobserved) anthropogenic deforestation. An unnamed 1957 college textbook described by Rosenblum "claimed that one of the basic reasons for the underdevelopment of the savanna region was 'the presence of the native population'" (Rosenblum 1963: 11). This blunt portrayal echoed earlier, subtler statements, and anticipated subsequent ones (e.g. Verstraete et al. 2009).

In Southern and East Africa, savanna was frequently understood as unpeopled, full of wildlife, and available to outsiders. Fictional portrayals of unbound "savanna Africa" stimulated tourism, and were prominent in White settler imaginations (Adams and McShane 1992; Akama 1996; Hughes 2011; Staples 2006). East African landscapes became the archetypal savanna, unchanged since the Pleistocene and the ultimate contrast to humanized Europe (Anderson and Grove 1987; Neumann 2011). Further, the (singular) African savanna became the archetypal wild environment (Preston-Whyte et al. 2006: 132), occupied only by wild animals and timeless people, like the Maasai stereotype above (Akama 2002; Galaty 2002; Norton 1996). Colonial and post-colonial authorities made this concept of savanna real by evicting people from protected areas (Neumann 1997), as African agriculturalists were considered inauthentic in or incompatible with the wild savanna.

Across the continent, socially constructed environmental "mismatches" allowed policymakers to claim and place blame for environmental changes, and to justify heavy-handed policies meant to either eliminate or promote savanna behaviors depending on context. Purportedly apolitical savanna imagery continues to justify resource dispossession (Robbins 2012: 12–13).

What/Where/Why Is Savanna?

It is neither necessary nor unnecessary that savanna exists as a category in Africa or elsewhere. Ever since Malte-Brun (1819), a small but persistent social group has argued against highly generalized environmental categories (Malte-Brun 1819). The specifying impulse might reflect desires to correct "geographer's fictions" (Gleave and White 1969), beliefs about the characteristics and meaning of biophysical conditions (Domínguez-Rodrigo 2014), efforts to improve resource management (Raynaut 1997), or new approaches to data collection and analysis (Sayre et al. 2013). Emphatically, narrower environmental categories are just as social as broad ones, and thus can support non-biophysical generalizations and political-economic imperatives.

The generalizing versus specifying quandary in environmental thought has existed at least since the 1690s (Locke 1706: 78). Savanna represents a generalizing worldview: "Despite their [...] differences, the savannas of the world are believed to share the same basic patterns of structure and function" (Scholes 1997: 259). In academic geography, debate about the value of general versus specific knowledge has been especially prominent since the 1950s, though not in physical geography (Cresswell 2013). Contrary to Miller and Goodchild (Miller and Goodchild 2015: 456), we would argue that it is not "perhaps wise" of physical geographers to avoid this debate, or other points of philosophical engagement. Geographic generalization is necessary and broad patterns and processes exist, but broad features arise through the spatial dependency and heterogeneity that exists in local contexts. Further, advances in complexity theory, data science, and biogeography underscore that interactions within particular environments produce emergent, system-wide behaviors that cannot be understood—or sometimes even observed—in examining either the local or the global alone (Graham and Shelton 2013; Gregory 2017; Miller and Goodchild 2015).

Instead of choosing either generalizations or specificities, we need both. The environmental categories with scientific lineages tracing to Humboldt (1819) are not suitable, and were not intended, for particular descriptions. As the spatial and conceptual center, savanna stabilizes the forest-savanna-desert ontology as fact without history, even though it was formalized in Europe in the 1810s: evident in *The Swiss Family Robinson*, it became scientific in Humboldt's equinoctial Americas. The concept of savanna approximates the global condition of semi-arid climate and low-latitude location. Due to the term's long history it is probably inescapable in labeling emergent behavior associated with this global condition. But importantly, spatial extent does not indicate complexity, globalness, or localness; forest, savanna, desert, and other highly generalizing categories should not be accepted as default descriptors of large areas. The specificities of locations grouped into the broad categories are important. Biogeographers have identified multivariate and multi-scalar patterns in semi-arid, low-latitude regions. Erstwhile savanna in Africa, for instance, can be meaningfully bifurcated based on various non-contiguous pairs of biophysical criteria: dystrophic versus eutrophic soils (Breman and Kessler 1995); bimodal versus unimodal precipitation seasonality (Ellis and Galvin 1994); Zambezian versus Sudanian floristic zones (White 1983);

elevation greater or less than 3000 feet (Stock 2004); or clustered versus dispersed tree distribution (Moore 1996). Many studies described as relating to African savanna do not justify such generalization (e.g. Asner et al. 2009; Pellegrini et al. 2017). More globally, scholars call semi-arid, low-latitude locations categorically savanna, even if the purpose of generalization is, ironically, to characterize expansive areas as not uniform (Furley and Metcalfe 2007; Lehman et al. 2014). The need to generalize should not be privileged over the need to describe locally specific environmental conditions.

Finally, the usefulness of environmental description depends upon social context (Robbins 2001; Simon 2010, 2016). There is no inherently correct or incorrect way to categorize environments. Practicing CPG means actively inquiring why a particular environment might or might not be considered to exist in a location (see Davis, Sayre, and Simon, this volume). Environmental categories are useful to most physical geographers, though much more widely useful elsewhere in society. The forest-savanna-desert categories allow immediate, if facile, generalizations, and inevitably evoke notions about object-context relationships, including environmentally deterministic ideas about human identity, authenticity, and capability that can have profound environmental and ethical consequences for how landscapes are managed. All categories simplify reality. Nonetheless, physical geographers have social responsibilities to avoid overgeneralizing about environmental conditions and to embrace specificities as well, in order to weaken facile characterizations that can sustain unjust political-economic relationships.

References

Adams, J.S., and T.O. McShane. 1992. *The myth of wild Africa*. New York: Norton.

Adanson, M. 1757. *Histoire naturelle du Sénégal*. Paris: Bauche.

Akama, J.S. 1996. Western environmental values and nature-based tourism in Kenya. *Tourism Management* 17 (8): 567–574.

———. 2002. The creation of the Maasai image and tourism development in Kenya. In *Cultural tourism in Africa*, ed. J.S. Akama and P. Sterry, 43–54. Arnhem, The Netherlands: ATLE.

Anderson, D., and R.H. Grove. 1987. Introduction. In *Conservation in Africa*, ed. D. Anderson and R.H. Grove, 1–12. Cambridge, UK: Cambridge University.

Anonymous. 1684. *Recueil de divers voyages faits en Afrique et en Amérique*. Paris: Veuve Cellier.

Arber, E. 1885. *The first three English books on America*. Birmingham, UK: [Publisher not named].

Archibold, O.W. 1995. *Ecology of world vegetation*. London: Chapman & Hall.

Asner, G.P., et al. 2009. Large-scale impacts of herbivores on the structural diversity of African savannas. *Proceedings of the National Academy of Sciences of the USA* 106 (12): 4947–4952.

Aubréville, A. 1949. *Climats, forêts, et désertification de l'Afrique tropicale*. Paris: Société d'Éditions Géographiques, Maritimes, et Coloniales.

———. 1962. Savanisation tropicale et glaciations Quaternaires. *Adansonia* 2 (1): 16–84.

Bassett, T.J., and D. Crummey, eds. 2003. *African savannas*. Oxford, UK: James Currey.

Bassett, T.J., and Z. Koli Bi. 2000. Environmental discourses and the Ivorian savanna. *Annals of the Association of American Geographers* 90 (1): 67–95.

Bennett, B. 2001. What is a forest? On the vagueness of certain geographic concepts. *Topoi* 20: 189–201.

Blaut, J.M. 1979. Some principles of ethnogeography. In *Philosophy in geography*, ed. S. Gale and G. Olsson, 1–7. Dordrecht, The Netherlands: D. Reidel.

Breman, H., and J.-J. Kessler. 1995. *Woody plants in agro-ecosystems of semi-arid regions with an emphasis on the Sahelian countries*. Berlin: Springer-Verlag.

Burenhult, N. (Ed.) 2008. Language and landscape: Geographical ontology in cross-linguistic perspective [Special Issue]. *Language Sciences*, 30 (2–3): 135–382.

Cinnamon, J. 2003. Narrating equatorial African landscapes: Conservation, histories, and endangered forests in northern Gabon. *Journal of Colonialism and Colonial History* 4 (2). https://doi.org/10.1353/cch.2003.0038.

Cole, M.M. 1963. Vegetation nomenclature and classification, with particular reference to the savannas. *South African Geographical Journal* 45 (2): 3–14.

———. 1986. *The savannas: Biogeography and geobotany*. London: Academic Press.

Correia, D. 2013. F**k Jared Diamond. *Capitalism Nature Socialism* 24 (4): 1–6.

Cresswell, T. 2013. *Geographic thought*. Chichester: Wiley-Blackwell.

Domínguez-Rodrigo, M. 2014. Is the "Savanna Hypothesis" a Dead Concept for Explaining the Emergence of the Earliest Hominins? *Current Anthropology* 55 (1): 59–81.

Dove, M.R. 1992. The dialectal history of "jungle" in Pakistan. *Journal of Anthropological Research* 48 (3): 231–253.

———. 1997. The epistemology of Southeast Asia's anthropogenic grasslands. *Southeast Asian Studies (Kyoto)* 35 (2): 223–239.

———. 2004. Anthropogenic grasslands in Southeast Asia. *Agroforestry Systems* 61: 423–435.

Drude, O. 1887. *Atlas der Pflanzenverbreitung*. Gotha: Perthes.

Eiten, G. 1986. The use of the term "savanna". *Tropical Ecology* 27 (1): 10–23.

Ellis, J., and K.A. Galvin. 1994. Climate patterns and land-use practices in the dry zones of Africa. *BioScience* 44 (5): 340–349.

Eltringham, N. 2006. The Hamitic hypothesis, race, and the Rwandan genocide. *Social Identities* 12 (4): 425–446.

Fairhead, J., and M. Leach. 1996. *Misreading the African landscape*. Cambridge: Cambridge University Press.

Forsberg, F.R. 1967. A classification of vegetation for general purposes. *Tropical Ecology* 2: 1–28.

Furley, P.A., and S.E. Metcalfe. 2007. Dynamic changes in savanna and seasonally dry vegetation through time. *Progress in Physical Geography* 31 (6): 633–642.

Galaty, J. 2002. Narrative inventions of the pastoralists in East Africa. *Visual Anthropology* 15 (3): 347–367.

Gleave, M.B., and H.P. White. 1969. The West African Middle Belt: Environmental fact or geographer's fiction? *Geographical Review* 59 (1): 123–139.

Graham, M., and T. Shelton. 2013. Geography and the future of big data, big data and the future of geography. *Dialogues in Human Geography* 3 (3): 255–261.

Gregory, K.J. 2017. Putting physical environments in their place: The next chapter? *The Canadian Geographer / Le Géographe canadien* 61(1): 11–18.

Grisebach, A. 1872. *Die Vegetation der Erde nach ihrer Klimatischen Anordnung, Band II*. Leipzig: Engelmann.

Gruley, J., and C.S. Duvall. 2012. The evolving narrative of the darfur conflict as represented in *The New York Times* and *The Washington Post*. *GeoJournal* 77 (1): 29–46.

Hacking, I. 2002. *Historical ontology*. Cambridge: Harvard University Press.

Harrison Church, R.J., et al. 1964. *Africa and the islands*. London: Longmans.

Hartwig, G., and A.H. Guernsey. 1876. *The polar and tropical worlds*. Columbus, OH: Watrous.

Helms, J.A. 2002. Forest, forestry, forester: What do these terms mean? *Journal of Forestry* 100 (8): 15–19.

Herbertson, A.J. 1905. The major natural regions. *Geographical Journal* 25: 300–312.

Hills, T.L. 1965. Savannas: A review of a major research problem in tropical geography. *Canadian Geographer* 9 (4): 216–228.

Hintjens, H.M. 1999. Explaining the 1994 genocide in Rwanda. *Journal of Modern African Studies* 37 (2): 241–286.

Huggan, G., and H. Tiffin. 2010. *Postcolonial ecocriticism*. London: Routledge.

Hughes, D.M. 2011. Whites lost and found: Immigration and imagination in savanna Africa. In *Environment at the margins*, ed. B. Caminero-Santangelo and G. Myers, 159–184. Athens: Ohio University Press.

Humboldt, A. 1819. *Voyage aux régions equinoxiales du nouveau continent, tome second*. Paris: Maze.

Ickowitz, A. 2006. Shifting cultivation and deforestation in tropical Africa. *Development and Change* 37 (3): 599–626.

Inkpen, R. 2005. *Science, philosophy and physical geography*. New York: Routledge.

Jefremovas, V. 1997. Contested identities: Power and the fictions of ethnicity, ethnography and history in Rwanda. *Anthropologica* 39 (1–2): 91–104.

Jeltsch, F., G.E. Weber, and V. Grimm. 2000. Ecological buffering mechanisms in savannas: A unifying theory of long-term tree-grass coexistence. *Plant Ecology* 150 (1/2): 161–171.

Köppen, W. 1900. Versuch einer Klassifikation der Klimate, vorzugsweise nach ihren Beziehungen zur Pflanzenwelt. *Geographische Zeitschrift* 6 (11): 593–611.

Körner, C. 2007. Climatic treelines: Conventions, global patterns, causes. *Erdkunde* 61 (4): 316–324.

Lawesson, J.E. 1994. Some comments on the classification of African vegetation. *Journal of Vegetation Science* 5: 441–444.

Lehman, C.E.R., et al. 2014. Savanna Vegetation-Fire-Climate relationships differ among continents. *Science* 343: 548–552.

Loarie, S., R. van Aarde, and S. Pimm. 2009. Elephant seasonal vegetation preferences across dry and wet savannas. *Biological Conservation* 142 (12): 3099–3107.

Locke, J. 1706. Of the conduct of the understanding. In *The Posthumous Works of Mr. John Locke*, 1–137. London: Churchill.

Locke, J.C. 1848. *The Swiss Family Robinson. second series, Vol. 1*. New York: Harper & Brothers.

Malte-Brun, C. 1819. Analyse critique: *Voyage aux régions équinoxiales du nouveau continent. Nouvelle Annales des Voyages, de la Géographie et de l'Histoire* 3: 202–222.

Mangin, A. 1872. *The desert world*. London: Nelson and Sons.

Mark, D.M., et al., eds. 2011. *Landscape in language*. Amsterdam: John Benjamins.

Mayhew, S. 2009. *Oxford dictionary of geography*. 4th ed. Oxford: Oxford University Press.

Menaut, J.C. 1983. The vegetation of African savannas. In *Tropical savannas*, ed. F. Bourlière, 109–150. Amsterdam: Elsevier.

Miller, H.J., and M.F. Goodchild. 2015. Data-driven geography. *GeoJournal* 80: 449–461.

Moore, F. 1738. *Travels into the inland parts of Africa*. London: Cave.

Moore, J.J. 1996. Savanna chimpanzees, referential models and the last common ancestor. In *Great ape societies*, ed. W.C. McGrew, L.F. Marchant, and T. Nishida, 275–292. Cambridge: Cambridge University.

Müller, K. 1857. *Das Buch der Pflanzenwelt: Eine botanische Reise um die Welt, Band 1*. Leipzig: Spamer.

Murray, J., ed. 1990. *The cultural atlas of the World: Africa*. Alexandria: Stonehenge Press.

Neumann, R.P. 1997. *Imposing wilderness: Struggles over livelihood and nature preservation in Africa*. Berkeley: University of California.

———. 2011. "Through the Pleistocene": Nature and race in Theodore Roosevelt's *African Game Trails*. In *Environment at the Margins*, ed. B. Caminero-Santangelo and G. Myers, 22–42. Athens: Ohio University Press.

Nix, H.A. 1983. Climate of tropical savannas. In *Tropical savannas*, ed. F. Bourlière, 37–61. Amsterdam: Elsevier.

Norton, A. 1996. Experiencing nature: The reproduction of environmental discourse through safari tourism in East Africa. *Geoforum* 27: 355–373.

Orians, G.H., and J.H. Heerwagen. 1992. Evolved responses to landscapes. In *The adapted mind*, ed. J.H. Barkow, L. Cosmides, and J. Tooby, 555–579. New York: Oxford University Press.

Oxford English Dictionary. 2006. *Oxford English dictionary online version*. 3rd ed. Oxford: Oxford University Press.

PBS.org. 2001. Africa: Explore the regions: Savanna. http://www.pbs.org/wnet/africa/explore/savanna/savanna_overview.html. Accessed 21 Jan 2017.

Pellegrini, A.F.A., et al. 2017. Woody plant biomass and carbon exchange depend on elephant-fire interactions across a productivity gradient in African savanna. *Journal of Ecology* 105 (1): 111–121.

Pinkerton, J. 1802. *Modern geography*. Vol. 2. London: Cadell and others.

Pratt, D.J., P.J. Greenway, and M.D. Gwynne. 1966. A classification of East African rangelands, with an appendix on terminology. *Journal of Applied Ecology* 3 (2): 369–382.

Preston-Whyte, R., S. Brooks, and W. Ellery. 2006. Deserts and savannah regions. In *Tourism and environmental change*, ed. S. Gössling and C.M. Hall, 128–141. London: Taylor & Francis.

Radcliffe, S.A. 2009. Environmentalist thinking and/in geography. *Progress in Human Geography* 33: 1–19.

Raleigh, C., A. Linke, and J. O'Loughlin. 2014. Extreme temperatures and violence. *Nature Climate Change* 4 (2): 76–77.

Raveneau de Lussan. 1690. *Journal du voyage fait à la mer de Sud*. Paris: Coignard.

Raynaut, C., ed. 1997. *Sahels: Diversité et dynamiques des relations sociétés-nature*. Paris: Karthala.

Reclus, É. 1889. *Nouvelle géographie universelle, tome 14: Océan et terres océaniques*. Paris: Hachette.

Richards, P. 1996. Forest indigenous peoples: Concept, critique and cases. *Proceedings of the Royal Society of Edinburgh, Section B* 104: 349–365.

———. 2001. Are "Forest" wars in Africa resource conflicts? In *Violent environments*, ed. N.L. Peluso and M. Watts, 65–82. Ithaca: Cornell University Press.

Ritter, C. 1822. *Die Erdkunde im Verhaltnis zur Natur und zur Geschichte des Menschen, Erster Theil: Afrika*. Berlin: Reimer.

Robbins, P. 2001. Fixed categories in a portable landscape: The causes and consequences of land-cover categorization. *Environment and Planning A* 33: 161–179.

———. 2012. *Political ecology*. 2nd ed. Chichester: Wiley-Blackwell.

Rosenblum, P. 1963. Africa as seen in American geography textbooks. *Africa Today* 10 (2): 4–5+11.

Rosindell, J., et al. 2012. The case for ecological neutral theory. *Trends in Ecology & Evolution* 27 (4): 203–208.

Rubenstein, J.M. 2008. *The cultural landscape*. 9th ed. Upper Saddle River: Pearson.

Salazar, N. 2009. Imaged or imagined? Cultural representations and the "tourismification" of peoples and places. *Cahiers d'Études Africaines* 49 (1–2): 49–71.

Sayre, R., et al. 2013. *A new map of standardized terrestrial ecosystems of Africa*. Washington, DC: AAG.

Schimper, A.F.W. 1903. *Plant geography on a physiological basis*. Oxford: Clarendon.

Scholes, R.J. 1997. Savanna. In *Vegetation of Southern Africa*, ed. R.M. Cowling, D.M. Richardson, and S.M. Pierce, 258–277. Cambridge: Cambridge University Press.

Shantz, H.L., and C.F. Marbut. 1923. *The vegetation and soils of Africa*. New York: American Geographical Society.

Simon, G.L. 2010. The 100th meridian, ecological boundaries, and the problem of reification. *Society & Natural Resources* 24 (1): 95–101.

———. 2016. How regions do work, and the work we do: A constructive critique of regions in political ecology. *Journal of Political Ecology* 23: 197–203.

Sloane, H. 1725. *A voyage to the Islands Madera, Barbados, Nieves, St. Christophers, and Jamaica, Vol. 2*. London: Sloane.

Smith, B. 2001. Fiat objects. *Topoi* 20 (2): 131–148.

Smith, B., and D.M. Mark. 2001. Geographic categories: An ontological investigation. *International Journal of Geographic Information Science* 15 (7): 591–612.

———. 2003. Do mountains exist? Towards an ontology of landforms. *Environment and Planning B* 30 (3): 411–427.

Smith, B., and A.C. Varzi. 2000. Fiat and bona fide boundaries. *Philosophy and Phenomenological Research* 60: 401–420.

Staples, A. 2006. Safari adventure: Forgotten cinematic journeys in Africa. *Film History* 18 (4): 392–411.

Stock, R. 2004. *Africa South of the Sahara*. 2d ed. New York: Guilford.

Stott, P. 1999. *Tropical rainforest: A political ecology of hegemonic myth-making*. London: IEA Environment Unit.

Tadaki, M., et al. 2015. Cultivating critical practices in physical geography. *The Geographical Journal* 181 (2): 160–171.

Tansley, A.G. 1920. The classification of vegetation and the concept of development. *Journal of Ecology* 8 (2): 118–149.

Tilley, H. 2011. *Africa as living laboratory: Empire, development and scientific knowledge, 1870–1950*. Chicago: University of Chicago Press.

Trochain, J.-L. 1957. Accord interafricain sur la définition des types de végétation de l'Afrique tropicale. *Bulletin de l'Institut des Etudes Centrafricaines* 13–14: 55–94.

UNESCO. 1973. *International classification and mapping of vegetation*. Paris: UNESCO.

Varzi, A.C. 2001. Vagueness in geography. *Philosophy & Geography* 4 (1): 49–65.

Verstraete, M.M. 1986. Defining desertification. *Climatic Change* 9 (1): 5–18.

Verstraete, M.M., R.J. Scholes, and M. Stafford Smith. 2009. Climate and desertification: Looking at an old problem through new lenses. *Frontiers in Ecology and the Environment* 7 (8): 421–428.

Walter, H. 1971. *Ecology of tropical and subtropical vegetation*. Edinburgh: Oliver & Boyd.

Warming, E. 1909. *Oecology of plants*. Oxford: Clarendon.

White, A.S. 1892. *The development of Africa*. 2nd ed. London: George Philip.

White, F.C. 1983. *The vegetation of Africa*. Paris: UNESCO/AETFAT/UNSO.

Winterbotham, W. 1799. *An historical, geographical, commercial, and philosophical view of the American United States, Vol. 2*. 2nd ed. London: Printed for the Compiler.

Wyss, J. 2007 [1816]. *The Swiss Family Robinson*. New York: Penguin Books.

7

Between Sand and Sea: Constructing Mediterranean Plant Ecology

Diana K. Davis

Introduction

The lands surrounding the Mediterranean Sea have one of the longest histori-cal records of continuous and intensive human use. The region has, for a variety of reasons examined here, long been perceived as a ruined environ-ment; an Eden degraded by anthropogenic deforestation, burning, and graz-ing for thousands of years (Attenborough 1987; Brandt et al. 1996; Thirgood 1981). A lively debate among scholars has been taking place, though, for the last couple of decades between those who continue to support this "ruined landscape" theory and those who champion the more recent theories of co-evolution, variability, and ecological resilience in the region. Much of the debate focuses on competing views of the amount and significance of defores-tation in the region and what these views imply for the alleged degradation, and hence sustainability, of many parts of the Basin.

Recent paleoecological research combined with current plant ecological studies in the Mediterranean region, much of which is arid to semi-arid, have shown that although human activities have certainly had an impact, especially in agricultural areas, much of the vegetation ecology of the region is not "degraded."[1] In fact, a growing number of scholars have concluded that tradi-tional (pre-industrial) land use systems in the Mediterranean region have sup-ported human populations for 5–10 millennia and thus, "in the long test of history, Mediterranean land-use systems have been [largely] sustainable"

D. K. Davis (✉)
Department of History, University of California at Davis, Davis, CA, USA

© The Author(s) 2018
R. Lave et al. (eds.), *The Palgrave Handbook of Critical Physical Geography*,
https://doi.org/10.1007/978-3-319-71461-5_7

(Butzer 1996, p. 145). The region also has some of the highest rates of biodiversity anywhere in the world, and it is widely believed that this is due in large part to the intense human use of the environment over thousands of years (the diversity-disturbance theory) (Blondel 2006; Blumler 1998; Huston 1994). Much of the vegetation is adapted to disturbance, and co-evolved with, both fire and grazing, for example, in addition to drought and aridity (Batanouny 2001; Blumler 1998; Di Castri 1981; Perevolotsky et al. 1998).

This chapter argues that the origins of the ruined landscape interpretation of plant ecology in the Mediterranean region are primarily found in the colonial period in the Maghreb (Algeria, Tunisia, and Morocco) and that this view was informed in important ways by French misunderstandings of arid and semi-arid lands. As a result largely of colonialism and existing political-economic and social relations in the Mediterranean Basin during the nineteenth century, what began as a pre-colonial environmental narrative (or discourse) of general dryland decline was further refined to a narrative of destruction and degradation during the period of French colonialism in the North Africa. Shortly thereafter, it was incorporated into the young science of plant ecology in early twentieth-century Morocco and was interwoven into the dominant paradigm of linear succession to climax vegetation with the use of a highly subjective sampling methodology. This narrative still influences a good deal of policy in the Basin today, resulting in questionable environmental and social outcomes.

The location of the colonial Maghreb between the sands of the Sahara and the Mediterranean Sea influenced, more than has been previously recognized, both the construction of Mediterranean plant ecology and its application in environment and development policies around the Basin during the twentieth century. Thus, this analysis demonstrates some of the ways that structural power relations shape not only the landscapes we study but also who studies them and why (King and Tadaki, this volume). It further shows how landscapes are often problematically shaped by human actions motivated by the results of such politically inflected studies and the knowledge they create. Finally, it suggests that knowledge construction and its application, while often quite political, are also sometimes results of certain forms of inertia and strong cultural norms outside the conventional spheres of politics and economics.

Early and Colonial Origins

The roots of much contemporary thinking about the Mediterranean landscape are very deep. Influenced by a variety of economic and political debates and events, understandings about the region's environment have changed a great deal over the last several hundred years. The Mediterranean Sea was, for instance, perceived in the West primarily as a border, a dividing line between Europe and the uncivilized and barbaric torrid zone of Africa for many centuries. This conception remained dominant into the early nineteenth century. The lands bordering the Sea, in southern Europe, North Africa and the Levant, were usually conceived in the early nineteenth century as three (or more) separate regions, each with differences in both physical and human geography.

The Swiss botanist Augustin Pyramus de Candolle (1778–1841) was almost certainly the first to conceive of a "Mediterranean region" in terms of vegetation and geographical botany in his influential 1820 essay (De Candolle 1977).[2] He basically saw the region as a botanical melting pot that reflected all three of the continents of Europe, Africa, and Asia, but not as a coherent, unified botanical region (Blais et al. 2012; Deprest 2002; Drouin 1998). Importantly, most botanical work up to about 1820 on North Africa and most of the surrounding Mediterranean territories did not portray the landscape as ruined or degraded. In fact, often these lands and their vegetation were lauded by botanists and geographers as either pleasant and esthetically pleasing or as very fertile.[3] This interpretation of the Mediterranean landscape began to change to one of degradation in the 1830s and 1840s.[4]

The reasons for this change in perception were complex and were as influenced by political economy as by studies in botany and natural history. Importantly, in the decades before ideas of the "Mediterranean landscape" began to take shape in Europe, notions of deserts, arid, and semi-arid lands as devastated landscapes were being developed. The French were likely the earliest of the western powers to develop the inaccurate thesis, articulated clearly as early as the mid-eighteenth century, that "Oriental despots" like the Ottomans, along with varieties of nomads, deforested the environment and created deserts such as those in "formerly flourishing" Mesopotamia, Nineveh, and the "Orient." This deforestation/desiccation thesis (an early articulation of desertification) was taken up and propagated by many influential French figures. It became, by the turn of the nineteenth century, the widespread belief that the hot, "oriental" countries from Turkey to Mesopotamia and the Levant, around the Mediterranean to Egypt and across North Africa to the Maghreb were ruined, deforested, desiccated lands (Davis 2016a).

Thus, by the time that the lands around the Mediterranean were beginning to be regarded as a distinctive botanical region early in the nineteenth century, the idea that arid and semi-arid lands were ruined landscapes had been established and was becoming widespread. As the majority of what we now consider Mediterranean landscapes are arid and semi-arid zones, much of this notion that drylands are degraded landscapes came to be applied to the region during the nineteenth century. It highlights the political nature of much of this narrative of deforestation, desiccation, and ruin that it became strongly dominant among politicians, physicians, foresters, and many others in Europe and the UK during the nineteenth century, a period of active imperial expansion. It did not become as dominant as quickly among botanists, geographers, and some other natural historians. However, as the nineteenth century progressed, "deforestation and consequent aridity became one of the greatest 'lessons of history' that every literate person knew" (Williams 2003, p. 430).

With the capture of Algiers by France in 1830, a 125-year colonial saga began that produced a great deal of colonially constructed knowledge about many different things from agriculture to health and medicine to the environment. Influenced by the growing belief that arid and semi-arid lands were the product of mismanagement and devastation, an inaccurate environmental history of the Maghreb was developed during the colonial period. It was first constructed in Algeria and incorrectly blamed the native Algerians for deforesting, overgrazing, and desertifying the land (Davis 2007). It was later applied to Tunisia and Morocco as they were occupied. This story was primarily constructed by powerful colonial actors including foresters, physicians, some large landowners, and many politicians, and it was deployed to disenfranchise a great many North Africans of their land and resources to be used for European colonization. It was also used to control populations like nomads, who were blamed for overgrazing, in order to further the colonial project and provide grazing land for European livestock production.

In Colonial Algeria, the deforestation/desertification narrative began to take coherent shape around mid-nineteenth century primarily in colonialist tracts but not yet in the botanical literature (Davis 2007). A few decades later, though, in the 1880s and especially the 1890s, the declensionist narrative did begin to be incorporated into botanical and ecological work on Algeria and the wider Maghreb as it became more dominant and pervasive in the colonial Maghreb and in France.

By this time, in the late nineteenth century, developments in botany, and what would become known as plant ecology, led to the strengthening of phytosociology in Europe. This approach attempted to define zones of vegetation in a country or region based on various physical factors including latitude, altitude, temperature, annual precipitation, and later, soil types.[5] Pioneered early in the nineteenth century as phytogeography, it was further refined into phytosociology primarily by German and French botanists, around the turn of the twentieth century. It entails the categorization of the one or two "dominant" species of each region and their associated plants. In effect, each region is defined by its dominant species, most often some kind of tree. It is firmly based on notions of equilibrium and also on succession to climax.

Many assumptions were made during these categorizations and frequently "relict" vegetation—vegetation presumed to be a remnant of previously widespread vegetation—was used to define the dominant species (Matagne 1999).[6] Moreover, in some cases, vegetation that was not even present in the landscape during observation was chosen as the dominant species, based on the assumption that it had been there before "degradation" had occurred and was thus "no longer present." Assumptions of deforestation were especially common and were very problematic as they then often led to inappropriate reforestation and other forms of environmental "improvement."

French botanist Charles Flahault was a vigorous proponent of phytosociology and was closely involved in its promotion and acceptance in Europe (Flahault 1896). It was later codified by his student, Swiss botanist Josias Braun-Blanquet with his *relevé* method of sampling in the 1920s; subsequently, it became dominant in France and much of Europe.[7] The *relevé* method, however, has been repeatedly criticized for its nearly "complete subjectivity" since the dominant species is determined by a visual inspection of the area by an expert botanist rather than by random or systematic sampling (Allen 2001, pp. 122–125; Barbour et al. 1999; Poore 1955). The *relevé* method, therefore, is particularly susceptible to bias and the incorporation of inaccurate histories of landscape change such as the assumption of the deforestation and degradation of the Maghreb. Nonetheless, phytosociology remains dominant in much of France, in many parts of Europe, and much of the rest of the world outside of Britain and N. America (Allen 2001; Barbour et al. 1999).

The French colonization of the Maghreb for about 125 years allowed botanists quite familiar with the Mediterranean sections of southern France (and sometimes other parts of the northern Mediterranean Basin) to make detailed comparisons with the vegetation of the southern Mediterranean in North Africa. It was in the Maghreb, specifically in Morocco, that the first "scientific" definition of Mediterranean vegetation was constructed, using the phytosociological approach, early in the twentieth century. This definition carried with it, though, most of the incorrect colonial assumptions about longstanding degradation in the Maghreb. Since that time, French research and French ecologists rooted in phytosociology's succession to climax approach have dominated the field of vegetation ecology of the Mediterranean Basin (Allen 2001, pp. 122–125; Roumieux et al. 2010). They have also had an influence in other Mediterranean climate regions around the world.

French Colonial Phytosociology and the Emergence of Mediterranean Plant Ecology

About a decade after the conquest of Morocco by the French in 1912, French botanist Réné Maire (1878–1949) published an influential phytogeographic paper on Moroccan vegetation. Based in Algeria, since 1911, as the chair of botany at the University of Algiers, Maire had worked with the two most prominent botanists in colonial Algeria, Louis Trabut (1853–1929) and Jules Battandier (1848–1922), and he had published an Algerian vegetation atlas as well as a phytogeographic map of Algeria and Tunisia.[8] Maire had also worked and published with Braun-Blanquet on the vegetation of the Maghreb and so was very familiar with his *relevé* method for phytosociology.

Drawing on this previous experience, Maire applied French phytogeographic methods to Moroccan vegetation in 1921 and published a paper on its 14 botanical zones in a government publication (Maire 1921). Fully persuaded by the French colonial narrative of environmental decline in the Maghreb, Maire noted that 9 of the 14 regions in Morocco were badly degraded. He had defined all 14 regions by their arboreal or potential arboreal vegetation, and thus the regions with few or no trees were classified as deforested. Many of his deductions of the "natural" vegetation were "based on the few relicts that have escaped grazing and cultivation" (Maire 1921, p. 60).

Two years later, in 1923, a young French botanist, Louis Emberger (1897–1969), arrived in Morocco. Emberger had been initiated into phytosociology by his father-in-law, famous botanist Charles Flahault, in Montpelier before arriving in Morocco. Armed with the education in France at that time which taught that the Maghreb had been degraded and desiccated for centuries, Emberger was fully convinced that Morocco was badly deforested and overgrazed. This conviction was evident in nearly all of his publications. In this, he was following a decades-long tradition of French botanical research in Algeria, Tunisia, and Morocco which had accepted the colonial environmental history of decline and ruin for the region.

Emberger was quickly embraced by the French botanists working in Algeria and Morocco. He was especially close to René Maire, who became his mentor, and they collaborated and published many articles and books together. It is with this complex background and intellectual heritage, then, that Emberger's work on the Mediterranean region must be understood. His research was part of an ongoing effort to try to delineate and define the Mediterranean environment which had first been attempted in the nineteenth century based on floristic inventories. Emberger, though, was the first to define and delimit Mediterranean climate together with vegetation zones in 1930 (Emberger 1930). Emberger was one of the earliest scholars to try to delineate the climate zones of the Mediterranean region, and he made several long-lasting and influential contributions (Emberger 1930).[9] His work on Mediterranean climate zones was part of a wider effort to understand and define the highly variable and complex Mediterranean region.[10] The fundamental basis of all of these definitions was the long summer dry period and the relatively wet but mild winters, with several more complex variations given by various researchers.

Emberger, though, was primarily interested in the vegetation of the Mediterranean which he believed "closely mirrored the climate" (Emberger 1930, p. 643). He termed his Mediterranean vegetation zones "étages" because he conceived them to be so tightly associated with the "étages climatiques," the climate zones.[11] His extensive research and publications are widely credited with constructing and defining Mediterranean bioclimate zones as well as Mediterranean vegetation zones (Allen 2001; Nahal 1981). In his detailed 1930 article "The Vegetation of the Mediterranean Region," accompanied with a map, Emberger used Morocco as the perfect example since he believed it alone had all the series of bioclimatic and vegetation zones found in the Mediterranean region.[12] He further claimed that the "vegetation of other countries around the Mediterranean may be examined and appreciated on the basis of that of the Cherifien Empire [Morocco]" (Emberger 1934, p. 152). His goal was to be able to identify and delimit all of the climate zones in the Mediterranean region and the "natural" vegetation associated with them using phytosociology, relicts, and notions of succession to climax vegetation. He came to call these "bioclimatic" zones, "étages bioclimatiques," a term that became widely adopted (Fig. 7.1).

Since he based his definitions of Mediterranean bioclimate zones on Morocco, employed the phytosociological method to determine the natural (potential) vegetation for each, and accepted the colonial environmental history, Emberger deduced that a great many of the Mediterranean bioclimate zones were badly degraded. In Morocco, for instance, using his bioclimatic calculations to deduce the natural (climax) vegetation, he calculated that 85% of the area that "should" be forested was deforested. "Only ruins" remain, he lamented (Emberger 1934, p. 163). As he explained another way in a key publication on Morocco, since "the physical equilibrium of a country requires a forest cover [taux de boisement] theoretically of 30%," Morocco, with its 3-million forested hectares, has "a deficit of nearly 4 million hectares of forest" (Emberger 1934, p. 163).[13] But even this additional nearly 4 million hectares

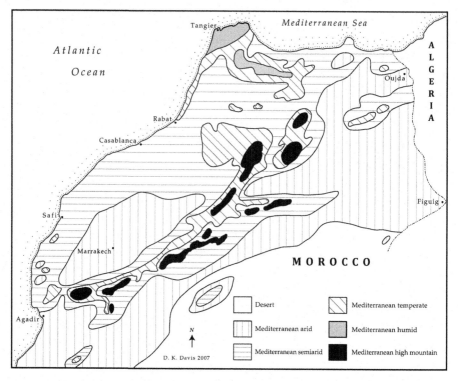

Fig. 7.1 The stages of vegetation in Morocco. This map illustrates how Emberger drew the main vegetation zones of Morocco as he conceived them in about 1934. Five of the six vegetation zones depicted here (excluding the desert zone) are defined by their trees or potential trees. This map is a precursor to the much more detailed phytogeographic maps of Morocco, the Maghreb, and the Mediterranean Basin that Emberger created in later. Created by Diana K. Davis. Source: D. K. Davis 2007. *Resurrecting the Granary of Rome: Environmental History and French Colonial Expansion in North Africa*, © 2007 Ohio University Press, p. 153. This material is used by permission of Ohio University Press

resulted in a total that was still significantly less than the ideal represented by his calculated potential (natural) vegetation of 20 million hectares of forest.

In a display of the arboreal chauvinism dominant in the nineteenth and early twentieth centuries, Emberger defined all of his vegetation zones, except the desert zone and the high alpine zone (above the tree line) by their dominant trees, considering them naturally forested zones. This included even the Mediterranean arid zone, although he conceded that the "forests there are very thin and comparable to savannas" (Emberger 1930, p. 708). Regions that did not conform to what the calculated climax predicted were automatically defined as degraded (Emberger 1930, pp. 713–714). Several regions were defined as badly degraded, especially grasslands. In the successional vegetation hierarchy dominant at the time, annual grasslands were usually considered the most degraded vegetation, and it was widely assumed that grasslands were the result of "retrogressive succession" away from the ideal climax of woody vegetation, usually a tree of some sort.[14]

Refined and given more details over the years, Emberger's definition of Mediterranean vegetation zones and "proper" forest cover became widely adopted. In 1933, Emberger published another influential article extending his bioclimatic and vegetation zones to all of the Mediterranean-type regions of the world, including California, Chile, S. Africa, and parts of Australia (Emberger 1933).[15]

By 1938, in addition to several articles refining these ideas, Emberger had produced his definitive phytogeographic map of Morocco which showed the potential or "natural" vegetation of the colony as deduced by the standard phytogeographic methods (Emberger 1939). This map formalized and institutionalized with great authority the deforestation/desertification narrative for Morocco and allowed environmental statistics, including inaccurate deforestation statistics, to be calculated. Such a calculation was easy because anywhere there were no trees in places that the map indicated the potential vegetation to be trees/forest, the area was simply classified as deforested. At the request of the Forestry department, Emberger made a similar potential vegetation map for the whole Maghreb in 1942 (Fig. 7.2), including Algeria and Tunisia, which facilitated similarly fraught deforestation calculations (Boudy 1948, pp. 170–173).[16] The available paleoecological evidence, however, does not support these colonial claims of massive deforestation of 66–85% in Morocco although it does show some deforestation in some areas at certain times (Davis 2007, pp. 9–10). These colonial deforestation statistics, however, justified numerous forest conservation and reforestation projects that were ecologically questionable and that caused many hardships for the local Moroccan people.

From Colonial Constructs to International Expertise

Emberger's definition of Mediterranean bioclimates and the related zones of vegetation quickly became very widely referenced sources on Mediterranean plant ecology and were dominant in most of the work by French and other European ecologists for at least half a century. Chair of Botany at the science faculty in Montpelier from 1937, Emberger established and directed the Center for Phytosociological and Ecological studies, (CEPE 1961) still

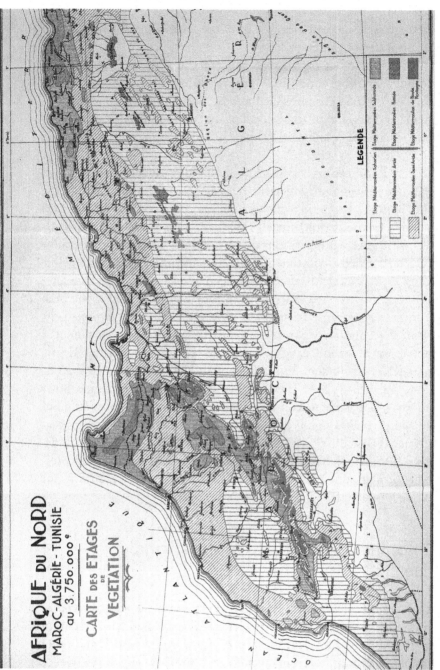

Fig. 7.2 The stages of vegetation in North Africa. Constructed on the same principles as Emberger's map of the stages of vegetation in Morocco and in the Mediterranean Basin, this map represents the culmination of Emberger's thinking about the stages of vegetation in the region. The six zones shown here are slightly different from his original 1930 conception, with the addition of the Saharan desert zone and a single high mountain zone encompassing what he had previously separated as two different high mountain zones (inferior

functioning today, and supervised dozens of doctoral students in their eco-logical studies. Many of these students, trained by Emberger in phytosocio-logical methods and notions of widespread deforestation and degradation, became prolific and widely published ecologists of the Mediterranean and dominated publishing in this area for decades.[17]

Several years later, when Emberger was employed as an expert by UNESCO, he was instrumental in the construction of two highly influential maps: the *Bioclimatic Map of the Mediterranean Zone* of 1962 (for UNESCO & FAO) and the *Vegetation Map of the Mediterranean Zone* of 1970 (also for UNESCO & FAO). Both maps were part of the Arid Zone Program of UNESCO which was a high-profile and globally influential program on the southern Mediterranean countries from Morocco to the Levant and on eastward to Pakistan. This UNESCO program took problematic colonial knowledge of arid, semi-arid lands and desertification, institutionalized it with great expert authority, and spread it around the world during the 1950s and 1960s (Davis 2016a). This program also exported the subjective phytosociological approach to vegetation study to the Middle East and North Africa via training pro-grams (several run by Emberger) where it became the most common approach to vegetation ecology for many years, as did conventional succession theory. The content for these Mediterranean maps was chosen primarily by Louis Emberger and three colleagues who also wrote the explanatory notices.[18] The maps were made and printed in France. They were based on a variety of pub-lished sources and on many estimates since there was a serious "lack of data for certain sectors of the map" (UNESCO 1963, p. 8).

The Bioclimatic map aimed to sketch a "picture of the major climatic com-plexes determining the different types of vegetation" relying "almost entirely on the two factors of temperature and amount of water available" (ibid., p. 11). Indicative of the political-economic motivations behind it, the Vegetation map was "designed to promote the rational use of land and to expand agricultural and forestry output" (UNESCO 1969/1970, p. 8). The general principal followed in making this map was

> to show the 'potential' vegetation, i.e. the vegetation as it would be without the intervention of man and animals. Accordingly, what is shown does not corre-spond to the vegetation actually found, which has been extremely modified by man … [especially] for many steppes and pseudosteppes where man-caused degradation has been in progress for thousands of years. (ibid., p. 47)

Determining the potential vegetation was fundamental in guiding environ-ment and development programs in the region as it often still is today.

This vegetation map and the explanatory notice are suffused with the colonial degradation narrative, a fixation on a single climax vegetation, and what the vegetation of the region "should" be—the potential vegetation. The overall picture drawn from this phytosociological vegetation map was of a Mediterranean region that was deforested and overgrazed and in need of significant restoration.[19] The concluding discussion highlights that it is important "not to forget the 'desertifying' action of man" (UNESCO 1969/1970, p. 74).[20]

This map, due to the international authority of the UN and of UNESCO's Arid Zone Program, became the standard reference map of Mediterranean vegetation for several decades. It was not questioned until the mid–1980s and then more widely in the 1990s, and yet it is still frequently referenced today.[21] The US Defense Mapping Agency, for example, had planned to use this UNESCO vegetation map for the Mediterranean section of its 1984 Vegetation Map of Africa but discovered that its classifications, "based largely on potential vegetation conditions, did not provide an adequate estimate of actual vegetative structure or crown cover," and thus their methodology had to be refined.[22] However, UNESCO's 1983 map of *The Vegetation of Africa* did use both the Mediterranean bioclimate and vegetation maps and relied heavily on Emberger's zones and classifications of the (degraded) vegetation (White 1983). More recently, these problematic UNESCO Maps of the Bioclimates and Vegetation of the Mediterranean Zone were used as the primary sources for the FAO's 2001 *Global Ecological Zoning for the Global Forest Resources Assessment* (map and database) for the section on the Middle East and North Africa (FAO 2001).

The "New" Ecology and the Policy Impasse

The 1980s saw a number of academic ecologists working on the Mediterranean who began to conceive of Mediterranean plant ecology differently. Two major schools began to develop about this time, the first of which followed the phytosociology tradition of Emberger, Braun-Blanquet, and Gaussen which incorporated strongly the degraded Eden narrative with its assumptions of succession to climax. The second school championed by the likes of Italian Francesco Di Castri and others, incorporated much more of a co-evolutionary approach that engaged the "new" ecology which did not assume universal equilibrium conditions and linear succession to climax (Blondel et al. 1999; Blumler 1998; Di Castri 1981; Perevolotsky et al. 1998). The second approach, although not without some aspects of the successional ruined landscape view, began to ask new questions about vegetation being co-evolved with and adapted to human activities over thousands of years.

This newer research also began to acknowledge the great resilience of much Mediterranean vegetation—in fact some were arguing in the 1990s that the vegetation of this zone is more robust and resilient than that in more cool temperate and tropical zones (Blumler 1993; Grove et al. 2001). As a result largely of Mediterranean vegetation's resilience in the face of variable precipitation, the region has recently been identified as one of the more robust and least sensitive of the terrestrial ecosystems to climate variability, along with many of the world's arid and semi-arid lands (Seddon et al. 2016). Contrary to common wisdom, Seddon's 14-year analysis demonstrates that the most ecologically sensitive regions are the tropical rainforests, the boreal forest regions, and alpine areas, among others, not the drylands.[23] This correlates strongly with research in the drylands over the last couple of decades which has demonstrated that low rainfall areas with a coefficient of variability of interannual precipitation (CV) above 30–33% generally contain vegetation highly resilient to drought and also to grazing pressure (Davis 2016a, ch. 1; Mortimore 2009; Reynolds et al. 2007) (Fig. 7.3).

These regions with large interannual precipitation variability are governed by ecological dynamics not at equilibrium, and they display the weaknesses of the conventional theory of linear succession to climax vegetation perhaps most clearly.[24] In these highly variable areas, there are often multiple stable states for the vegetation (rather than a single "climax"), and these stable states may be changed by things like drought, wildfire, or various anthropogenic disturbances (Briske et al. 2005; Mortimore 2009). Thus, annual grasslands and garrigue/mattoral, for example, are normal, healthy vegetation communities in many parts of the Mediterranean, not degraded landscapes undergoing vegetation regression as often presumed (Blumler 1993, 1998; Perevolotsky et al. 1998).

Primarily relevant in zones around or below the 300–350-mm rainfall isohyets, this non-equilibrium area applies to most of the warm deserts of the world as well as to large portions of the semi-arid lands (Davis 2016a). It also applies to a significant area of the eastern and southern Mediterranean and to some southerly areas of the northern Mediterranean. Importantly, due in part to the rugged, mountainous nature of some of the Mediterranean region, the 200-, 300-, and 400-mm isohyets as well as the 30–33% CV isolines may fall quite close together so that significant climatological and ecological conditions can and do change a great deal over quite small areas of land.

Despite this research over the last two decades, the degraded Eden view of Mediterranean landscapes prevails in policy circles that control agricultural

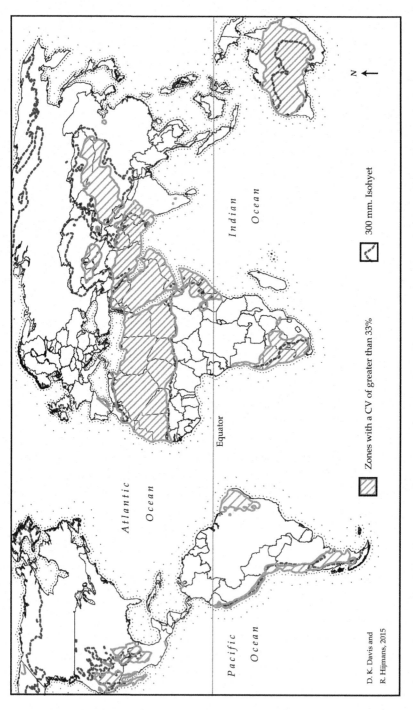

Fig. 7.3 Drylands variability map. This map shows the variability of drylands by demonstrating the 300-mm. isohyet and the 33% CV (coefficient of variation of interannual rainfall). Created by Diana K. Davis and Robert Hijmans. Source: D. K. Davis 2016. *The Arid Lands: History, Power, Knowledge*, The MIT Press, p. 16. Reproduced with the permission of The MIT Press. Higher resolution map available at: http://www.geovet.org/DrylandsMaps.html

and environmental projects on the ground including those in many national governments and in the work of international organizations such as the United Nations. In Europe, the degraded Eden interpretation has permeated much of the research of Mediterranean Desertification and Land Use (MEDALUS) operationalized by the European Union (EU) and a great many of the policies it, and other similar projects, have generated for Mediterranean Europe. Recent analyses have demonstrated just how little degradation there is in the region and how complex and locally specific the question of degradation is (Grove et al. 2001; Mulligan et al. 2016). Mulligan et al., for example, argue that whereas land degradation is commonly claimed to be affecting "80% of the arid and dry areas of the Mediterranean ... negative trends in vegetation cover (possibly reflecting degradation) [affect] only 1% of the study region" (p. 441).

Much of the degradation debate hinges on whether the currently widespread xeric, sclerophyllous vegetation (drought resistant) typified by mattoral, maquis, or garrigue-type plant associations is primarily (or solely) the result of human intervention in otherwise "natural" disturbance regimes and whether the entire region should "naturally" be more wooded as assumed by much phytosociological research.[25] A significant amount of new paleoecological research over the last decade or so is showing strongly; however, that xeric, sclerophyllous vegetation has existed for at least 6000 years since the mid-Holocene, and in some places even longer, predating significant human impact on the regional environment (Collins et al. 2012; Roberts et al. 2011). Moreover, the region has experienced significant but variable aridification during that time which could partly explain the frequency of mattoral (Allen 2001; Roberts et al. 2011). A perusal of the contemporary ecological literature on the Mediterranean today reveals a great many scholars who are engaging with the co-evolution and resilience approach to vegetation change in the region and do not believe that it is significantly deforested or degraded.

The stubbornly tenacious degraded Eden view of the Mediterranean has prevailed in policy circles more often than not, though, and has led to the implementation of a variety of unsuccessful programs such as fire and grazing suppression that have marginalized traditional livelihoods and that are correlated with an increase in destructive wildfires (Butzer 1996; Lloret et al. 2009). Pastoralist livestock producers have been especially hard hit by these programs which have usually enforced settlement and the lowering of livestock numbers. Afforestation, a nearly ubiquitous policy to "restore" the allegedly deforested environment (especially with conifers), may be contributing to global warming rather than ameliorating it (Naudts et al. 2016) as well as depleting groundwater and expropriating smallholder land.

The degraded Eden theory appears to still be most influential primarily in the work of various UN agencies and national governments, as well as in the work of large, pan-European organizations such as MEDALUS and others who directly inform EU environmental and agricultural policy (Mulligan et al. 2016). It has been argued that new science/knowledge can be very slow to penetrate when it does not serve an organization's larger political-economic goals (Bauer et al. 2009; Davis 2016b). This particular environmental narrative of a degraded Eden is useful to mobilize attention and concern over the environment as well as to justify and obtain power and/or funding by scientists, national governments, and international organizations, including authority in policy development and implementation. Scholars have begun to recognize that the old degraded Eden view needs to be reconsidered or discarded and that the new views of variability and resilience must be incorporated. Mulligan et al., for instance, have recently argued that "the legislative and institutional framework must recognize this need for spatio-temporal sophistication to build robust and sustainable agriculture" (Mulligan et al. 2016, p. 444). Policymakers have so far been slow to listen.

Conclusion

The analysis provided in this chapter points in part to the power of potential vegetation maps, the ideologically informed methods at the heart of some of the most influential of these maps, and the outdated understandings of vegetation ecology that accompany some of them. The power relations at play in the development of Mediterranean plant ecology were deep and complex, showing how important it is to consider knowledge politics together with material landscapes and social dynamics (Lave et al. this volume). Like much work in Critical Physical Geography, this analysis also suggests, however, the pitfalls of institutional rigidity and some cultural norms as have been identified, for example, in France. The long dominance of the Braun-Blanquet phytosociological approach in French vegetation ecology and the reluctance of many scholars to engage with other, newer ideas and debates in ecology, have been attributed in part to the tendency of university and governmental research programs to be inflexible and retain existing programs and ideas rather than to explore new ones (Golley 1993; Lefeuvre 1990, 2003). This perhaps explains why discussions of rainfall and climate variability by Emberger and others did not lead later to any substantial engagements with what has become known as the "new" theories of non-equilibrium ecology, especially in arid and semi-arid lands around the Mediterranean region.[26]

A better understanding of the deep history of Mediterranean plant ecology as delineated in this chapter may help us to understand the deeply political and colonial nature of much of this knowledge which has led to policies and programs that have generated inequitable social outcomes as well as to problematic ecological results. The decolonization of this knowledge may lead to better environmental management policies with less "reforestation," more grazing, less stringent fire suppression, and ultimately a more sustainable Mediterranean environment with more equitable social outcomes in the future.

Notes

1. Perceptions and interpretations of "degradation" are, of course, highly subjective and relate in no small part to what the "norms" of a proper landscape are (Behnke et al. 2002; Davis 2007; Sprugel 1991).
2. This famous 1820 essay drew on his earlier, less well-known essay on the same subject in which he discussed the vegetation found in the Mediterranean region and noted that much of it was endemic to Africa (Drouin 1998, p. 153). De Candolle also drew on earlier work by his mentor, French botanist Réné Desfontaines, on North Africa.
3. See, for example, the joyous account of the pleasing vegetation, "the garden of Europe," and the "genial climate" Malte-Brun describes (Malte-Brun 1829 [1810]).
4. By 1832, for example, the French explorer Bory de Saint-Vincent was describing parts of Greece as destroyed by deforestation and fire caused by human use in the French *Exploration scientifique de Morée* (1832–1838) (Drouin 1998, p. 151). There are similar lamentations of environmental ruin in the French *Description de l'Egypte* (1809–1829) and in the *French Exploration scientifique de l'Algérie* (1844–1867).
5. For more details on phytosociology, see (Davis 2007, pp. 144–146), and (Barbour et al. 1999). The term phytosociology was first used in 1896 by the Polish scientist Josef Paczoski who was drawing on a long European tradition of phytogeography reaching all the way back to Alexandre von Humboldt. It was not formally adopted until 1910 at the International Congress of Botany attended by Flahault and other proponents.
6. The serious problems with the utilization of relict vegetation have been widely discussed (Davis 2007; Fairhead et al. 1998).
7. The kind of phytosociology promoted and taught by Braun-Blanquet and his French colleagues, including Louis Emberger, is often termed the "Zurich-Montpelier school." Phytosociology is underpinned by the concept of plant associations, whereas the other main approach to vegetation analysis utilizes

the continuum (or individualistic) concept of community; this has been much more common in the Anglophone world of the UK and North America (Barbour et al. 1999).

8. Louis Trabut had been one of the first to apply phytogeographic methods, including the extensive use of relict vegetation, in Algeria and one of the first botanists to incorporate the deforestation/degradation narrative in his botanical research and publishing in the late 1880s.

9. One of these was Emberger's influential pluviometric quotient which was quite innovative since it takes into account the effects of temperature, precipitation, and evaporation on plant associations.

10. Many scholars were working on these topics and his work complemented that of several others including Henri Gaussen, another French botanist, as well as the French geographer Emmanuel de Martonne and German geographer Wladimir Koppen.

11. The use of the term "*étage*" reveals the history of the idea of vegetation groups being related to zones of altitude going back at least as far as von Humboldt and his influential writings on plant geography.

12. This map may be viewed at: http://gallica.bnf.fr/ark:/12148/bpt6k5449747p/f769.item.r=emberger. Last accessed 26 December 2017.

13. For a discussion of the significance of the *taux de boisement*, see Davis, D.K. and P. Robbins (under review) "Ecologies of the Colonial Present: Pathological Forestry from the '*Taux de Boisement*' to Contemporary Plantations," *Environment and Planning E: Nature and Space.*

14. Emberger accepted the ideas of succession to climax vegetation as did most botanists of the period. For a helpful discussion of these ideas of "Clementsian succession" and their problems in a Mediterranean context, see (Allen 2001, pp. 162–164).

15. In California, for example, French phytosociologists including some of Emberger's students consider chaparral a degraded form of forest, whereas most American ecologists consider it a stable "climax" formation (Barbero et al. 1982, pp. 72–73).

16. The Directory of forestry in Morocco, Paul Boudy lauded this map and Emberger's method for the "precise scientific base" it provided which was a great help to foresters in the region (Boudy 1948, p. 170).

17. A list of Emberger's 82 students and their thesis titles may be found in a volume dedicated to him, see (Emberger 1971, pp. 509–512). Most of these students, while conducting independent research on the Mediterranean and often heading in new directions, carried with them Emberger's underlying assumptions of a ruined landscape as is evident in the majority of their publications.

18. These were Henri Gaussen, the French botanist, phytosociologist, and vegetation cartographer; Egyptian ecologist Mohamed Kassas; and an Italian ecologist (de Phillipis). All were trained in phytosociology and conventional succession theory.

19. The descriptions of the stages and zones of vegetation are particularly enlightening in this respect as at least one-third of them are described as degraded or overgrazed (UNESCO 1969/1970, pp. 60–73).

20. The western section was based on Emberger's studies and maps as well as a variety of other sources on the western end of the Basin. The eastern section was derived from Kassas' estimates using a very similar methodology, a lot of guesswork and a variety of secondary sources, but the data gaps were large as is made clear in the text. Gaussen contributed to southern European vegetation and to parts of the eastern Basin as well as directing and coordinating the cartography.

21. It is interesting to note that one of Emberger's later students gently questioned Emberger's definition of the Mediterranean bioclimate and suggested that it might not be the best for diachronic studies of vegetation change (Daget 1977).

22. See: http://www.grid.unep.ch/data/data.php?category=biosphere and then click on "GNV33, Vegetation Map of Africa, U.S. Defense Mapping Agency (1984)" (www.grid.unep.ch/data/summary.php?dataid=GNV33&category= biosphere&dataurl=&browsen=). Last accessed 6 September 2017.

23. However, a few of the higher elevation mountainous parts of the Mediterranean do show a greater vegetation sensitivity than the surrounding lower elevation areas. See Seddon, 2016.

24. Much of the pathbreaking research behind these new understandings has been conducted in range science/ecology and thus is not as widely appreciated in policy circles as it should be (Behnke et al. 2016; Behnke et al. 1993; Sayre 2017; von Wehrden et al. 2012). See also Sayre this volume.

25. There is also a related debate about erosion in the Mediterranean basin and whether it is accelerated beyond a negative threshold or primarily a natural phenomenon in this hilly and mountainous region. For enlightening discussions of erosion, see (Blumler 1998) and (Stocking 1996).

26. For an example of puzzling over variability, see (Emberger et al. 1962, pp. 203, 206). Emberger even noted the "abundance of annual plants" and their seed production here as well as their importance for pasturelands. A few French ecologists have engaged with the newer theories of non-equilibrium ecology but primarily in the sub-Saharan African context (Hiernaux et al. 2002).

References

Allen, H.D. 2001. *Mediterranean ecogeography*. London: Prentice Hall.

Attenborough, D. 1987. *The first eden: The mediterranean world and man*. London: Fontana/Collins.

Barbero, M., and P. Quezel. 1982. *Classifying mediterranean ecosystems in the mediterranean Rim Countries and in the Southwestern USA*. Berkeley, CA: Pacific Southwest Forest and Range Experiment Station, Forest Service, USDA.

Barbour, M.G., J.H. Burk, W.D. Pitts, F.S. Gilliam, and M.W. Schwartz. 1999. *Terrestrial plant ecology*. 3rd ed. Menlo Park, CA: Addison, Wesley, Longman.

Batanouny, K.A. 2001. *Plants in the deserts of the Middle East*. Berlin: Springer-Verlag.

Bauer, S., and L.C. Stringer. 2009. The role of science in the global governance of desertification. *The Journal of Environment & Development* 18 (3): 248–267.

Behnke, R.H., and M. Mortimore, eds. 2016. *The end of desertification? disputing environmental change in the drylands*. Dordrecht: Springer.

Behnke, R.H., I. Scoones, and C. Kerven, eds. 1993. *Range ecology at disequilibrium: New modes of natural variability and pastoral adaptation in African Savannas*. London: Overseas Development Institute.

Behnke, R.H., P.M. Döll, J.E. Ellis, and P.A. Harou. 2002. Responding to desertification at the national scale. In *Global desertification: Do humans cause deserts?* ed. J.F. Reynolds and D.M. Stafford Smith, 357–385. Berlin: Dahlem University Press.

Blais, H., and F. Deprest. 2012. The mediterranean, a territory between France and Colonial Algeria: Imperial Constructions. *European Review of History* 19 (1): 33–57.

Blondel, J. 2006. The 'Design' of mediterranean landscapes: A millennial story of humans and ecological systems during the historic period. *Human Ecology* 34 (5): 713–729.

Blondel, J., J. Aranson, and J.-Y. Bodiou. 1999. *The mediterranean region: Biological diversity in space and time*. Oxford: Oxford University Press.

Blumler, M.A. 1993. Successional pattern and landscape sensitivity in the mediterranean and near East. In *Landscape sensitivity*, ed. D.S.G. Thomas and R.J. Allison, 287–305. Chichester: John Wiley & Sons Ltd.

———. 1998. Biogeography of land-use impacts in the near East. In *Nature's geography: New lessons for conservation in developing countries*, ed. K. Zimmerer and K. Young, 215–236. Madison: The University of Wisconsin Press.

Boudy, P. 1948. *Économie forestière nord-africaine: Milieu physique et milieu humain*. Vol. I. Paris: Éditions Larose.

Brandt, C.J., and J.B. Thornes, eds. 1996. *Mediterranean desertification and land use*. Chichester: John Wiley & Sons.

Briske, D.D., S.D. Fuhlendorf, and F.E. Smeins. 2005. State-and-transition models, thresholds, and rangeland health: A synthesis of ecological concepts and perspectives. *Rangeland Ecology and Management* 58 (1): 1–10.

Butzer, K.W. 1996. Ecology in the long view: Settlement Histories, agrosystemic strategies, and ecological performance. *Journal of Field Archaeology* 23 (1): 141–150.

de Candolle, A. 1977. Géographie botanique. In *Ecological phytogeography in the nineteenth century*, ed. F.N. Egerton. New York: Arno Press.

di Castri, F., ed. 1981. *Mediterranean-Type Shrublands*. Amsterdam: Elsevier Scientific Publishing Company.

Collins, P., B. Davis, and J. Kaplan. 2012. The mid-holocene vegetation of the mediterranean region and Southern Europe, and comparison with the present day. *Journal of Biogeography* 39 (10): 1848–1861.

Daget, P. 1977. Le bioclimat méditerranéen: Analyse des formes climatiques par le système d'Emberger. *Vegetatio* 34 (2): 87–103.

Davis, D.K. 2007. *Resurrecting the granary of Rome: Environmental history and French Colonial Expansion in North Africa*. Athens: Ohio University Press.

———. 2016a. *The arid lands: History, power, knowledge*. Cambridge, MA.: The MIT Press.

———. 2016b. Deserts and drylands before the age of desertification. In *The end of desertification? Disputing environmental change in the drylands*, ed. R.H. Behnke and M. Mortimore, 203–228. Dordrecht: Springer.

Deprest, F. 2002. L'Invention géographique de la méditerranée: Éléments de réflexion. *L'Espace Geographique* 1 (1): 73–92.

Drouin, J.-M. 1998. Bory de Saint-Vincent et la géographie botanique. In *L'Invention scientifique de la Méditerranée*, ed. M.-N. Bourget, B. Lepetit, D. Nordman, and M. Sinarellis, 139–157. Paris: École des Hautes Études en Science Sociales.

Emberger, L. 1930. La Végétation de la région méditerranéene: Essai d'une classification des groupements végétaux. *Revue Générale de Botanique* 42 (503): 641–662, 705–721.

———. 1933. Nouvelle Contribution à l'étude de la classification des groupements végétaux. *Revue Générale de Botanique* 45 (1): 473–486.

———. 1934. Aperçu générale [de la végétation]. In *La Science au Maroc*, ed. P. Boudy and P. Despujols, 149–182. Casablanca: Imprimeries Réunies.

———. 1939. Aperçu générale sur la végétation du Maroc. Commentaire de la carte phytogéographique du Maroc. *Veröffentlichungen des Geobotanischen Institutes Rübel in Zürich* 14 (1): 40–157.

———. 1971. *Travaux de botanique et d'écologie*. Paris: Masson et Cie., Éditeurs.

Emberger, L., and G. Lemée. 1962. Plant ecology. In *The problems of the arid zone: Proceedings of the paris symposium*. Paris: UNESCO.

Fairhead, J., and M. Leach. 1998. *Reframing deforestation: Global analyses and local realities: studies in West Africa*. London: Routledge.

FAO. 2001. *Global ecological zoning for the global forest resources assessment 2000, final report*. Rome: UNFAO Forestry Department.

Flahault, C. 1896. Au Sujet de la carte botanique, forestière et agricole de France. *Annales de Géographie* 5 (15 October): 449–457.

Golley, F.B. 1993. *A history of the ecosystem concept in ecology: More than the sum of the parts*. New Haven: Yale University Press.

Grove, A.T.D., and O. Rackham. 2001. *The nature of mediterranean Europe: An ecological history*. New Haven: Yale University Press.

Hiernaux, P., and M.D. Turner. 2002. The influence of farmer and pastoralist management practices on desertification processes in the Sahel. In *Global desertifica-*

tion: Do humans cause deserts? ed. J.F. Reynolds and D.M. Stafford Smith, 135–148. Berlin: Dahlem University Press.

Huston, M. 1994. *Biological diversity: The coexistence of species on changing landscapes.* Cambridge: Cambridge University Press.

Lefeuvre, J.-C. 1990. La recherche en écologie en France: Heur et malheur d'une discipline en difficulté. *Aménegement et. Nature* 91 (1): 1–4.

———. 2003. Science et éducation dans le domaine de l'environnement. In *La Charte de l'environnement: Enjeux scientifiques et juridiques.* Paris: Ministère de l'Écologie et du Développement Durable.

Lloret, F., J. Pinol, and M. Castellnou. 2009. Wildfires. In *The physical geography of the mediterranean,* ed. J. Woodward, 541–558. Oxford: Oxford University Press.

Maire, R. 1921. Coup d'oeil sur la végétation du Maroc. In *Sur les productions végétales du Maroc,* ed. É. Perrot and L. Gentil, 59–71. Paris: Larose.

Malte-Brun, C. 1829 [1810]. *Universal geography, or a description of all the parts of the world on a new plan, according to the great natural divisions of the globe.* 6 vols. Vol. 4. Philadephia: John Laval and SF Bradford.

Matagne, P. 1999. *Aux Origines de l'écologie: Les naturalistes en France de 1800 à 1914.* Paris: Éditions du CTHS.

Mortimore, M. 2009. *Dryland opportunities: A new paradigm for people, ecosystems and development.* Gland: IUCN.

Mulligan, M., S. Burke, and A. Ogilvie. 2016. Much more than simply 'Desertification:' Understanding agricultural sustainability and change in the mediterranean. In *The end of desertification? Disputing environmental change in the drylands,* ed. R.H. Behnke and M. Mortimore, 427–450. Dordrecht: Springer.

Nahal, I. 1981. The mediterranean climate from a biological viewpoint. In *Mediterranean-Type Shrublands,* ed. F. di Castri and D.W. Goodall, 63–86. New York: Elsevier Scientific Publishing Co.

Naudts, K., Y. Chen, M. McGrath, J. Ryder, and A. Valade. 2016. Europe's forest management did not mitigate climate warming. *Science* 351 (6273): 597–600.

Perevolotsky, A., and N. Seligman. 1998. Role of grazing in Mediterranean rangeland ecosystems. *BioScience* 48 (12): 1007–1017.

Poore, M.E. 1955. The use of phytosociological methods in ecological investigations: The Braun-Blanquet system. *Journal of Ecology* 43 (1): 226–244.

Reynolds, J.F., D.M. Stafford Smith, and E.F. Lambin. 2007. Global desertification: Building a science for Dryland Development. *Science* 316 (5826): 847–850.

Roberts, N., D. Brayshaw, C. Kuzucuoglu, R. Perez, and L. Sadori. 2011. The mid-holocene climatic transition in the mediterranean: Causes and consequences. *The Holocene* 21 (1): 3–13.

Roumieux, C., G. Raccasi, E. Franquet, A. Sandoz, F. Torre, and G. Metge. 2010. Actualisation des limites de l'aire du bioclimat méditerranéen selon les critères de Daget (1977). *Écologia Méditerranéa* 36 (2): 17–24.

Sayre, N. 2017. *The Politics of Scale: A History of Rangeland Science.* Chicago: The University of Chicago Press.

Seddon, A.W., M. Marcias-Fauria, P.R. Long, D. Benz, and K.J. Willis. 2016. Sensitivity of Global Terrestrial Ecosystems to climate variability. *Nature* 531 (7593): 229–232.

Sprugel, D.G. 1991. Disturbance, equilibrium, and environmental variability: What is 'Natural' vegetation in a changing environment? *Biological Conservation* 58 (1): 1–18.

Stocking, M. 1996. Soil erosion: Breaking new ground. In *The lie of the land: Challenging received wisdom on the African Environment*, ed. M. Leach and R. Mearns, 140–154. London: The International African Institute.

Thirgood, J.V. 1981. *Man and the Mediterranean Forest: A history of resource depletion*. New York: Academic Press.

UNESCO. 1963. *Bioclimatic map of the mediterranean zone. Explanatory notes*. 30 vols. Vol. 21, *Ecological study of the mediterranean zone*. Paris: UNESCO—FAO.

———. 1969/1970. *Vegetation map of the mediterranean zone. Explanatory notes*. 30 vols. Vol. 30, *Ecological study of the mediterranean zone*. Paris: UNESCO—FAO.

von Wehrden, H., J. Hanspach, P. Kaczensky, J. Fischer, and K. Wesche. 2012. Global assessment of the non-equilibrium concept in Rangelands. *Ecological Applications* 22 (2): 393–399.

White, F. 1983. *The vegetation of Africa: A descriptive memoir to accompany the UNESCO/AETFAT/UNSO vegetation map of Africa*. Paris: UNESCO.

Williams, M. 2003. *Deforesting the earth: From prehistory to global crisis*. Chicago: The University of Chicago Press.

8

How the West Was Spun: The De-politicization of Fire in the American West

Gregory L. Simon

Introduction

In late Spring 2015, after yet another wildfire threatened yet another suburban Southern California settlement, a gaggle of media outlets were quick to report that the already vexing problem of costly wildfires in the American West was only getting worse. A report in *Scientific American*, for example, described how the weather and landscapes of the American West were expected to "usher in regular wildfires" around the region; "drought and heat wrought by stubborn ocean conditions have left great stretches of it dryer and more combustible than usual this year", the report told us. According to this article (and many others like it), the western United States is becoming more and more 'combustible' with each passing—and increasingly hot and arid—season. Provocatively, the article suggested that "vulnerable residents" now find themselves "staring down the barrel of a torturous fire season…" (Upton 2016). While the report also noted that keys for adapting to increased combustibility "lie in how fires and the lands that fuel them are managed", there is no mistaking the primary culprit for these stubborn and "rattling" wildfire threats: our changing and increasingly inhospitable climate. Another widely circulated news article describing a deadly fire in the Sierra Nevada foothills a few days later conveyed a similar story: "scorching heat and tinder-dry conditions across the West" are contributing to "massive wildfires in the past week

G. L. Simon (✉)
Department of Geography and Environmental Sciences, University of Colorado, Denver, CO, USA

© The Author(s) 2018
R. Lave et al. (eds.), *The Palgrave Handbook of Critical Physical Geography*,
https://doi.org/10.1007/978-3-319-71461-5_8

that have destroyed properties and sent residents to seek shelter..." (Associated Press 2016).

In each case, the causes of increased combustibility are portrayed as a by-product of warming weather, stubborn high-pressure zones, and increasingly desiccated western landscapes. But what about the institutions, policies, and billions of (US) dollars worth of financial incentives that help produce human settlements and immense social risks on these landscapes? In our list of common explanatory variables, where are these powerful social forces that turn historically active fire regimes into a string of deadly and costly firestorm events? (Fig. 8.1) I argue here that, unfortunately, these important expressions of material accumulation and risk are all too frequently (and conveniently) ignored within mainstream scientific and media reporting. Indeed the afore-mentioned "down the barrel of a gun" metaphor seems more apt if we are willing to admit that residential communities—and the planning and construction industry that creates them—are also holding the weapon (Upton 2016).

In a similar fashion, "the flammable West" is a phrase that gets used often by media and public policy outlets. It depicts a region that seems, almost like clockwork, to 'catch fire' and go 'up in smoke' each year. A 2013 northern California public television news article titled *The Flammable West: Mega Fires in the Age of Climate Change* is one example of such journalism. The article provides a useful, if startling, description of fire trends in the western United States. It tells us that compared to annual averages during the 1970s, the

Fig. 8.1 An all too familiar scene. Are wildfires threatening homes in wildland-urban interface areas of the US West? Or are homes impinging on natural fire regime events?

period 2002–2011 contained twice as many fires larger than 1000 acres, seven times more fires exceeding 10,000 acres, nearly five times more fires greater than 25,000 acres, and an average fire season lasting two-and-a-half months longer (Green 2013). Like many other similar reports, *The Flammable West* provides an important public service announcement on the importance of fire mitigation and adaptation policies. But like the *Scientific American* report above, it also reproduces and fortifies a troubling trend within the suburban and exurban fire discourse: the persistent focus on the region's tendency to burn, as if this were the natural order of things. As if flammability *was* the *problem* rather than the *symptom* of a larger, engrained, and more pernicious underlying set of social-economic processes (Fig. 8.2).

The de-politicization of these human and financial drivers was on full display in early 2016 when California Governor Jerry Brown introduced a US$719 million one-time funding package and an extra $215 million to the state's emergency fund to assist efforts to fight the state's next round of large wildfires. A spokesperson for the governor explained how "conditions have changed in California" while noting an increase in devastating wildfires in recent years due to persistent drought conditions linked to climate change and its effects across the state (Mai-Duc 2016). Given the governor's office's climate-centric description of destructive fires and their causes, the state's budget earmark is, quite fittingly, called the 'drought package'. But this type of policy framing and budget justification further obfuscates the other important 'condition' that has dramatically changed around the region: the steady encroachment of human settlements into formerly undeveloped areas at the urban fringe. This undeniably massive modification to the California landscape is conspicuously left out of the public conversation. Although the bill could more accurately be called the 'drought and urban encroachment package', government officials and other special interest groups seem quite content with the current, non-controversial title.

A Critical Physical Geography of Fire

The time has come to illuminate flammability. In mainstream reporting, scientific research, and ongoing policy debates, the term 'flammable' (or 'combustible', 'tinderbox', etc.) is often deployed in a manner that naturalizes costly fires while obfuscating influential, shortsighted, and sometimes-reckless development histories and regional growth policies. This chapter marks a Critical Physical Geography (CPG) intervention in two crucial ways. First, it explains how suburban landscapes and associated fire risks and costs are produced dialectally

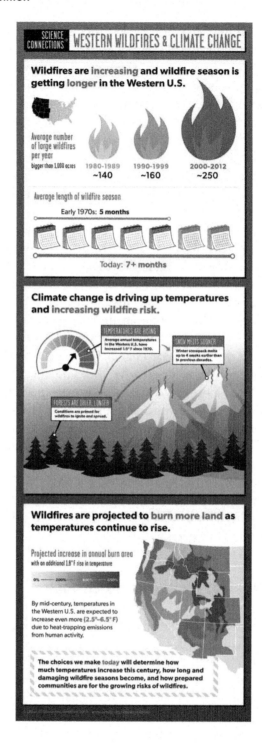

through a powerful and self-sustaining positive feedback loop; physical landscapes stand as both an artifact of diverse profitable development incentives and also a lucrative arena within which future opportunities to extract profits and immense wealth are activated—in both pre- and post-fire conditions (Simon 2014). Contemporary fire-prone suburban landscapes of the American West are decidedly neoliberal landscapes—profits in production, profits in protection. Second, this chapter interrogates various ways the simultaneous production of risk and profits is obscured within mainstream fire science, environmental management, and urban development policy-making. I illustrate here the limited (and thus limiting) ways civil society and policy-makers come to know and debate hazardous environments in the West (which in turn influences how we modify and manage the physical landscape).

The following 'Illuminating Flammability' section introduces the concept of 'the Incendiary' as a way of describing how landscapes of the American West (and other fire-prone areas) are produced over time through capitalist growth imperatives—a recursive process that generates both immense wealth and risks for diverse parties (Simon and Dooling 2013). The affluence-vulnerability interface is then presented as an alternative to the wildland-urban interface as an analytic framework that better elucidates the underlying socio-economic drivers of rapidly changing suburban landscapes. The 'Lucrative Landscapes' and 'A Character Profile' sections further illustrate how areas at the urban periphery are lucrative landscapes and briefly outline how development pressures are altering large portions of the region while generating unprecedented fire activity, risks, and costs.

While these two sections explain fire and its production as a dynamic set of socio-physical *conditions*, the subsequent two sections explore fire and its production as a set of contested and continually evolving *ideas*. The 'Smoke Screen' section introduces the concept of de-politicization—a process through which issues (i.e., high-risk, costly fires) are systematically stripped of one or more of their important and politically provocative foundations—in this case the economic incentives and avarice that produce expensive, injurious wildfires.

Fig. 8.2 An informational panel developed by the Union of Concerned Scientists depicting the relationship between fire and climate change in the US West. The panel offers many important and revealing statistics. But this image also reveals something else: the minimization of profitable land use planning decisions and the privileging of climatic forces when explaining the "growing risks of wildfires" in the West. The only reference to residential developments is in the context of adaptation strategies, thus portraying homes as passive victims and not as part of a larger structure of "risk"-producing suburbanization. (Photo Credit: Union of Concerned Scientists 2013)

This section demonstrates how mainstream and scientific reporting on wildfires de-politicizes fire in the American West, most notably by naturalizing costly wildfires and privileging climate change as an explanatory variable. The 'Debates of Distraction' section discusses the related process of re-politicization. This process arises when banal narratives, contentious debates, and the pursuit of 'relevant' science become mired in various alternative disagreements (frequently in the form of contested, place-specific issues) or proxy debates (often manifest in larger ideological disagreements such as the appropriate role of government in regulating individual and community uses of natural resources on public and private lands).

This chapter suggests that critical physical geographers will need to play an important role in reshaping how we study, know, and manage wildfire risks around the region. Through their research and outreach, CPGers can help infuse the public's understanding of fire activity around the West with a clear sense that many wildfire risks, costs, and vulnerabilities at the urban periphery are profoundly social in nature. Infusing Physical Geography's already strong understanding of physical fire-climate dynamics (e.g., Westerling and Bryant 2008, Peterson 2010, Smithwick et al. 2009, Hessl 2011, Westerling et al. 2014) with a robust appreciation for important social processes and land use policy dynamics will set CPGers apart from many other physical scientists. In so doing, researchers will be able to present policy-makers and the media with a diverse suite of ecological *and* social factors to help explain the rise and implications of dangerous wildfires. This should help temper the inclination for popular and scientific media outlets to understate (or simply ignore) these important social drivers of risk (i.e., the financial incentives spurring increased suburbanization and land use/cover changes at the urban fringe) in favor of more narrowly focused, climate change-centric explanations. A Critical Physical Geography approach to fire will thus challenge normative accounts of social-environmental change in the West that de-politicize society's unflagging pursuit of suburban development, and instead inform a new set of land use management practices and perspectives about how we want to coexist with fire in the future.

Illuminating Flammability: Introducing 'The Incendiary'

In its common usage, 'flammability' connotes the physical symptom of a landscape but not the root causes behind its making. The term 'flammable' implies that an entity, such as a landscape, holds qualities that make it susceptible to fire. It is an adjective used to describe an object that just *happens to have* the capacity to easily go up in flames. Consider instead the term 'Incendiary', which in noun form implies that an object (or person, place) is an agent that actively produces and incites fire. It *makes things* flammable, much like an arsonist.

Imagine a network of elusive, brazen, and dangerous arsonists afflicting a series of towns and cities around the American West. Every few weeks, these individuals randomly ignite one or two fires. Some of the fires are controlled with only minor damage while others quickly spread and endanger nearby communities, resulting in lost lives, considerable private property damage, and millions of (US) dollars in firefighting and rebuilding costs. This problem could be confronted through a series of adaptive measures, which might include rapid emergency response efforts or direct mitigation of flammable land features through vegetation clearing and building code modifications. One could argue however that a more effective and long-lasting approach would be to also directly confront the source of the problem itself, that is, investigate the incendiaries and undercut the arsonist cell. Why are they lighting the landscape on fire? How are they getting the necessary money and resources? And what is it about their environment, funding, background, character, and psychology that lead them to perpetrate such acts? To address these questions is to grapple with the root causes of the problem. This approach accepts that while it is important to treat the source of fire—flammability—it is also important to treat the source of flammability, the Incendiary. Confronting the Incendiary means closely examining its history, engrained foundations, essential nature, and core qualities.

As a society, we would never accept the first option of simply reacting and adapting to an arsonist. It is thus puzzling that we accept it with wildfires. If we understand the landscape as a troublesome individual, as 'the Incendiary', then the best way to substantively reduce the symptom of flammability is to engage in appropriate fire reaction and mitigation activities *while also* confronting their root causes: the political economic structures, planning policies, socio-cultural behaviors, and environmental systems that continue to produce, support, and enrich the Incendiary. If fire can be understood as a symptom of a flammable landscape, then flammability exists as one symptom of a landscape that is an Incendiary. Like the arsonist, it is the landscape as Incendiary that should receive our direct and critical inquiry.

The Affluence-Vulnerability Interface

In order to excavate and treat 'the Incendiary', managers, planners, and scientists will need to move beyond analysis that conforms to—and is bound spatially by—the wildland-urban interface (WUI). The WUI is one the most ubiquitous phrases circulating through the suburban and exurban wildfire management discourse. It is *the* land designation used to connote the uneasy overlap of human settlements with traditionally undeveloped or wild (and oftentimes already fire-prone) environments. The WUI is a rather recent concept and geographic construct and is described by the National Wildfire Coordinating Group as "…the zone of transition between unoccupied land and human development" (National Wildfire Coordinating Group 2014). The establishment of a WUI land designation—despite its somewhat malleable definition—has substantial policy consequence. This designation is easy to map and has thus made legible the geographical area supporting the structured implementation of a number of land use and forest management practices. These include early efforts to extend the US Forest Service's 'fire exclusion paradigm' into developed areas through dedicated fire suppression-based home protection (Coehn 2008) and more recent 'Fire Adapted Communities' approaches premised on providing services that increase community education, preparedness, and resilience to periodic fire events (FAC 2014). Over the past 30 years, the WUI has emerged as a useful land classification—a conceptual container within which we can study, interpret, and manage the messy and complex transition from non-urban to urban, and public to private.

A shift in perspective is in order. This chapter argues for a move away from the wildland-urban interface as the central organizing framework guiding the management of wildfires (and the *symptoms* of flammability) at the urban periphery. Instead, it suggests the adoption of an affluence-vulnerability interface (AVI) approach. This approach encourages decision-makers to pay greater attention to the systemic causes of change, risk, and vulnerability, factors that are quite often implicated in policies that generate profit opportunities for stakeholders in urban and exurban settings (including landowners, the construction industry, individual homeowners, private fire services, and cities in search of new tax revenues—see below for more details). Critically examining the AVI therefore signals a conceptual shift from the management of particular *areas*, to the management of social-ecological *processes*. Analyzing the AVI also means closely assessing various ways the simultaneous production of risk and profits is concealed within mainstream fire and urban development discourse. This conceptual tack will entail analyzing policies, social norms, economic incentives, and environmental changes that produce both increased profits and risks in areas currently recognized as the WUI (Simon 2016).

Of course, it would be unwise and irresponsible to just do away with the WUI all together. The wildland-urban interface can certainly function as one useful organizing principle since it does hold a level utility in day-to-day land management activities. The WUI characterizes a land designation and set of material conditions that are grounded in a particular time and space. The inadequacy of the WUI as a concept however lies in its inability, by itself, to reveal the forces behind its own creation, the same forces driving an increase in wildfire disasters. The AVI, on the other hand, is valuable for illustrating complex economic, social, and environmental drivers (i.e., the Incendiary)— across multiple spatial and temporal scales—that inform the development of the WUI.

Lucrative Landscapes at the Urban Periphery: Taking Profits, Adding Risk

Suburban landscapes of the US West are lucrative landscapes. They are areas— converted into various forms of capital and surplus value—that generate high levels of profit and revenue for interested parties near and far (see Table 8.1 for examples of these profit-seeking opportunities.) From early land use extraction activities to contemporary private fire mitigation services, diverse groups extract wealth from these regions, thus leveraging the suburban landscape as a source of prosperity and increased affluence.

In many areas of the West, the formation of lucrative fire-prone landscapes begins when previously undeveloped areas fall under the speculative eye of resource extraction industries. Profitable mining, timber, agriculture, and other extractive activities allow parties to take profits from the land while introducing basic infrastructure (water, electricity, graded roads, etc.) that are later used to justify and enable cost-efficient entry points for eventual suburban developments. Private and public landowners benefit financially from the eventual subdivision and sale of these landscapes as land values increase with the arrival of new amenities. Meanwhile, various development interests in the home and municipal infrastructure construction industries procure large contracts in fast growing urban peripheries around the West. (See the following section for figures illustrating the size and scope of this immense suburban transformation.) These suburban developments present opportunities for lucrative *post-fire* construction contracts as well. According to one wildfire

Table 8.1 Fire-prone areas of the US West are highly lucrative landscapes. For well over a century, many groups and individuals have benefited financially from these land-scapes. In a dialectical fashion, profitable activities produce communities with high exposure to wildfires, which in turn spur opportunities for wealth accumulation in response to fire risks and events. Several examples of these profitable activities and associated risks are listed in this table (Simon 2014)

Lucrative landscapes: Profitable activity	Extracting profits: Specific example	Risky real estate: New exposures and risks
Pursuit of profits leading to increased social risks		
Resource extraction	Historical logging and mining activities, including large-scale removal of valuable timber	Introduced municipal infrastructure such as graded roads enabling further growth
Land subdivisions and real estate syndicates	Conversion of open space into developable neighborhoods and profitable housing tracks	further paved the way for new residential developments in the area
Home construction industry	New lucrative home and municipal infrastructure construction opportunities	Introduced thousands of new homes and residents to the landscape
Re- and afforestation activities	New vegetation cover (e.g., eucalyptus) increases property values in new neighborhoods	New and arguably more dense and flammable vegetation
City and county property tax revenues	High fire risk area houses produce millions in tax revenue annually for many cities and counties	Pursuit of new tax base introduce high-density housing developments
Pursuit of profits in response to increased social risks		
Insurance company profit potential	Company fails to meet claim payouts despite customer payments and substantial government support	Financial vulnerabilities add to composite household-level risks
Private firefighting services	Private sector fire companies charge for concierge-level fire services and product sales	Responders unfamiliar with the area, adding confusion to scene
Home protection entrepreneurship	Creation of market opportunities for new products like buffer mulch, fire foams, fireproof features	Generates a sense of security and sustained home demand in fire-prone landscapes
Post-disaster home reconstruction	Homes in fire areas are often much bigger, closer, and more valuable after the reconstruction process	Adds to overall landscape fuel load and assists fire spread

analysis, "there are 897,102 residential properties in the western U.S. that are currently located in High or Very High wildfire-risk categories, with a reconstruction value of more than US$237 billion" (Botts et al. 2015).

Landowners and the construction industry are hardly the only ones to profit from the development of sub- and exurban landscapes that are historically prone to wildfires. Cities and their oftentimes-overburdened budgets can be some of the largest beneficiaries. If developed and financed efficiently, the development of land at the urban periphery can generate an extraordinary boost in property tax revenues for cities. Consider the case of Oakland, California, where property tax revenues generated in very high fire risk areas are 57 percent higher per unit compared to the rest of the city (US$6650/unit and $4798/unit, respectively). Despite only containing 23 percent of the total taxable units in the city, very high fire risk areas account for 33 percent of the property tax revenue (Simon 2014, 2016). For the City of Oakland, the decision to permit massive development projects in this area—like other similar landscapes around the region—was indeed a financially lucrative decision.

While the occupants of these residential developments at the urban fringe may be exposed to periodic fire activity and potentially catastrophic losses, there are also distinct financial benefits associated with homeownership for those willing to remain and rebuild. Analysis from Colorado Springs, Colorado, and Oakland, California, shows that after major firestorm events, home rebuilds were 14 percent and 11 percent larger than original home structures, respectively. In Oakland, the construction of new, bigger, and better homes translated into an increase in home values (in the ten years after the fire) that was nearly double the rate of home value increases in non-impacted parts of the city (Simon 2016).

The development of homes in fire-prone areas also presents new profitable opportunities for a fast emerging private firefighting industry. While firefighting activities have historically been operated by public agencies, today, the United States is witnessing the rapid privatization of the residential fire response sector. In 2012 there were already 256 private firefighting companies in the United States—a number industry forecasters expect will grow to more than 320 by 2017. Over the same period the number of private firefighters is expected to increase from 16,880 to 27,200. As the website of a leading community fire information portal put it, although private firefighters "make up just 4.3% of the nation's total firefighters … this is an industry on the verge of catching fire because of growing trend towards privatization" (WildfireX 2015, in Simon 2016). Along with a vast array of new consumer products such as fire mulch and home spray kits, more homes at risk means more homes to protect and still more opportunities for private sector profits.

Revenue-generating activities at the city's edge are certainly not benign. Over time, the generation of financial benefits has coincided with the production and maintenance of social risks, vulnerabilities, and costs. This is the nature of urban growth under capitalism—it produces both beneficiary and disadvantaged groups, simultaneously. And as the examples above illustrate, in many instances, we see that one outcome co-constitutes the other—in a dialectical fashion, efforts to increase affluence oftentimes necessitate elevating levels of fire risk, and higher levels of social risk and vulnerability frequently spur opportunities to generate further financial gains.

Factors influencing increased social vulnerability and higher-risk mitigation costs are inextricably tied to ever-changing profit-seeking practices and diverse forms of economic opportunism. Understanding the AVI and the root causes of fire risk is an important first step toward substantively reducing future costs associated with patterns of material accumulation and seemingly unfettered urban expansion into this risky real estate—as the old adage goes, 'you have to understand the problem before you can find the solution'.

A Character Profile of the Incendiary: The Changing American West

Processes supporting the Incendiary have led to unmistakable population growth—and increased fire risks and costs—around the American West over the past several decades. Suburbanization has increased the number of houses in urban areas by as much as 27 percent from 1970 to 2000, with approximately 38 percent of this new development occurring near or within the WUI (FEMA 2002). Between 1990 and 2000 alone, more than one million homes in total were introduced to the WUI in the states of California, Oregon, and Washington (Hammer et al. 2007). Across the western United States, WUI areas have seen a 300 percent population growth rate in the past 50 years, which outpaces overall regional population growth rates for the same time period (IAWF 2013). Spatially, these areas of the western United States have experienced 60 percent expansion since 1970 (Theobold and Romme 2007), with traditional wildlands converted to wildland-urban interface designated areas at a rate of 400 acres per day, an equivalent of close to two million acres per year (IAWF 2013). The most alarming suburbanization statistic, however, concerns what *hasn't* been developed. As of 2008, only 14 percent of private land in WUI areas of the western United States had actually undergone land conversion. By 2013, this number increased to 16 percent (Gorte 2013) (Fig. 8.3). These numbers reveal something quite startling: over 80 percent of the WUI environment remains eligible for further growth, increased social vulnerability, and higher firefighting costs. As of 2012, 46 million homes were located in the WUI. Based on current trends, that number is expected to increase to 54 million by 2022 (United States Forest Service 2015).

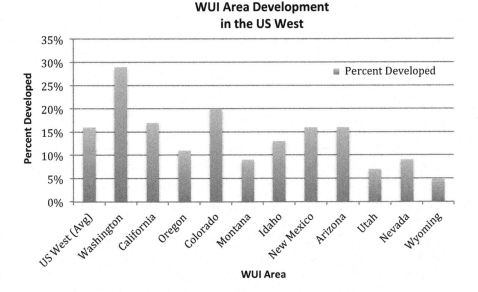

Fig. 8.3 Percentage of WUI area developed in the US West as of 2013. (Source: Headwaters Economics)

The growing number of structures destroyed by wildfires illustrates the damaging impact of wildfires on human populations. In total, from 2000 to 2012, the United States lost 38,701 structures to wildfires, an average of 2977 structures per year (IAWF 2013). In California, for example, since 1923, 15 of the most damaging 20 fires (in number of structures destroyed) have occurred within the past 25 years; nine of these fires have occurred over the last ten years (CalFire 2013). This means that in California's modern history, about 75 percent of the largest and most destructive wildfires have occurred in the past 25 years, and nearly 50 percent have taken place in the last decade alone. Death, injury, and long-lasting health problems are other well-documented negative outcomes resulting from wildfires. From elderly community members unable to flee fast moving flames to emergency first responders (such as the 19 Prescott City firefighters who lost their lives in the tragic 2013 Arizona Yarnell Hill Fire), bodily harm and trauma as a result of destructive fires are constant concerns in the region.

The implications of wildfires go beyond structural damage and bodily harm and include immense financial commitments by city, state, and federal agencies to fight fires at the WUI—cost burdens that displace other, arguably more essential needs such as health care, education, and environmental conservation. Over the past 50 years, the cost of fire mitigation activities has grown dramatically in the United States. In the 1970s, the federal budget allocated to firefighting wildfires averaged US$420 million. This figure jumped to $1.4 billion by 2000 (Ingalsbee 2010) and increased again to $2.5 billion by 2012. Estimates place the total fire mitigation budget in 2012 at a lofty US$4.7 billion when inclusive of federal, state ($1.2 billion), and local ($1 billion) governments in the United States (IAWF 2013). These costs have risen primarily as a result of increased fire mitigation requirements due to several factors. First is a buildup of fuels resulting in part from past fire suppression policies. For several decades ending in the 1970s, forest policy mandated a strict commitment to fire prevention. Prescribed and controlled burns were banned due to their perceived threat to the surrounding environment. This policy, we now know, led to a steady accumulation of forest materials and an increased likelihood of larger, more intense, and more dangerous wildfires. Other influential factors increasing mitigation costs include a warming climate, persistent drought conditions in the West, and, I would argue most importantly, the development of residential communities adjacent to already fire-prone public lands.

To be sure, wildfires are common occurrences in the US West even in the absence of human activity due to normal climate variability and frequent and sometimes-prolonged droughts. Wildfires have occurred for millennia and provide crucial ecological services required to recycle nutrients, improve soil condition, and initiate plant succession. Despite this active fire history, wildfire trends are changing because of a dramatically altered western climate, a climate now characterized by higher regional average temperatures, increased rates of evapotranspiration, and more pronounced levels of aridity (at least as compared to recent history). These emerging conditions are, in turn, resulting in longer and more active fire seasons. But make no mistake, while climate change itself is certainly generating environmental conditions favorable to higher-frequency and intensity fires, it is the region's long history of fire suppression and, most notably, the widespread encroachment of human populations into already high fire risk areas that are most responsible for increased fire exposure, risk, and mitigation costs across the region (Moritz et al. 2014). The effects of climate change on the US West are a lot like adding fuel to an already burning fire.

Smoke Screen: When Explaining Wildfires Obscures the Incendiary

Contemporary management and scientific discourses on wildfires *de-politicize* 'The Incendiary' and the political economic root causes of fire disasters. De-politicization refers to the process of stripping an issue or event of one or more of its important and politically provocative foundations. This allows particular foundational explanations of social-environmental change—in this context, processes related to the AVI and its associated controversies—to go unnoticed and unchallenged. Because popular media and policy-makers tend to overlook the AVI when reporting on increased fire risk around the region, critical physical geographers (and physical scientists more generally) will need to more directly reference the role of urban sprawl, and the financial incentives that support it, when informing these public outlets. Moreover, this 'critical' engagement by physical geographers should also closely evaluate the use and development of scientific fire categories that tend to treat human-caused fires as if they were natural, inevitable, and unavoidable, and thus outside the influence of urban planning and development decisions. The following sections outline these concerns in greater detail.

Naturalizing Wildfire Hazards: 'Firestorm' as a Scientific Category

A 'firestorm' is one of the many frequently used fire classifications. The largest urban wildfire in modern history, for example—the Oakland Hills Firestorm—was labeled in this manner because of its immense size, heat intensity, and high winds (FEMA 1992). But this label raises an interesting question about the meaning and legitimacy of environmental categories such as 'firestorms'. What exactly are they? And what differentiates a devastating firestorm from a seasonal wildfire or a run-of-the-mill fire? Upon investigating the term's origins, one important issue becomes immediately clear: there is no such thing as a *natural* firestorm. Quite the contrary, 'fires' are only 'firestorms' when society says they are. Firestorms are social constructs that we have, for many decades now, defined, classified, suppressed, created, feared, and managed.

A firestorm is defined by the American National Fire Protection Association as "a fire which creates its own weather" (Ewell 1995). This occurs "when the heat, gases, and motion of a fire build up", pulling "air into the base of the fire", leading to towering convection columns that "result in long-distance spotting and tornado-like vortices" (NFPA 1992). For a firestorm to be generated, sufficient fuel load is required that will ignite several adjacent fires in a large area (Fig. 8.4). When these multiple sites of ignition coalesce, they become a single firestorm, generating sufficient updraft to create swirling winds and large pyrocumulus cloud formations overhead.

This firestorm definition and its widespread use as a conceptual construct, scientific category, and distinct and observable 'thing' have occurred because

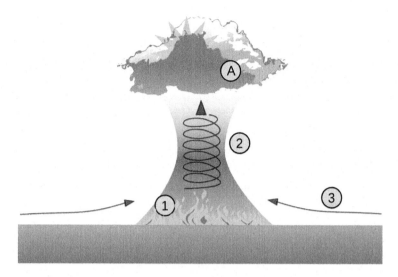

Fig. 8.4 The scientific classification of a firestorm. Decidedly unnatural firestorms appear to be part of the scientifically legitimized and inexorable natural order of things. (1) Large fire area. (2) Updraft and thermal column. (3) Strong winds generated by updraft. (A) Pyrocumulus cloud

social risk thresholds are constantly exceeded in various socio-natural landscapes. The rhetorical fine-tuning from fire to firestorm thus emerged within a particular social context where fires were (and continue to be) deemed 'out of control' and a threat to nearby social assets. And although an 'out-of-control' fire could be viewed as perfectly normal in other historical contexts, fire scientists and management officials continue to elevate the significance of the condition formerly known as fire in response to society's growing anxiety with it. Firestorms threaten our viewsheds and the aesthetic appeal of our natural surroundings. They get too close to us. They burn our property. And they threaten our lives and livelihoods. Thus they are not simply fires. They are menacing firestorms. We define them. We fear them. We often create them. And we certainly make them more costly. They are human-made disasters.

For a fire to be a firestorm it must be sufficiently large and intense. We map onto our firestorm designations' particular measurable attributes such as fire size, wind speed and direction, pace of spread, and vertical development. However, this classification process does something else, something rather more powerful than produce a neat delineation and classification of fire. Presenting the term 'firestorm' as a scientifically objective category has the surreptitious effect of cloaking it with a sense of authenticity, as if it were something natural and inexorable. In truth, such efforts to classify 'fires' simply

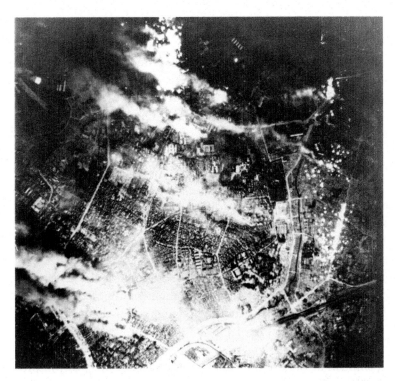

Fig. 8.5 By their very etymological origins, firestorms are social constructs. With their early usage describing the conflagrant outcomes of WWII air raids, the catalyzing source of firestorms has always been exogenous, intractable, and 'out of local control'—whether from bombing campaigns or the threat of global climate change

reflect society's increased proximity to them, a sense of threat from them, and need to order and retain control over them.

The Tyranny of Climate Change When Explaining Wildfire Hazards

The vernacular shift toward 'firestorm' has emerged over time. Although the precise origins of the term remain difficult to pin down, many historical records show that the term was used frequently during WWII to describe the conflagrant outcomes of massive firebombing campaigns across Europe and Japan (Fig. 8.5). This military origin has consequences, which can be traced through to contemporary fire terminology that connotes the catalyzing source of fire as menacing and exogenous. Today, the threat of falling bombs onto target landscapes and the resulting 'firestorms' they create is replaced, discursively, by the imposed, out-of-our-control threat of climate change, increasing aridity, and lack of falling rain throughout large portions of the US West. The *causes* of WUI fires that are internal to impacted environments, such as the presence of extensive home developments, are instead rendered as *victims* of these external threats (Davis 1998). This framing results in the *de-politicization* of pernicious urban sprawl and the profitable industry standing behind it, and the *natural-ization* of wildfires (and firestorms) as simply an unfortunate by-product of global climate change.

But I cannot emphasize this point enough: *there is nothing disastrous about fire in and of itself.* For areas of the US West, firestorms and wildfires are disasters because of human actions. We insert private properties and construct flammable assets. We impose market values. We increase exposure. We up the cost of fire. We create fire victims. We cultivate loss. In short, we produce the disaster.

Worse yet, most residential structures placed in areas already susceptible to fire could hardly be more inappropriate for their environment. Like Duraflame logs, they are composed primarily of wood and petroleum products (although in the case of homes the petroleum is in furniture, carpets, paints, staining materials, water sealants, etc. rather than in paraffin wax). They are both highly combustible once ignited, and they both assist the growth, spread, and duration of a fire (Fig. 8.6).

Fires are only disasters when human populations and all our trappings are placed within the eventual (and oftentimes historical) spatial extent of fires. We exacerbate fires and oftentimes increase their geographic extent and intensity by introducing more combustible material on the landscape. We then naturalize these events, obscuring our role in causing them, by developing labels and empirically supported (i.e., scientifically credible) categories such as 'firestorm'. This scientific and mainstream labeling diminishes the very political role humans play in creating these events and crises. The decidedly *unnatural* condition of damaging and costly fire events appears to simply be a part of the *natural* order of things when, in fact, there exist many financial incentives and social demands (see Table 8.1 and accompanying text) that facilitate their formation. The systematic production of economic benefits from attempts to mitigate these risky landscapes—through, for example, the recent proliferation of private firefighting agencies and do-it-yourself fire safety kits—is thus able to proceed as simply a logical response to these 'flammable' landscapes and seemingly inevitable disasters.

The tyranny of climate change as a dominant explanatory variable in media and policy-making circles suppresses public awareness of the ever-changing profit-seeking practices and diverse forms of economic opportunism that help

Fig. 8.6 In many ways, homes are a lot like compressed/extruded fire logs. They are both heavily composed of petroleum and wood products, highly combustible once ignited, and assist fire growth and spread. Wildfire disasters are manufactured through the construction and placement of these flammable, Duraflame-like objects on the landscape

produce increased social vulnerability and higher fire risk mitigation costs. Consider the 2016 Fort McMurray fire, which burned hundreds of thousands of acres in Alberta, Canada. While the massive fire still burned, a chorus of articles covered the fire using titles such as *We Need to Talk About Climate Change: Tragedies Like the Fort McMurray Fires Make it More Important, Not Less.* This article, like many others, ties the massive blaze to the impacts of climate change and points to the clear and present dangers of our now drier, longer, and more disastrous fire seasons. The author notes that the cause of the fire is indeed a "messy mix of factors" including forest management practices, urban encroachment, and the effects of El Niño. But the article also singles out climate change as the topic (and causal factor) that we have failed to adequately grapple with at the policy level (Holthaus 2016). That we need to address the elephant in the room—climate change—is true, to an extent. Climate change is extremely important and not adequately accounted for in many policy circles. But a quick read of fire reporting, including another article titled, *Fort McMurray and The Fires of Climate Change*, leads one to wonder just how marginalized the issue of climate change really is within the media (Kolbet 2016).

Much more importantly, the leap to illuminate (and implicate) climate change has the simultaneous effect of concealing the important role urban expansion and lucrative developments have in creating this tragedy. The McMurray fire would surely have received much less coverage if it seared through only the surrounding boreal forest. What gets overlooked in this climate-frenzied coverage is Fort McMurray's development history: rapid growth in population and size over the past several decades supporting large-scale oil extraction from an enormous subterranean tar sands deposit. When only focusing on the fire's impacts or the influential role of climate change, the actions of corporations and governments seeking to exploit this lucrative landscape fade into the explanatory background. City inhabitants are rightfully portrayed as the victims; but quite erroneously, so too are the city officials and oil industry players that continue to fuel this regional growth. Moreover, the fact that the Fort McMurray area was developed *in pursuit of fossil fuels that in turn drive anthropogenic climate change* is also rendered marginal to the story. Not only are patterns of regional oil development crucial to explaining this wildfire, they are also central to explaining the additional burden of climate change. If we were to drill down in search of the structural root causes of fire disasters like Fort McMurray, what we would find would be patterns of rapacious urban and regional development. When the American West is spun as a 'flammable' landscape it tells a very different and far less controversial story.

Debates of Distraction: Our Inability to See the Incendiary for the Spark

Our difficulty addressing the underlying social causes of increased wildfire risk and costs can be explained in part by a myriad of distracting alternative and proxy debates. Despite their diversity these corollary disputes hold a similar quality: each functions as a spark that ignites disputes at neighborhood, city, and regional levels. Once communities, managers, scientists, and politicians become mired in these debates, the Incendiary becomes less visible, less acknowledged, and seemingly less important. As we labor to put out small fires, we fail to see the whole wildfire complex. We may understand this as a process of *re-politicization*, where public conversations on the social causes and implications of fire risk (as well as strategies to destabilize such trends) are replaced by other, seemingly more contentious debates of distraction.

In this process of re-politicization, arguments over landscapes and land features oftentimes serve as convenient and tractable sites for engaging with, and ostensibly "settling", broader disagreements and social tensions such as the proper role of the government or the importance of private property rights in land management (Alagona 2013). This chapter contributes to this discussion by suggesting that not only are broad debates fought in small arenas, but, in fact, the acrimony found in these small arenas can distract us from addressing larger disagreements, tensions, and contradictions. These alter-debates may actually *prevent* us, for example, from directly confronting the social drivers of fire risk. We are left tinkering around the edge of the problem, constantly putting out little fires, instead of grappling with the root cause of the major blaze itself.

The Confounding Debate over How to Measure Vegetation Flammability

One such example concerns eucalyptus management around the West, particularly in coastal areas that support large stands of eucalypt species. Eucalypts, according to a University of California professor of forestry and conservation, have been described as "the worst tree anywhere as far as fire hazard is concerned". The Oakland/Berkeley Hills area provides a microcosm of the debate over the flammability and relative danger of these prevalent yet contested trees. Here, two factions have fought for many decades over the suitability of eucalypts in this densely populated, hilly area containing a historically active fire regime. For one side of the debate, eucalyptus trees represent a highly flammable and thus dangerous tree cover. For others, eucalypts represent a highly aesthetic and ecologically valuable species that is conveniently and unfairly blamed for the spread of recent wildfires. Over time, the debate over eucalyptus (and landscape flammability) has, in large measure, been contested around how best to enumerate and thus 'prove' its contribution to the overall landscape fuel load. This is a scientific process that is fraught with inaccuracies and subjectivities. How much should the leaf litter, branches, and trunk, respectively, contribute to overall combustibility calculations? Should replacement vegetation cover be subtracted from the total? If measuring correlations between eucalyptus groves and historical burn area, how many trees constitute a grove? How contiguous must trees and groves be to assist fire spread? As the community continues to chase scientific clarity on these and other fundamental yet elusive questions of flammability, all parties involved have descended into a 'debate of distraction' vortex.

The tone and content of this particular debate obfuscates the fact that eucalyptus trees actually accompanied home construction in these residential neighborhoods. Thus if one is to talk honestly about eucalyptus, one must speak directly to its residential landscape counterpart, the home. Rather than contemplate eucalyptus and fire in relation to homes, it would seem more productive to consider eucalypts and homes in relation to fire. Instead, within these controversies over how to protect residents, the homes themselves are rarely controversial. Disagreements over the flammability of eucalyptus divert attention away from broader mechanisms of real estate development that have produced increased fire risk in the first place. Flare-ups such as those associated with the relative flammability of local vegetation surreptitiously naturalize residential fire (and our concerns over fire *risk*) as inexorable and simply 'the way things are'.

Wood Shingles as Distracting Political Objects

Another debate of distraction concerns the deeply political and protracted process of challenging the powerful wood shingle and cedar shake industry. As early as 1959 a report by the National Fire Protection Association encouraged officials in California and Texas to limit the use of wood products on home exteriors (FEMA 1992). Not only are wood shingles and shake roofs prone to easy ignition (compared to fire-resistant alternatives), they also have a tendency to produce flaming brands that start new spot fires well ahead of the main fire front. For many decades beginning in the 1960s, politicians around the West lobbied to enact strict state and city legislation mandating the use of fire-resistant roofing materials. Despite isolated pockets of success, this form of fire-safe home construction remained an elusive goal, in large part because of a powerful triumvirate comprised by the home construction building industry, the Cedar Shake and Shingle Bureau and the Forest Products Association.

By the early 1990s, many cities including Los Angeles finally passed ordinances preventing the use of wood shingle materials on new building construction. The Cedar Shake and Shingle Bureau quickly labeled such legislation as "unwarranted and discriminatory" and "unconstitutional". Amidst these still ongoing flare-ups in many parts of the West, wood shingles have become political objects that seemingly come to represent a choice between the destruction of cedar shingle homes or the destruction of the cedar shingle industry. This important yet distracting public dispute has led discussions over residential fire risk to begin not by asking *whether* to build more homes but rather by debating *how* to build them. By placing the focus of the debate on home materials and not the homes themselves, the inevitability of home construction—and the seemingly unfettered path to increased fire risks and costs—goes largely unquestioned.

Conclusion

Suburban landscapes of the American West are lucrative landscapes. For well over a century, diverse stakeholders have extracted profits and surplus value from already fire-prone areas at the urban periphery while simultaneously inserting considerable social risks and mitigation costs back on the landscape. This persistent process of wealth accumulation drives environmental transformations and rampant suburbanization around the West. It occurs both in the production of residential developments and again in their protection. I have argued that we should understand this process as 'the Incendiary' because much like an arsonist, these economic incentives and patterns of development do not just reflect the region as a flammable landscape but rather reveal the oftentimes reckless forces producing that very flammability. I have suggested that critical physical geographers are well suited to excavate and address these powerful drivers of social-ecological change and lead a shift from the study of wildland-urban interface (WUI) *areas* to the study of affluence-vulnerability interface (AVI) *processes*.

This shift is needed because hazardous resource management and planning histories are concealed behind a series of scientific framings, policy debates, and community disagreements that de-politicize the Incendiary and divert our attention away from the affluence-vulnerability interface. Suburban and exurban areas of the American West—and their injurious and costly wildfires—are 'spun' as strangely natural and inviolable. They are nearly always portrayed as the inevitable by-product of climatic changes and are rarely characterized as the catastrophic outcomes of profit-seeking urban and regional developments. Moreover, when decidedly unnatural urban firestorms are classified neatly as a scientific category, they are also legitimized as simply part of the 'natural order of things'. This process of de-politicization is supported by a secondary process of re-politicization, which occurs as the arena for debate is filled with other ideological disputes (such as the appropriate role of government in regulating land use) or micro debates (such as what is the most appropriate roofing material or residential landscaping vegetation). The debate over wildfires in the West thus all too frequently ignores the structural root causes of fire disasters.

References

Alagona, P. 2013. *After the grizzly: Endangered species and the politics of place in California*. Berkeley, CA: University of California Press.

Associated Press. 2016. Possible human remains found in deadly California wildfire. June 25.

Botts, H., T. Jeffrey, S. McCabe, B. Stueck, and L. Suhr. 2015. Wildfire hazard risk report: Residential wildfire exposure estimates for the Western United States. Report by CoreLogic.

CalFire. 2013. State of California Wildfire Incident Information. http://cdfdata.fire.ca.gov/incidents/incidents_statsevents. Accessed 17 July 2013.

Cohen, J. 2008. The wildland-urban interface fire problem: A consequence of the fire exclusion paradigm. *Forest History Today*, Fall. pp. 20–26.

Davis, M. 1998. *Ecology of fear: Los Angeles and the imagination of disaster*. New York: Metropolitan Books.

Ewell, P.L. 1995. The Oakland-Berkeley Hills Fire of 1991. USDA Forest Service Gen. Tech. Rep. PSW-GTR-158.

FEMA. 1992. The East Bay Hills Fire Oakland-Berkeley, California. U.S. Fire Administration/Technical Report Series USFA-TR-060, FEMA, Washington, DC.

———. 2002. Fires in the Wildland/Urban Interface. U.S. Fire Administration, Topical Fire Research Series 2(16).

Fire Adapted Communities. 2014. Guide to Fire Adapted Communities. Fire Adapted Communities Organization

Gorte, R. 2013. The rising cost of Wildfire Protection. *Headwaters Economics*. http://headwaterseconomics.org/wildfire/fire-costs-background/

Green. M. 2013. The flammable west: Mega-Fires in the age of climate change (with real-time fire map). *KQED News*. 9, Aug

Hammer, R.B., V.C. Radeloff, J.S. Freid, and S.I. Stewart. 2007. Wildland-urban interface housing growth during the 1990s in California, Oregon, and Washington. *International Journal of Wildland Fire*. 16: 255–265.

Hessl, A. 2011. Pathways for climate change effects on fire: Models, data, and uncertainties. *Progress in Physical Geography* 35 (3): 393–407.

Holthaus, E. 2016. We need to talk about climate change: Tragedies like the Fort McMurray fire make it more important, not less. *Slate*. 6 May.

IAWF. 2013. International Association of Wildland Fire. http://www.iawfonline.org/pdf/WUI_Fact_Sheet_08012013.pdf. Accessed 25 May 2015.

Ingalsbee, T. 2010. *Getting burned: A taxpayer's guide to wildfire suppression costs.* Eugene, OR: Firefighters United for Safety, Ethics and Ecology (FUSEE).

Kolbert, E. 2016. Fort McMurray and the fires of climate change. *The New Yorker.* 5 May.

Mai-Duc, C. 2016. Brown's budget earmarks big money for natural disasters. *Los Angeles Times*. 8, January.

Moritz, M.A., E. Batllori, R.A. Bradstock, A.M. Gill, J. Handmer, P.F. Hessburg, J. Leonard, S. McCaffrey, D.C. Odion, T. Schoennagel, and A.D. Syphard. 2014. Learning to coexist with wildfire. *Nature* 515: 58–66.

National Wildfire Coordinating Group. 2014. Wildland urban interface wildfire mitigation desk reference guide. PMS 051. August.

NFPA. 1992. *The Oakland/Berkeley hill fire*. National Wildland/Urban Interface Fire Protection Initiative.

Peterson, D.L. 2010. Managing fire and fuels in a warmer climate. *Northwest Woodlands* 26 (16–17): 28–29.

Simon, G.L. 2014. Vulnerability-in-production: A spatial history of nature, affluence, and fire in Oakland, California. *Annals of the Association of American Geographers* 104 (6): 1199–1221.

———. 2016. *Flame and fortune in the American West: Urban development, environmental change, and the great Oakland Hills Fire*. Berkeley, CA: University of California Press.

Simon, G.L., and S. Dooling. 2013. Flame and fortune in California: The material and political dimensions of vulnerability. *Global Environmental Change* 23 (6): 1410–1423.

Smithwick, E.A.H., M.G. Ryan, D.M. Kashian, W.H. Romme, D.B. Tinker, and M.G. Turner. 2009. Modeling the effects of fire and climate change on carbon and nitrogen storage in lodgepole pine (Pinus contorta) stands. *Global Change Biology* 15 (3): 535–548.

Theobald, D.M., and W.H. Romme. 2007. Expansion of the US wildland-urban interface. *Landscape and Urban Planning* 83: 340–354.

United States Forest Service. 2015. National forests on the edge: Development pressures on America's National Forests and Grasslands http://www.fs.fed.us/openspace/fote/national_forests_on_the_edge.html. Accessed 25 May 2015

Upton, J 2016. How the U.S. West Can Live With Fire. Scientific American. Found online at: http://www.scientificamerican.com/article/how-the-u-s-west-can-live-with-fire/. (First appearing on Climate Central on May 28, 2015).

Westerling, A.L., and B.P. Bryant. 2008. Climate change and wildfire in California. *Climatic Change* 87 (1): 231–249.

Westerling, A., T. Brown, T. Schoennagel, T. Swetnam, M. Turner, and T. Veblen. 2014. Briefing: Climate and wildfire in western US forests. In *Forest conservation and management in the Anthropocene: Conference proceedings. Proceedings. RMRS-P-71*, ed. V. Alaric Sample and R. Patrick Bixler, 81–102. Fort Collins, CO: US Department of Agriculture, Forest Service. Rocky Mountain Research Station.

WildfireX. 2015. www.wildfirex.com/private-firefighting/. Data based on a 2012 study conducted by market research firm IBISWorld. Accessed 30 Mar 2015.

9

Critical Physical Geography in Practice: Landscape Archaeology

Daniel Knitter, Wiebke Bebermeier, Jan Krause, and Brigitta Schütt

Introduction

"Critical Physical Geography allows us to investigate material landscapes, social dynamics, and knowledge politics together, as they co-constitute each other" (Lave et al. this volume). In order to answer critical physical geographic research questions, physical as well as social analyses are required. The results of these analyses produce new insights regarding physical and social aspects. They help to state new questions, rethink the use of available data and develop new methodological approaches for both the physical- and social-oriented, analyses. This practice of conducting critical physical geographic research is focused on the present, but we could ask similar questions about people, societies and their relation to the environment for eco-social systems that flourished thousands of years ago. Landscape archaeology is the interdisciplinary field that tries to do this, linking numerous disciplines from the humanities and natural sciences, such as archaeology, geography, philosophy, biology, (geo)physics and linguistics.

> Landscape archaeology is an archaeology of how people visualized the world and how they engaged with one another across space, how they chose to manipulate

D. Knitter (✉)
Department of Geography, Christian-Albrechts-Universität zu Kiel, Kiel, Germany

W. Bebermeier • J. Krause • B. Schütt
Institute of Geographical Sciences, Freie Universität Berlin, Berlin, Germany

© The Author(s) 2018
R. Lave et al. (eds.), *The Palgrave Handbook of Critical Physical Geography*,
https://doi.org/10.1007/978-3-319-71461-5_9

their surroundings or how they were subliminally affected to do things by way of their located circumstances. It concerns the intentional and the unintentional, the physical and the spiritual, human agency and the subliminal (David and Thomas 2008, p. 38).

For Kluiving and Guttmann-Bond (2012, p. 15): "Landscape archaeology is the science of material traces of past peoples within the context of their interactions with the wider natural and social environment they inhabited."

Both definitions have in common, that landscape archaeologists are specifically interested in the different ways by which (a) humans and societies have created their landscapes, and (b) landscapes influence settlement behaviour and adaptation strategies. Human perceptions, conditions and norms of culture and society as well as the natural environmental setting are the building blocks of what can be referred to as landscape. The integrative research questions that follow include (1) how humans utilized the different environmental prerequisites, for example, how did they cope with strong rainfall variabilities in semiarid regions causing irregular crop failures?; and (2) what kind of adaptation strategies they developed, in which geographic areas and under which cultural and societal conditions?

In order to investigate the sustainability of these eco-social systems, we follow a diachronic and comparative approach. By describing these systems as *landscapes* we make sure that analyses and conclusions always consider the etic (i.e. the perspective of an outside observer; what do *we* think they as a group considered important?) as well as the emic (i.e. the perspective of the subject; the perspective from within the social group; how did *they* think, perceive and categorize their world?) viewpoint of the research object (Pike 2015, Ingold 1993, Nakoinz and Knitter 2016, pp. 10–13). Landscapes are a product of natural features in an area, for example, specific bedrocks, climate characteristics, flora and fauna. But at the same time landscapes are also the result of human actions. Due to actively changing their environment, humans *created* their landscapes following certain reasons, aims or expectations within a specific natural setting.

Hence, landscapes are archives of eco-social systems. In order to understand them, we need social and physical perspectives that iteratively advance their research questions and methodologies based on the results and findings of the social and physical investigations. In this regard landscape archaeology is similar to Critical Physical Geography. However, there is one big difference: in landscape archaeology, the research object is solely based in the material world (representing a hybrid in the terminology of Latour 1993), and insights about past societal conditions are often based on this material evidence, referred to as "material culture" (Renfrew and Bahn 2012, p. 12). A landscape archaeologist that works with the material remains of societies is forced to integrate natural dynamics, since they influence the material culture before, during and after its extraction, production, utilization and abandonment by humans. All these aspects form the observable *past human-environmental system*: the reference frame for landscape archaeological research.

Due to the particular characteristics of material culture, a landscape archaeologist has two perspectives on his/her research object: (a) a natural scientific perspective, which perceives material culture as physical objects, governed by natural laws and integrated in a (complex) system of many interrelated parts; and (b) a social scientific perspective, which considers material culture as a means to understand past societies and cultures. Based on a continuous integration of these two perspectives, the researcher creates different societal constructs that are interpreted.

While in the natural-science perspective material culture is a passive or non-autonomous object and part of the natural world, in the social science perspective, material culture is an active agent created and constructed within a certain socio-cultural context—a context that is generated by the researcher's interpretation and imagination. The landscape archaeological challenge is to simultaneously take both perspectives into account because: (1) without considering the natural influence on material culture, that is, the processes that changed the properties of material culture after it was deposited, we (re) construct incorrect ideas regarding past societies; (2) without considering the socio-cultural influence on material culture, we are not able to understand the role of culture and the specific relationships and interdependencies between human and nature; and (3) without considering the individual characteristics of natural environments, that is, the different natural limits to human actions, societal activities cannot be understood comprehensively.

Hence, the prerequisites of critical physical geographic research hold true also for landscape archaeology: in order to do it, one has to try to be holistic and to take the whole past eco-social system into consideration.

There are different strategies to understanding past conditions. The attempt of physical geographers often follows the principle of uniformitarianism: Processes that take effect today (and thus are measurable) have also taken effect in the past, that is, processes of the past acted in the same way as today (Slaymaker 2006, p. 1080). This perspective allows to draw conclusions about the formation of phenomena that were created in the past. Uniformitarianism does not require presuppositions about the rates of operation of processes (Kennedy 2000, 503). However, "uniformitarianism is [only] a guiding tenet and *not* a rule of nature. As theories about the operation of nature change, so it is possible [...] that one 'Uniformitarian' explanation will come to replace another" (Kennedy 2000, 504; emphasis in the original). This indicates that there are different possibilities for how to think of and (re)construct phenomena. Schumm (1991) illustrated this in his book *To Interpret the Earth* where he highlights different problems regarding the scientific method and epistemology of Earth Sciences. For him, the reconstruction of past phenomena is an approach rather than a (simple) method (Schumm 1991, preface). By introducing different challenges of investigation and interpretation, Schumm shows how a concept of multiple hypotheses and continuous questioning of the results might help to arrive at better conclusions and new questions. This way of doing research sounds familiar to a critical physical geographer. Therefore, in the following, we adapt Schumm's approach of reconstructing past phenomena with a focus on landscape archaeological questions in order to show the different facets and challenges when employing Critical Physical Geography in landscape archaeology.

Scale and Place

When conducting landscape archaeological research in the field, the first questions that need to be answered are: Where is the research area and how is it delimited? What is the time period of interest? These questions refer to time, space and location. They create a common framework that enables comparative analyses of what is the prerequisite for discussions about process-related phenomena (Schumm 1991, p. 36).

Time

Time is a measure of changes in a system (Schumm 1991, p. 36). Since we cannot measure past circumstances directly, time functions as a scale of reference for changes and can be used as a surrogate for variables that are no longer active (Schumm 1991, p. 36).

When thinking time in a past human-environmental system, it is important to consider (1) how time is measured, (2) how the former people perceived time and (3) how the researchers interpret time:

(1) There are at least four different approaches that are employed to measure time and to derive chronological information (see Bell 2005, pp. 52–62 for general information and further references): (a) radiometric methods that are based on the radioactive decay of isotopes, for example, radiocarbon dating; (b) incremental methods, such as tree-ring dating; (c) age

equivalence methods that use marker horizons or sedimentary sequences in order to establish an event chronology; and (d) artefact dating, where the relative position of archaeological artefacts within a sequence and knowledge about archaeological periods is used to establish a relative chronology. The potential pitfall in measuring time lies in the different basic principles and presumptions underlying these four methods which produce different interpretations of measured time (Fig. 9.1).

(2) Once time is measured and a chronology is established, that chronological signature has to be interpreted. Time is a surrogate for former processes, their frequency and magnitude. Thanks to our chronology we can derive information about the duration of processes, but were the former people able to perceive them? Did they recognize, for example, a decrease in precipitation due to climate change; degradation of their soils due to insufficient manuring; erosion along pathways across slopes? Some of these processes are not directly observable. Hence, we have to question if, how and where such processes were recognized or interpreted by the former people and whether some of their material traces can be considered as adaptation or mitigation measures.

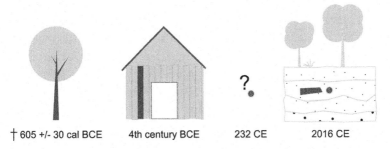

† 605 +/- 30 cal BCE   4th century BCE   232 CE   2016 CE

Fig. 9.1 We started in year 2016 CE with our excavation to clarify when and how people lived in the research area. In a certain layer we found the wooden remains of a house, next to pottery remains and a coin. The imprint on the coin stated that it was created in 232 CE under emperor X of state Y. The pottery on the other hand had the stylistic characteristics of culture Z that occurred throughout the fourth-century BCE. Lastly, we used a sample of the wood for radiocarbon dating. We know that the age will correspond to the time when the tree died. At this point it stops to integrate 14C isotopes from the atmosphere. Based on the known rate of radioactive decay of this isotope, it is possible to use the amount of the remaining 14C isotopes to assess the age of the tree (however, since the amount of 14C in the atmosphere changes, the date has to be calibrated according to a calibration curve; the calibration curve is created using an incremental method of age detection). In the end we received an age of 605 ± 30 calibrated years BCE. The results of the different dating approaches seemed to be a mess and very contradictory. This is usually the case in landscape archaeology and necessitates a very careful investigation of all the different aspects that might influence our archive

(3) The third challenge related to time is the link that researchers draw between landscape change and societal developments. A classic example, still debated today, is the relationship between the dynamics of climate and society (see Huntington 1915 as an early example; Clarke et al. 2016 as data-informed state of the art). For instance, in Mesopotamia, researchers link specific climatic characteristics, such as decreasing precipitation rates to phases of cultural instability (Clarke et al. 2016, pp. 97–98). Is this accurate, or the coincidence simply a result of the different resolutions of the chronological records, addressed above? Does this temporal coincidence enable us to infer that climate is a determining factor of societal organization? Certainly not, since (a) knowing when a process took place does not tell us what triggered it, and (b) the different lengths of these physical and social processes prohibit us from inferring causal relationships between them (see Hacking 1996, pp. 66–70 on the difficulties of correlation and causation).

Space

"Space is the three dimensional field in which natural phenomena function and occur, and in which the subject of an investigation exists" (Schumm 1991, p. 47). This definition is already controversial since the perspectives on and investigations of "space" differ strongly (see Thrift 2009, Kent 2009). It is not surprising that this heterogeneity of perspectives within geography is also common in (landscape) archaeology (see Müller-Scheeßel 2013), and also essential in Critical Physical Geography.

In Schumm's (1991, p. 47) definition, space is regarded as something real; it is absolute and measurable in terms of Euclidean geometry. Space itself influences the occurrence and functioning of processes. The processes that can be investigated and the conclusions that can be drawn depend on the scale of investigation: as spatial scale changes, the status of potential explanatory variables changes (Fig. 9.2): At the macro-scale, details are not accessible but large degrees of the research object are visible (Fig. 9.2, left). For instance, the integration of a topographic map and data about the location of archaeological sites allows to identify potential patterns of location. Nevertheless, assessments about the *specific* local characteristics are not considered. Interpretations that ignore this fact—for instance, due to the assumptions of *general* laws of settlement distribution—might reconstruct misleading formation processes and weigh factors wrongly.

At the micro-scale, details are visible and open to investigations, but the broader picture and interrelations of different parts of the system in general

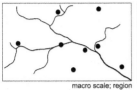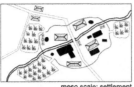

macro scale; region   meso scale; settlement   micro scale; excavation trench/ drilling

Fig. 9.2 (left) On the macro-scale, settlements are shown as small dots; their patterning and their relation to environmental features can be investigated, but the potential internal processes that lead to their patterning cannot be investigated or proved. (centre) On the meso-scale, details of the settlement plan and its utilization of specific environmental features get obvious and can be analysed; the availability of data is greatly increased, what necessitates a first selection of potentially important features. (right) On the micro-scale, the natural dynamics can be reconstructed in detail, and specific contexts of artefact distribution and usage are open for detailed investigations based on, for example, excavations. The resolution is so high, that larger objects, for example, floodplain terraces or features of settlement plans, cannot be recognized anymore and the conclusions drawn only on this scale are prone to miss important aspects from the meso- and macro-scales

are not observable. Thus, spatial interrelations cannot be evaluated (Fig. 9.2, right). A sampling profile offers detailed stratigraphic information about the sedimentation characteristics at a certain location. But without knowing where and to which degree former people influenced the natural dynamics in the catchment, the drilling cannot yield complete answers about the complex causes of the formation of the sedimentary archive. Reconstructions of large-scale environmental dynamics and their multiple triggers might be misleading. The same holds true for societal dynamics: an excavation trench yields high-resolution information about local characteristics, but the broader picture of its formation conditions remains unresolved as long as it is not integrated into the larger-scale context of societal and natural dynamics.

Location

Location refers to the places where observations are made (Schumm 1991, p. 54) and summarizes aspects related to (a) the spatial dependency of phenomena, (b) the spatial extrapolation of results and (c) the comparability of research results with spatial significance.

In landscape archaeology the "problem" of location is frequently linked to administrative problems like spatial or methodological restrictions of work permits. These restrictions may lead to differences in the data acquired at the investigated sites, impeding their comparison. For instance, in a landscape archaeological project dealing with iron smelting in lower Silesia (Poland), two sites are under investigations (see Thelemann et al. 2016; Thelemann et al. 2015). In the hinterland of settlement A, promising sediment archives are located, which are investigated by several percussion drillings. The settlement itself is located on an agricultural field, whose owner denied permission for further investigations. Further downstream at settlement B, the full methodological toolkit (drillings, excavations and geophysical prospection methods) is employed. Coarse sandy deposits of a braided river represent at this site the main sediment archive. Sediment characteristics did not support a high-resolution environmental reconstruction (see Thelemann et al. 2016). Thus the data from each site have notably different strengths and weaknesses. The difficulty of integrating these disparate results is a common locational challenge in landscape archaeological research (see Thelemann et al. 2016).

In this example, we know why different methods were applied. Nevertheless, when searching the literature for additional settlements with comparable characteristics, the absence of certain methods or data-types due to locational issues is not written or mentioned. The most appropriate methods for certain locations, and thus notable holes in the evidence go unremarked; we do not know what we are missing.

Cause and Process

The trajectories of processes and their causes are coupled in numerous, non-linear ways and simple analogies are not sufficient to gain a comprehensive understanding of the research object. When we refer to causes and processes, it is of crucial importance that we use multiple, competitive hypotheses as well as composite hypotheses, that is, hypotheses that complement each other (Schumm 1991, p. 13, pp. 33–34; Fig. 9.5).

Convergence

Convergence, or equifinality, "(…) refers to a situation when different processes and different causes produce similar effects" (Schumm 1991, p. 58). Under conditions of equifinality, the principle of analogy breaks down, making it difficult to infer processes and causes from effects. Comparisons of locations and their characteristics are complicated and may be misleading.

In a perfect world, that is, a world where we know all the process and their causes that act upon a location, equifinality would not exist. However, in general, we do not have information about *all* the processes and their causes that acted upon and created the characteristics of a location. The fragmentary nature of data in landscape archaeology produces phenomena that exhibit equifinal characteristics.

In the following brief example, we demonstrate two different causal links that could explain the creation of a specific phenomenon. To deduce testable hypotheses, we have taken into account the differing processes and causes that could trigger the development of the phenomenon in question.

In the semiarid north central part of Sri Lanka, man-made reservoirs, so-called tanks or *wewas*, have been utilized since the fourth-century BCE to store runoff for paddy irrigation (Fig. 9.3). A very high density of these reservoirs occurs in the hinterland of the city of Anuradhapura, which is regarded as the first capital of Sri Lanka (fourth-century BCE to eleventh-century CE; Schütt et al. 2013). The Malwathu Oya river passes the city in the east. An ancient stone bridge, still located at the present-day river, indicates that the river course has been surprisingly stable over time. Which processes inhibit the river from meandering through the floodplain? To answer this question, at least two hypotheses can be formulated. The first is that the reservoirs function as sediment traps. Their construction affected not only the water cycle

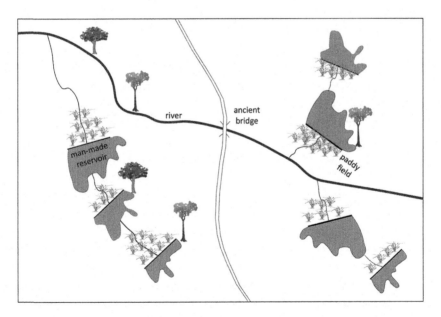

Fig. 9.3 A typical cultural landscape in north central Sri Lanka, composed of man-made ancient reservoirs to store rainfall and runoff for paddy irrigation (Tree and plant symbols used with courtesy of the Integration and Application Network, University of Maryland, Center for Environmental Sciences, http://ian.umces.edu/symbols/)

but also the sediment budget, since a reasonable share of eroded sediments is stored in the tank bodies. These sediments are absent from the bedload of the Malwathu Oya river resulting in increased erosion and vertical incision of the river. As a consequence of this down cutting, the river channel shifts less frequently, or not at all, since effective flood events with the required energy to shift the incised channel are less frequent. Second, the construction of the tanks coincided with an intensification of agricultural practices in the hinterland leading to an increase in soil erosion. The described sediment storage capacity of the tanks was exceeded, and another chain of processes occurred: the higher erosion rates resulted in higher bedload for a given Malwathu Oya river discharge. During regular flood events the increased bedload resulted in the development of levees at the river banks, and these eventually stabilized the river's course.

Both hypotheses may explain the stability of the Malwathu Oya river over millennia. The task of a landscape archaeologist is to carefully investigate the different routes of explanation via an interdisciplinary methodological approach that incorporates multiple hypotheses.

Divergence

In contrast to convergence, divergence refers to situations where similar causes and processes produce different results (Schumm 1991, p. 62). If different effects are produced by similar causes, it is difficult to assess the ultimate cause of a phenomenon (Schumm 1991, p. 63).

An illustrative example is the landscape archaeological system of north-east Jordan during the Early Bronze Age (3500–3000 BCE). The Early Bronze Age settlement of Jawa is located in this basalt desert steppe. Dry conditions can trigger different processes and lead to different effects in this region:

First, dry conditions can lead local societies to develop new adaptation techniques, such as irrigation terraces as water management systems (Fig. 9.4a). In the surroundings of the ancient settlement of Jawa, a highly sophisticated water distribution and storage system was established, which appears to be the earliest hydraulic system of its kind in the world (see Whitehead et al. 2008). This complex system combines surface and floodwater runoff with agricultural terrace systems, located on slopes, small plateaus and valleys in the direct vicinity of Jawa. These systems prove the existence and the functioning of floodwater irrigation measures in the Early Bronze Age and show that agriculture at Jawa was an important part of the settlement's economy (Meister et al. 2017, Meister et al. in press).

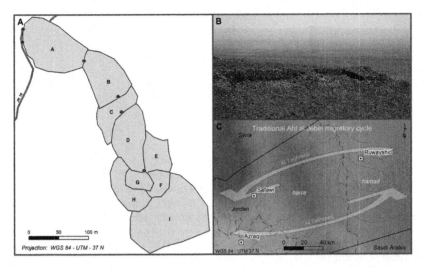

Fig. 9.4 (a) Ancient water management systems in the surroundings of Jawa, NE Jordan (after Meister et al. 2017); (b) remnants of the ancient Chalcolithic to Early Bronze Age settlement of Jawa, north-east Jordan (Helms 1981, Müller-Neuhof 2015); (c) sketch of annual pastoral migration routes after Roe (2000) in the Jordanian basalt desert steppe (Meister et al. Submitted)

Second, dry conditions can also lead to migration or abandonment (Fig. 9.4b). For the fortified ancient settlement of Jawa, by far the largest and one of the best-preserved prehistoric sites in the region (Helms 1981), the main occupation phase is documented according to the pottery remains from the Late Chalcolithic to the beginning of the Early Bronze Age (Levantine Early Bronze Age IA, c. 3500–3000 BCE; Helms 1981). After this, the flourishing settlement was abandoned. Only from the transition from the Early Bronze Age IV to the Middle Bronze Age I (around 2000 BCE), a minor reoccupation is documented.

Third, next to the implementation of local-scale techniques, dry conditions can also lead to societal innovations and the development of new forms of subsistence, such as mobile pastoralism (Fig. 9.4c). In the Badia (desert steppe in north-east Jordan, east of Jawa), herders traditionally follow a common pattern of annual migration including two constituent movement patterns which are described as *al tashreeq* and *al taghreeb*, meaning "the easting" and "the westing," respectively (Roe 2000, p. 56). New studies reveal that this circle of pastoral migration routes presumably has been consistent through prehistoric, historic and modern times (Meister et al. in press).

Three different societal strategies can be observed in this example that are initiated by one potential cause. This implies that in order to understand the past eco-social dynamics, further parameters have to be taken into account, and new hypotheses have to be developed.

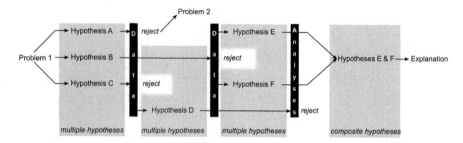

Fig. 9.5 In landscape archaeology, multiple and composite hypotheses are necessary in order to arrive at a holistic explanation. The complexity of the research problems necessitates multiple hypotheses that are competitive and that are tested against data. Some hypotheses will be rejected and lead to new problems. Based on this, new hypotheses are developed, and together with already tested ones they are evaluated against other data. In the end we hopefully arrive at an explanation based on composite hypotheses of complementing ideas that developed throughout the research process (after Schumm 1991, p. 13)

Multiplicity

Multiplicity refers to multiple causes that act simultaneously and in combination in order to produce a phenomenon (Schumm 1991, p. 70). As is obvious in the examples of convergence and divergence, mono-causal explanations of complex landscape archaeological problems are not sufficient, since the number of interrelations between the variables is numerous, not linear and complex. Analogies might be misleading, prohibiting deterministic explanations. Accordingly, a framework of multiple and composite hypotheses is required to explain phenomena based on different variables and to illustrate the converging and diverging characteristics of the acquired explanations (Schumm 1991, p. 71; Fig. 9.5).

Efficiency

Efficiency refers to the impact of an event or a series of events on a system (Schumm 1991, p. 66). The simple assumption that the stronger the impact, the larger the response only holds true for single-variable systems under controlled (preferably laboratory) conditions. This is not the case in the real world, where more than one variable affects a system and different effects might cancel or strengthen each other. This especially holds true where, for example, a society starts settling in an area and changes the environmental conditions.

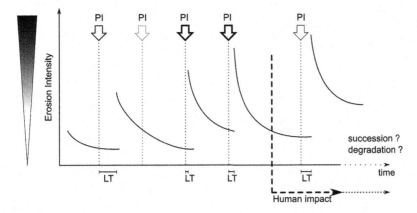

Fig. 9.6 The efficiency of processes as a function of process intensity (PI) and pre-event conditions in an area. The example is based on the assumption that in a runoff-controlled land surface, erosional processes are triggered by rainfall events (LT = lag-time)

This can be exemplified by a simplified process-response system of rainfall and erosion (Fig. 9.6). After the onset of a precipitation event, erosional processes start with a lag-time, encompassing the time needed to make the land surface ready to generate surface runoff; this includes splash effect of raindrops and infiltration processes which control the generation of Hortonian overland flow and saturation overland flow (Horton 1933, Horton 1945, Ahnert 2003). If a precipitation event is strong enough that either Hortonian overland flow or saturation overland flow is generated, erosion processes start. After the rainfall event, with decreasing surface runoff, erosion rates decrease. As processes like infiltration and splash effect are physically controlled, precipitation rates and erosion rates behave proportionally: the stronger the rainfall event, the higher the erosion rate. However, proportionality of the relation does not go along with linearity, as multiple other factors affect the described processes. This might be the case when several precipitation events of high magnitude occur repeatedly so that soil conditions have not completely settled to their pre-event stage. In consequence, a subsequent rainfall event of identical magnitude may have a shorter lag-time and a higher erosion rate. Hence, knowledge about preconditions is required to conduct a valid assessment of the relation between rainfall magnitude and erosion intensity.

This relation gets even more complex when focusing on a past eco-social system where humans actively change the environment. For instance, initial settlement activities frequently went along with clear-cutting and soil cultivation such as ploughing and harrowing (Kalis et al. 2003). These processes directly affect the exposure of a land surface to erosion processes, that is, they change the landscapes sensitivity (see section "Sensitivity"):

- Clear-cutting increases landscape sensitivity to erosion by surface runoff as the loss of vegetation cover, that is, the loss of interception, increases the share of rainfall that reaches the land surface and becomes effective as surface runoff; simultaneously the loss of roots causes reduced soil stability.
- Harrowing increases soil erodibility due to the mechanical destruction of soil aggregates.
- Ploughing disturbs soil cohesion and thus exposes soil to erosion; concurrently ploughing increases surface roughness and pore volume, both supporting infiltration processes and reducing surface runoff generation.

All these activities influence the efficiency of processes taking place. Hence, if settlement expansion dates to a period of moderate to low precipitation intensities, the onset of erosion processes will not necessarily occur simultaneously with the onset of settlement but is delayed until a rainfall event occurs that exceeds the critical threshold for surface runoff generation.

The efficiency of an event triggering a process strongly depends on the sensitivity and exposure of a system. A valid assessment, for example, of the efficiency of a rainfall event triggering erosion processes is only possible if the landscape character and the environmental preconditions are known. A lack of this knowledge might lead to a misinterpretation of contemporaneity and lag-time of different processes and, thus, to a misinterpretation of causalities.

System Responses

The spatial and temporal boundaries of the past human-environmental systems and the processes occurring within it produce complex questions whose solution necessitates the consideration of diverse standpoints and the application of numerous methods. In this section on system responses, we further complicate this picture by focusing on the non-deterministic components of the landscape, that is, singularity, sensitivity and complexity. These terms describe the autonomy and self-organizing characteristics of the eco-social system.

Singularity

Singularity or indeterminacy describes the characteristics that make one thing different from others (Schumm 1991, p. 75). Singularity can be considered as the unexplained variation in a data set. For instance, although Greek poleis were mostly built with a common layout and structure of their chora (hinterland), each polis functioned and reacted differently to internal and external dynamics that affected it (see Kirsten 1956, Bintliff 2012; Fig. 9.7).

Singularity is closely linked to the scale of investigation and the amount of available data. "(…) [W]e can predict for a population based upon a large sample, but only order of magnitude estimates can be made for an individual" (Schumm 1991, p. 76). Hence, singularity causes each landscape to respond to changes with different processes, varying in type, duration and intensity (Schumm 1991, p. 76). A striking example is the development of central places throughout history. Antique city competitors like Antiochia (modern Antakya, Turkey) and Aleppo (Syria) or Ephesos (modern Selcuk, Turkey) and Pergamon (modern Bergama, Turkey), although comparable in many aspects, followed very different paths and attracted different interactions. For instance, sediments brought by the streams silted up the harbours of Pergamon and Ephesos but only in Ephesos did the people invest the resources to clean and rebuild it—at least once (Stock et al. 2013, Knitter

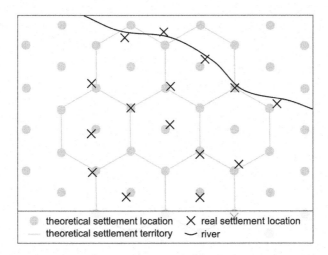

theoretical settlement location   X real settlement location
theoretical settlement territory   river

Fig. 9.7 If we try to explain the pattern of settlements, we can use theoretical models like central place theory (Christaller 1933); singularity in this case describes the shift of the expected settlement location from the observed one due to different reasons, for example, the accessibility to water. A shift from the expected pattern due to singularity does not reject the theoretical model

et al. 2013). Environmental disasters and earthquakes shocked the inhabitants of Antiochia and Aleppo but only the latter was able to revive (Knitter et al. 2014). All four cities had supra-regional importance that resulted from very different functions, caused by specific natural and social conditions. Their singularity is the result of an integration of these at the micro-, meso- and macro-scale. Hence, singularity forces the landscape archaeologist to follow a comparative approach in order to be able to separate singular features from general patterns.

Sensitivity

Landscape sensitivity refers to the capacity of landscapes to resist the forces of change, independent of whether these arise from the variability of climate, endogenous triggers or human impacts. A change occurs when a threshold is passed and adjustment occurs; if systems respond to a minor external change, that is the threshold is low, they are referred to as sensitive (Schumm 1991, p. 78). Sensitive systems respond with short lag-time and visible reactions to changing external influences, even when these changes are very small. In contrast, systems are frequently categorized as insensitive when they seemingly do not respond to changes of environmental factors. This lack of response might be caused by high thresholds required to trigger a process. Furthermore, a large lag-time or a low magnitude of resulting processes might not have been perceivable by the people of that time, for example, due to insufficient measuring devices. This would prohibit them from inferring a causal relation.

The most well-known landscape engineering measure to reduce landscape sensitivity is the construction of terraces on sloped terrain, a technique spread globally and in its basic structures traceable to the Bronze Age (Radkau 2000). Terraces aim to reduce runoff and average slope on a particular cultivated area upslope of a contour parallel slope-break (Fig. 9.8). The slope-break can be created, for example, by erecting a dry-stone wall. This induces a reduction in runoff rate and velocity on the slope, fostering its infiltration and the deposition of wash load. As a result, the area upslope of the contour parallel slope-break gets filled in by sediments, levelling the surface. In consequence, the sensitivity of the slopes to erosion is reduced and the exposure of the terraces' cultivable area to edaphic drought is reduced. The result is a morphodynamically very stable type of eco-social system (Beckers et al. 2013a, b). Terraced landscapes require continuous maintenance, including raising of terrace walls, draining excess runoff and repairing damage (Schütt et al. 2005). Due to the

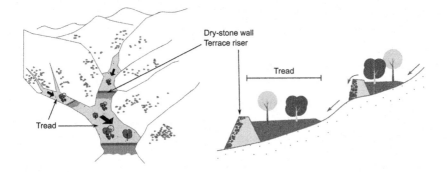

Fig. 9.8 An exemplary agricultural terrace system (after Frederick and Krahtopoulou 2000, El Amami 1983)

double function of agricultural terraces to prevent soil erosion and to foster soil water availability by water harvesting, they are particularly widespread in drylands.

A consequence of terracing slopes is thus a significant decrease of surface runoff and transported sediments into receiving streams. If the thickness of the alluvium of these receiving streams is taken as a proxy for settlement activities, chronological or process-oriented misinterpretation may occur due to the lack of expected sediments eroded from the slopes (see Chap. 3).

Complexity

"When something is complex, it is composed of numerous interconnected parts. (…) The complex system when interfered with or modified is unable to adjust in a progressive and systematic fashion, and its response can be complex" (Schumm 1991, p. 85).

Complex systems are composed of various structurally coupled parts. Due to this, the dynamics of such systems are sensitive and dependent on initial conditions. They possess patterns of emergence and self-organization and are often autopoietic, that is, the system functions as a whole to continually produce its own components (see Gershenson 2008). Accordingly, these systems always have a history, such that similar processes occurring in two systems would result in completely different patterns (e.g. Nicolis and Prigogine 1977). If one element of a landscape is changed, for example, a dam is built in order to supply drinking water for a settlement, all other aspects of the fluvial system change as well (numerous examples in Winiwarter and Bork 2014). Hence, landscape archaeological investigations have to consider all the composing parts of a system in order to reconstruct the processes that occurred at a certain time in a certain area. With this step, all the other points mentioned throughout our contribution come into play. This will inevitably further complicate the picture and change the way how the complexity of the specific past human-environmental system is faced by the investigating researcher. A comprehensive reconstruction of a past eco-social system is of course an impossible (but very exciting) endeavour that necessitates an open mind for new ideas, hypotheses, concepts and explanations.

Synthesis

The potential pitfalls in landscape archaeological research presented here can be seen from two perspectives: on the one hand they point to the necessity of avoiding simple or mono-causal explanations that exclude relevant social or natural aspects of the studied object. On the other hand they can be seen as a guideline to develop hypotheses and to conduct joined, integrative research that is aware of its different discipline-specific limitations.

Taken together three main fields of landscape archaeologically relevant questions can be answered (see Schumm (1991), pp. 98–100; Fig. 9.9), that is, (1) what was the observed past human-environmental system and how did it develop? (2) How did the system operate and what controlled the different processes that took place? (3) Can the investigated eco-social system be compared to other systems? What is specific about the system and where does it follow general patterns?

The landscape archaeologist faces a variety of challenges when trying to reconstruct the eco-social systems that operated thousands of years ago. Only with close collaboration of researchers focusing on social and physical questions including past and present states is it possible to detect the potential pitfalls arising from the nature of the research subject and its integration in natural dynamics. Such a collaboration allows us to understand the past human-environmental system and to answer questions such as: What can we learn from the patterns and structures excavated and extracted? How can the different readings of the subject, that is, one from an etic, the other from an emic viewpoint, fruitfully be integrated? What does the implementation of specific measures, like terraces, tell us about the society, and how can the influence of such features on the societal conditions be analysed? Finally, what can we learn from the strategies, techniques and behavioural patterns of former people and their landscape about current issues in our own eco-social systems?

Fig. 9.9 The different questions and their corresponding problems and pitfalls can guide an investigation of a landscape archaeological system (LAS) and help to develop research questions. The pictures as well as the system description and research questions only show the physical geographic perspective. Nevertheless, referring to what we presented throughout this chapter, this perspective cannot be thought of in isolation, and none of the stated questions can be answered by a physical perspective alone. Integrative thinking, interdisciplinary collaboration and an open mind are necessary

All these questions illustrate the importance and necessity of developing an awareness of the pitfalls in interdisciplinary landscape archaeological research, that is, in historically oriented critical physical geographic research. The systematic analysis of misleading interpretations in our studies might give some support to avoid these pitfalls. However, success of landscape archaeological research deeply depends on respect and mutual understanding between the social and natural scientific disciplines involved and the willingness of the scientists at least to learn some of the co-operating discipline's languages and concepts.

Acknowledgements The authors are grateful to the Excellence Cluster Topoi (EXC 264)—The Formation and Transformation of Space and Knowledge in Ancient Civilizations—for supporting this study. Furthermore, we want to thank landscape archaeologists Thusitha Wagalawatta and Julia Meister for drawing Figs. 9.2 and 9.4. We want to thank Rebecca Lave and Stuart Lane for their helpful and constructive comments that improved this contribution.

References

Ahnert, F. 2003. *Einführung in die Geomorphologie.* Stuttgart: Eugen Ulmer.

El Amami, S. 1983. *Les aménagements hydrauliques traditionnels de Tunisie.* Tunis: Centre de Recherche du Genie Rural.

Beckers, B., J. Berking, and B. Schütt, 2013a. Ancient water harvesting methods in the Drylands of the Mediterranean and Western Asia. *eTopoi. Journal for Ancient Studies.*

Beckers, B., B. Schütt, S. Tsukamoto, and M. Frechen. 2013b. Age determination of Petra's engineered landscape—Optically stimulated luminescence (OSL) and radiocarbon ages of runoff terrace systems in the Eastern Highlands of Jordan. *Journal of Archaeological Science* 40 (1): 333–348.

Bell, M. 2005. *Late quaternary environmental change: Physical and human perspectives.* 2nd ed. Harlow, England and New York: Pearson/Prentice Hall.

Bintliff, J.L. 2012. *The complete archaeology of Greece: From hunter-gatherers to the twentieth century AD.* Chichester, West Sussex and Malden, MA: Wiley-Blackwell.

Christaller, W. 1933. *Die zentralen Orte in Süddeutschland—Eine ökonomisch-geographische Untersuchung über die Gesetzmäßigkeiten der Verbreitung und Entwicklung der Siedlungen mit städtischer Funktion.* Jena: Gustav Fischer.

Clarke, J., et al. 2016. Climatic changes and social transformations in the Near East and North Africa during the "long" 4th millennium BC: A comparative study of environmental and archaeological evidence. *Quaternary Science Reviews* 136: 96–121.

David, B., and J. Thomas. 2008. *Handbook of landscape archaeology.* Walnut Creek, CA: Left Coast Press.

Frederick, C., and A. Krahtopoulou. 2000. Deconstructing agricultural terraces: Examining the influence of construction method on stratigraphy, dating and archaeological visibility. In *Landscape and Land Use in Postglacial Greece,* ed. P. Halstead and C. Frederick, 79–93. London: Continuum International Publishing Group.

Gershenson, C. 2008. *Complexity: 5 questions.* Automatic Press/VIP.

Hacking, I. 1996. *Einführung in die Philosophie der Naturwissenschaften.* Stuttgart: Reclam.

Helms, S.W. 1981. *Jawa: Lost city of the black desert.* London: Methuen & Co. Ltd.

Horton, R.E. 1933. The Rôle of infiltration in the hydrologic cycle. *Eos, Transactions American Geophysical Union* 14 (1): 446–460.

———. 1945. Erosional development of streams and their drainage basins; Hydrophysical approach to quantitative morphology. *Geological Society of America Bulletin* 56 (3): 275–370.

Huntington, E. 1915. *Civilization and climate.* New Haven, London and Oxford: Yale University.

Ingold, T. 1993. The Temporality of the landscape. *World Archaeology* 25 (2): 152–174.

Kalis, A.J., J. Merkt, and J. Wunderlich. 2003. Environmental changes during the Holocene climatic optimum in Central Europe—human impact and natural causes. *Quaternary Science Reviews* 22 (1): 33–79.

Kennedy, B.A. 2000. Uniformitarianism. In *The dictionary of physical geography*, ed. D.S.G. Thomas and A. Goudie, 502–504. Malden, MA: Blackwell Publishers.

Kent, M. 2009. Space: Making room for space in physical geography. In *Key concepts in geography*, ed. N.J. Clifford, S.L. Holloway, S.P. Rice, and G. Valentine, 97–118. Los Angeles: SAGE.

Kirsten, E. 1956. *Die Griechische "Polis" als historisch-geographisches Problem des Mittelmeeraumes.* Bonn: E. Dümmler.

Kluiving, S.J., and E.B. Guttmann-Bond. 2012. LAC 2010: First International Landscape Archaeology Conference. In *Landscape archaeology between art and science from a multi- to an interdisciplinary approach*, ed. S.J. Kluiving and E.B. Guttmann-Bond. Amsterdam: Amsterdam University Press.

Knitter, D., H. Blum, B. Horejs, O. Nakoinz, B. Schütt, and M. Meyer. 2013. Integrated centrality analysis: A diachronic comparison of selected Western Anatolian locations. *Quaternary International* 312: 45–56.

Knitter, D., O. Nakoinz, R. Del Fabbro, K. Kohlmeyer, M. Meyer, and B. Schütt. 2014. The Centrality of Aleppo and its environs. *eTopoi. Journal for Ancient Studies* 3: 107–127.

Latour, B. 1993. *We have Never been Modern.* Cambridge: Harvard University Press.

Meister, J., J. Krause, B. Müller-Neuhof, M. Portillo, T. Reimann, and B. Schütt. 2017. Desert agricultural systems at EBA Jawa (Jordan): Integrating archaeological and paleoenvironmental records. *Quaternary International* 434: 33–50.

Meister, J., D. Knitter, J. Krause, B. Müller-Neuhof, and B. Schütt in press. A Pastoral for Millennia: Investigating pastoral mobility in NE Jordan using quantitative spatial analyses. *Journal of Archaeological Science.*

Müller-Neuhof, B., A. Betts, and G. Willcox. 2015. Jawa, Northeastern Jordan: The First 14C Dates for the Early Occupation Phase. *Zeitschrift für Orientarchäologie* 8: 124–131.

Müller-Scheeßel, N. 2013. Mensch und Raum: Heutige Theorien und ihre Anwendung. In *Theorie in der Archäologie: Zur jüngeren Diskussion in Deutschland. Münster: Waxmann*, ed. M.K.H. Eggert and U. Veit, 101–138.

Nakoinz, O., and D. Knitter. 2016. *Modeling human behaviour in landscapes.* Switzerland: Springer International Publishing.

Nicolis, G., and I. Prigogine. 1977. *Self-organization in nonequilibrium systems: From dissipative structures to order through fluctuations*. First ed. Hoboken: John Wiley & Sons.

Pike, K.L. 2015. *Language in relation to a unified theory of the structure of human behavior*. Berlin and Boston: De Gruyter Mouton.

Radkau, J. 2000. *Natur und Macht. Weltgeschichte der Umwelt*. München: C.H.Beck.

Renfrew, C., and P. Bahn. 2012. *Archaeology: Theories, methods and practice*. London: Thames & Hudson Ltd.

Roe, A.G. 2000. *Pastoral livelihoods: Changes in the role and function of livestock in the northern Jordan Badia*. Doctoral thesis, Durham University.

Schumm, S.A. 1991. *To interpret the earth: Ten ways to be wrong*. Cambridge: Cambridge University Press.

Schütt, B., S. Thiemann, and B. Wenclawiak. 2005. Deposition of modern fluvio-lacustrine sediments in Lake Abaya, South Ethiopia—A case study from the delta areas of Bilate River and Gidabo River, northern basin. *Zeitschrift für Geomorphologie N. F., Suppl.-Bd* 138: 131–151.

Schütt, B., W. Bebermeier, J. Meister, and C.R. Withanachchi. 2013. Characterisation of the Rota Wewa tank cascade system in the vicinity of Anuradhapura, Sri Lanka. *DIE ERDE—Journal of the Geographical Society of Berlin* 144 (1): 51–68.

Slaymaker, O. 2006. Uniformitarianism. In *Encyclopedia of geomorphology*, ed. A. Goudie, 1080–1081. London: Taylor.

Stock, F., A. Pint, B. Horejs, S. Ladstätter, and H. Brückner. 2013. In search of the harbours: New evidence of late Roman and Byzantine harbours of Ephesus. *Quaternary International* 312: 57–69.

Thelemann M., E. Lehnhardt, W. Bebermeier, and M. Meyer. 2015. Iron, human and landscape—Insights from a micro-region in the Widawa catchment area, Silesia. In: D. Knitter, W. Bebermeier, and O. Nakoinz (Eds.): *Bridging the gap—Integrated Approaches in Landscape Archaeology. eTopoi Journal for Ancient Studies*. Special Volume 4. 109–138.

Thelemann M., W. Bebermeier, P. Hoelzmann, and B. Schütt. 2016. Landscape history of the Widawa catchment area—A case study of long-term landscape changes since the Saalian Drenthe stadial in Silesia, Poland, Quaternary International. https://doi.org/10.1016/j.quaint.2016.09.015.

Thrift, N. 2009. Space: The fundamental stuff of geography. In *Key concepts in geography*, ed. N.J. Clifford, S.L. Holloway, S.P. Rice, and G. Valentine, 85–96. Los Angeles: SAGE.

Whitehead, P.G., S.J. Smith, A.J. Wade, S.J. Mithen, B.L. Finlayson, B. Sellwood, and P.J. Valdes. 2008. Modelling of hydrology and potential population levels at Bronze Age Jawa, Northern Jordan: A Monte Carlo approach to cope with uncertainty. *Journal of Archaeological Science* 35 (3): 517–529.

Winiwarter, V., and H.-R. Bork. 2014. *Geschichte unserer Umwelt: Sechzig Reisen durch die Zeit*. Darmstadt: Primus.

10

Shifting Climate Sensitivities, Shifting Paradigms: Tree-Ring Science in a Dynamic World

Christine Biermann and Henri D. Grissino-Mayer

The uniformitarian principle is assumed in all dendrochronological inferences, and, as in all sciences of the past, if this principle does not hold, no conclusions regarding the past can be made.
—Harold C. Fritts 1976, *Tree Rings and Climate*

Dendrochronology, the science of tree-ring dating, has been used to study numerous types of environmental and social phenomena, from rainfall in the Amazon basin (Brienen et al. 2012) to the historical timber trade in Northern Europe (Bridge 2012). The annual growth rings of trees are of particular value as natural archives, or sources of information about past environments, because ring width, structure, and chemical composition are all influenced by the environmental conditions under which the tree ring was formed. Using large datasets that document tree growth patterns over centuries to millennia, dendroclimatologists attempt to reconstruct environmental conditions (e.g. temperature, precipitation, stream flow, snow pack, wildfires, and hurricanes) of the past. Because of this versatility, tree rings and the researchers that analyze them play crucial roles in contextualizing climate change and rendering

C. Biermann (✉)
Department of Geography, University of Washington, Seattle, WA, USA

H. D. Grissino-Mayer
Department of Geography, University of Tennessee, Knoxville, TN, USA

© The Author(s) 2018
R. Lave et al. (eds.), *The Palgrave Handbook of Critical Physical Geography*,
https://doi.org/10.1007/978-3-319-71461-5_10

it visible and comprehensible to policymakers and the public. Deep knowledge of past relationships between tree growth and environmental conditions also allows researchers to investigate questions about the future of forest ecosystems in a warming world. The scholarly contributions of tree-ring scientists, however, have been contested, and over the past two decades tree-ring based climate research has witnessed controversy (e.g. McIntyre and McKitrick 2005; Wahl and Ammann 2007; Mann et al. 2012; Anchukaitis et al. 2012).

In this chapter, we explore tree-ring science as a field of study that exemplifies both the already existing practices of Critical Physical Geography (CPG) "in the wild" and the potential for even more integrative, reflexive, and engaged scholarship. As authors, we present original research on tree growth-climate interactions but also consider the science of dendrochronology itself as our object of analysis. In other words, we are concerned both with material relationships among climate, trees, forests, and people *and* with the ways in which scientific ideas about these relationships are being tackled, challenged, and reformulated. We specifically interrogate the principle of uniformitarianism that undergirds tree-ring-based climate research and call attention to a growing body of work that finds tree growth-climate relationships that fluctuate over time.

Inspired by CPG's call to integrate diverse methodologies and ways of knowing, we bring together quantitative analyses of relationships between climate and tree growth in Great Smoky Mountains National Park (GSMNP), Tennessee, with qualitative data on the practices, opinions, and perspectives of tree-ring scientists. Unless otherwise noted, all quoted material is excerpted from responses to an anonymous survey of tree-ring scientists (n = 48) administered by the first author in 2016. We employ multiple methods here not to triangulate results or scrutinize a singular research object but to bring into conversation strands of scholarship that might otherwise be pursued separately, placed in different academic books or journals, presented in separate conference sessions, and read by audiences with little overlap. This integration allows us to contribute to scholarship on tree growth and climate while simultaneously considering how this body of knowledge shapes and is shaped by the politics and discourses of climate change and the social identities of tree-ring researchers.

Ultimately, we find that temporal instability in climate-tree growth relationships might be addressed by fostering what Jasanoff (2003, 2007) calls "technologies of humility," in which scientists are compelled to continually "reflect on the sources of ambiguity, indeterminacy, and complexity" (Jasanoff 2007, p. 33) in their research and to recognize that even the most ostensibly robust scientific work is insufficient to achieve generalizable solutions to pressing socioecological concerns such as climate change. The survey results discussed here indicate that a culture of humility is already being actively fostered by many tree-ring scientists, but that significant obstacles remain to developing and implementing new patterns of thought, methods, and technologies.

Tree Rings, Climate Science, and Controversy

Extracting climate information from tree rings seems relatively straightforward at first glance: long growing seasons and abundant energy and water tend to coincide with wide annual rings, while seasons in which resources are more limited are marked by narrow annual rings (Speer 2010). But the apparent simplicity that makes tree-ring research so compelling and accessible belies the methodological complexity of climate reconstruction and the changing assumptions that underpin it. Over the past two decades, the methods for achieving tree ring-based climate reconstructions have come under heightened scrutiny by climate change skeptics and deniers, with highly publicized controversies materializing around the "hockey stick" graph (Mann et al. 1998, 1999) published in the Intergovernmental Panel on Climate Change's (IPCC) assessment reports and the 2009 hacking of a server of the University of East Anglia's Climatic Research Unit ("Climategate") (Holliman 2011; Anderegg and Goldsmith 2014). Such controversies highlight the high political stakes of climate reconstruction and the powerful but precarious position of tree-ring scientists as scientific authorities on climate. As one researcher and survey respondent noted, tree ring-based climate reconstructions "seem to have a target on their back."

Even as dendroclimatology has weathered a storm of attacks from outside the field, it has also experienced a gradual upwelling of concern from within regarding the reliability of reconstructions and the assumption of consistent, linear relationships between tree growth and climate variables over time. Climate reconstruction is made possible through a number of assumptions and principles, two of the most basic being the principles of uniformitarianism and limiting factors. Together, these principles assert that tree growth is limited by the factor in shortest supply relative to demand (e.g. moisture, energy, sunlight, etc.), and crucially, that the factors that limited tree growth during the past century also limited tree growth in prior centuries and will likely continue to limit tree growth in the future. While it has long been known that growth is a function of many interacting factors other than just climate (e.g. genetics, age, competition, and various disturbance processes), the general consensus in tree-ring science has been that, at sites where growth is strongly affected by climate, the climate-tree growth relationship can be isolated and the influences of other factors can be minimized to enhance the climate signal (i.e. increase the signal-to-noise ratio). After identifying and isolating the relationship between climate and tree growth in the historical period for which climate data are available, scientists then use tree-ring data to reconstruct how climate has varied over a longer time scale, before the collection of instrumental data. In short, climate reconstruction from tree-ring proxy data has proceeded under the assumption that an essential climate-tree growth relationship exists and can be isolated for a given species at a given site that functions relatively consistently over time.

Recently, however, findings of time-varying responses by trees to climate have shown stable, linear relationships between tree growth and climate variables to be less common than previously believed. The identification of a "divergence problem" in dendroclimatology in the mid- to the late 1990s drew attention to the lack of temporal stability in climate-tree growth relationships. "Divergence" refers to a weakening of the relationship between temperatures and tree growth beginning in the 1960–1970s at high northern latitude and high-elevation sites, where trees are generally expected to be more responsive to changes in temperature than moisture (D'Arrigo et al. 2008). At many sites, trees that previously responded positively to temperature appear no longer to be temperature-sensitive (Jacoby and D'Arrigo 1995; Briffa et al. 1998; Barber et al. 2000; Driscoll et al. 2005). Instrumental temperature records and the predicted temperature values derived from tree-ring chronologies begin to diverge around the mid-twentieth century, with actual temperatures increasing while tree growth has remained stable or declined.

The past decade has seen a surge of published research on the temporal stability of climate-tree growth relationships, with a large number of studies finding unstable relationships for numerous tree species worldwide. These species include white spruce (*Picea glauca*) in the Yukon Territory (Porter and Pisaric 2011), Alaska yellow cedar (*Xanthocyparis nootkatensis*) in southern Alaska (Wiles et al. 2012), pines and European larch (*Larix decidua*) in the European Alps (Carrer and Urbinati 2006; Oberhuber et al. 2008), mountain hemlock (*Tsuga mertensiana*) in the North Cascades (Marcinkowski et al. 2015), eastern hemlock (*Tsuga canadensis*) in the central Appalachians (Saladyga and Maxwell 2015), and hardwoods in the central USA (Maxwell et al. 2016). No consensus, however, has been found regarding the cause(s) of shifting climate-tree growth relationships, the degree to which such instability is the norm, and the implications that such findings have for climate reconstruction, tree-ring science, and models of ecosystem dynamics in a warming world. Some cases of instability are thought to be directly or indirectly related to anthropogenic warming over the past half-century, while other studies have found multi-century patterns of instability that appear unrelated to recent climatic changes (Frank et al. 2007). Research has also linked the changing sensitivities of trees to localized anthropogenic forcings. In Bavaria, for example, changes in the ways silver firs respond to climate appear to be related to local SO_2 emissions (Wilson and Elling 2004). Still others suggest that instability may be a product of the methodological choices (e.g. artifacts from imprecise detrending) made by researchers rather than a material phenomenon inherent within the affected trees (D'Arrigo et al. 2008; Esper and Frank 2009). More research is needed, however, to determine the frequency, extent, and causes of unstable climate-tree growth relationships, and many tree-ring researchers are pursuing this line of inquiry.

Findings of time-varying responses by trees to climate are not new (Cook and Johnson 1989; Van Deusen 1990), but this recent emphasis calls into question the basic assumption of uniformitarianism in all tree ring-based climate reconstructions. The accuracy and reliability of reconstructions are dependent on relatively consistent climate-growth relationships over time, especially during the twentieth century when tree-ring data are calibrated with instrumental climate data. Hence, the implications of temporal instability in climate-growth relationships are potentially vast. If trees do not consistently respond through time to one or more climate variable, knowledge of past climates might be compromised. Tree-ring science has thus come to a crossroads. The assumptions that underpin dendroclimatology are increasingly being interrogated, and the precepts upon which the field was established are yielding to a new formulation of climate-tree growth relationships as fluid and non-linear. It is at this crossroads that our chapter is situated.

Climate Sensitivity of Pines in Great Smoky Mountains National Park, Tennessee

We now turn to our research on climate-growth relationships in mid-elevation pine-oak woodlands in GSMNP. A conservation priority in the southern Appalachians, pine-oak forests have declined over the past century due to a combination of factors that include fire exclusion, southern pine beetle (*Dendroctonus frontalis*) outbreaks, climate change, and timber harvesting (Harrod et al. 2000; South and Buckner 2003; Dale et al. 2010; Coyle et al. 2015). Using tree-ring data, we analyze the responses of pines to climate over the twentieth century, with the goal of understanding which specific climatic factors drive pine growth and if these factors have remained consistently significant over the twentieth century. We have two broad aims. First, we demonstrate how changes in climate-growth relationships are analyzed and how "instability" comes to be defined and recognized. Second, we use this case to reflect on some of the possible mechanisms for temporal instability in climate-tree growth relationships. A definitive explanation for the shifts we identify is certainly beyond the scope of the chapter. We suggest, however, that the responses of pines growing in GSMNP to climate are unstable and were derived through interactions via both positive and negative biotic and abiotic feedbacks, including physiological mechanisms, effects of micro- and macro-topography, possible factors related to anthropogenic activities, and climate state. This has significant normative implications, which we explore here as well.

Data and Methods

We assess the climate sensitivity of pines using a composite tree-ring chronology of pines growing at mid-elevation sites (400–750 m) in GSMNP (Fig. 10.1). To build this chronology we targeted mature canopy-dominant shortleaf pines (*Pinus echinata*) growing in pine-oak woodlands on xeric

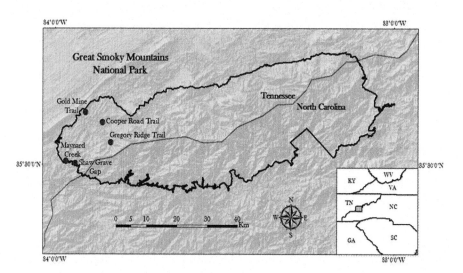

Fig. 10.1 Map of study sites in Great Smoky Mountains National Park, Tennessee. The tree-ring chronology we analyze is a composite of five individual chronologies developed in the westernmost portion of the park

south- or southwest-facing slopes in the westernmost region of GSMNP. Because shortleaf pine is known to hybridize in the wild with pitch pine (*Pinus rigida*) (Smouse and Saylor 1973), we also sampled hybrids and pitch pine individuals at one of the five sites.

The composite tree-ring chronology was built from five individual site chronologies, encompassing a total of 390 core samples from 245 trees. Two cores were extracted from each tree at breast height (1.3 m), air-dried, and sanded using standard dendrochronological methods (Orvis and Grissino-Mayer 2002). Ring widths were measured to 0.001 mm on a Velmex measuring system interfaced with Measure J2X software. We cross-dated all tree rings to ensure the correct calendar year was assigned to every tree ring by identifying common patterns of wide and narrow rings among the tree-ring series using the list method (Yamaguchi 1991), and then statistically verified our cross dating using the program COFECHA (Holmes 1983; Grissino-Mayer 2001). All ring-width series were then detrended in the program ARSTAN (Cook 1985) to allow the tree-ring series to be averaged into a single index chronology. We chose relatively conservative forms of standardization, using negative exponential curves or linear regression lines with the goal of removing age-related growth trends but retaining possible decadal-scale climate signals in the data.

We analyzed the relationship between tree growth and climate over the twentieth century using historical precipitation, temperature, and Palmer Drought Severity Index (PDSI) data from the National Centers for Environmental Information (NCEI). For precipitation and PDSI, we used the NCEI Eastern Tennessee divisional data over the period 1895 to 2007, the last year corresponding to the last year of the tree-ring data. For temperature, we analyzed pine sensitivity to monthly average minimum temperature rather than mean temperature because preliminary analyses found that minimum temperatures were strongly correlated with shortleaf pine growth in the southern Appalachians. We obtained instrumental minimum temperature records (1910–2007) from a single weather station (McGhee Tyson Airport, Alcoa, Tennessee), as this station began recording daily minimum temperatures earlier than other stations in the region and is the nearest weather station offering 100% data coverage.

We calculated correlations between annual tree growth and monthly climate parameters, spanning from June of the previous growing season to October of the current growing season, recognizing that growth may be influenced by previous year conditions as well as current season conditions (Fritts 1976; Speer 2010). To test the possible changing relationship between pine growth and climate over time, we performed moving correlation analysis at

45-year intervals (e.g. 1910–1954, 1911–1955, etc.) using the program DendroClim2002 (Biondi and Waikul 2004). This technique indicates the periods during which climate variables were significantly correlated with tree growth. Using the program treeclim (Zang and Biondi 2015; version 2.0.0 released 5 September 2016), we then plotted the results of moving correlation analysis on correlation evolution graphs to visually highlight periods in which significant shifts occurred in the climate-tree growth relationship. In all analyses, statistical significance is reported at $p < 0.05$.

We emphasize that the methods we employed have become standard procedures in tree-ring science over the past decade because these new technologies can account for the dynamic nature in tree growth responses to climate over time. DendroClim2002 and its companion program treeclim were created by and for tree-ring scientists in response to concerns about temporal (in)stability and represent significant additions to dendroclimatological research practices since the mid- to the late 1990s. The software offers a user-friendly way to test and visualize climate responses over time and has helped researchers to recognize and contend with complexity and instability in climate-growth relationships. Both programs have added a higher level of accuracy in the selection of climate variables to be reconstructed.

Our survey results reflect these perceptions well. Of our 48 respondents, 35 (72.9%) believe that temporal instability affects the reliability of climate reconstructions, and many report that they have altered their own research practices in light of this issue. For example, one researcher noted that (s)he abandoned conducting a climate reconstruction from a high-elevation chronology in the western USA after performing moving correlation analysis and finding that relationships among climate and growth appeared highly unstable throughout the short period of instrumental data. Another response echoed this, stating that "if the DendroClim correlation graphs aren't a big straight band of color across the twentieth century, the chronology should absolutely not be used for climate reconstruction." For some respondents, testing temporal stability also served to demonstrate the credibility of tree-ring science in the broader climate change research and policy community, thus preventing future attacks by climate change deniers: "if analyses are not done properly, things like 'hide the decline' (Climategate) could happen again." In our analyses, we perform moving correlation analysis in DendroClim2002 not to reconstruct climate or demonstrate credibility but to illuminate the "ragged fringes" of scientific understandings of climate-growth relationships (Jasanoff 2003, p. 227).

Major Findings

Our analyses indicate that rather than being limited by a single factor, annual growth of pines in GSMNP is limited by multiple, interacting climatic variables. We identified significant positive correlations between growth and three groupings of climatic variables: (a) average minimum temperature for three winter months (January, February, and March) as well as October of the current growing season (Fig. 10.2a); (b) precipitation for February, May, and

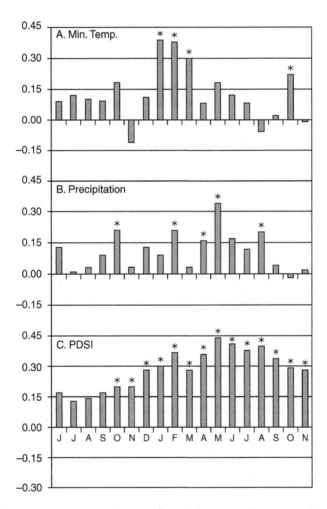

Fig. 10.2 Bootstrapped correlation coefficients between the composite chronology and monthly **(a)** average minimum temperature, **(b)** precipitation, and **(c)** PDSI, from previous June (left) to current growing season November (right). Asterisks indicate statistically significant correlations ($p < 0.05$)

August of the current growing season and October of the previous growing season (Fig. 10.2b); and (c) moisture availability (via PDSI) throughout the current growing season (Fig. 10.2c). Looking for seasonal groupings of significant monthly parameters, we conclude that low winter temperatures limit annual radial growth of shortleaf and pitch pine in GSMNP during our study period.

This response to winter minimum temperature is noteworthy because other studies in the southeastern USA have found mid-elevation conifers to be predominantly precipitation or moisture sensitive (Friend and Hafley 1989; Copenheaver et al. 2002). Although some studies have noted a weak winter temperature signal (Stambaugh and Guyette 2004) or even a negative summer temperature signal (Grissino-Mayer and Butler 1993), no prior studies have investigated the effects of *minimum* temperatures on pine growth in the southeastern USA. In our analysis, we found that when winter minimum temperatures are warmer, radial growth tends to be greater in the following growing season. Warmer winters may allow pines to photosynthesize during winter thaws or break dormancy earlier in the spring, leading to above average annual growth. Minimum temperatures may be more important than average temperatures because a certain temperature threshold must be reached for the trees to remain photosynthetically active or break dormancy (Perry 1971).

We also noted a significant positive correlation between growth and growing season PDSI, suggesting that long-term moisture availability may be more important for pine growth than total precipitation, emphasizing the important role of temperature and soil conditions working together with rainfall to moderate tree growth. In general, years with high annual growth corresponded with low drought stress and abundant soil moisture. These conditions are particularly beneficial in the late spring and summer months, when southern pines put on most of their annual cambial growth (Zahner 1962; Dougherty et al. 1994; Emhart et al. 2006).

Despite the significant relationships identified between pine growth and monthly climate, the sensitivity of pine growth to climate has fluctuated over the past century (Fig. 10.3). Moving correlation analysis performed over 45-year windows indicated that no single monthly climate parameter was significantly related to annual growth for the entire duration of the study period (1910–2007). PDSI during February is the one month that came closest to being temporally stable for the full length of the 98-year-long period we examined, but showed a weakened relationship for 45-year periods ending around 1962 to 1972 (Fig. 10.3c). Because trees did not respond consistently to one or more dominant environmental factor over time, the factors that

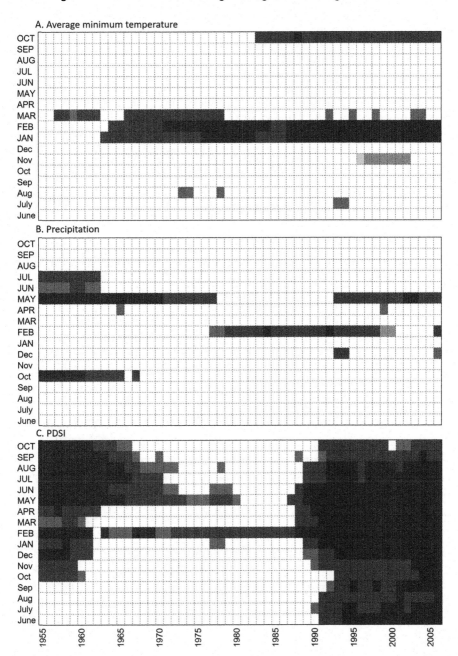

Fig. 10.3 Results of moving correlation analysis between the composite tree-ring chronology and monthly climate: (**a**) average minimum temperature, (**b**) precipitation, and (**c**) PDSI, from previous June (bottom of y-axis) to October of the current growing season (top of y-axis). Years shown are the last years of the 45-year moving intervals, that is, correlations plotted for 2005 represent correlations calculated from 1961 to 2005. All shading indicates statistically significant correlations. The darker the shading, the stronger the correlation ($p < 0.05$)

limit growth of pines in GSMNP in one year, decade, or even century, should not be assumed to limit growth in another period.

Before mid-twentieth century, winter minimum temperatures appear to have had little influence on pine growth, but when data after 1964 are added to our analysis, the relationship between winter temperature and growth becomes significant (Fig. 10.3a). Similarly, current October temperature was significantly positively correlated with growth in the initial monthly correlation analysis (Fig. 10.2a), but the moving analysis shows that the correlation was not significant until the latter part of the century (Fig. 10.3a). Had our analyses stopped before 1983, the results would suggest no relationship between October temperature and growth. Precipitation during May, June, and July precipitation were also inconsistently correlated to growth (Fig. 10.3b). Had we simply used the results of the correlation analysis, we would have inferred that May precipitation was a likely candidate for reconstruction, but the relationship between pine growth and May precipitation was statistically insignificant for 45-year periods ending from 1978 to 1992. While the pine response to winter temperature has strengthened over the twentieth century, the response to growing season precipitation actually weakened. The most dramatic changes occurred in the relationship between pine growth and drought (as measured by PDSI), when correlations dropped precipitously in the 45-year periods ending 1972 to 1990 but increased again ca. 1990.

Inspection of the changes in climate-tree growth relationships over time via correlation evolution graphs was particularly instructive (Fig. 10.4). We observed that the responses to average minimum temperatures were split among positive and negative correlations for 45-year periods ending in 1955, but by the 45-year period ending in 2007, nearly all monthly responses were positive or near positive (Fig. 10.4a), with the exception of the strong negative correlation with previous November temperature (Fig. 10.2a). This demonstrates an overall strengthening of the pine growth response to temperatures over the twentieth and early twenty-first centuries. We also identified key transition periods of temporal instability. For example, a perturbation in the climate-pine growth relationship occurred in the 45-year periods ending in the late 1950s, seen primarily in the weakening response to PDSI (Fig. 10.4c) as well as strengthening response to minimum temperatures (Fig. 10.4a). The relationship between pine growth and PDSI remained weak until the 45-year periods ending in the mid- to late 1980s when the correlations increased dramatically, stabilizing in the early 1990s to correlations similar to those we observed for the most recent 45-year period ending in 2007. Curiously, no overall trend or transition periods could be identified in the response by pines

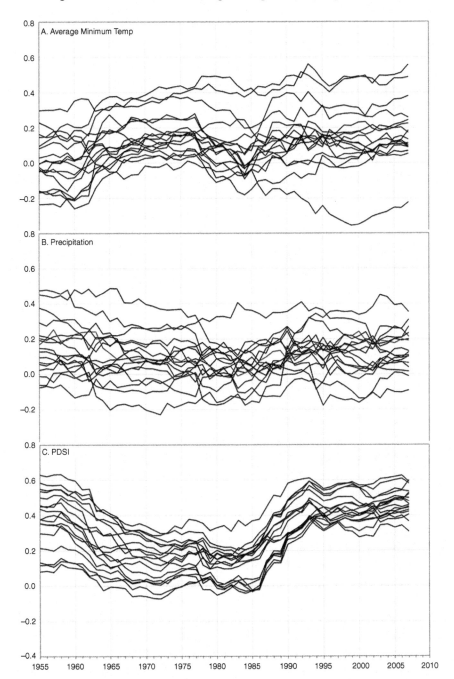

Fig. 10.4 Correlation evolution graphs showing evolving correlation patterns over time, beginning with the 1910–1955 base interval, for all months of the three climate variables analyzed in this study. The coefficients are plotted according to the last year of the 45-year moving interval, that is, correlations plotted for 2005 represent correlations calculated from 1961 to 2005

to precipitation (Fig. 10.4b). Based on these findings, we classify the climate-tree growth relationship in this study as unstable.

Causes and Implications of Instability of Climate Responses in GSMNP and Beyond

Our designation of this relationship, however, raises questions about what constitutes and causes such instability of climate responses, both for our site specifically and for climate and trees more generally. At present, dendrochronologists have not established a standard protocol for distinguishing between a "stable" climate response and an "unstable" one. Our designation is based on the fact that the parameters most strongly related to growth change over time, but other definitions of instability are possible as well. Several survey responses address this issue:

> In my opinion, the most urgent change to be made is a more statistically founded discussion of what instability actually means, and how to quantify and compare it.

> [H]ow do we determine what level of inconsistency is acceptable for a chronology to be used for a reconstruction? And furthermore, how do we communicate the complexity of all of this to the public without allowing a few people to pick this up and run with it to discredit climate science?

Our survey revealed a wide range of views exist as to the extent of temporal instability in tree growth-climate relationships. While some responses suggest that growth-climate relationships are inherently variable because "trees are not thermometers," and "a linear, univariate relationship for a biological organism is a silly assumption," others imply that "trees probably still react to climate in the same way as in previous millennia," but that "what has changed is rather the yearly 'composition' of the climate." A subset of those that attribute temporal instability to climatic changes view the issue as possibly unique to the Anthropocene and a function of anthropogenic climate change. In this view, if climate was relatively stable, we would expect the responses of trees to climate to be stable as well.

This range of perspectives is also found in the published literature, where numerous causal mechanisms have been put forth to explain apparent instability in the sensitivities of trees to climate. Possible causes include methodological factors such as the reliability of weather station data (Frank et al. 2007) and the type of detrending pursued in chronology development

(D'Arrigo et al. 2004), as well as factors considered "internal" to the tree, such as tree genetics and aging (Szeicz and MacDonald 1994). Other hypothesized mechanisms focus on climate itself, attributing shifts in sensitivity to climatic extremes or regime shifts in climate, possibly linked to oceanic-atmospheric climate oscillations or recent anthropogenic warming. For example, extreme high temperatures may induce moisture stress, causing trees to exhibit a stronger response to moisture than temperature. Furthermore, trees may exhibit complex non-linear or threshold responses. For example, trees that responded positively to temperature in the past may begin to respond to factors other than temperature once a certain temperature threshold is passed. Still other factors that may contribute to temporal instability include acclimation of trees to changing conditions, atmospheric pollution, insect outbreaks, logging, fire (and fire suppression), and land use change.

While we do not aim to and indeed cannot identify one or more definitive causes for instability in pine responses to climate in GSMNP, here we expand on the ways in which intertwined human and ecological processes influence tree sensitivity to climate. We posit that tree sensitivity to climate should be understood as a relational and contingent outcome—not inherent to a given site, species, or climate regime, but produced through a suite of interacting factors. This conceptualization reinterprets the principle of limiting factors, one of the established principles of the field. Instead of focusing attention primarily on climate parameters as limiting factors, researchers also must consider how tree-climate relationships are influenced by land use history and changes in ecosystem structure, composition, and function caused by both human and natural disturbances. Such factors do not merely confound or disturb an underlying climate response but may indeed enable it in the first place.

Recent tree-ring studies support this conceptualization of responses to climate as a relational outcome, influenced by clear cutting (White et al. 2014), earthworm invasion (Larson et al. 2010), landscape development (Wilmking and Myers-Smith 2008), and anthropogenic changes to the water table (Smiljanić et al. 2014). In North Carolina, for example, red spruce experienced a shift in sensitivity to temperature post 1930, coinciding with clear-cut harvesting (White et al. 2014). In the north central USA, the climate responses by hardwoods have been altered by invasive earthworms that dramatically reduce leaf litter on the forest floor (Larson et al. 2010). In survey responses, dendroecologists note that oftentimes "basic ecological knowledge is not incorporated into dendroclimatology," and that "most ecologists have no problem" understanding time-varying responses to climate by trees, because they recognize that limiting factors are variable over time and space. Together, these studies and responses remind us that tree-climate relationships are embedded in broader landscapes, and that considering these landscapes and their dynamics can aid in understanding how tree growth responds variably to climate.

Southern Appalachian pine-oak woodlands, including the stands we sampled, have changed dramatically over the past two centuries. Low intensity, frequent fire played a significant role in shaping the landscape through a combination of natural and anthropogenic ignitions (Grissino-Mayer 2016). From the mid-nineteenth to early twentieth centuries, Euro-American settlers used fire to clear land for grazing, agriculture, and timber harvesting (Pyle 1988), but fire has been actively suppressed since ca. 1940 (Grissino-Mayer 2016). The forests of the westernmost portion of GSMNP were a mosaic of open woodlands and closed canopy forests at the start of the twentieth century (Ayres and Ashe 1905) but have changed in composition and structure between the 1920s and today. In particular, fire-tolerant pines are failing to regenerate since ca. 1940, while both canopy density and basal area have increased (Harrod et al. 1998; Harrod and White 1999). In our study sites, mesic, fire-intolerant species such as red maple (*Acer rubrum*), eastern white pine (*Pinus strobus*), and mountain laurel (*Kalmia latifolia*) comprised the understory, with very few shortleaf and pitch pines present as seedlings or saplings, suggesting an ongoing shift in species composition.

Over the past century, the pine-oak woodlands of GSMNP have experienced a number of other changes as well. Leaf litter and duff have increased at many sites, as the exclusion of fire has allowed the surface layer of soil organic matter to thicken over time rather than be consumed by fire as fuel. The chestnut blight (*Cryphonectria parasitica*) also affected the region, wiping out all chestnut trees (*Castanea dentata*) and dramatically changing the composition of the forests in GSMNP. At our study sites, chestnut seedlings and saplings were present in the understory growing from old root stock and stumps, indicating the former presence of chestnuts. Our sites also showed evidence of widespread southern pine beetle infestation and mortality, a native beetle that has contributed to declines of pine timberland throughout the southeastern USA. The weakening relationship of pine growth with precipitation in the mid-twentieth century, and strengthening relationship to winter temperature, may indeed be related to the changing ecosystem of which they are part, and in particular the build-up of leaf litter and duff, the establishment and growth of shade-tolerant and fire-sensitive species, the park's policy of fire suppression, and other major disturbances such as the southern pine beetle outbreak. By thickening the surface layer of soils and increasing the density of understory vegetation, such factors may have also moderated the effect of drought on pines, leading to a diminished relationship between growth and moisture conditions (precipitation and PDSI). Essentially, such factors would be considered "noise" that potentially can mask or disrupt the climate "signal" in tree growth.

In summary, the pine trees we sampled may have been growing for more than two centuries, but the forest ecosystems of which they are a part have changed dramatically in the lifetime of individual trees. As one survey response notes, "How can we expect trees to behave the same way if ecosystems bear little resemblance to the places that they once were?" In GSMNP, many of the pines we sampled likely established and grew in very different conditions—soils with a moderated organic horizon, a less dense understory, with recurring fire—than what they experience now.

Not only have GSMNP forests themselves changed dramatically over the past two centuries, but the broader socioecological landscape in which they are embedded has also transformed, with implications for tree growth responses to climate. Atmospheric pollution from fossil fuel-burning power plants, mining and smelting operations, and vehicular emissions has been a particular problem in the southern Appalachians, with prevailing winds carrying pollutants to the region from throughout the Midwestern and South Central USA (Ke et al. 2007). GSMNP experiences higher levels of air pollution than any other US national park, and visibility in the park decreased 40–80% between 1948 and 2002 (National Park Service 2002). Pollutants (nitrous oxides, sulfur dioxide, carbon, and mercury) in the air are deposited by precipitation, and nitric and sulfuric acids are particularly harmful for forest ecosystems (McLaughlin and Percy 1999; Tomlinson 2003). High-elevation forests are often the most drastically and visibly impacted by acidic deposition, but pine species that characterize lower-elevation pine-oak woodlands also can be affected (Allen and Gholz 1996; Flagler and Chappelka 1996). In GSMNP, tree-ring analysis of shortleaf pines found increased trace metals and suppressed growth beginning in 1970 caused by air pollution and acidic deposition from copper smelting operations upwind (Baes and McLaughlin 1984; Shaver et al. 1994). Similar growth-trend declines have been identified in conifers elsewhere in eastern North America (Adams et al. 1985; LeBlanc et al. 1987). By altering tree growth patterns, atmospheric pollution and acidic deposition may contribute to instability in the climate-tree growth relationship over time.

Of course, understanding the histories of acidic deposition and changes in forest structure and composition does not allow us to conclusively determine one or more mechanisms for temporal instability in the growth response of pine trees to climate. We explore these factors not to develop firm conclusions but to focus on site-specific factors that are often overlooked in considerations of temporal instability. In bringing to light these possible mechanisms, we are reminded that the forests of GSMNP are products of multiple pressures that interact in complex ways. Despite the apparent absence of sustained human activity or impact, the pine-oak woodlands in GSMNP observed today are fundamentally different from the woodland ecosystems that existed when the park was created in 1934. Given such a dynamic landscape—notwithstanding variation in climate and other factors—it is not surprising that tree growth does not correspond with climatic factors in a predictable, linear fashion. While our dataset is inappropriate for climate reconstruction, it may offer an opportunity to analyze more specifically how historical changes in ecosystem structure, composition, and function influence relationships between climate and tree growth in hybrid socioecological landscapes. Additionally, instability in the response by trees to climate may provide a sense of the multiple possible trajectories that the pine-oak woodland ecosystems may take in a changing climate, thus countering the idea that the fate of an ecosystem is determined by climate and climate alone.

A Culture of Humility in Tree-Ring Science

The conceptualization of tree growth-climate relationships that we offer here is not solely our own. Many dendrochronologists are likewise grappling with the plasticity of environment-organism relationships, the limits of uniformitarianism, and the role of tree-ring science in climate change policy and politics. Not only is tree-ring science experiencing a shift in thought in which longstanding precepts are being questioned and reconfigured, but many researchers are also preaching broader changes to the culture of the field and to science in general. We conclude this chapter by reflecting on the normative implications of tree growth-climate relationships as dynamic, relational, and unstable, and explore how and why some tree-ring scientists are fostering a culture of humility in their science.

By casting doubt on tree-ring-based climate reconstructions, temporal instability also stands to threaten the authority of science and scientists in international climate change policy and "the hegemony exercised by the predictive natural sciences over contingent, imaginative, and humanistic … visions of the future" (Hulme 2011). Certainly, reconstructions may be less accurate than previously believed if trees do not respond consistently to climate over time. This realization has generated concern among researchers about climate policy that is based on climate reconstructions and models of future climate scenarios: "From my perspective, addressing the issue of temporal stability in tree-ring research is crucial to building better GCMs [global climate models]... Better models leads to better informed climate policy." Others fear that "future decisions based on 'wrong' reconstructions and false assumptions may cause harm," and note that "if we cannot trust the results from research, then we cannot have trust in the policies that are created from these results." These responses portend an ambivalence toward the "epistemological authority over the future claimed, either implicitly or explicitly, by modeling activities" (Hulme 2011).

But even as instability of growth-climate relationships might undermine hegemonic global climate policy and politics, it provides support for climate change adaptation at the local or micro-scale. Understanding responses by trees to climate as fundamentally dynamic and shaped through complex webs of interrelationships implies also that the consequences of climate change will be, as one response states, "context sensitive to the local conditions that mediate how the global change impacts the micro-environment where organisms actually live." Some researchers therefore see climate-growth research as necessary for developing management strategies that promote resilience not only for forests but also people: "Trees are so sensitive to climate changes and often are a gateway to explaining animal-climate relationships. Adaptations of trees are reflective of ecosystem changes and ultimately necessary human adaptations." In short, the plasticity of tree response to climate suggests that both human and nonhuman responses to climate change are contingent and variable rather than determined by climate alone.

While many researchers recognize the problems with policies informed by flawed scientific assumptions, they simultaneously fear that public awareness of these issues will encourage climate change denial and erode scientific authority. Responses state that "climate change deniers are always looking for ammunition and will likely use this (temporal instability)" and that tree-ring scientists should expect "further attacks from climate deniers and public officials that mistrust science." Some suggest that the only way to prevent this is to make sure claims are "absolutely waterproof," or else "the general public will lose faith in science and not care about any findings anymore." In this modernist view of science, uncertainty is viewed as "threat to collective action" and "a disease that knowledge must cure" (Jasanoff 2007, p. 33). Science must therefore be cleansed of any flawed assumptions in order to retain its privileged position as an authority on climate change.

Another position has emerged, however, among tree-ring scientists, as scientists not only interrogate long-established principles and methods but also promote broader changes in the culture of the field, embracing humility about "both the limits of scientific knowledge and about when to stop turning to science to solve problems" (Jasanoff 2007, p. 33). Our survey demonstrates that a small but vociferous subset of tree-ring scientists view temporal instability not merely as a source of uncertainty to be resolved but as a phenomenon that exemplifies the dynamic interactions among climate, landscape, trees, and people. In this perspective, temporal instability is "a positive trend," and respondents note that "perhaps there is value from understanding unstable climate relationships... Why did that happen? What caused that relationship to break down?" In addition, some researchers note that "being too definitive" is problematic in a world made of contingency and interconnection and express discomfort with the expectation that science can or should provide certainty about complex socioecological issues such as climate change. In summary, findings of temporal instability have spurred self-reflection in the tree-ring science community. This self-reflection has taken many forms. We see the most promising among them as the fostering of a culture of humility among tree-ring scientists—a culture in which researchers are "acknowledging the limits of prediction and control" and "[confronting] head-on the normative implications of our lack of perfect foresight" (Jasanoff 2003, p. 227).

References

Adams, H., S. Stephenson, T. Blasing, and D. Duvick. 1985. Growth-trend declines of spruce and fir in mid-appalachian subalpine forests. *Environmental and Experimental Botany* 25 (4): 315–325.

Allen, E.R., and H.L. Gholz. 1996. Air quality and atmospheric deposition in southern U.S. forests. In *Impact of air pollutants on Southern Pine Forests*, ed. S. Fox and R. Mickler, 83–170. New York: Springer Verlag.

Anchukaitis, K.J., P. Breitenmoser, K.R. Briffa, A. Buchwal, U. Büntgen, E.R. Cook, R.D. D'Arrigo, et al. 2012. Tree rings and volcanic cooling. *Nature Geoscience* 5 (12): 836–837.

Anderegg, W., and G.R. Goldsmith. 2014. Public interest in climate change over the past decade and the effects of the 'climategate' media event. *Environmental Research Letters* 9 (5): 054005.

Baes, C., and S. McLaughlin. 1984. Trace elements in tree rings: Evidence of recent and historical air pollution. *Science* 224: 494–497.

Barber, V., G. Juday, and B. Finney. 2000. Reduced growth of Alaskan white spruce in the twentieth century from temperature-induced drought stress. *Nature* 405: 668–673.

Biondi, F., and K. Waikul. 2004. DendroClim2002: A C++ program for statistical calibration of climate signals in tree-ring chronologies. *Computers and Geoscience* 30 (3): 303–311.

Bridge, M. 2012. Locating the origins of wood resources: A review of dendroprove-nancing. *Journal of Archaeological Science* 39 (8): 2828–2834.

Brienen, R.J.W., G. Helle, T.L. Pons, J.L. Guyot, and M. Gloor. 2012. Oxygen isotopes in tree rings are a good proxy for Amazon precipitation and El Nino-Southern Oscillation variability. *Proceedings of the National Academy of Sciences of the United States of America* 109 (42): 16957–16962.

Briffa, K., F. Schweingruber, P. Jones, T. Osborn, S. Shiyatov, and E. Vaganov. 1998. Reduced sensitivity of recent tree-growth to temperature at high northern latitudes. *Nature* 391: 678–682.

Carrer, M., and C. Urbinati. 2006. Long-term change in the sensitivity of tree-ring growth to climate forcing in *Larix decidua*. *New Phytologist* 170 (4): 861–872.

Cook, E. 1985. *A time series analysis approach to tree-ring standardization*. Ph.D. Tucson, University of Arizona School of Renewable Natural Resources.

Cook, E.R., and A.H. Johnson. 1989. Climate change and forest decline: A review of the red spruce case. *Water, Air and Soil Pollution* 48 (1): 127–140.

Copenheaver, C., L. Grinter, J. Lorber, M. Neathour, and M. Spinney. 2002. A dendroecological and dendroclimatic analysis of *Pinus virginiana* and *Pinus rigida* at two slope positions in the Virginia Piedmont. *Castanea* 67: 302–315.

Coyle, D.R., K.D. Klepzig, F.H. Koch, L.A. Morris, J.T. Nowak, S.W. Oak, W.J. Otrosina, W.D. Smith, and K. Gandhi. 2015. A review of southern pine decline in North America. *Forest Ecology and Management* 349: 134–148.

D'Arrigo, R., R. Kaufmann, N. Davi, G. Jacoby, C. Laskowski, R. Myeni, and P. Cherubini. 2004. Thresholds for warming-induced growth decline at elevational tree line in the Yukon Territory, Canada. *Global Biogeochemical Cycles*, 18 (3).

D'Arrigo, R., R. Wilson, B. Liepert, and P. Cherubini. 2008. On the 'divergence problem' in northern forests: A review of the tree-ring evidence and possible causes. *Global Planetary Change* 60 (3): 289–305.

Dale, V.H., M.L. Tharp, K.O. Lannom, and D.G. Hodges. 2010. Modeling transient response of forests to climate change. *Science of the Total Environment* 408 (8): 1888–1901.

Dougherty, P.M., D. Whitehead, and J.M. Vose. 1994. Environmental constraints on the structure and productivity of pine forest ecosystems: A comparative analysis. *Ecological Bulletins* 43: 64–75.

Driscoll, W., G. Wiles, R. D'Arrigo, and M. Wilmking. 2005. Divergent tree growth response to recent climate warming, Lake Clark National Park and Preserve, Alaska. *Geophysical Research Letters*, 32 (20).

Emhart, V.I., T.A. Martin, T.L. White, and D.A. Huber. 2006. Genetic variation in basal area increment phenology and its correlation with growth rate in loblolly and slash pine families and clones. *Canadian Journal of Forest Research* 36 (4): 961–971.

Esper, J., and D. Frank. 2009. Divergence pitfalls in tree-ring research. *Climatic Change* 94 (3): 261–266.

Flagler, R.B., and A.H. Chappelka. 1996. Growth response of southern pines to acidic precipitation and ozone. In *Impact of Air Pollutants on Southern Pine Forests*, ed. S. Fox and R. Mickler, 388–424. New York: Springer Verlag.

Frank, D., U. Büntgen, R. Böhm, M. Maugeri, and J. Esper. 2007. Warmer early instrumental measurements versus colder reconstructed temperatures: Shooting at a moving target. *Quaternary Science Reviews* 26 (25): 3298–3310.

Friend, A., and W. Hafley. 1989. Climatic limitations to growth in loblolly and shortleaf pine: A dendroclimatological approach. *Forest Ecology and Management* 26 (2): 113–122.

Fritts, H. 1976. *Tree rings and climate*. New York: Academic Press.

Grissino-Mayer, H.D. 2001. Evaluating crossdating accuracy: A manual and tutorial for the computer program COFECHA. *Tree-Ring Research* 57: 205–221.

———. 2016. Fire as a once-dominant disturbance process in the yellow pine and mixed pine-hardwood forests of the Appalachian Mountains. In *Natural disturbances and historic range of variation: Type, frequency, severity, and post-disturbance structure in Central Hardwood Forests*, ed. C. Greenberg and B. Collins, 123–146. Berlin: Springer Verlag.

Grissino-Mayer, H.D., and D.R. Butler. 1993. Effects of climate on growth of short-leaf pine in northern Georgia: A dendroclimatic study. *Southeastern Geographer* 33 (1): 65–81.

Harrod, J., and R. White. 1999. Age structure and radial growth in xeric pine-oak forests in western Great Smoky Mountains National Park. *Journal of the Torrey Botanical Society* 126: 139–146.

Harrod, J., P. White, and M. Harmon. 1998. Changes in xeric forests in western Great Smoky Mountains National Park, 1936–1995. *Castanea* 63: 346–360.

Harrod, J., M. Harmon, and P. White. 2000. Post-fire succession and twentieth century reduction in fire frequency on xeric southern Appalachian sites. *Journal of Vegetation Science* 11 (4): 465–472.

Holliman, R. 2011. Advocacy in the tail: Exploring the implications of 'climategate' for science journalism and public debate in the digital age. *Journalism* 12 (7): 832–846.

Holmes, R. 1983. Computer-assisted quality control in tree-ring dating and measurement. *Tree-Ring Bulletin* 43: 69–78.

Hulme, M. 2011. Reducing the future to climate: A story of climate determinism and reductionism. *Osiris* 26 (1): 245–266.

Jacoby, G.C., and R.D. D'Arrigo. 1995. Tree ring width and density evidence of climatic and potential forest change in Alaska. *Global Biogeochemical Cycles* 9 (2): 227–234.

Jasanoff, S. 2003. Technologies of humility: Citizen participation in governing science. *Minerva* 41 (3): 223–244.

———. 2007. Technologies of humility. *Nature* 450: 33.

Ke, L., X. Ding, R.L. Tanner, J.J. Schauer, and M. Zheng. 2007. Source contributions to carbonaceous aerosols in the Tennessee Valley Region. *Atmospheric Environment* 41: 8898–8923.

Larson, E., K. Kipfmueller, C. Hale, L. Frelich, and P. Reich. 2010. Tree rings detect earthworm invasions and their effects in northern hardwood forests. *Biological Invasions* 12 (5): 1053–1066.

LeBlanc, D., D. Raynal, and E. White. 1987. Acidic deposition and tree growth I: The use of stem analysis to study historical growth patterns. *Journal of Environmental Quality* 16 (4): 325–333.

Mann, M.E., R.S. Bradley, and M.K. Hughes. 1998. Global-scale temperature patterns and climate forcing over the past six centuries. *Nature* 392: 779–787.

———. 1999. Northern hemisphere temperatures during the past millennium: Inferences, uncertainties, and limitations. *Geophysical Research Letters* 26 (6): 759–762.

Mann, M.E., J.D. Fuentes, and S. Rutherford. 2012. Underestimation of volcanic cooling in tree-ring-based reconstructions of hemispheric temperatures. *Nature Geoscience* 5 (3): 202–205.

Marcinkowski, K., D. Peterson, and G. Ettl. 2015. Nonstationary temporal response of mountain hemlock growth to climatic variability in the North Cascade Range, Washington, USA. *Canadian Journal of Forest Research* 45 (6): 676–688.

Maxwell, J., G. Harley, and S. Robeson. 2016. On the declining relationship between tree growth and climate in the Midwest United States: The fading drought signal. *Climatic Change* 138: 127–142.

McIntyre, S., and R. McKitrick. 2005. Hockey sticks, principal components, and spurious significance. *Geophysical Research Letters,* 32 (3).

McLaughlin, S., and K. Percy. 1999. Forest health in North America: Some perspectives on actual and potential roles of climate and air pollution. *Water, Air and Soil Pollution* 116: 151–197.

National Park Service. 2002. *Air Quality in the National Parks.* 2nd ed. Washington, DC: U.S. Department of the Interior.

Oberhuber, W., W. Kofler, K. Pfeifer, A. Seeber, A. Gruber, and G. Wiesner. 2008. Long-term changes in tree-ring-climate relationships at Mt. Patscherkofel (Tyrol, Austria) since the mid-1980s. *Trees-Structure and Function* 22 (1): 31–40.

Orvis, K., and H.D. Grissino-Mayer. 2002. Standardizing the reporting of abrasive papers used to surface tree-ring samples. *Tree-Ring Research* 58: 47–50.

Perry, T. 1971. Dormancy of trees in winter. *Science* 171: 29–36.

Porter, T.J., and M. Pisaric. 2011. Temperature-growth divergence in white spruce forests of Old Crow Flats, Yukon Territory, and adjacent regions of northwestern North America. *Global Change Biology* 17 (11): 3418–3430.

Pyle, C. 1988. The type and extent of anthropogenic vegetation disturbance in the Great Smoky Mountains before National Park Service acquisition. *Castanea* 53: 183–196.

Saladyga, T., and R. Maxwell. 2015. Temporal variability in climate response of eastern hemlock in the Central Appalachian Region. *Southeastern Geographer* 55 (2): 143–163.

Shaver, C.L., K.A. Tonnessen, and T.G. Maniero. 1994. Clearing the air at Great Smoky Mountains National Park. *Ecological Applications* 4 (4): 690–701.

Smiljanić, M., J.W. Seo, A. Läänelaid, M. van der Maaten-Theunissen, B. Stajić, and M. Wilmking. 2014. Peatland pines as a proxy for water table fluctuations: Disentangling tree growth, hydrology and possible human influence. *Science of the Total Environment* 500: 52–63.

Smouse, P., and L. Saylor. 1973. Studies of the *Pinus rigida-serotina* complex I. A study of geographic variation. *Annals of the Missouri Botanical Garden* 60: 174–191.

South, D., and E. Buckner. 2003. The decline of southern yellow pine timberland. *Journal of Forestry* 101 (1): 30–35.

Speer, J.H. 2010. *Fundamentals of Tree-Ring Research*. Tucson: The University of Arizona Press.

Stambaugh, M., and R. Guyette. 2004. Long-term growth and climate response of shortleaf pine at the Missouri Ozark Forest Ecosystem Project. In *Proceedings of the 14th Central Hardwood Forest Conference*, ed. D.A. Yaussy, D.M. Mix, R.P. Long, and P.C. Goebel, 448–458. Newtown Square, PA: USDA Forest Service.

Szeicz, J., and G. MacDonald. 1994. Age-dependent tree-ring growth responses of subarctic white spruce to climate. *Canadian Journal of Forest Research* 24 (1): 120–132.

Tomlinson, G.H. 2003. Acidic deposition, nutrient leaching and forest growth. *Biogeochemistry* 65 (1): 51–81.

Van Deusen, P.C. 1990. Evaluating time-dependent tree ring and climate relationships. *Journal of Environmental Quality* 19 (3): 481–488.

Wahl, E., and C. Ammann. 2007. Robustness of the Mann, Bradley, Hughes reconstruction of Northern Hemisphere surface temperatures: Examination of criticisms based on the nature and processing of proxy climate evidence. *Climatic Change* 85 (1): 33–69.

White, P., P. Soulé, and S. van de Gevel. 2014. Impacts of human disturbance on the temporal stability of climate–growth relationships in a red spruce forest, southern Appalachian Mountains, USA. *Dendrochronologia* 32 (1): 71–77.

Wiles, G., C. Mennett, S. Jarvis, R. D'Arrigo, N. Wiesenberg, and D. Lawson. 2012. Tree-ring investigations into changing climatic responses of yellow-cedar, Glacier Bay, Alaska. *Canadian Journal of Forest Research* 42 (4): 814–819.

Wilmking, M., and I. Myers-Smith. 2008. Changing climate sensitivity of black spruce in a peatland-forest landscape in Interior Alaska. *Dendrochronologia* 25 (3): 167–175.

Wilson, R., and W. Elling. 2004. Temporal instability in tree-growth/climate response in Lower Bavarian Forest region: Implications for dendroclimatic reconstructions. *Trees-Structure and Function* 18 (1): 19–28.

Yamaguchi, D.K. 1991. A simple method for cross-dating increment cores from living trees. *Canadian Journal of Forest Research* 21 (3): 414–416.

Zahner, R. 1962. Terminal growth and wood formation by juvenile loblolly pine under two soil moisture regimes. *Forest Science* 8 (4): 345–352.

Zang, C., and F. Biondi. 2015. Treeclim: An R package for the numerical calibration of proxy-climate relationships. *Ecography* 38 (4): 431–436.

11

Forest Land-Use Legacy Research Exhibits Aspects of Critical Physical Geography

David Robertson, Chris Larsen, and Steve Tulowiecki

The natural agency of trees and other forest organisms can give the impression that forests are completely natural. Most forests, however, are influenced by past human activities. For example, soil and vegetation conditions in forested sites within northern France still bear the influence of Gallo-Roman occupation 1600 years ago (Plue et al. 2008). Similarly, the predominance of white oak forests in eastern USA may be due to land-use practices of Native Americans and Euro-American settlers (McEwan et al. 2011). These forests thus contain land-use legacies (LULs): biotic (e.g. vegetation) and abiotic (e.g. soil) features of ecosystems influenced by land-use history (Perring et al. 2016).

Research frameworks similar to Critical Physical Geography (CPG) have been identified in physical geography, political ecology, environmental history, and other areas of research (Lave et al. 2014), but no systematic evaluation of the pervasiveness of CPG approaches in any specific field of inquiry has been conducted. This chapter explores the relationship between CPG and forest LUL research, an area of interdisciplinary scholarship well suited to a CPG framework. The chapter begins with an exploration of the peer-reviewed literature on LULs and their relevancy to CPG. This is followed by a focused analysis of studies examining LULs in forest ecosystems, which answers the following questions. From which disciplines are the journals and authors?

D. Robertson (✉) • S. Tulowiecki
Department of Geography, SUNY Geneseo, Geneseo, NY, USA

C. Larsen
Department of Geography, University at Buffalo, Buffalo, NY, USA

© The Author(s) 2018
R. Lave et al. (eds.), *The Palgrave Handbook of Critical Physical Geography*,
https://doi.org/10.1007/978-3-319-71461-5_11

227

How varied are the data (biophysical and human, quantitative and qualitative) employed? How prevalent is the use of CPG intellectual tenets of *transdisciplinarity, reflexivity,* and *power and justice*? Our analysis is premised on the observation, articulated by Lave et al. (2014), that existing bodies of integrative research demonstrate the potential of a CPG framework. Arguing that LUL scholarship is one such precedent, this systematic analysis of forest LUL research and its relationship to the CPG approach reinforces this claim. In addition, analysis of forest LUL research through a CPG lens highlights strengths and weaknesses in this body of work.

Land-Use Legacies and Critical Physical Geography

A term introduced to the academic literature in 1994, LULs are the focus of a growing body of research in the human-ecological sciences.[1] The earliest scholarly use of the term "land-use legacies" was by Wallin et al. (1994), in a study of forest LULs created by past logging. However, the term "land-use legacies" was used only in the article's title. Similar incidental use of the term characterizes subsequent papers until LULs received a systematic presentation by Foster et al. (2003).

The most frequently cited article in LUL research, the Foster et al. 2003 paper emphasizes that consideration of historical land use and its legacies "adds explanatory power to our understanding of modern conditions at scales from organisms to the globe and reduces missteps in anticipating or managing for future conditions (77)." They explain that the focus of LUL research is not the immediate effects of human activities (which differentiates it from land change science), but rather "their enduring consequences on ecosystem structure and function decades or centuries or longer after they have occurred and natural processes have been operative (2003, 78)." The paper reviews how forestry, agriculture, modification of natural disturbance regimes (e.g. fire), and manipulation of animal populations has had enduring effects on ecosystem features such as vegetation structure and composition, soil structure and chemistry, and carbon and nitrogen dynamics. They show how LUL effects are relevant in wildland conservation, and resource and restoration management. Providing an influential introduction to LUL research, Foster et al. (2003) outlined a research agenda increasingly pursued by investigators working at the intersection of land-use history and ecological and environmental science. Growth in LUL research mirrors growth of a broader interest in the Anthropocene and an understanding, as Lave et al. (Chap. 1 this volume) state, that "the material world is now shaped by deeply intermingled social and physical processes."

General Trends in Forest LUL Research

To assess the growth of LUL research, we identified LUL studies using *Web of Science*, an online citation indexing service utilizing multidisciplinary databases.[2] A total of 155 articles have been published on LULs since 1994 (Fig. 11.1), and the volume of research has increased through time with highs of 25 articles in 2014 and 2015, and nearly 50% of articles (76 of 155 articles) published in the last four years (2013–2016). The breadth of representation in both human-ecological and environmental sciences in the articles confirms the observation that human activity is now so varied and widespread that "consideration of land-use legacies could be boundless (Foster et al. 2003, 78)." The articles include LUL studies on many ecosystems (e.g. grasslands and wetlands), species (e.g. fish and amphibians), ecosystem functions (e.g. soil and aquatic chemistry), and land cover types (e.g. agricultural and urban landscapes).

The 155 articles were published in 85 different journals, with the top 15 listed in Table 11.1. Natural science and ecology-oriented journals including *Ecological Applications* and *Landscape Ecology* are dominant publication out-

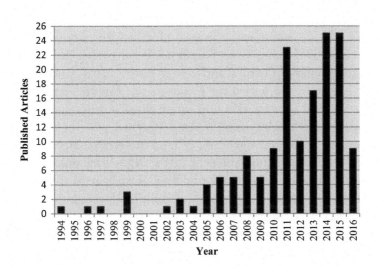

Fig. 11.1   Peer-reviewed articles published on land-use legacies. A total of 155 articles have been published on land-use legacies since 1994 (Source: *Web of Science*, 6 May, 2016)

Table 11.1 The 15 journals that were the most frequent outlets for the 155 land-use legacy articles (Source: *Web of Science*, 6 May, 2016)

Journal	Articles
Forest Ecology and Management	11
Ecological Applications	9
Landscape Ecology	8
Applied Vegetation Science	5
Ecological Monographs	4
Restoration Ecology	4
Ecosphere	4
Global Change Biology	4
Agriculture, Ecosystems & Environment	3
Applied Geography	3
PLOS One	3
Biological Conservation	3
Journal of Applied Ecology	3
Landscape and Urban Planning	3
Soil Biology and Biochemistry	3

lets but indicative of the fact that LUL research spans many knowledge domains, the list also includes social science journals such as *Applied Geography* and *Landscape and Urban Planning*. Moreover, many of the journals are multidisciplinary in nature and emphasize applied research outcomes, including the leading journal outlet for LUL research, *Forest Ecology and Management*, which published 11 of 155 articles.

The prominence of *Forest Ecology and Management* is also an indicator of the dominant topical focus of LUL research: forests. A plurality of all LUL articles (72 of 155) had a topical focus on forests.[3] The fact that forests are a central focus in LUL scholarship should not be surprising as forests are typically a complex tangle of natural and human histories: past land-use and land-clearance activities like agricultural conversion, forestry, and fire use can produce ecological legacies affecting forest characteristics such as function, size-structure, and species composition. Moreover, forests are particularly likely to exhibit ecological signatures of land use because trees are long-lived elements of landscape (Munteanu et al. 2015). More fundamentally, LULs are evident in forests because of the central role forests play in society (Williams 1989). Forests are affected by economy, politics, and culture, and in turn, affect these aspects of society. For example, forests may hold cultural meaning, and legacies of their use (e.g. effects of indigenous or historical land-use practices) may provide cultural ecosystem services (e.g. serve as cultural heritage).

Given their status as socio-biophysical entities, comprehensive analysis of forest LULs requires integration of data and approaches from the natural and social sciences. As a result, forest LULs are well suited to investigation using a

CPG approach, which brings together social and physical science to understand socio-biophysical landscapes and eco-social transformations (Lave et al. 2014, Lave et al. Chap. 1 this volume).

Although LUL research and CPG have yet to be explicitly linked in empirical analyses, LUL scholarship represents an existing body of integrative research that demonstrates the potential of CPG (see Lave et al. 2014). How forest LULs have been analyzed at the study level, and how these studies relate to CPG, is the focus of the remainder of this chapter.

Assessing Critical Physical Geography Practices in Forest Land-Use Legacy Research

To perform a study-level evaluation of the relationship between forest LUL research and CPG, *Web of Science* was utilized to extract 40 forest-related studies from the larger list of 155 LUL articles. To increase the relevancy of this analysis, the 40 studies were composed of two categories of peer-reviewed articles: the 20 most frequently cited and the 20 most recently published.[4] Study-level evaluation of CPG practices in the 40 forest LUL articles included three areas of assessment: (1) an article overview including discussion of publication sources, authorship, and topical foci; (2) evaluation of biophysical and human data use; and (3) incorporation of CPG intellectual tenets. To assess the latter two characteristics, a set of article evaluation guidelines and a visual model for conveying results were derived from CPG and related literatures.

The use of biophysical and human data in forest LUL articles is represented in the four-square portion of the article analysis diagram (Fig. 11.2), adapted

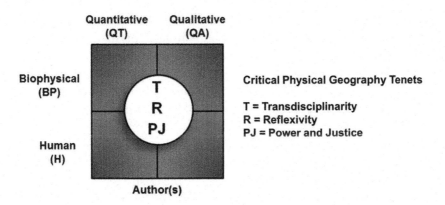

Fig. 11.2 Article analysis diagram

from Lave et al.'s "methods four-square diagram" (Chap. 1 this volume). Data was categorized as biophysical or human, and quantitative or qualitative, according to the following definitions: (1) *biophysical quantitative* data are quantitative expressions of objects or variables that are physical (e.g. percent bedrock) or biological (e.g. area of forest), regardless of whether they were influenced by natural or anthropogenic factors; (2) *biophysical qualitative* data are qualitative expressions (with no accompanying quantification) of objects and variables that are physical (e.g. limestone bedrock) or biological (e.g. red maple forest), regardless of whether they were influenced by natural or anthropogenic factors; (3) *human quantitative* data are quantitative expressions about humans (e.g. a population of 1000 people) or their actions (e.g. ten years of occupancy); and (4) *human qualitative* data are qualitative expressions (with no accompanying quantification) about humans (e.g. a population) or their actions (e.g. forest cleared for farmers). Quantitative data was defined as structured numerical data (usually aggregated or statistically analyzed). Qualitative data was defined as less structured non-numerical data (usually text), or nominal-value numerical data holding no quantitative meaning (Montello and Sutton 2006).

The prevalence of three intellectual tenets of CPG in the articles is represented in the circle overlaying the center of the article analysis diagram with the absence of a circle representing absence of tenet use. Distilled primarily from foundational CPG commentaries including Lave et al. (2014), Lave (2015), and Lave et al. (Chap. 1 this volume), the three intellectual tenets are *transdisciplinarity*, *reflexivity*, and *power and justice*. *Transdisciplinary* research demonstrates substantive interweaving of physical and social science (Lave et al. Chap. 1 this volume), integrating biophysical and human data and producing understanding of biophysical landscapes as eco-social systems. Transdisciplinary analysis may include: triangulation, whereby researchers explore the convergence and corroboration, or paradox and contradiction, of findings derived from biophysical and human data (Bryman 2006); complementarity, whereby researchers seek "elaboration, enhancement, illustration, or clarification" of findings derived from biophysical and human data (Greene et al. 1989, 259); iterative analysis, whereby researchers "work back and forth" between findings derived from biophysical and human data, "modifying research plans in one area in response to new data or questions in another" (Lave et al., Chap. 1 this volume); and, expansion, whereby researchers seek to "extend the breadth and range of inquiry" by using biophysical and human data "for different inquiry components" (Greene et al. 1989, 259).

Reflexive research demonstrates critical awareness that science is "inextricably imbricated in social, cultural, and political-economic relations" affecting research questions asked, methods employed, and findings (Lave et al. Chap. 1 this volume). Reflexivity is demonstrated when researchers engage in substantive introspection regarding knowledge production including consideration of how the background and perspective of the researcher shapes questions asked, methods employed, findings considered most consequential, and conclusions communicated (Malterud 2001; Tadaki et al. 2015). Reflexive research involves explicit reporting of researchers' beliefs, values, perspectives, and assumptions in the articles and discussion of how they may have influenced the research process.[5]

Research embodying the *power and justice* tenet demonstrates critical awareness of the relationship between biophysical landscapes and social power. It also promotes social and/or environmental justice (Lave 2015). Specifically, the research shows awareness of the cultural, economic, or political factors, structural and/or locally contingent, that shape, and are shaped by, the biophysical landscape. This awareness includes consideration of power relationships rooted in colonial history, or governed by factors such as economy and class, politics and policy, or gender, race and ethnicity. The research also delivers applied outcomes in the form of knowledge providing services to society or environment (Lave et al. 2014).

Results

Article Overview

Studies of forest LULs have diverse publication outlets. The 40 forest LUL articles extracted from the larger database were published in 30 journals. The leading five journals were *Ecological Applications* (five articles); *Ecological Monographs* and *Forest Ecology and Management* (three articles each); and, *BioScience* and *Regional Environmental Change* (two articles each). Ecology-oriented journals dominate the source list, but not exclusively. Multidisciplinary journals (e.g. *Agriculture, Ecosystems & Environment*), applied journals (e.g. *Applied Geography*), and policy- and management-oriented journals (e.g. *Land Use Policy* and *Global Environmental Change: Human and Policy Dimensions*) also published studies.

The cross-disciplinary and collaborative nature of forest LUL scholarship is revealed through authorship. The 40 articles had 185 authors and only two were single-authored. The median author count per article was four, showing that forest LUL research has been a collaborative endeavor dominated by team-science approaches. The authors also have diverse academic and professional affiliations. Although the epistemological background of authors was not surveyed, research approaches and department affiliations indicate that the majority are natural scientists. Of the authors, 78% were university-based researchers and the leading departmental affiliations were biology (25%), forestry (20%), geography (18%), and environmental science (11%). Researchers from the USA (63% of authors) and Europe (30% of authors) conducted the majority of forest LUL research.

The topical focus of the 40 forest LUL studies varied substantially. The majority of studies (22 of 40) focused on forest legacy effects produced by multiple human activities with general land-use and/or land cover data used as an analytic variable (e.g. Rhemtulla et al. 2007). In other studies, forest LULs resulting from agriculture (e.g. Mattingly et al. 2015), most notably farming-related land clearance (e.g. Kepfer-Rojas et al. 2015), were studied (ten of 40 articles). Other activities serving as topical foci included grazing (e.g. Ponette-González et al. 2016) and forestry (e.g. Blixt et al. 2015). In addition, legacy effects and influences related to climate change (e.g. Ameztegui et al. 2016), anthropogenic fire use or altered natural fire regimes (e.g. Hahn and Orrock 2015), and precolonial Native American land use (e.g. Tulowiecki and Larsen 2015) also received consideration.

The studies examined a range of forest attributes across a relatively small range of forest biomes, with only a few studies conducted beyond the Western world. Multiple ecological effects of past land use were explored in the studies including change in forest function (e.g. carbon sequestration), forest structure (e.g. forest cover and fragmentation), and forest composition (e.g. the presence or abundance of tree species). However, alteration in tree species composition was the dominant variable measured, with regional analyses of US forests (13 of 40 articles) and European forests (8 of 40 articles) accounting for more than half of all studies conducted. Including dependencies, 60% of the articles sampled (24 of 40) investigated forest LULs in US territory and more than three-quarters (31 of 40) analyzed temperate forests.

Biophysical and Human Data Use

The results of the article analysis are presented in Figs. 11.3 and 11.4. All 40 forest LUL studies (100%) utilized quantitative biophysical data including measures such as area of forest cover or percent species distribution. Qualitative biophysical data, employed in 29 of 40 studies (73%), was the second most common type of data utilized. This included textual descriptions of forest types, soil series, and climate conditions. Use of human data in the research

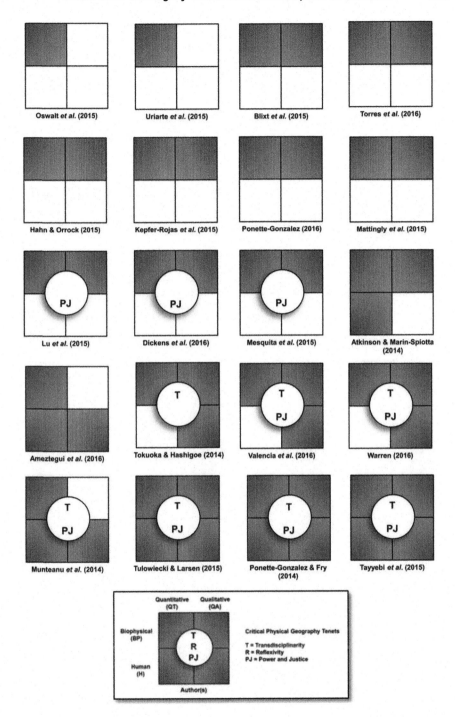

Fig. 11.3 Article analysis diagrams of the 20 most recent forest land-use legacy articles (Source: *Web of Science*, 6 May, 2016)

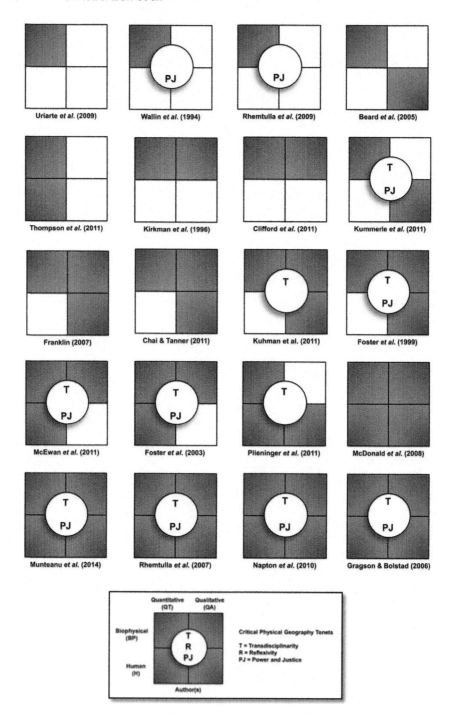

Fig. 11.4 Article analysis diagrams of the 20 most-cited forest land-use legacy articles (Source: *Web of Science*, 6 May, 2016)

was less prevalent, appearing in 24 of 40 studies (60%). Intriguingly, qualitative human data such as narrative land-use histories or overviews of evolving land management policies were more common in the research, appearing in 20 of 40 studies (50%), than was quantitative human data like demographic or socio-economic statistics, which appeared in only 15 of 40 articles (38%). This result may seem counterintuitive given that forest LUL research is dominated by natural scientists who are presumably more familiar with quantitative analyses. Perhaps natural scientists are more comfortable using everyday forms of information like narratives and anecdotes, than numerical demographic and socio-economic data more clearly linked to social science inquiry.

Overall, forest LUL research has utilized diverse forms of data with 35 of 40 studies (88%) employing more than one of the four forms of data: five studies (12%) utilized one data type, 14 studies (35%) utilized two, 13 studies (33%) utilized three, and nine studies (23%) utilized all four forms of data (Figs. 11.3 and 11.4). Moreover, the majority of articles, 24 of 40 (60%), showed substantive use of both biophysical and human data suggesting that the majority of forest LUL researchers recognize the need for data use across natural and social science knowledge domains. These numbers also show that an essential precursor for transdisciplinary research exists in forest LUL research—the combined use of biophysical and human data. However, the fact that 40% of studies do not utilize biophysical and human data together raises questions about the thoroughness of some forest LUL research, as LULs are eco-social phenomena requiring comprehensive analysis of both biophysical and human objects and variables.

Relatedly, it is notable that data diversity and combined use of biophysical and human data are both significantly higher in the 20 most-cited LUL articles (Fig. 11.4) than in the 20 most recent articles (Fig. 11.3). Of the 20 most-cited articles, 65% utilized more than two forms of data and 75% used both biophysical and human data. Comparatively, the 20 most recent articles registered 45% on both measures; further, in some cases, the focus on LULs may reflect adoption of an increasingly popular academic phrase rather than an attempt to unearth the eco-social complexities of forest LULs. The differences in data use between these samples suggest that integration of biophysical and human data produces more citation-worthy and consequential results in LUL scholarship. It also suggests that future adoption of a CPG framework explicitly emphasizing the interweaving of natural and social science could advance LUL scholarship.

An article appearing in the most-cited list that effectively used all four forms of data (i.e. quantitative and qualitative, biophysical and human) to study LUL effects in forests is Napton et al. (2010) (Fig. 11.4). They assess and compare driving forces of landscape change in four forested ecoregions of the southeastern USA. The project utilized probability sampling of Landsat imagery, manual interpretation of US Geological Survey land cover classes, and diverse thematic literatures dealing with regional history, population, and economy. A convergence of evidence approach was used to determine the driving forces of land change, which was found to be affected most by commercial forestry and agriculture, economic and population growth, and changes in transportation and technology. They also found that driving forces of forest change were modified through time by the legacies of past land decisions which in turn were the product of earlier driving forces interacting with environment in each ecoregion.

Transdisciplinarity

Forest LUL research is characterized by diverse data use. Combined use of biophysical and human data, however, is a relatively straightforward task in research when these data are segregated in a study. Common uses of human qualitative data in natural science research, for example, include its use in a descriptive and introductory capacity early in a research project with quantitative approaches later applied in an analytical and confirmatory way. Qualitative data is also commonly used to exemplify quantitative results and conclusions (Montello and Sutton 2006). Alternatively, more substantive integration of biophysical and human data in research, what Lave et al. (Chap. 1 this volume) calls "transdisciplinary" research, is more challenging; as a central intellectual tenet of CPG, transdisciplinary research utilizes biophysical and human data *and* interweaves that data analytically.

Of the 40 forest LUL articles sampled, only 17 (43%) were categorized as transdisciplinary in that they showed analytical integration of biophysical and human data and generated understanding of forests as eco-social systems (Figs. 11.3 and 11.4). Unsurprisingly, studies utilizing more forms of data had higher incidence of transdisciplinarity. No studies utilizing one form of data were classified as transdisciplinary, whereas one of 14 studies (7%) utilizing two forms of data, nine of 13 studies (69%) utilizing three forms, and seven of eight studies (88%) utilizing four forms of data were identified as transdisciplinary.

Transdisciplinary studies had authors whose professional affiliations leaned toward interdisciplinary academic fields: 13 of 17 transdisciplinary articles (76%) had first authors affiliated with forestry, geography, or environmental science programs. These results confirm Lave et al.'s observation (Chap. 1 this volume) that "cross-training" in physical and social science facilitates transdisciplinarity, and conversely, that the rigid knowledge boundaries that exist in many disciplines are a barrier to transdisciplinary research. It is interesting to note, however, that a logistical barrier to transdisciplinarity—the tendency of journals to either publish physical or social science but not both (Lave et al., Chap. 1 this volume)—may be less of an obstacle in LUL research. The 17 articles classified as transdisciplinary were published in 16 journals including high-impact ecology journals like *Ecological Applications, Global Change Biology,* and *Ecological Monographs.*

Transdisciplinary research was somewhat more common in the top-cited articles (10 of 20), than in the most recent articles (7 of 20), suggesting that the interweaving of biophysical and human data in forest LUL research also produces more notable scholarship. However, with less than half of the 40 articles classified as transdisciplinary, it appears that a comprehensive understanding of forest LULs as eco-social phenomena is being overlooked in much of the literature. While 24 of 40 studies (60%) use biophysical and human data of various forms, only 17 of 40 (43%) interweave those data analytically.

An article that effectively interweaves biophysical and human data is Tulowiecki and Larsen (2015). A study drawn from the list of 20 most recent forest LUL articles, the authors evaluate forest LULs of Native American societies by examining the extent to which the geographic distributions of tree species (ca. 1800 CE) were attributable to Seneca Iroquois land use in Chautauqua County, New York State. The study demonstrates transdisciplinarity in that it integrates biophysical data, such as vegetation data from original land survey records and environmental data (e.g. climate, soil, and topographic GIS data) with complementary human data, such as the settlement locations of Native American societies collected from the archaeological and historical record. The transdisciplinarity of this article extends into its methodology, where quantitative methods from biogeography (i.e. species distribution modeling) are expanded using quantitative methods from archaeology (i.e. caloric cost modeling), disentangling biophysical and human impacts on tree species distributions.

Reflexivity

Authors of the 40 forest LUL articles included no substantive reflexive discussion in their research (Figs. 11.3 and 11.4), since none consciously engaged in critical self-reflection of how their personal perspectives and biases influenced the research. Given that natural scientists dominate authorship of the articles, and that reflexivity is a less-familiar practice in the natural sciences, this finding is predictable. Of the three CPG intellectual tenets analyzed in this chapter, reflexivity is the tenet most deeply embedded in critical social theory (e.g. Tadaki et al. 2015). Moreover, reflexivity confronts fundamental assumptions of positivistic science: the existence of external realities and value-free knowledge production (Cloke et al. 1991). As such, reflexivity is an epistemologically distant concept for most natural scientists.[6] Although a single scientific method is not employed in the natural sciences, and while adoption of critical rationalist and other post-positivistic perspectives has led many natural scientists to question the ability of scientific explanation to obtain truths about reality, a positivistic logic privileging scientific empiricism remains dominant (Inkpen 2005). This holds true in forest LUL research.

This is not to say that the 40 forest LUL articles lack critical or contextual thinking. Natural scientists are increasingly trained in exploring multiple lines of explanation through integrative inquiry (Tadaki et al. 2015), and many studies reviewed herein reflect these qualities. Rather, forest LUL researchers have not developed a mindset for explicit reflexive commentary, as called for in CPG (e.g. Lave et al. 2014). Yet, expectations regarding reflexive practice in natural science research should be checked by noting that reflexivity is not universally practiced, nor is it uncontested, even in the social sciences (see e.g. Lynch 2000). While social scientists may be more likely to recognize that knowledge is culturally situated and relative, they often remain reluctant to apply reflexivity to their own knowledge. As noted by Breuer and Roth (2003, N.P.), social scientists have a tendency to "write widely about the constructed nature of knowledge without accepting that their knowledge bears all the marks of construction and subjectivity."

Power and Justice

To meet the power and justice tenet, forest LUL articles must demonstrate an awareness of how relationships of social power shape, or are shaped by, the biophysical landscape. The articles were also required to deliver findings with the potential to service society and/or environment, thus fulfilling the CPG goal of research committed to eco-social transformation (Lave et al. 2014).

Power and justice is the most common CPG tenet in forest LUL research, identifiable in 19 of 40 articles (48%) (Figs. 11.3 and 11.4). In 14 of those 19 articles (74%), research addressing power and justice issues was also transdisciplinary, suggesting that research integrating biophysical and human data was more likely to generate knowledge pertinent to social or ecological causes. Moreover, unlike measures of data use and transdisciplinarity, the power and justice tenet was almost equally prevalent in the 20 most recent articles (9) as in the 20 most-cited articles (10). Together, these results show that colonial history, and the socio-economic, political, and cultural drivers of forest change have received considerable attention in forest LUL research, and that the field has a considerable applied dimension, particularly in the areas of resource and environmental management. It is interesting to note, however, that the relationship between social power and the biophysical landscape was limited to analysis of the effects of the former on the latter: no transdisciplinary articles substantively explored how social relations are affected by the biophysical landscape.

Demonstrating the power and justice tenet are Ponette-González and Fry (2014) from the most recent articles (Fig. 11.3), and Kuemmerle et al. (2011) from the most-cited articles (Fig. 11.4). Ponette-González and Fry (2014) spatially analyze data drawn from historical sources including land-tenure maps and landscape histories, and modern land cover data, to examine the legacy effects of past land tenure on the contemporary landscape in the Xalapa-Coatepec region of Mexico. The exploitation of land and labor by colonial powers and discussion of forest degradation and land reform initiatives provide context for interpreting LULs, with the researchers showing that a better understanding of land-use and forest history produces benefits for market-based environmental protection programs in the region and elsewhere in Latin America. Kuemmerle et al. (2011) examine how the collapse of socialism, which resulted in widespread farmland abandonment and forest expansion, affected net carbon fluxes in western Ukraine. They used satellite-based forest disturbance rates, historic forest resource statistics, and a carbon bookkeeping model, to reconstruct carbon fluxes in Eastern Europe from land-use change in the twentieth century and assess future carbon fluxes in forest expansion and logging scenarios.

Forest Land-Use Legacy Research and Critical Physical Geography

Legacies of past land use can be persistent, with broad-reaching impacts on the contemporary and future state of ecosystems and implications for environmental management (Foster et al. 2003). As this chapter shows, these characteristics apply to forests, which comprise a dominant focus in LUL scholarship and a growing area of academic inquiry. As socio-biophysical entities rooted in complex human histories, comprehensive understanding of forest LULs requires use of diverse forms of biophysical and human data. These qualities make forest LULs a topic well suited to an emergent CPG aimed at drawing together physical and social science in the service of eco-social transformation (Lave et al. Chap. 1 this volume). Although a CPG framework has yet to be explicitly adopted in forest LUL research, studies of legacy effects on forest ecosystems operationalize many CPG practices. Forest LUL research thus demonstrates the feasibility of a CPG framework with its emphasis on cross-disciplinary inquiry, biophysical and human data use, transdisciplinarity, and critical inquiry.

For example, forest LUL scholarship engages researchers from various disciplines and team-science approaches dominate. Although ecology is the foundational science, and natural scientists dominate authorship, forest LUL research is cross-disciplinary, involving scholars from biology, forestry, geography, environmental science, anthropology, and a range of other fields. Most studies involve multiple researchers collaborating and publishing in journals with interdisciplinary and applied research missions. In addition, the use of diverse forms of data, particularly the combined use of biophysical and human data, is a characteristic of CPG that forest LUL scholarship best exemplifies. The vast majority of forest LUL studies analyzed in this chapter employed more than one form of data, and although quantitative and qualitative forms of biophysical data were most commonly used, the majority of articles demonstrated substantive use of both biophysical and human data. Relatedly, a significant number of forest LUL studies were also transdisciplinary, substantively interweaving biophysical and human data at the analytical level. Another strength that parallels concerns in CPG is a commitment to power and justice issues. Almost half of the forest LUL studies analyzed paid substantive attention to socio-economic, political and cultural drivers of landscape change, and had applied research outcomes.

Collectively, these research commitments not only demonstrate the utility of the CPG perspective in analyzing socio-biophysical landscape entities, but also confirm CPGs suitability as a methodological framework for advancing forest LUL scholarship. Reinforcing this claim is the tendency of the 20 most-cited and assumedly more consequential articles in the sample to use more diverse forms of data and have a higher incidence of CPG tenet use.

This review of forest LUL scholarship through a CPG lens also reveals areas for improvement in this body of work, as well as challenges for the emergent sub-discipline. To a significant degree, forest LUL scholarship already cuts across the natural and social sciences. However, explicit adoption of a CPG framework would better emphasize that LULs are eco-social entities whose understanding *must* include rigorous analysis of human data. It would also draw greater numbers of social scientists into collaborative LUL research improving transdisciplinarity. Collaboration with social sciences would enhance the interweaving of human data and methods in forest LUL research, particularly socio-economic, demographic, or other forms of quantitative human data and analysis useful for complementing, expanding, or triangulating biophysical results.

Greater collaboration with social scientists would also improve LUL research by enhancing adoption of critical concerns and practices, which were found to be absent (in the form of CPG "tenets") from almost half of all forest LUL studies analyzed. The influence of the biophysical landscape on societal relations, for example, appears to be uncharted territory in forest LUL research that could be addressed through collaboration with human geographers trained in critically analyzing the role landscape plays in social and cultural reproduction. The absence of reflexive research practices is also notable, and adoption of a CPG framework would enhance understanding of the complexities of knowledge production in LUL research. But, as the CPG tenet most deeply embedded in critical social theory, the absence of reflexivity may be a cautionary sign for CPG: adoption of critical methodologies epistemologically distant from traditional approaches to natural science, and in the case of reflexivity not consistently utilized in social science itself, may be particularly difficult to achieve. Conversely, critical practices that natural scientists are already familiar with may be more readily embraced. For example, for researchers studying forest LULs from interdisciplinary fields like forestry, geography, and environmental science, consideration of socio-economic and political relationships driving ecological change and the generation of applied research outcomes are established traditions. There is thus much potential for CPG to inform LUL research on legacy effects of humans on ecosystems.

Notes

1. Although scholarly use of the term "land-use legacies" is relatively new, interest in human-landscape agency has long been pursued in environmental history, ecological anthropology, paleoecology, and geography. Awareness that environmental conditions are products of past societal interactions is also a unifying theme in historical ecology (McClenachan et al. 2015) and land change science (Turner et al. 2007).

2. All *Web of Science* article searches were conducted on May 6 2016. Studies were defined as having a substantive focus on land-use legacies if the term and its

variants (e.g. ["land-use" OR "land use"] AND ["legacies" OR "legacy"]) appeared in the article title or topic.

3. Studies were determined to have a forest focus if the study site was currently or formerly heavily treed, vegetation status was part of the study, or forest was a major land cover assessed.

4. No overlap existed in the 20 most frequently cited articles which were published between 1994 and 2014, and the 20 most recent articles which were published between 2014 and 2016.

5. Criteria for analyzing reflexive research practices in LUL scholarship are restricted to the researchers' positionality. This is a narrower conceptualization of reflexivity than that discussed by Tadaki et al. (2015), who appeal for the development of a reflexive critical disposition toward the politics of knowledge production in physical geography more broadly, and not just within an integrative Critical Physical Geography subfield.

6. Although positivistic assumptions also underpin research in the social sciences (della Porta and Keating 2008), post-positivistic practices like reflexivity emerged as a response to positivistic research practices. Social theory-informed approaches are now common in the social sciences where the degree to which the world, "is real and objective, endowed with an autonomous existence outside the human mind and independent of the interpretation given to it by the subject (della Porta and Keating 2008, 22)," has received greater attention than in the natural sciences.

References

Ameztegui, A., L. Coll, L. Brotons, and J.M. Ninot. 2016. Land-use legacies rather than climate change are driving the recent upward shift of the mountain tree line in the pyrenees. *Global Ecology and Biogeography* 25: 263–273.

Atkinson, E.E., and E. Marín-Spiotta. 2014. Land use legacy effects on structure and composition of subtropical dry forests in St. Croix, U.S. Virgin Islands. *Forest Ecology and Management* 335: 270–280.

Beard, K.H., K.A. Vogt, D.J. Vogt, F.N. Scatena, A.P. Covich, R. Sigurdardottir, T.G. Siccama, and T.A. Crowl. 2005. Structural and functional responses of a subtropical forest to 10 years of hurricanes and droughts. *Ecological Monographs* 75: 345–361.

Blixt, T., K.-O. Bergman, P. Milberg, L. Westerberg, and D. Jonason. 2015. Clear-cuts in production forests: From matrix to neo-habitat for butterflies. *Acta Oecologica* 69: 71–77.

Breuer, F., and W. Roth. 2003. Subjectivity and reflexivity in the social sciences: Epistemic windows and methodical consequences. *Forum: Qualitative Social Research* 4: Article 25. http://nbn-resolving.de/urn:nbn:de:0114-fqs0302258

Bryman, A. 2006. Integrating quantitative and qualitative research: How is it done? *Qualitative Research* 6: 97–114.

Chai, S.-L., and E.V.J. Tanner. 2011. 150-year legacy of land use on tree species composition in old-secondary forests of Jamaica. *Journal of Ecology* 99: 113–121.

Clifford, M.J., N.S. Cobb, and M. Buenemann. 2011. Long-term tree cover dynamics in a pinyon-juniper woodland: Climate-change-type drought resets successional clock. *Ecosystems* 14: 949–962.

Cloke, P., C. Philo, and D. Sadler. 1991. *Approaching human geography: An introduction to contemporary theoretical debates.* New York: The Guilford Press.

Della Porta, D., and M. Keating, eds. 2008. *Approaches and methodologies in the social sciences: A pluralistic perspective.* Cambridge: Cambridge University Press.

Dickens, S.J.M., S. Mangla, K.L. Preston, and K.N. Suding. 2016. Embracing variability: Environmental dependence and plant community context in ecological restoration. *Restoration Ecology* 24: 119–127.

Foster, D.R., M. Fluet, and E.R. Boose. 1999. Human or natural disturbance: Landscape-scale dynamics of the tropical forests of Puerto Rico. *Ecological Applications* 9: 555–572.

Foster, D.R., F. Swanson, J. Aber, I. Burke, N. Brokaw, D. Tilman, and A. Knapp. 2003. The importance of land-use legacies to ecology and conservation. *BioScience* 53: 77–88.

Franklin, J. 2007. Recovery from clearing, cyclone and fire in rain forests of Tonga, South Pacific: Vegetation dynamics 1995–2005. *Austral Ecology* 32: 789–797.

Gragson, T.L., and P.V. Bolstad. 2006. Land use legacies and the future of Southern Appalachia. *Society and Natural Resources* 19: 175–190.

Greene, J.C., V.J. Caracelli, and W.F. Graham. 1989. Toward a conceptual framework for mixed-method evaluation designs. *Educational Evaluation and Policy Analysis* 11: 255–274.

Hahn, P.G., and J.L. Orrock. 2015. Land-use legacies and present fire regimes interact to mediate herbivory by altering the neighboring plant community. *Oikos* 124: 497–506.

Inkpen, R. 2005. *Science, philosophy and physical geography.* New York: Routledge.

Kepfer-Rojas, S., K. Verheyen, V.K. Johannsen, and I.K. Schmidt. 2015. Indirect effects of land-use legacies determine tree colonization patterns in Abandoned Heathland. *Applied Vegetation Science* 18: 456–466.

Kirkman, L.K., R.F. Lide, G. Wein, and R.R. Sharitz. 1996. Vegetation changes and land-use legacies of depression wetlands of the Western Coastal Plain of South Carolina: 1951–1992. *Wetlands* 16: 564–576.

Kuemmerle, T., P. Olofsson, O. Chaskovskyy, M. Baumann, K. Ostapowicz, C.E. Woodcock, R.A. Houghton, P. Hostert, W.S. Keeton, and V.C. Radeloff. 2011. Post-soviet farmland abandonment, forest recovery, and carbon sequestration in Western Ukraine. *Global Change Biology* 17: 1335–1349.

Kuhman, T.R., S.M. Pearson, and M.G. Turner. 2011. Agricultural land-use history increases non-native plant invasion in a southern Appalachian forest a century after abandonment. *Canadian Journal of Forest Research* 41: 920–929.

Lave, R. 2015. Introduction to special issue on Critical Physical Geography. *Progress in Physical Geography* 39: 571–575.

Lave, R., M.W. Wilson, E.S. Barron, C. Biermann, M.A. Carey, C.S. Duvall, L. Johnson, et al. 2014. Intervention: Critical Physical Geography. *Canadian Geographer* 58: 1–10.

Lu, X., D.W. Kicklighter, J.M. Melillo, J.M. Reilly, and L. Xu. 2015. Land carbon sequestration within the conterminous United States: Regional- and state-level analyses. *Journal of Geophysical Research G: Biogeosciences* 120: 379–398.

Lynch, M. 2000. Against reflexivity as an academic virtue and source of privileged knowledge. *Theory, Culture and Society* 17: 26–54.

Malterud, K. 2001. Qualitative research: Standards, challenges and guidelines. *The Lancet* 358: 483–488.

Mattingly, W.B., J.L. Orrock, C.D. Collins, L.A. Brudvig, E.I. Damschen, J.W. Veldman, and J.L. Walker. 2015. Historical agriculture alters the effects of fire on understory plant beta diversity. *Oecologia* 177: 507–518.

McClenachan, L., A.B. Cooper, M.G. McKenzie, and J.A. Drew. 2015. The importance of surprising results and best practices in historical ecology. *BioScience* 65: 932–939.

McDonald, R.I., G. Motzkin, and D.R. Foster. 2008. Assessing the influence of historical factors, contemporary processes, and environmental conditions on the distribution of invasive species. *Journal of the Torrey Botanical Society* 135: 260–271.

McEwan, R.W., J.M. Dyer, and N. Pederson. 2011. Multiple interacting ecosystem drivers: Toward an encompassing hypothesis of oak forest dynamics across Eastern North America. *Ecography* 34: 244–256.

Mesquita, R.D.C.G., P.E.D.S. Massoca, C.C. Jakovac, T.V. Bentos, and G.B. Williamson. 2015. Amazon rain forest succession: Stochasticity or land-use legacy? *BioScience* 65: 849–861.

Montello, D.R., and P.C. Sutton. 2006. *An introduction to scientific research methods in geography*. London: Sage Publications.

Munteanu, C., T. Kuemmerle, M. Boltiziar, V. Butsic, U. Gimmi, Lúboš Halada, D. Kaim, et al. 2014. Forest and agricultural land change in the carpathian region-A meta-analysis of long-term patterns and drivers of change. *Land Use Policy* 38: 685–697.

Munteanu, C., T. Kuemmerle, N.S. Keuler, D. Müller, P. Balázs, M. Dobosz, P. Griffiths, et al. 2015. Legacies of 19th century land-use shape contemporary forest cover. *Global Environmental Change* 34: 83–94.

Napton, D.E., R.F. Auch, R. Headley, and J.L. Taylor. 2010. Land changes and their driving forces in the Southeastern United States. *Regional Environmental Change* 10: 37–53.

Oswalt, C.M., S. Fei, Q. Guo, B.V. Iannone III, S.N. Oswalt, B.C. Pijanowski, and K.M. Potter. 2015. A subcontinental view of forest plant invasions. *NeoBiota* 24: 49–54.

Perring, M., Pieter De Frenne, Lander Baeten, Sybryn Maes, Leen Depauw, Haben Blondeel, María Mercedes Carón, and Kris Verheyen. 2016. Global environmental change effects on ecosystems: The importance of land-use legacies. *Global Change Biology* 22: 1361–1371.

Plieninger, T., H. Schaich, and T. Kizos. 2011. Land-use legacies in the forest structure of silvopastoral oak woodlands in the Eastern Mediterranean. *Regional Environmental Change* 11: 603–615.

Plue, J., M. Hermy, K. Verheyen, P. Thuillier, R. Saguez, and G. Decocq. 2008. Persistent changes in forest vegetation and seed bank 1,600 years after human occupation. *Landscape Ecology* 23: 673–688.

Ponette-González, A.G., and M. Fry. 2014. Enduring footprint of historical land tenure on modern land cover in Eastern Mexico: Implications for environmental services programmes. *Area* 46: 398–409.

Ponette-González, A.G., H.A. Ewing, M. Fry, and K.R. Young. 2016. Soil and fine root chemistry at a tropical andean timberline. *Catena* 137: 350–359.

Rhemtulla, J.M., D.J. Mladenoff, and M.K. Clayton. 2007. Regional land-cover conversion in the U.S. Upper Midwest: Magnitude of change and limited recovery (1850-1935-1993). *Landscape Ecology* 22: 57–75.

———. 2009. Legacies of historical land use on regional forest composition and structure in Wisconsin, USA (Mid-1800s-1930s-2000s). *Ecological Applications* 19: 1061–1078.

Tadaki, M., G. Brierley, M. Dickson, R. Le Heron, and J. Salmond. 2015. Cultivating critical practices in physical geography. *Geographical Journal* 181: 160–171.

Tayyebi, A., B.C. Pijanowski, and B.K. Pekin. 2015. Land use legacies of the Ohio river basin: Using a spatially explicit land use change model to assess past and future impacts on aquatic resources. *Applied Geography* 57: 100–111.

Thompson, J.R., D.R. Foster, R. Scheller, and D. Kittredge. 2011. The influence of land use and climate change on forest biomass and composition in Massachusetts, USA. *Ecological Applications* 21: 2425–2444.

Tokuoka, Y., and K. Hashigoe. 2014. Effects of stone-walled terracing and historical forest disturbances on revegetation processes after the abandonment of mountain slope uses on the Yura Peninsula, Southwestern Japan. *Journal of Forest Research* 20: 24–34.

Torres, I., B. Pérez, J. Quesada, O. Viedma, and J.M. Moreno. 2016. Forest shifts induced by fire and management legacies in a *Pinus pinaster* woodland. *Forest Ecology and Management* 361: 309–317.

Tulowiecki, S.J., and C.P.S. Larsen. 2015. Native American impact on past forest composition inferred from species distribution models, Chautauqua county, New York. *Ecological Monographs* 85: 557–581.

Turner, B.L., II, E.F. Lambin, and A. Reenberg. 2007. The emergence of land change science for global environmental change and sustainability. *PNAS* 104 (52): 20666–20671.

Uriarte, M., C.D. Canham, J. Thompson, J.K. Zimmerman, L. Murphy, A.M. Sabat, N. Fetcher, and B.L. Haines. 2009. Natural disturbance and human land use as determinants of tropical forest dynamics: Results from a forest simulator. *Ecological Monographs* 79: 423–443.

Valencia, V., S. Naeem, L. García-Barrios, P. West, and E.J. Sterling. 2016. Conservation of tree species of late succession and conservation concern in coffee agroforestry systems. *Agriculture, Ecosystems and Environment* 219: 32–41.

Wallin, D.O., F.J. Swanson, and B. Marks. 1994. Landscape pattern response to changes in pattern generation rules: Land-use legacies in forestry. *Ecological Applications* 4: 69–580.

Warren, R.J. II. 2016. Ghosts of cultivation past—Native American dispersal legacy persists in tree distribution. *PLoS One* 11 (3): e0150707.

Williams, M. 1989. *Americans and their forests: A historical geography*. New York: Cambridge University Press.

12

Critical Invasion Science: Weeds, Pests, and Aliens

Christian A. Kull

Introduction

In July 2011, *Nature* magazine printed several irate responses to an article about the science of biological invasions by Mark Davis entitled "Don't judge species on their origins". The first response, led by the eminent scholar Daniel Simberloff, was titled threateningly "Non–natives: 141 scientists object". The spat has since widened. Science writers have published books with titles like *The New Wild: Why Invasive Species Will Be Nature's Salvation*, *Where Do Camels Belong? Why Invasive Species Aren't All Bad*, and *Rambunctious Garden: Saving Nature in a Post-Wild World*, while invasion scientists have defended their field with journal articles such as "Misleading criticisms of invasion science: a field guide" and "The rise of invasive species denialism" (Fig. 12.1).[1]

What is going on here? The movements of plants and animals from one part of the world to another, their establishment and success in new environments, and their impacts on host communities would appear to be a fascinating, yet solidly scientific endeavor. A glance at the titles above, however, shows that the debate is more than scientific—it is about terminology, about values, about politics. It appears, then, that calls for a "Critical Physical Geography" (CPG) have emerged at the right time for studies of invasive species, whether in biogeography or elsewhere across the natural and social sciences.

C. A. Kull (✉)
Institut de géographie et durabilité, Université de Lausanne, Lausanne, Switzerland

© The Author(s) 2018
R. Lave et al. (eds.), *The Palgrave Handbook of Critical Physical Geography*,
https://doi.org/10.1007/978-3-319-71461-5_12

Fig. 12.1 Texts, reviews, and popular science books involved in recent debates over the terminology, politics, and values in the study of biological invasions.

According to the introductory chapter in this handbook, CPG has three core intellectual tenets: (1) that landscapes are not just biophysical but deeply shaped by human actions and structural inequalities, (2) that power relations affect who studies landscapes and how, and (3) that the resulting knowledge has deep impacts on lives and landscapes. These tenets are strongly applicable to the science of biological invasions. Let me illustrate with what might be one of the most striking examples: South Africa (Fig. 12.2).

Tenet 1: Invasion landscapes deeply shaped by human actions and structural inequalities. Most problematic alien invasive species in South Africa were introduced in the colonial era, with goals related to utility (e.g., timber resources), land rehabilitation (e.g., dune stabilization), or science and aesthetics (e.g., botanic gardens and personal gardens). Colonization- and Apartheid-era policies not only separated people but also created starkly disjunctive landscapes whose legacies endure today (from peri-urban townships to rural "homelands"; large properties for farming, forestry, and game ranching; suburban estates; conservation zones) and which form the matrix across which invasive species spread and are managed (van Wilgen et al. 2011; van Wilgen and Richardson 2012).

Tenet 2: Power relations affect who studies invasions, how, and the questions asked. South Africa has been a global leader in invasion biology, with strong

Fig. 12.2 Scenes of plant invasion in South Africa: (a) dense acacia brush in the foothills above Muizenberg and False Bay, Western Cape province; (b) lone black wattle shrub and woodlot behind hut near the Swazi border, Mpumalanga province; (c) cattle enclosure made from black wattle near Butterworth, former Transkei, Eastern Cape province; (d) public works labourer controlling lantana infestation, also near Butterworth.

historical roots in Cape Town-based circles of botanists, naturalists, and foresters (Pooley 2014). I may be sticking my foot in it to say this, but it is largely a "white" science—see for instance the core staff of the world-renowned Centre for Invasion Biology at Stellenbosch University. To be clear, this is not an accusation as I recognize the historical path dependency, the very engaged stances of many of these researchers, and the structural difficulties of attracting students from previously disadvantaged backgrounds to this field of study (but, this last element is perhaps precisely the point I am making). A generation has passed since the end of Apartheid, but according to some observers, "environmental engagement does not transcend but rather pronounces ecological and social inequities" (Carruthers et al. 2011; Bennett 2014; Green 2014; Lidström et al. 2015, p. 21). Significant research has been undertaken on the impacts of biological invasions for poorer, more marginal South Africans (e.g., Shackleton et al. 2007; Shackleton et al. 2015; Mukwada et al. 2016), yet problem framings until more recently largely started with the biological and hydrological impacts of invasions, not with the concerns of rural people.

Tenet 3: Resulting knowledge has deep impacts on invasion landscapes and lives of people. The science produced on invasions in South Africa has numerous direct impacts. Biological control programs (releasing insects and pathogens to control invasions) date back to early twentieth-century struggles with prickly pear cactus (Beinart and Wotshela 2011) and continue today, sometimes dramatically shaping the ecology of invasive plant communities and leading to complaints by local users (Shackleton et al. 2007). The Working for Water program, a major post-Apartheid government program for job creation and ecological restoration paid and trained tens of thousands of poor black South Africans in mechanical and chemical control of invasive species (Turpie et al. 2008; Neely 2010; van Wilgen et al. 2011; Lidström et al. 2015). There is no doubt that the science has contributed to important impacts on many plants (uprooted), people (employed), resource-based livelihoods (disrupted), and landscapes (transformed).

In this chapter, I argue for the necessity of a critical approach to the study of invasive plants and animals. I first explore what a "critical" invasion science means. Then the bulk of the chapter investigates four main aspects of invasion science ripe for critical analysis: the history of the science, the terminology, the categories, and the social-political-ethical context. I conclude with four proposals for further work in critical invasion science and a review of the questions it might ask.

The Emergence of a Critical Approach

First, however, I pass by a necessary detour. What is "Critical Physical Geography", or why have I titled this chapter "critical invasion science"? Like many a good term, "critical" can mean different things in different contexts. As the *Oxford English Dictionary* notes, while "critical" may be defined most commonly as "given to judging, esp. given to adverse criticism", it has also meant "exercising careful judgment or observation" and has been specifically associated with the Frankfurt School of social theory and philosophy and its engagement with thinkers like Kant, Hegel, Marx, Weber, and Freud. This is the sense informing the use of the term in geography, where "critical" has gone on to signify scholarship informed by social theories, particularly with a sensibility to emancipatory forms of politics and also a "deconstructive impulse" with respect to scientific knowledge (Forsyth 2003; Gregory et al. 2009, p. 121).

"Critical Physical Geography" was proposed by Lave et al. (2014). The fact that this idea took hold reflects, in my opinion, three broader phenomena. The first is the commitment by practitioners to an integrative discipline, a rear-guard defense against trends of splitting human from physical geography. Such trends are deeply ironic given endless calls for more interdisciplinarity, especially at the boundaries of nature and society (cf Kull and Rangan 2016). In this sense, CPG emerges as a call for revitalizing some of the integration across physical and human that makes geography whole.

Second, political ecologists are relatively prominent among early enthusiasts for CPG. My intuition is that this enthusiasm by political ecologists for CPG is due to the fact that CPG lays claim to and gives visibility to territory that has been increasingly marginalized in political ecology. To put it bluntly: for people with a foot in both the natural sciences and critical social sciences, political ecology has been a productive home. Yet political ecology has over time more and more emphasized the social side over the natural science. So CPG gives an alternative home for people of a political ecological spirit with a real commitment to the natural sciences. As such, CPG shares many key elements with political ecology, including the three tenets referred to in my introduction, but also an epistemological "double posture" (cf. Gautier and Benjaminsen 2012; Robbins 2012). That is, it takes seriously the knowledge created by the natural sciences at the same time as deconstructing the categories and the authority of these sciences.

The third phenomenon I see as contributing to the resonance of Critical Physical Geography might be the rise of the label *science*. In my view, "critical" is a necessary antidote to "science", to question the modernist scientific separation of nature and society and the power relations in the production of knowledge. Let me explain. In the past decade or two, it has become trendy to re-label various domains of inquiry with the epithet "science".[2] We now have, for instance, *conservation science* (Kareiva and Marvier 2012), *land change science* (Gutman et al. 2004), *sustainability science* (Kates et al. 2001), *resilience science* (Leslie and Kinzig 2009), *vulnerability science* (Cutter 2003), and, of course, *invasion science* (Richardson 2011b). Two contradictory trends seem to be pushing this fad. One is the use of "science" to replace "biology" or other disciplinary epithets in order to represent an interdisciplinary spirit, particularly across a natural-social divide. The second trend is the recourse to "Science-with-a-capital-S" to assert a sense of authority, and sometimes to draw a line between "sciency" epistemologies (whether natural science or social science) and other interpretive or critical approaches.

The interdisciplinarity across the nature-society divide in these "sciences" is often couched in the language of "coupled systems". In many cases, there is an implicit assumption that the natural sciences will set the "factual parameters", whereas the humanities and social sciences will assess impacts, develop solutions, and convince people of the issue at hand (Demeritt 2009, p. 128; Lidström et al. 2015, p. 9). In invasion science, in particular, the feeling that sometimes emerges is that social scientists are invited to collaborate only to help invasion scientists find out how to cope with attitudes, values, and perceptions among the public which conflict with the biologists' goals of managing invasive species (Lidström et al. 2015). For instance, Estévez et al.'s (2015) excellent review of socio-cultural factors in conflicts over invasions nonetheless concludes that more effective risk communication from scientists to the public will help avoid conflict (and, implicitly, allow the fight against invasives to continue). This model of knowledge creation and action—where science creates privileged knowledge that then calls on social sciences to help apply this knowledge—is the opposite of a "critical" approach. The consequence is that certain forms of knowledge production and communication are excluded, and such coupled approaches can become apolitical, technical approaches indisposed to interrogating the social assumptions, values, and power relations that underlie them. Hence the need for a "critical" approach to counter-balance the "science" approach, not only to do science but also to interrogate unstated power relations, categories, and ideologies.

In the next four sections, I seek to do exactly this for the study of biological invasions. I begin by briefly placing the field in its context: where it comes from, and what inherited assumptions or path dependencies it gains from its particular historical roots. Then I investigate a series of debates in invasion science (including those mentioned in the introduction) that are ripe for CPG-style reflection and critique.[3] For each, I demonstrate what is at stake and why it matters.

Invasions: History of a Science

The ability to take a step backwards, to gain perspective, is crucial to a critical reading of what a particular science is doing and why (see Davis, Marchesi, and Sayre, this volume). For instance, Thomas Malthus's widely known theories on population growth take on a different significance if one takes into account the fact that he developed his ideas as a politically engaged actor in the context of a crowded, burgeoning eighteenth-century London experiencing the birth pains of the industrial revolution and an associated urban proletariat. The same goes for the study of invasive species. Indeed, perhaps tellingly, in the nineteenth century, a dominant scientific approach to many plants and animals now considered invasive was *acclimatization*. Associated with colonialism and settler societies, acclimatization sought to "improve" environments by purposefully introducing and propagating alien plants and animals: rabbits, willows, and trout in Australia and eucalypts and acacias around the Mediterranean (Osborne 2000).

The study of plants, animals, and other organisms that are "out of place"—their characteristics, the causes of their displacement, their behavior in their new environments, their impacts on ecology and economy, people's reactions, and management strategies—not coincidentally goes back to the same historical period. Histories of invasion biology and allied fields have already been written (Davis 2009; Johnson 2010; Chew and Hamilton 2011; Richardson 2011a; Frawley and McCalman 2014; Vaz et al. 2017); here I highlight some key factors shaping the nature of the field and its assumptions.

The modern field of invasion biology dates to the 1980s. Large international research consortia served as catalysts, such as the international "SCOPE 37"[4] research program launched in 1982 (Drake et al. 1989; Simberloff 2011a, 2013; Kull and Rangan 2015). In the 1990s, the field was institutionalized into science, policy, and programs. Publications on invasions grew tenfold; new specialized journals like *Biological Invasions* and *Diversity and Distributions* were founded under field-leading editors and SCOPE participants Daniel Simberloff and David Richardson; governments funded programs like the European Commission's project to inventory invasive species (DAISIE) or the Global Invasive Species Programme (GISP), as well as diverse national and international legislation (Vaz et al. 2017).

The invasion biology field builds on a variety of practical and intellectual heritages. Some of the categories and terminologies of invasion—which as we will see later, are highly contested—draw on ideas of European naturalists working in peri-urban countrysides in the 1800s and early 1900s, such as Hewett Watson and Albert Thellung (Chew and Hamilton 2011; Kowarik and Pyšek 2012). The field of weed science, which crystallized with the edition of a field-defining textbook in 1942 categorizing weed types as well as focusing on practical control strategies, was explicitly designated as a stepping-stone for the SCOPE program (Kull and Rangan 2015). In the post-war period, weed sciences and weed services boomed with the conversion of wartime industries into the production of fertilizer and chemical herbicides; this martial legacy shadows invasion biology to this day (Atchison and Head 2013; Tassin 2017). Finally, it is common to refer to Oxford biologist Charles Elton, who published the prescient *Ecology of Invasions by Plants and Animals* in 1958, as the field's father or prophet, despite the 30-year gap between his work and the efflorescence of the field (Simberloff 2011a; Chew 2015; Vaz et al. 2017). Elton's work, publicized via BBC radio broadcasts, set a pattern of use of military metaphors in describing invasions.

The SCOPE program re-directed and applied these inherited concepts and approaches to the study of natural ecosystems and biodiversity (e.g., Drake et al. 1989; Cronk and Fuller 1995). From the 1980s, studies of invasion in natural areas took off. It was increasingly informed by the broader field of ecology's relative disengagement from anthropic landscapes. Most invasion biology work in the 1990s and 2000s, for instance, largely ignored cities and other strongly humanized areas, despite the irony that the objects of study were human-introduced species (Salomon Cavin and Kull 2017).

Invasions: Words and Labels

An aspect of invasion biology that has already received a lot of critical attention is the vocabulary of "alien invasive species". A lot of ink has been spilt about the war metaphors of enemy *invasion*, such as those used by Elton in his 1958 tome, and the term *alien* with its resonance in both pop culture and immigration politics. My point is not to rehash these debates (see, for instance, Peretti 1998; Comaroff and Comaroff 2001; Subramaniam 2001; Simberloff 2003; Larson 2005; Warren 2007; Davis 2009; Kull and Rangan 2015) but to point out the importance of attention to labels and language in highlighting potential underlying assumptions, potential foregone conclusions, in scientific studies. Humans are of course used to words having different meanings in different contexts (e.g., *invasion* as a military term vs. as a medical term); any other term one could choose would come with its own baggage (for instance, *colonizing* or *pioneer* species carry their own metaphorical echoes). But a critical approach might follow the practical advice of Kueffer and Larson (2014) in evaluating metaphors in terms of factual correctness, socially acceptable language, neutrality, and transparency. It would then go further to evaluate what kinds of concrete impacts the choice of terms actually has on the conduct of science, on the framing of policy debates, and on practical management actions (Larson 2011).

Practitioners of invasion biology are highly aware of the rhetorical and ideological importance of labels. This applies even to the name of the field itself. While the field has for several decades passed under the label invasion biology (or invasion ecology), two new field names were recently proposed. The first is "species redistribution ecology", or SPRED ecology, proposed in a textbook titled *Invasion Biology* that bravely sought to abolish its own name (Davis 2009). Davis's proposal is based on the argument that the fundamental object of study of invasion biology—how and why species spread and move—falls within community ecology and biogeography and does not merit a different field.[5] He notes that the term "invasion" is too emotive and that too much unfounded stock is put on the distinction between native and alien. The neologism SPRED ecology has, however, not taken off.[6] It is, perhaps, hampered by its narrow focus and by the fact that it was promoted by an author somewhat outside the mainstream (Davis was the author to whom 141 scientists objected in this chapter's opening paragraph).

In contrast, the successful term "invasion science" comes from a top scientist in the field, David Richardson, visionary leader of the world-renowned Centre for Invasion Biology at Stellenbosch University, alumnus of the SCOPE program, long-time editor of *Diversity and Distributions*, and editorial board member of *Biological Invasions* (Richardson 2011b). As noted in the section "The Emergence of a Critical Approach", the epithet "science" appears to carry a particular strategic ambit of legitimacy attached to the authority of science. In addition, the replacement of "biology" with "science" also tries to signal a broadening of the scope of the field from just biological aspects of invasions to concerns with costs and benefits and human value systems (Richardson and Ricciardi 2013). The ecumenical focus of the title "invasion science", its links to institutional centers of power, and its shying away from thorny debates over terminology would seem to explain the label's success. Thus the contrast between the stories of the two neologisms—one stuck in the starting blocks, the other running to an early lead—merits CPG-style attention, as it demonstrates not only the importance of power and networks in the production of scientific knowledge but also the stickiness of paradigmatic concepts.

Invasions: Categories

Defining *what* one is studying is crucial to any science. Yet, *how* this object is defined involves choices and boundaries, and these have consequences—on the science that is done, and on the ways in which it is relevant to policy and management. As Nathan Sayre has noted, scientific categories should not be taken for granted (Sayre 2015). *Invasive species* is the central category for the type of research this chapter engages with. But the definition of this term is far from settled. There are three ideal-type concepts that appear singly or in various combinations in most definitions of invasive species (cf. Williamson 1996, p. 58–59; Boonman-Berson et al. 2014; Kull et al. 2014; Tassin 2014).

First, some definitions emphasize *origins*. In this model, an invasive species is an *alien*, that is, a species that comes from elsewhere. This definition emphasizes the crossing of some biogeographical barrier (Richardson et al. 2000). This definition carries an unexamined ideology of natural purity and nativeness and is troubled by a black-and-white dichotomy between alien and native that in many cases is not so clear—there are quite a few species whose origins or original distributions are unclear (Bean 2007). To overcome these issues some definitions further specify that transport of the species has to be at the hands of humans (Richardson et al. 2000); if a species arrives naturally, then it does not count. While practical, this potentially introduces an ideology that humans are separate from nature.

A second set of definitions emphasize *behavior*. In this model, an invasive species is an *invader*, one that gains terrain, spreads quickly, and becomes dominant in a given ecosystem (Valéry et al. 2008). On its own, this definition poses the problem of distinctions of temporal and spatial scale: what distinguishes an invader, then, from a pioneer species or a colonizer? (Hoffmann and Courchamp 2016a).

The third set of definitions emphasize *impacts*. In this model, an invasive species is a *weed* or a *pest*, one that has negative impacts on native vegetation or on society, public health, or the economy (McNeely 2001; Simberloff et al. 2013). This is a value judgment, which raises the question of how this value is determined, by whom, or from what perspective. It also predisposes the field toward an investigation of only the negative impacts and not the positive (Tassin and Kull 2015).

Definitions of the invasive species concept have been hotly debated in the field (Colautti and Richardson 2009; Blackburn et al. 2011); in that sense, CPG-style work has been initiated. But definitions are often not made explicit in studies, with consequences on the types of conclusion that become possible and the implicit judgments behind them. For instance, a study based on the assumption that invasives must be alien might miss a native species that—for whatever reason (climate change, human disturbance)—currently acts as a landscape transformer. Or, an article surveying a taxa or a region to establish an inventory of invasive species might include in this inventory any species listed by a scientific study or expert opinion as "invasive" without regard to the definition used (as I did myself in a survey of introduced plants in Madagascar: Kull et al. 2012). This potentially mixes together plants from elsewhere with noxious weeds and those that spread quickly, hiding large differences in ecological processes and human interactions.

In addition to defining what *invasive* means, one must also consider what the implications are of selecting *species* as the central unit of analysis. A critical approach contributes to highlighting the advantages and disadvantages, winners and losers, or hidden assumptions behind the choice of units of analyses. In invasion studies, it has long been noted, for instance, that it is particular *populations* of a species, in specific contexts, which are invasive, not the species itself (Colautti and MacIsaac 2004). The Monterey pine, or radiata, is a case in point: it is endangered in its native habitat in California, but invasive in numerous places around the southern hemisphere where it has been grown for forestry. The impression given in study after study is that it is the *biological species* that is invasive, not particular populations in particular contexts. This results in online lists and databases of flora and fauna that typically list invasive species abstracted from their geographic context. Regional listings are often agglomerated to larger regional or national scales (a plant exhibiting weedy behavior in Miami is listed as invasive in Florida, and thus in the United States). So in many cases one can quickly find on the Internet or in scientific publications that species A is "reported invasive in country X, Y, and Z" even though the inclusion of some of those countries might involve very minor populations. This results in lists of invasive species in online databases or legislative appendices that forbid transport, restrict cultivation, or mandate eradication with sometimes little attention to context. What appears as a precautionary principle to some might constrain legitimate choices for others.

A further consequence of the focus on species is that it distracts from the processes favoring invasion. To illustrate, take the case of a variety of often thorny American bushes—such as *Lantana camara, Acacia farnesiana, Mimosa pigra, Leucaena leucocephala, Prosopis* spp.—that are widely seen as invasive species across the sub-humid and semi-arid tropics of the eastern hemisphere. With numerous publications and reports listing these and other species, the implicit message is that it is their fault and that they are the entities that must be controlled. Yet these species were transported (originally) by humans, and they tend to be present in environments rendered "invasible" by human actions: by our lighting of grass or forest fires, by our grazing practices, or by our introduction of seed dispersers, like the common mynah bird. The outcome is that invasion biologists and environmental managers address invasive *species* more than arguably more relevant populations, human disturbances, or specific places.

An alternative to species-based approaches—an alternative that should be of particular interest to geographers—is a place-based approach. Together with Jacques Tassin, I suggested in an earlier article that:

> Instead of using an *a priori* judgment to call for a blanket ban of a wide array of plant species, the focus should be on the processes that societies (communities, governments, agencies) use to anticipate and debate the changes to landscapes and human lives that are possible outcomes of specific plant introductions and diffusion in specific places. Who are the winners and losers, now and in the foreseeable future? ... Who has the right to decide, and the might to enforce? .... We suggest an evidence-based, context-specific, socially-negotiated approach .... The judgment of 'weed versus useful' should not be

made at a global level, it should remain contextual to local and regional scales, to particular ecosystems and landscapes, particular economies and socio-political situations. (Kull and Tassin 2012, pp. 2230 and 2232)

Recent work in the Australian outback shows how prioritizing "place-based" management over the species-based management imposed by government interpretations of invasion science could better address Aboriginal cultural issues, budgetary constraints, and on-the-ground outcomes (Bach 2015; Bach and Larson 2017). Critical work could further question the choice of scales and units of analyses and how they shape scientific, social, and practical outcomes.

Invasions: Social, Political, Ethical Dimensions

Efforts to prevent, control, or eradicate particular invasions can be embroiled in a variety of conflicts. These include struggles over priorities, funding, responsibility, worldviews, ethics, and more. As with any intervention, there will be winners and losers. In a number of cases, for instance, the livelihoods of certain members of rural communities have become dependent on invasive species, whether for fodder, woodfuel, or food, to the point that the removal of the invader would have negative livelihood outcomes (Shackleton et al. 2007; Ellender et al. 2010; Kull et al. 2011; Middleton 2012). In other contexts, the invasive species is more broadly disliked for its negative impacts on livelihoods (Awanyo 2001; Mwangi and Swallow 2008). A critical invasion science engages with these conflicts and builds on them to guide research. I illustrate this with two examples: the question of toxic chemicals and the question of labor.

First, a major conundrum in invasion science is the battling of one environmental evil (invasive species) with another (chemical poisons). How does one balance a desire to combat invasions using herbicides and pesticides with the resultant environmental pollution, and with the mortality and suffering of sentient beings (in the case of invasive animals)? According to Jacques Tassin (2017), this ethical quandary has not been adequately and openly addressed in invasion studies (cf. Orion 2015). This is all the more pressing given the entanglement of weed science with the post-war chemical industry, as noted earlier. As Paul Robbins (2007) noted in his analysis of the American lawn, the chemical industry played a far-from-neutral role in the development of the cultural ideal of a perfect green suburban lawn, creating the demand for their products. A similar role in terms of invasive species management is not farfetched—as biologist and historian Matt Chew notes, invasive species are marketing opportunities for pesticide manufacturers.[7]

The second example is related: the control and eradication of invasive species is difficult work, potentially involving elements such as exposure to toxic chemicals, hard physical labor, and the killing of living things. These consequences are more commonly borne by certain sectors of society than others—a laboring class characterized by relative poverty, migrant status, or indigeneity (cf. Murray 1994; Atchison and Head 2013; Head et al. 2015). In northern Australia, for instance, Aboriginal rangers submit to difficult, hot, and poisonous weed work largely following the exigencies of state agency lists and contractor funding incentives. These tasks are, according to the rangers, the most unsatisfying of their job, and the most distant from their official mandate to be doing work related to "caring for country" (Bach 2015; Bach and Larson 2017). A similar disconnect and dissatisfaction was noticed among park rangers whose jobs over a 30-year period centered on killing goats, cats, and other feral animals in the Galápagos Islands (Hennessy 2014). Similarly, South Africa's Working for Water alien management program has been criticized for risk exposure and low pay (Lidström et al. 2015, p. 23).

The scientific literature on invasions quite often frames conflicts over the management of invasive species as "conflicts of interest" (Cullen and Delfosse 1984; Shackleton et al. 2007; Estévez et al. 2015). A critical perspective on invasion studies suggests that this literature often takes on an overly simplistic "us-and-them" framing. It tends to view conflicting interests in relatively straightforward ways: for instance, community Y opposes control of species X because of cultural belief Z, or because Y makes money selling the products from X. Such a framing suggests that there are relatively clearly bounded interests, implying that they might be resolved through approaches such as cost-benefit analyses (Le Maitre et al. 2002) or conflict resolution and negotiation (van Wilgen and Richardson 2012; Mukwada et al. 2016). From a critical perspective, what is missing is a more complete sense of the complex historical and current entanglements that have dialectically shaped the invasion problem in different locales. Conflicts over *Prosopis* in Rajasthan, for instance, cannot be understood without reference to questions of land access and institutional incentives to state foresters (Robbins 2001) just as conflicts over *Acacia* in Portugal must grapple with rural depopulation, outmigration, the history of plantation forestry, and perceptions of wildlife danger (Kull et al. 2017).

Instead of taking a black-and-white approach to conflicts over invasive species, a critical approach might engage more deeply with the complicated, rough-and-tumble, unpredictable, and practical necessity to "live with", accommodate, or coexist with invasive species. This is not only because full eradication and even partial control is often unrealistic, but also because control efforts are politically or socially untenable in some contexts (Atchison and Head 2013; Chandrasena 2014; Head et al. 2015). Rangan et al. (2014, p. 124) cite a struggling cattle and sheep rancher in Australia who is constrained by invasive species policies, and who says "I'm sick and tired of poisoning the things that want to live here, and trying to raise the things that want to die". These kinds of questions around adaptation, winners and losers, and unintended consequences are rich in critical opportunities that could push invasion scientists to pose their questions and frame their approaches differently.

Toward a Critical Invasion Science

The above discussions have hinted at some directions for a "critical invasion science". In this concluding section, I build on the previous sections and formulate four proposals for what a critical approach to invasion science might do, and what questions it might ask.

   (1) Questioning words and labels. The terminology used in research is powerful, as it can reflect assumptions and beliefs and thus frame research questions and interpretations of results. What are the concepts used in posing questions and guiding analysis, where do they come from, what do they show, and what do they hide? A critical approach would encourage invasion scientists to ask whether the use of different labels might lead to different research questions, and whether certain labels reflect the worldview (or political stance) of a particular interest group (perhaps more socially dominant) and thus might miss alternative framings and conclusions. *Specifically, for invasives,* one might begin with the name "invasion science" or the term "invasives", as I have already done above. More specifically, one can ask how the use of terminology affects research. A study I supervised in eastern Madagascar can illustrate the need for this approach. Posing the research question as "is *Grevillea banksii* invasive in eastern Madagascar" required the researcher to present criteria of what it means to "be invasive" (which, as we saw above, involves consequential choices between competing definitions regarding origins, impacts, and behavior) and then data to assess whether the plant meets the selected criteria. A different question, such as "why is *Grevillea banksii* spreading in eastern Madagascar" would have focused the research on different processes and different data. Each word selected for a research question—invasive, spread, alien, native, neophyte, naturalized, transformer species, adventive, feral—constrains the kind of information that will be sought. The

terminology can be questioned before a study is undertaken, for instance, when a scientists asks whether her research question should be framed using the concept of "invasion" as opposed to "colonization" (Hoffmann and Courchamp 2016a). Or it can be questioned afterward, as Larson (2011) does when he asks what the impacts are of terms and metaphors like "invasion meltdown" used to communicate research results.

(2) Questioning scale and its impacts. A critical approach would ask how it matters that research is framed at a particular temporal, spatial, or organizational scale. Does it change the questions that are asked, the evidence that is applied, or the analytical connections that can be made? *Specifically, for invasives,* an important scalar consideration I mentioned above is the way in which the category of *biological species* has become the object of analysis and communication, rather than particular populations of particular species in particular places. One crucial project for a critical invasion science would thus be to assess invasive species databases, the institutional and sociological process of their creation, and their impacts (Kull and Rangan 2015; Lidström et al. 2015), and to evaluate the benefits and consequences of a more "place-based" approach to invasion science.

(3) Caring explicitly how the science is used, who wins, who loses. Of course scientists care about these matters, but a critical approach would be explicit about it. *Specifically, for invasives,* the dominant discourse of the science of invasion biology is of the urgency and importance of the issue, incessantly promoted as the "second greatest threat" to biodiversity (Chew 2015). This leads to an under-exploration of opposing views—those of rural residents whose livelihoods are based on the abundant and robust growth of certain invaders, or of people who labor in chemical protection suits in the tropical sun to poison invasive plants, or of advocates for amphibians made sick by toxic chemicals. That is, in caring narrowly for the protection of biodiversity or certain suites of ecosystem services, the broader impacts of this science are downplayed. For instance, Courchamp et al. (2017, p. 13) state that criticisms, internal strife, and an unaware society "hinder the progress of invasion biology". Similarly, van Wilgen and Richardson (2012, p. 56) basically say that opposing voices do not matter: their proposals regarding the problem of pine invasions in South Africa "will require political commitment to policies that could be unpopular in certain sectors of society". A critical approach would interest itself more in the impact of these conclusions and in opposing views. It might, for instance, seek to co-construct research questions with different interest groups (perhaps resulting in questions like: "how would eradication of species X affect ecological dynamics and the provision of wood-fuel in this region", or "given that local stakeholders are not keen on full eradi-

cation, what are the impacts of partial control via bio-control agents on livelihoods and novel ecosystem dynamics").

*(4) **Questioning the voice of expertise.*** By this I do not mean questioning the expertise of scientists and their research outcomes. Instead, I mean questioning the voice, or attitude, or posture whereby science has a *monopoly* on expertise and on translating that expertise into action. ***Specifically, for invasives,*** more attention could be paid to the embedded landscape knowledge of local people about weeds and pests (Bentley et al. 2005; Vaarzon-Morel and Edwards 2012). This could be an intellectually and practically significant shift. Intellectually, because, for instance, it might contribute to a reconsideration of the kinds of questions asked. For instance, *Plantago major* was known by Native Americans as "white man's footprint", a name that usefully directs attention to the society that transported the plant and aided its spread through ecological disturbance. And yet while it may be seen as an invader, it has not displaced other species and became widely appreciated in Native American communities for its different uses (Kimmerer 2013). This is also a practically important shift, because local knowledge based on landscape experience might reveal patterns and processes not easily noticed by transient field workers. More fundamentally, the sharing of knowledge and co-produced questioning might lead to better appreciation of the social complexity inherent to ecologically dynamic situations and orient research toward solutions acceptable to all parties.

In the introductory chapter, Lave et al. suggest that a critical approach leads to the asking of new questions, or to adding layers to questions we already ask, and provide an example of a case of a soil scientist working in Oakland, California. This applies well to the case for a "critical invasion science". Let me illustrate with a final set of examples from a research project in which I am currently involved—the rapid expansion of potentially invasive Australian *Acacia* plantations in Vietnam (Richardson et al. 2015; Cochard et al. 2017). An invasion biologist might start and end their study with mention of widespread commercial plantations of this species, followed by investigations of dispersal mechanisms, seed banks, soil allelopathy, and spatial spread. On top of this, a critical invasion scientist might add additional layers of inquiry:

- How do political-economic factors shape the distribution of acacia plantations and thus "propagule pressure"?
- To what extent does strong government policy favoring tree cover and economic interest in acacia plantations reduce local scientific attention to potential invasive behavior?

- How does the introduction of tree breeding, and notably hybrid strains of *A. mangium* crossed with *A. auriculiformis*, affect seed viability, dispersal, and invasive behavior?
- To what extent is the spread of acacia constrained by dense human land use outside plantation areas?
- For which people, and in what contexts, is spontaneous acacia spread beneficial, or harmful, or irrelevant?
- How might these peoples' concerns and experiential knowledge affect the construction of research questions regarding acacia in the landscape?

A critical "spirit" is of course already widely held by many natural and social scientists. Much scientific training promotes, somewhere along the line, attention to the construction of categories, to things that do not fit pre-existing models, and to the implications of one's research. But this is far from universal, and often not explicit. Furthermore, in research at the interface of society and environment, it needs to go much further, as this Handbook's introduction suggests. Because we live in a post-natural world where social processes profoundly affect almost all landscapes and environmental processes (Urban, this volume), a critical spirit is needed to incorporate attention to these social processes, often deeply structured, from the get-go, and not treat them as add-ons to the natural science problem. A critical spirit also involves holding a mirror up to science: what are the ideologies, power relations, and social legacies that shape how we produce knowledge, and what are the effects of that knowledge on the eco-social landscapes we study and the people that live in them? This kind of critical approach could benefit from broader training, reflection, encouragement, and attention.

Notes

1. The publications cited in this paragraph are, in order, Davis et al. (2011), Simberloff (2011b), Pearce (2015), and Thompson (2014) for which, interestingly, the American edition subtitle is *Why Invasive Species Aren't All Bad*, but Britain it is *The Story and Science of Invasive Species*, Marris (2011), Richardson and Ricciardi (2013), and Russell and Blackburn (2017).
2. For a more in-depth discussion of the "Science" phenomenon, see my blog: https://christiankull.net/2013/11/25/is-everything-a-science/.
3. Courchamp et al.'s (2017) list of 24 issues in invasion science provides more inspirations for critical enquiry.

4. SCOPE is the Scientific Committee on Problems of the Environment, established by the International Council for Science in 1969 (http://www.scopenvironment.org). It has sponsored over 70 authoritative investigations of particular topics, including biological invasions (number 37).
5. Interestingly, a similar assertion that biological invasions and natural colonization were not that different recently sparked a vehement debate (Hoffmann and Courchamp 2016a, b; Wilson et al. 2016).
6. However, interestingly, a workshop involving a number of invasion scientists is advertised for 2018 without using the word "invasion" (the title is "Species range extensions and local adaptation"). See http://andina4argentina.weebly.com (accessed 7 April 2017).
7. See https://milliontrees.me/2017/04/01/ecological-restorations-follow-the-money/, accessed 4 April 2017.

References

Atchison, J., and L. Head. 2013. Eradicating bodies in invasive plant management. *Environment and Planning D: Society and Space* 31 (6): 951–968.

Awanyo, L. 2001. Labor, ecology, and a failed agenda of market incentives: The political ecology of agrarian reforms in Ghana. *Annals of the Association of American Geographers* 91 (1): 92–121.

Bach, T.M. 2015. "All about healthy country": Aboriginal perspectives of weed management in the Kimberley, Western Australia. PhD, Monash University.

Bach, T.M., and B.M.H. Larson. 2017. Speaking about weeds: Indigenous elders' metaphors for invasive species and their management. *Environmental Values* 26 (5): 561–581.

Bean, A.R. 2007. A new system for determining which plant species are indigenous in Australia. *Australian Systematic Botany* 20: 1–43.

Beinart, W., and L. Wotshela. 2011. *Prickly pear: The social history of a plant in the Eastern Cape.* Johannesburg: Wits University Press.

Bennett, B.M. 2014. Model invasions and the development of national concerns over invasive introduced trees: Insights from South African history. *Biological Invasions* 16: 499–512.

Bentley, J.W., M. Webb, S. Nina, and S. Pérez. 2005. Even useful weeds are pests: Ethnobotany in the Bolivian Andes. *International Journal of Pest Management* 51 (3): 189–207.

Blackburn, T.M., P. Pyšek, S. Bacher, J.T. Carlton, R.P. Duncan, V. Jarosìk, J.R.U. Wilson, and D.M. Richardson. 2011. A proposed unified framework for biological invasions. *Trends in Ecology & Evolution* 26 (7): 333–339.

Boonman-Berson, S., E. Turnhout, and J. van Tatenhove. 2014. Invasive species: The categorization of wildlife in science, policy, and wildlife management. *Land Use Policy* 38: 204–212 https://doi.org/10.1016/j.landusepol.2013.11.002.

Carruthers, J., L. Robin, J.P. Hattingh, C.A. Kull, H. Rangan, and B.W. van Wilgen. 2011. A native at home and abroad: The history, politics, ethics and aesthetics of Acacia. *Diversity and Distributions* 17 (5): 810–821.

Chandrasena, N. 2014. Living with weeds—A new paradigm. *Indian Journal of Weed Science* 46 (1): 96–110.

Chew, M.K. 2015. Ecologists, environmentalists, experts, and the invasion of the 'second greatest threat'. *International Review of Environmental History* 1: 7–4.

Chew, M.K., and A.L. Hamilton. 2011. The rise and fall of biotic nativeness: A historical perspective. In *50 years of invasion ecology: The legacy of Charles Elton*, ed. D.M. Richardson. London: Blackwell.

Cochard, R., D.T. Ngo, P.O. Waeber, and C.A. Kull. 2017. Extent and causes of forest cover changes in Vietnam's provinces 1993–2013: A review and analysis of official data. *Environmental Reviews* 25 (2): 199–217.

Colautti, R.I., and H.J. MacIsaac. 2004. A neutral terminology to define 'invasive' species. *Diversity and Distributions* 10: 135–141.

Colautti, R.I., and D.M. Richardson. 2009. Subjectivity and flexibility in invasion terminology: Too much of a good thing? *Biological Invasions* 11 (6): 1225–1229.

Comaroff, J., and J.L. Comaroff. 2001. Naturing the nation: Aliens, apocalypse and the postcolonial state. *Journal of Southern African Studies* 27 (3): 627–651.

Courchamp, F., A. Fournier, C. Bellard, C. Bertelsmeier, E. Bonnaud, J.M. Jeschke, and J.C. Russell. 2017. Invasion biology: Specific problems and possible solutions. *Trends in Ecology & Evolution* 32 (1): 13–22.

Cronk, Q.C.B., and J.L. Fuller. 1995. *Plant invaders: The threat to natural ecosystems*. London: Chapman & Hall.

Cullen, J.M., and E.S. Delfosse. 1984. Echium plantagineum: Catalyst for conflict and change in Australia. Proc. VI Int. Symp. Biol. Contr. Weeds, Vancouver, Canada, Agriculture Canada.

Cutter, S.L. 2003. The vulnerability of science and the science of vulnerability. *Annals of the Association of American Geographers* 93 (1): 1–12.

Davis, M.A. 2009. *Invasion biology*. Oxford: Oxford University Press.

Davis, M.A., M.K. Chew, R.J. Hobbs, A.E. Lugo, J.J. Ewel, G.J. Vermeij, J.H. Brown, et al. 2011. Don't judge species on their origins. *Nature* 474: 153–154.

Demeritt, D. 2009. Geography and the promise of integrative environmental research. *Geoforum* 40: 127–129.

Drake, J.A., H.A. Mooney, F. di Castri, R.H. Groves, F.J. Kruger, M. Rejmánek, and M. Williamson, eds. 1989. *Biological invasions: A global perspective. SCOPE 37*. Chichester: John Wiley & Sons.

Ellender, B.R., O.L.F. Weyl, H. Winker, and A.J. Booth. 2010. Quantifying the annual fish harvest from South Africa's largest freshwater reservoir. *Water SA* 36 (1): 45–52.

Estévez, R.A., C.B. Anderson, J.C. Pizarro, and M.A. Burgman. 2015. Clarifying values, risk perceptions, and attitudes to resolve or avoid social conflicts in invasive species management. *Conservation Biology* 29 (1): 19–30.

Forsyth, T. 2003. *Critical political ecology: The politics of environmental science.* London: Routledge.

Frawley, J., and I. McCalman, eds. 2014. *Rethinking invasion ecologies from the environmental humanities.* London: Routledge.

Gautier, D., and T.A. Benjaminsen, eds. 2012. *Environnement, discours et pouvoir: L'approche Political ecology.* Versailles: Éditions Quæ.

Green, L. 2014. Ecology, race, and the making of environmental publics: A dialogue with Silent Spring in South Africa. *Resilience: A Journal of the Environmental Humanities* 1 (2): 20.

Gregory, D., R. Johnston, G. Pratt, M. Watts, and S. Whatmore, eds. 2009. *The dictionary of human geography.* London: Wiley-Blackwell.

Gutman, G., A.C. Janetos, C.O. Justice, E.F. Moran, J.F. Mustard, R.R. Rindfuss, D. Skole, B.L. Turner II, and M.A. Cochrane, eds. 2004. *Land change science: Observing, monitoring and understanding trajectories of change on the Earth's surface.* Dordrecht: Kluwer Academic Publishers.

Head, L., B.M.H. Larson, R.J. Hobbs, J. Atchison, N. Gill, C.A. Kull, and H. Rangan. 2015. Living with invasive plants in the Anthropocene: The importance of understanding practice and experience. *Conservation and Society* 13 (3): 311–318.

Hennessy, E.A. 2014. On the backs of tortoises: Conserving evolution in the Galápagos Islands. PhD, University of North Carolina.

Hoffmann, B.D., and F. Courchamp. 2016a. Biological invasions and natural colonisations: Are they that different? *NeoBiota* 29: 1–14.

———. 2016b. When similarities matter more than differences: A reply to Wilson et al. *NeoBiota* 31: 99–104.

Johnson, S. 2010. *Bioinvaders.* Cambridge: White Horse Press.

Kareiva, P., and M. Marvier. 2012. What is conservation science? *BioScience* 62 (11): 962–969.

Kates, R.W., W.C. Clark, R. Corell, J.M. Hall, C.C. Jaeger, I. Lowe, J.J. McCarthy, et al. 2001. Sustainability science. *Science* 292 (5517): 641–642.

Kimmerer, R.W. 2013. *Braiding sweetgrass: Indigenous wisdom, scientific knowledge, and the teachings of plants.* Minneapolis: Milkweed Editions.

Kowarik, I., and P. Pyšek. 2012. The first steps towards unifying concepts in invasion ecology were made one hundred years ago: Revisiting the work of the Swiss botanist Albert Thellung. *Diversity and Distributions* 18: 1243–1252.

Kueffer, C., and B.M.H. Larson. 2014. Responsible use of language in scientific writing and science communication. *BioScience* 64: 719–724.

Kull, C.A., and H. Rangan. 2015. The political ecology of weeds: A scalar approach to landscape transformation. In *The international handbook of political ecology*, ed. R.L. Bryant, 487–500. Cheltenham: Edward Elgar.

———. 2016. Political ecology and resilience: Competing interdisciplinarities? In *Interdisciplinarités entre Natures et Sociétés: Colloque de Cerisy*, ed. B. Hubert and N. Mathieu, 71–87. Bruxelles: P.I.E. Peter Lang.

Kull, C.A., and J. Tassin. 2012. Australian acacias: Useful and (sometimes) weedy. *Biological Invasions* 14 (11): 2229–2233.

Kull, C.A., C.M. Shackleton, P.J. Cunningham, C. Ducatillion, J.-M. Dufour-Dror, K.J. Esler, J.B. Friday, et al. 2011. Adoption, use and perception of Australian acacias around the world. *Diversity and Distributions* 17 (5): 822–836.

Kull, C.A., J. Tassin, S. Moreau, H. Rakoto Ramiarantsoa, C. Blanc-Pamard, and S.M. Carrière. 2012. The introduced flora of Madagascar. *Biological Invasions* 14 (4): 875–888.

Kull, C.A., J. Tassin, and S.M. Carrière. 2014. Approaching invasive species in Madagascar. *Madagascar Conservation and Development* 9 (2): 60–70.

Kull, C.A., C. Kueffer, D.M. Richardson, A.S. Vaz, J.R. Vicente, and J.P. Honrado. 2017. Using the 'regime shift' concept in addressing social-ecological change. *Geographical Research*. https://doi.org/10.1111/1745-5871.12267.

Larson, B.M.H. 2005. The war of the roses: Demilitarizing invasion biology. *Frontiers in Ecology and Environment* 3 (9): 495–500.

———. 2011. *Metaphors for environmental sustainability: Redefining our relationship with nature*. New Haven: Yale University Press.

Lave, R., M.W. Wilson, E.S. Barron, C. Biermann, M.A. Carey, C.S. Duvall, L. Johnson, et al. 2014. Intervention: Critical Physical Geography. *The Canadian Geographer/Le Géographe canadien* 58 (1): 1–10.

Le Maitre, D.C., B.W. Van Wilgen, C.M. Gelderblom, C. Bailey, R.A. Chapman, and J.L. Nel. 2002. Invasive alien trees and water resources in South Africa: Case studies of the costs and benefits of management. *Forest Ecology and Management* 160: 143–159.

Leslie, H.M., and A.P. Kinzig. 2009. Resilience science. In *Ecosystem-based management for the oceans*, ed. K. McLeod and H.M. Leslie, 55–73. Washington: Island Press.

Lidström, S., S. West, T. Katzschner, M.I. Pérez-Ramos, and H. Twidle. 2015. Invasive narratives and the inverse of slow violence: Alien species in science and society. *Environmental Humanities* 7: 1–40.

Marris, E. 2011. *Rambunctious garden: Saving nature in a post-wild world*. New York: Bloomsbury.

McNeely, J.A., ed. 2001. *The great reshuffling: Human dimensions of invasive alien species*. Gland: IUCN.

Middleton, K. 2012. Renarrating a biological invasion: Historical memory, local communities and ecologists. *Environment and History* 18: 61–95.

Mukwada, G., W. Chingombe, and P. Taru. 2016. Strifes of the frontier: An assessment of Acacia mearnsii related park-community conflicts in the Golden Gate Highlands National Park, South Africa. *Journal of Integrative Environmental Sciences* 13 (1): 37–54.

Murray, D.L. 1994. *Cultivating crisis: The human cost of pesticides in Latin America*. Austin: University of Texas Press.

Mwangi, E., and B. Swallow. 2008. *Prosopis juliflora* invasion and rural livelihoods in the Lake Baringo Area of Kenya. *Conservation and Society* 6 (2): 130–140.

Neely, A.H. 2010. Blame it on the weeds: Politics, poverty, and ecology in the New South Africa. *Journal of Southern African Studies* 36 (4): 869–887.

Orion, T. 2015. *Beyond the war on invasive species: A permaculture approach to ecosystem restoration.* White River Junction, VT: Chelsea Green Publishing.

Osborne, M.A. 2000. Acclimatizing the world: A history of the paradigmatic colonial science. *Osiris* 15: 135–151. 2nd series.

Pearce, F. 2015. *The new wild: Why invasive species will be nature's salvation.* Boston: Beacon Press.

Peretti, J.H. 1998. Nativism and nature: Rethinking biological invasion. *Environmental Values* 7: 183–192.

Pooley, S. 2014. *Burning table mountain: An environmental history of fire on the Cape Peninsula.* Basingstoke: Palgrave Macmillan.

Rangan, H., A. Wilson, and C.A. Kull. 2014. Thorny problems: Industrial pastoralism and managing 'country' in Northwest Queensland. In *Rethinking invasion ecologies from the environmental humanities*, ed. J. Frawley and I. McCalman, 116–134. London: Routledge.

Richardson, D.M., ed. 2011a. *Fifty years of invasion ecology: The legacy of Charles Elton.* Oxford: Wiley-Blackwell.

———. 2011b. Invasion science: The roads travelled and the roads ahead. In *Fifty years of invasion ecology: The legacy of Charles Elton*, ed. D.M. Richardson, 397–407. Oxford: Wiley-Blackwell.

Richardson, D.M., and A. Ricciardi. 2013. Misleading criticisms of invasion science: A field guide. *Diversity and Distributions* 19: 1461–1467.

Richardson, D.M., P. Pyšek, M. Rejmánek, M.G. Barbour, F.D. Panetta, and C.J. West. 2000. Naturalization and invasion of alien plants: Concepts and definitions. *Diversity and Distributions* 6 (2): 93–107.

Richardson, D.M., J.J. Le Roux, and J.R.U. Wilson. 2015. Australian acacias as invasive species: Lessons to be learnt from regions with long planting histories. *Southern Forests: A Journal of Forest Science* 77 (1): 31–39.

Robbins, P. 2001. Tracking invasive land covers in India, or why our landscapes have never been modern. *Annals of the Association of American Geographers* 91 (4): 637–659.

———. 2007. *Lawn people: How grasses, weeds, and chemicals make us who we are.* Philadelphia: Temple University Press.

———. 2012. *Political ecology: A critical introduction.* 2nd ed. Oxford: Wiley.

Russell, J.C., and T.M. Blackburn. 2017. The rise of invasive species denialism. *Trends in Ecology & Evolution* 32 (1): 3–6.

Salomon Cavin, J., and C.A. Kull. 2017. Invasion ecology goes to town: From disdain to sympathy. *Biological Invasions* 19 (12): 3471–3487.

Sayre, N.F. 2015. The Coyote-proof pasture experiment: How fences replaced predators and labor on US rangelands. *Progress in Physical Geography* 39 (5): 576–593.

Shackleton, C.M., D. McGarry, S. Fourie, J. Gambiza, S.E. Shackleton, and C. Fabricius. 2007. Assessing the effects of invasive alien species on rural livelihoods: Case examples and a framework from South Africa. *Human Ecology* 35: 113–127.

Shackleton, R.T., D.C. Le Maitre, and D.M. Richardson. 2015. Stakeholder perceptions and practices regarding *Prosopis* (mesquite) invasions and management in South Africa. *Ambio* 44 (6): 569–581.

Simberloff, D. 2003. Confronting introduced species: A form of xenophobia? *Biological Invasions* 5: 179–192.

———. 2011a. Charles Elton—Neither founder nor siren, but prophet. In *Fifty years of invasion ecology*, ed. D.M. Richardson, 11–24. New York: Wiley.

———. 2011b. Non-natives: 141 scientists object. *Nature* 475: 36.

———. 2013. *Invasive species: What everyone needs to know*. Oxford: Oxford University Press.

Simberloff, D., J.-L. Martin, P. Genovesi, V. Maris, D.A. Wardle, J. Aronson, F. Courchamp, et al. 2013. Impacts of biological invasions: What's what and the way forward. *Trends in Ecology & Evolution* 28 (1): 58–66.

Subramaniam, B. 2001. The aliens have landed! Reflections on the rhetoric of biological invasions. *Meridians: Feminism, race, transnationalism* 2 (1): 26–40.

Tassin, J. 2014. *La grande invasion: qui a peur des espèces invasives?* Paris: Odile Jacob.

———. 2017. User de pesticides pour contrôler les espèces invasives: les facettes d'un paradoxe éthique. *Revue d'Écologie (Terre et Vie)* 72 (4): 425–438.

Tassin, J., and C.A. Kull. 2015. Facing the broader dimensions of biological invasions. *Land Use Policy* 42: 165–169.

Thompson, K. 2014. *Where do Camels belong?* Vancouver: Greystone Books.

Turpie, J.K., C. Marais, and J.N. Blignaut. 2008. The working for water programme: Evolution of a payments for ecosystem services mechanism that addresses both poverty and ecosystem service delivery in South Africa. *Ecological Economics* 65 (4): 788–798.

Vaarzon-Morel, P., and G. Edwards. 2012. Incorporating Aboriginal people's perceptions of introduced animals in resource management: Insights from the feral camel project. *Ecological Management and Restoration* 13 (1): 65–71.

Valéry, L., H. Fritz, J.-C. Lefeuvre, and D. Simberloff. 2008. In search of a real definition of the biological invasion phenomenon itself. *Biological Invasions* 10: 1345–1351.

van Wilgen, B.W., and D.M. Richardson. 2012. Three centuries of managing introduced conifers in South Africa: Benefits, impacts, changing perceptions and conflict resolution. *Journal of Environmental Management* 106: 56–68.

van Wilgen, B.W., A. Khan, and C. Marais. 2011. Changing perspectives on managing biological invasions: Insights from South Africa and the Working for Water Programme. In *Fifty Years of Invasion Ecology: The Legacy of Charles Elton*, ed. D.M. Richardson, 377–393. Oxford: Wiley-Blackwell.

Vaz, A.S., C. Kueffer, C.A. Kull, D.M. Richardson, S. Schindler, A.J. Muñoz-Pajares, J.R. Vicente, et al. 2017. The progress of interdisciplinarity in invasion science. *Ambio* 46 (4): 428–442.

Warren, C.R. 2007. Perspectives on the 'alien' versus 'native' species debate: A critique of concepts, language and practice. *Progress in Human Geography* 31 (4): 427–446.

Williamson, M. 1996. *Biological Invasions*. London: Chapman & Hall.

Wilson, J.R.U., P. García-Díaz, P. Cassey, D.M. Richardson, P. Pyšek, and T.M. Blackburn. 2016. Biological invasions and natural colonisations are different—The need for invasion science. *NeoBiota* 31: 87–98.

13

Mapping Ecosystem Services: From Biophysical Processes to (Mis)Uses

Simon Dufour, Xavier Arnauld de Sartre, Monica Castro, Michel Grimaldi, Solen Le Clec'h, and Johan Oszwald

Introduction

> Ecosystem services are the benefits people obtain from ecosystems. These include provisioning services such as food and water; regulating services such as regulation of floods, drought, land degradation, and disease; supporting services such as soil formation and nutrient cycling; and cultural services such as recreational, spiritual, religious and other nonmaterial benefits. (MEA 2005)

In 2005, the Millennium Ecosystem Assessment (MEA) drafted a global overview of the state of ecosystems and threats to the benefits they provide to humans (MEA 2005). This assessment, based on an extensive scientific literature review, clearly shows that ecosystems have been intensely modified over the last century, more than in any other period, and that this modification had some paradoxical impacts on human well-being. On the one hand, ecosystem modification has allowed average well-being to increase, but on the

S. Dufour (✉) • J. Oszwald
Department of Geography, Université Rennes 2/CNRS LETG, Rennes, France

X. Arnauld de Sartre • M. Castro
Institut Claude Laugénie, CNRS PASSAGES, Pau, France

M. Grimaldi
IRD BIOEMCO, Bondy, France

S. Le Clec'h
ETH Zurich, Zurich, Switzerland

© The Author(s) 2018
R. Lave et al. (eds.), *The Palgrave Handbook of Critical Physical Geography*,
https://doi.org/10.1007/978-3-319-71461-5_13

other hand, this increase has not been shared equally. Moreover, current trends in ecosystem changes indicate that degradation of services provided by ecosystems will mainly affect poor people (MEA 2005). To combat these trends, the MEA promotes deeper integration of ecosystem services in development and economic policies. Thus, over the last decade, ecosystem services have become (or aimed to become) a crucial component for justifying and implementing environmental policies.

However, despite the apparent simplicity of the ecosystem services concept, it is extremely complex (see Norgaard 2010; Lamarque et al. 2011; Ernstson and Sörlin 2013; Arnauld de Sartre et al. 2014; Barnaud and Antona 2014). The concept is used in many contexts or categories of discourse (political, social, heuristic, economics, etc.) that can be linked to two uses: global pedagogic use (to illustrate how society depends on natural objects and processes) and applied use (to improve practices for conserving and managing ecosystems) (Lamarque et al. 2011; European Commission/Directorate General for the Environment 2013; Hauck et al. 2013; Dufour et al. 2014). Although the pedagogic dimension has become well established since the publication of the MEA in 2005, the applied dimension remains a source of debate. These debates have structured evolution of the concept after publication of the MEA. Some scientists and institutions view the concept as a concrete opportunity for improvement in environmental management and advocate for development of methods and tools for mapping ecosystem services as a priority, to render the concept operational (Daily and Matson 2008; European Commission/Directorate General for the Environment 2013). There are examples of concrete methods and tools to evaluate ecosystem services (Waage et al. 2011) developed by scientific teams and/or NGOs such as InVEST (Natural Capital Project, www.naturalcapitalproject.org) and the follow-up of REDD+ programs (Hewson et al. 2013). Some private consortiums have begun to get into the business of ecosystem service valuation using mapping tools and geospatial information (e.g. EO Services for Ecosystem Valuation project, www.space4ecosystems.com). But the ecosystem services concept has also been highly criticized. Most critiques are ethical (Maris 2014) and/or economic (Milanesi 2010; Gomez-Baggethun and Ruiz-Perez 2011; Karsenty and Ezzine de Blas 2014; Kronenberg 2015).

Steering between these two approaches (i.e. acceptance and rejection), we aim to illustrate the relevance of a Critical Physical Geography framework for improving identification of the potentials and limitations of the concept of ecosystem services (Potschin and Haines-Young 2011; Tadaki et al. 2015a). We focus on a specific type of object, maps, which is used to operationalize the concept of ecosystem services. We believe that analyzing a tool is a good way to analyze the framework in which the tool is situated (Lascoumes and Le Galès 2004; Tadaki et al. 2015a). First, we illustrate how the diversity and complexity of biophysical processes that are intertwined in the production of certain ecosystem services strongly limit the production of simple and accurate maps of them. Second, we analyze how maps of ecosystem services have been used and justified within the scientific literature and the MEA. We show that, despite methodological limits, ecosystem service maps are produced and used with little critical consideration, despite the fact that the political dimension of maps is already well known (Crampton and Krygier 2001). Beyond the case of ecosystem service maps, this chapter aims to combine some physical geography and critical cartography perspectives to illustrate why we need to increase reflexivity in scientific practice and how a thorough understanding of biophysical processes can help develop a critical perspective on environmental management, which are among the goals of Critical Physical Geography (Lave et al. 2014; Tadaki et al. 2015b).

(The Difficulty in) Mapping Ecosystem Services

Potential uses of ecosystem service maps are to enable quantitative and spatially explicit arguments to raise awareness, to negotiate management plans, or to assess stocks, loss, or restoration of services (Costanza et al. 1997; Hauck et al. 2013). These maps are built using spatial methods that predict ecosystem service values in a given area. But the credibility of pedagogic use and the efficiency of applied uses depend on scientists' ability to produce reliable methods for the quantification and spatial location of ecosystem services. As with all quantification and cartographical methods, this requires simplification of complex entities, and that simplicity is achieved through methodological choices. For example, Eigenbrod et al. (2010) demonstrate that "land cover based proxies provide a poor fit to primary data surfaces for biodiversity, recreation and carbon storage" and that proxies are "unsuitable for identifying hotspots or priority areas for multiple services". The reliability of maps thus depends not only on biophysical processes themselves but on the methodological choices made to map them.

What Is Mapped in Ecosystem Service Maps

Ecosystems constitute physical, chemical, and biological processes, and they exist with or without humans; such processes become services by being appropriated by societies. Thus, there are two approaches to studying ecosystem services: analyze bio-chemico-physical functions of ecosystems (Eigenbrod et al. 2010); and consider the beneficiaries of ecosystem services (Zhu et al. 2010; Burkhard et al. 2012; Wolff et al. 2015). From a spatial perspective, processes that generate services are generally easier to characterize, whereas identifying beneficiaries is usually much more complex because the latter require not only the identification of processes, also their impacts (Wolff et al. 2015). For example, it is (relatively) easier to map carbon stocks or storage in a forest than to determine who benefits from climate regulation provided by this forest. Which indicator of benefits should be chosen? The difference in temperature? The decrease in climate-change risk? Moreover, at which scale should the benefits of ecosystem services be mapped (local, regional, or global)? Is a map of global differences in temperature even interesting? Thus, while identifiable, these benefits remain much harder to map systematically. Thus, ecosystem services are mainly modeled and mapped using indicators of ecosystem composition, patterns, processes, and functions (Portela and Rademacher 2001; Boyd and Banzhaf 2007; Davies et al. 2011; Dymond et al. 2012; Nemec and Raudsepp-Hearne 2012; Le Clec'h et al. 2014), especially patterns such as land cover (Eigenbrod et al. 2010).

These approaches circumvent questions raised by the differences in individual, social, or cultural perceptions of services.

Behind Ecosystem Services, Some Complex Biophysical Processes

While it is common to map biophysical objects and processes, it does not mean that the task is easy, because ecological functions are complex attributes of socio-ecological systems. To illustrate this issue, we take the example of services linked to soils. Soils play direct and indirect roles in providing many ecological functions and associated services, such as primary production, nutrient cycling, the water cycle, and surface runoff erosion. Following Dominati et al. (2010), reliable characterization of soils must consider interrelated inherent properties, such as slope, orientation, texture, and aggregate size, and properties and components that respond to management, such as soluble phosphate, pH, and bulk density (Fig. 13.1). Most of these properties influence and can be influenced by multiple and interacting processes, such as water absorption, symbioses, organic matter decomposition, aggregation and disaggregation, water infiltration, and nutrient loss or assimilation, and each service depends on several properties. For example,

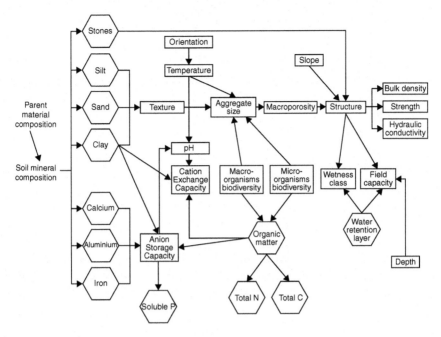

Fig. 13.1 Simplified relationships between soil components (hexagons) and properties (rectangles). Source: Dominati et al. (2010)

soil contributes to biological control of pests and diseases:by providing habitat to beneficial species, soils can support plant growth (rhizobium, mycorrhizae) and control the proliferation of pests (crops, animals or humans pests) and harmful disease vectors (e.g. viruses, bacteria). Soil conditions (e.g. moisture, temperature) determine the quality of the soil habitat and thereby select the type of organisms present. This service depends on soil properties and the biological processes driving inter- and intra-specific interactions (symbiosis, competition) (Dominati et al. 2010).Each service depends on several properties (of the soil, landscape, and climate), but each property also influences several services. For example, soil biodiversity (e.g. microflora, fungi, micropredators, engineer organisms) influences all soil functions, such as primary production, nutrient cycling, and climate regulation (Lavelle et al. 2006). The complexity of links between soil properties and services provided by soils makes quantification and mapping of indicators of services difficult because it is a challenge to capture the spatial dimension of all relevant properties in simple indicators (Grimaldi et al. 2014), even if data on these properties are available. Thus, the reliability of these maps should be thoroughly analyzed.

Reliability of Maps of Ecosystem Service Indicators

Although uncertainty is inherent in mapping techniques, this is rarely integrated in practices. In a review of 50 cases, Pagella and Sinclair (2014) point that uncertainty was explicitly discussed in relation to underlying data in only 6% of cases. To discuss the reliability of ecosystem service indicator maps, we illustrate the influence of methodological choices on map quality by a study case. Indeed, we analyzed how the choice of statistical method significantly affects the quality of maps produced using an empirical approach in the context of the Amazonian pioneer front.

Using data collected during a research project (AMAZ ES) that analyzed the influence of landscape changes on ecosystem services provided in the state of Pará, Brazil, we measured eight indicators of ecosystem services (a biodiversity index, Sphingidae richness, soil-engineer (termite and earthworm) richness, soil chemical quality index, soil carbon stock, water available for plants, water infiltration rates into the soil, and vegetation carbon stock) in three locations in the Brazilian Amazon rainforest (Maçaranduba, Pacajá, and Palmares II; see Grimaldi et al. (2014) and Le Clec'h et al. (2016) for details). These eight indicators differed in terms of relevant spatial scale and underlying biophysical processes. For each indicator, we compared two statistical methods that predicted the indicator as a function of predictive (explanatory) variables: decision trees and linear regressions. Following current practices in the scientific literature, the predictive variables were produced by remote sensing and GIS using Landsat data (NDVI, NDWI, land-cover classes, and trajectory of land cover over the last two decades) and ASTER DEM (topographic position index, slope, elevation, and distance to a hydrographic network). We also added a variable for site identification.

The analysis demonstrated three crucial aspects. First, the quality of the method (and thus of the resulting maps) varied among ecosystem service indicators, with coefficients of determination (R^2) of 0.74–0.46 for decision-tree methods and 0.75–0.18 for linear-regression methods (Table 20.1). The highest quality ($R^2 = 0.74$–0.75) was obtained for vegetation carbon stock. All others had an $R^2 < 0.7$. For several indicators, at least one of the two methods yielded an $R^2 < 0.5$ (i.e. soil-engineer richness, Sphingidae richness, biodiversity index, water available for plants, soil carbon stock). Variability in the quality of methods was partly related to the choice of indicators and the processes underlying them: vegetation carbon stock can be mapped accurately using land-cover classes that give a good indication of aboveground biomass, whereas indicators of soil services and soil carbon stock are more difficult to map due to the complexity of biophysical processes described previously and the difficulty in assessing them using basic remote-sensing data (Grimaldi et al. 2014).

Second, for a given indicator, the quality of the map depends on the statistical method used (Table 13.1). For the eight indicators, decision trees always predicted as well as (for carbon stocks) or better than linear regressions. Thus, the relative differences between the two methods ranged from 0 to 68%, with mean R2 of 0.60 and 0.43 respectively for decision trees and linear regressions. Even when the methods had equal R^2 values, their maps were significantly different (e.g. for soil carbon stock, Fig. 13.2). Differences between maps were visible both at the site and the intra-site scales, indicating several patterns, such as the influence of the hydrographic network on vegetation carbon stock in the linear-regression map (Fig. 13.3a). The difference between the methods is due to the statistics that underline the method but, for all indicators except vegetation carbon stock, also because of the variables considered as relevant by the algorithm in each one.

Third, the number and nature of predictive variables differ among indicators (Table 13.1). For most, the number of variables was high (6–8), which highlights the complexity of the underlying processes. This result indicates the need for a rich predictive dataset (to adapt to each indicator), a lack of which can limit the ability to render the ecosystem service framework operational (Le Clec'h et al. 2014). The indicators requiring only two variables (related to land cover) were vegetation carbon stock and rates of water infiltration into the soil. This result is consistent with the underlying processes, which are relatively less difficult to capture using remote-sensing data than those that generate other soil function such as biodiversity patterns and processes (Unwin and Kriedemann 2000).

This example underlies that biophysical processes that generate the service are complex (e.g. soil-related services) and/or related to properties other than land cover (e.g. soil type). Thus, the accuracy of ecosystem service indicator maps can be low and maps should be carefully used.

How Maps of Ecosystem Service Indicators Are (Mis)Used and Considered

In the previous section, we illustrated that mapping ecosystem service indicators can be difficult, especially when the objective is to consider several services together (Le Clec'h et al. 2016). It is possible to provide maps of certain

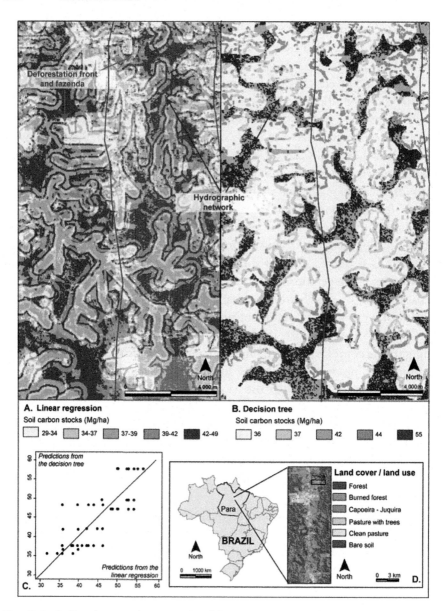

Fig. 13.2 Soil carbon stock maps using (**a**) linear regression (selected variables: land-cover classes and site) or (**b**) a decision tree (selected variables: land-cover classes and site). The methods are also compared using (**c**) a biplot graph. Maps correspond to Pacajà site, Brazilian Amazon, in 2007 (**d**)

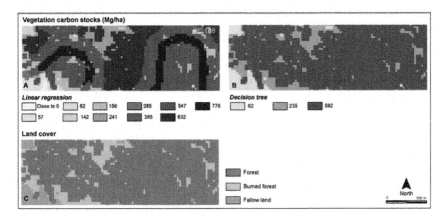

Fig. 13.3 Vegetation carbon stock maps (2007, Brazilian Amazon, Pacajá site, farm #108, see location in Fig. 13.2) using (**a**) linear regression (selected variables: land-cover classes and site) or (**b**) a decision tree (selected variables: land-cover classes and site). Land-cover map of the farm based and Landsat images (**c**)

services, but with a high degree of uncertainty. We believe that this limitation can have social and political implications because of the various ways that maps are used and considered. Thus, we argue that scientists should develop a more critical practice when mapping ecosystem services, which has not been the trend in the initial scientific literature dedicated to this subject.

The True Nature of Maps

To analyze how ecosystem service maps are used and considered, first we need to return to the nature of a map. Cartography is often considered a technique for representing data in a spatially explicit manner. Within this context, the goal of map making—a common activity in modern science—is to produce knowledge by using apparently neutral observation instruments that create a distance between the observer and the observed, which is considered essential to obtain objective results. However, the vision of maps, and more generally geomatic approaches (GIS, remote sensing), as neutral tools has been the subject of much theoretical and practical criticism by geographers and other social scientists. This critical dimension is long-established and has developed within critical cartography for the past 30 years (Wood 1992; Pickles 1994; Crampton and Krygier 2001). This body of work shows that maps, from an operational perspective, fail to produce complex representations of socio-ecological dynamics and, from a political standpoint, can help in controlling local populations and natural resources (Pickles 1994, 2004; Rajão 2013).

Table 13.1 Quality (coefficient of determination, R^2) of the methods used to map ecosystem service indicators

Service	Indicator	R^2 Decision tree	Linear regression	Number and nature of predictive variables selected for each method
Cultural, supporting	Biodiversity index	0.65	0.34	8: Land-cover classes, topographic position, site, slope, NDVI, elevation, NDWI, and trajectory of land cover
Production, pollination	Sphingidae richness	0.57	0.18	6: Site, distance to forest, elevation, land-cover classes, NDVI, and NDWI
Nutrient cycle, soil formation	Soil-engineer richness	0.46	0.26	NC
Production	Soil chemical quality index	0.67	0.50	7: NDWI, land-cover classes, site, elevation, slope, trajectory of land cover, and distance to hydrographic network
Climate regulation, soil fertility	Soil carbon stock	0.52	0.47	6: Site, distance to hydrographic network, trajectory of land cover, topographic position, NDWI, and land-cover classes
Production	Water available for plants	0.55	0.39	NC
Water cycle regulation, soil erosion control	Water infiltration rates into soil	0.66	0.57	2: Trajectory of land cover and land-cover classes
Climate regulation	Vegetation carbon stock	0.74	0.75	2: Land-cover classes and site

NC: not calculated

Like other scientific tools, maps are created using data that can only partially reproduce environmental complexity (Hausermann 2012). Further, not only are there technical limitations but also political and economic issues that explain and justify the conditions under which maps are made and used (Wood and Fels 1986; Pickles 1994; Crampton and Krygier 2001; Harley 2001; Wood and Krygier 2009). Similarly, the intensive use of remote sensing to describe environmental processes is not neutral: it excludes certain actors, provides limited understanding of environmental phenomena, and can produce controversial data (Fairhead and Leach 1996; Harwell 2000; Rajão 2013).

Lastly, while a map mirrors power relations, it also produces power; therefore, it contains a performative dimension (Crampton 2009). For example, Harris and Hazen (2006) show that maps, by indicating particular values for a given ecosystem, can lead to protection of the ecosystem by establishing boundaries for it, which can consequently shape the landscape. Maps are also potential tools for empowerment and can serve as a political tool. For example, studying local opposition to government practices in Indonesian forests, Peluso (1995) speaks of "counter mapping", in which maps are produced and used in protests and political demands. The use of mental maps and the emergence of participatory mapping reinforces "undisciplined mapping" because it allows stakeholders to guide the production of maps (Gould and White 1974; Alcorn 2000; Crampton and Krygier 2001; Del Casino and Hanna 2006; Noucher 2013).

Maps in the Ecosystem Service Literature

To assess whether the critical dimension is integrated into the creation of maps or not, we analyze how cartography is considered and practiced in studies of ecosystem services. To do so, we analyzed MEA reports, because the MEA is a crucial body of documents that promotes using the ecosystem service concept and maps, and the scientific literature on ecosystem service cartography from 1998 to 2012.

MEA reports were systematically analyzed, looking for (1) the words *map**, *carto**, *GIS*, and *remote sensing*; (2) all maps, regardless of subject, scale, and so on; and (3) citations of key authors of critical cartography. The reports included the three volumes of the MEA (*Current States & Trends, Scenarios, Policy Responses*), the report *Multiscale Assessments*, and the book *Bridging Scales and Knowledge Systems: Concepts and Applications in Ecosystem Assessment* authored by the *Sub-Global Working Group*.

Scientific literature concerning ecosystem service cartography was organized into three sets of documents: (1) 57 scientific articles published from 1998 to 2012 that address the mapping of ecosystem services (identified in the Web of Science database); (2) an edited volume: *Natural Capital: Theory and Practice of Mapping Ecosystem Services* (Kareiva 2011); and (3) scientific publications produced by the Ecosystem Services Partnership (ESP) (i.e. a special issue of *Ecosystem Services* published in 2013 following ESP's 2011 international conference, and abstracts from the short session "Mapping, visualisation and data access tools of ecosystem services" at ESP's 2015 international conference). All documents were studied using the same methods as the MEA reports.

Maps as Neutral Tools of Modern Science

There are at least two ways to see maps: as a neutral tool and a multiform object that contains and reflects certain socio-political issues. How maps are considered and used within the MEA documents is a good example of the "map as neutral tools" perspective, although this assertion changes based on the spatial scale at which the evaluation was conducted. Thus, in global assessment reports, maps have well-defined and limited objectives (Report 1; Chopra et al. 2005): assess land-cover changes, model ecological processes at a large scale or map risks. Maps are used as an assessment tool. For example, in the *Current States & Trends* report, 52% of the maps represent environmental degradation, 21% show the limits of biomes/ecosystems, 21% use economic and cultural indicators and 6% use physical data (e.g. albedo, evapotranspiration). The preferred scale is global: 70% of MEA maps represent the world. The remaining 30% represent mainly Africa and the tropics.

The MEA's use of the global scale and the subjects that are mapped are in line with the objective of assessing the world's ecosystems and their functions. Nonetheless, these choices are not neutral (Harris and Hazen 2006; Reid and Millennium Ecosystem Assessment (Program) 2006; Rajão 2013). Indeed, by showing environmental degradation as a global phenomenon, the MEA indirectly legitimizes global assessments, setting aside local struggles, solutions, and knowledge. Upscaling problems to the global scale implies that situations can be understood in the same way in very different regions and that these problems can be solved by coordinated, homogenous global action. In global maps, all regions of the world are treated as equal, but they do not provide detailed explanations of local spatial variations or causes, since the validity of data at the local level tends to be poor. Certain regions (and actors) are pinpointed as the source of problems, without considering macroeconomic factors and actors that help create local environmental problems. Thus, it is interesting that only one flux map exists in MEA reports (*Current States & Trends*, Chap. 7). In addition, global representation of environmental problems influences the subjects treated. For example, global representation of deforestation can give the idea that global climate change is the key environmental issue rather than the provision of environmental services to local inhabitants. When the global scale is favored, local phenomena difficult to map at a global scale can be easily ignored (e.g. bacteriological pollution, cultural values).

The MEA has no ecosystem service maps, but from 1998 to 2012 a growing number of scientific articles produced maps of ecosystem services using the same technical approach as in the MEA. This is probably because many of the authors of these articles were also involved in the MEA. Therefore, as in the MEA, location, distribution, and spatial variability of ecosystem services are considered as key elements to help institutions and decision makers in charge of ecosystems management (National Research Council 2005; Daily and Matson 2008; Naidoo et al. 2008; Eigenbrod et al. 2010). Following this logic, it is necessary to continually improve the methods behind and resolution and diversity of the data used to build ecosystem service cartography (Kandziora et al. 2013). Most scientists pinpoint the problem of the low accuracy of data used to produce the types of ecosystem service maps, but the limits identified are always linked to technical problems, which must be overcome by developing new methods and/or approaches, not to problems that result from researchers' own decisions, such as choice of scale. Furthermore, although recent articles clearly denounce the lack of map validation, since the problem of uncertainty is known (Crossman et al. 2013; Lavorel et al. 2014), the social and political dimensions of the maps produced are never discussed.

The lack of a critical approach when producing ecosystem services maps can also be demonstrated by analyzing the references that articles cited. While some publications in the field of critical cartography are cited in MEA's *Sub-Global Assessments* (e.g. Crampton 2001), they are absent from other reports and articles.

Maps as an Empowerment Tool

A second and less well-known way to consider and use ecosystem service maps exists in the literature. The conception of maps as multiform objects that contain and reflect certain socio-political issues is present in the MEA. Indeed, the text *Bridging Scales and Knowledge Systems* contains a critical perspective that highlights the complex scale issue in assessment processes and the need to combine multiple sources of information (Reid and MEA 2006). This text is unique in the MEA (Castro Larrañaga and Arnauld de Sartre 2014) because it contains a critical dimension and argues that environmental assessment (such as the MEA) and decision making should be based on multiple knowledge sources. It does not contain many maps, but geomatic tools are presented as non-neutral and useful tools that cannot, however, collect all appropriate information. It is the only text in the MEA that cites authors working on participatory mapping (Alcorn 2000) and critical cartography (Crampton and Krygier 2001).

MEA's *Sub-Global Assessments* include two dimensions not explored in the cartographic approach used for global-scale MEA reports. From an applied perspective, maps can be used as a tool for gathering other types of knowledge (e.g. lay, local). The role of participatory approaches in producing new knowledge is clearly highlighted in *Sub-Global Assessments* (Chaps. 5, 8, and 11) and *Bridging Scales and Knowledge Systems* (Chaps. 2, 6, 9, and 10). Maps are also mentioned as tools for improving communication between scientists and local stakeholders. Participatory mapping not only improves the quality of the information gathered, particularly local knowledge, but also helps empower local populations.

In the scientific articles on ecosystem service cartography that we analyzed, mapping as a process to upscale local information and empower local populations through participatory approaches was rare, and few articles used data collected from surveys of local stakeholders (Table 13.2). The empowerment dimension of mapping practices and references to critical cartography studies were absent (Table 13.2). Nonetheless, the idea of including as many actors as possible when mapping ecosystem services is growing within the scientific community. For example, the authors of *Natural Capital: Theory and Practice of Mapping Ecosystem Services* (Kareiva 2011) wrote a review article (Ruckelshaus et al. 2013) based on local studies in which they conclude that assessment of ecosystem services needs to improve integration of the variety of ecosystem valuation perspectives. They also argue for the need to increase empowerment of local experts (Ruckelshaus et al. 2013), even though this empowerment is probably not the same as that defended by

Table 13.2 Maps as a political tool in the scientific literature

Set of documents	Percentage of studies using bottom-up data (e.g. local knowledge) to map processes	Number of studies mentioning maps as an empowerment tool
Scientific articles (source: WOS 1998–2012; *n* = 57)	3	0
Abstracts in the 2015 ESP conference special session "Mapping, visualisation and data access tools of ecosystem services"	25	1
Case studies in *Natural Capital: Theory and Practice of Mapping Ecosystem Services* (Kareiva 2011)	10	0
Articles in special issue of *Ecosystem Services* (*n* = 13)	15	0

political ecologists or critical cartographers who generally consider the entire local community and/or poorest people and not only on local expert.

Conclusion

Mapping ecosystem services is difficult for certain specific services or when the objective is to consider several services together (Le Clec'h et al. 2016). When the underlying biophysical processes that generate the service are complex (e.g. soil-related services) or related to properties other than land cover (e.g. soil type), the accuracy of ecosystem service indicator maps can be low, and the methodological effort necessary to reach an acceptable level of accuracy is not realistic for applied uses. Indeed, remote sensing and GIS approaches are weak at getting at the subsurface, so its use in ecosystem services mapping privileges certain kinds of services over others. A Critical Physical Geography perspective on making/using such maps leads to two recommendations. It is crucial to consider (1) the uncertainty derived from the choice of statistical technique and the best indicators from a physical geography or ecological standpoint and (2) the social relevance or impact of technical choices. The accuracy issue recently became an important research topic with some studies about relevance and robustness of maps (Willemen et al. 2015). But most practices reduce mapping to a technological issue that can be solved by additional technology-based methods rather than an unavoidable limitation of using ecosystem service maps. Critical cartography studies clearly demonstrate that map production and use have sociological and political implications for resource and ecosystem management and, thus, contribute to power relationships. Indeed, a map is a rhetorical tool and not only a factual representation of reality; it functions within a specific spatial discourse, as a text. Currently, this critical perspective is nearly absent in the ecosystem service framework. Maps are commonly considered a neutral tool; when a viewpoint that is critical of maps is expressed, it is usually in publications at the margins of core science. Yet, we believe that this perspective is more crucial than technical questions in defining the real utility of the concept of ecosystem services. Thus, we argue that an urgent need exists to develop more critical studies of ecosystem service concepts, both in the scientific community and in application. Critical Physical Geography should play a large role in this work. Doing so could help natural scientists pay attention to socio-political issues related to the themes and tools in their work. It could also help social scientists understand the variable reliability of the maps. Indeed, the different origins, uses, and social effects of maps must be studied in a variety of geographical contexts. But to improve this analysis, which is the traditional area of work in critical cartography approaches, attention should focus on the diversity of the underlying biophysical processes that significantly affect map production. Detailed analysis of the socio-political implications of ignoring the complexity of biophysical processes when using ecosystem service maps, such as focusing on specific scales, themes, groups of people, and policy responses, still needs to be developed. How the misuse of the models in mapping exercise would have an effect on the possible beneficiaries of the map, considering the current or proposed policies, should be analyzed in the forthcoming years. Maps are a means to think, study, and debate, but must be analyzed in the political framework in which they originate (Bryan 2011) and not hide this framework. Mapping accurately all services is impossible because measuring related biophysical functions is complex. Thus an ecosystem services map value depends on its sociopolitical use as much as on its accuracy (Primmer and Furman 2012).

Acknowledgments We would like to thank N. Jégou for a very helpful advice on statistical analysis, the AMAZ ES team (P. Lavelle, T. Decaens, etc.) for field data and anonymous reviewers. This research was funded by the French ANR project "Approche géographique des services écosystémiques" (2011–2013) and by a grant to S. Le Clec'h from the Institut des Amériques.

References

Alcorn, J.B. 2000. Keys to unleash mapping's good magic. *Participatory Learning and Action Notes* 39: 10–13.

Arnauld de Sartre, X., M. Castro Larrañaga, B. Hubert, and C. Kull. 2014. Modernité écologique et services écosystémiques. In *Political Ecology Des Services Écosystémiques*, ed. X. Arnauld de Sartre, M. Castro, S. Dufour, and J. Oszwald, 31–47. Bruxelles: PIE Peter Lang.

Barnaud, C., and M. Antona. 2014. Deconstructing ecosystem services: Uncertainties and controversies around a socially constructed concept. *Geoforum* 56: 113–123.

Boyd, J., and S. Banzhaf. 2007. What are ecosystem services? The need for standardized environmental accounting units. *Ecological Economics* 63: 616–626.

Bryan, J. 2011. Walking the line: Participatory mapping, indigenous rights, and neoliberalism. *Geoforum* 42: 40–50.

Burkhard, B., F. Krolla, S. Nedkov, and F. Müller. 2012. Mapping ecosystem service supply, demand and budgets. *Ecological Indicators* 21: 17–29.

Castro Larrañaga, M., and X. Arnauld de Sartre. 2014. De la biodiversité aux services écosystémiques. Approche quantitative de la généalogie d'un dispositif. In *Political*

Ecology Des Services Écosystémiques, ed. X. Arnauld de Sartre, M. Castro, S. Dufour, and J. Oszwald, 49–81. Bruxelles: PIE Peter Lang.

Chopra, K.R., Millennium Ecosystem Assessment (Program), and Responses Working Group. 2005. In *Ecosystems and human well-being: Policy responses: Findings of the Responses Working Group of the Millennium Ecosystem Assessment*, ed. Kanchan Chopra et al. Washington, DC: The Millennium Ecosystem Assessment Series, Island Press.

Clec'h, S.L., J. Oszwald, T. Decaens, T. Desjardins, S. Dufour, M. Grimaldi, N. Jegou, and P. Lavelle. 2016. Mapping multiple ecosystem services indicators: Toward an objective-oriented approach. *Ecological Indicators* 69: 508–521.

Costanza, R., R. d'Arge, R. de Groot, S. Farber, M. Grasso, B. Hannon, S. Naeem, et al. 1997. The value of the world's ecosystem services and natural capital. *Nature* 387: 253–260.

Crampton, J. 2001. Maps as social constructions: Power, communication and visualisation. *Progress in Human Geography* 25: 235–252.

Crampton, J.W. 2009. Cartography: Performative, participatory, political. *Progress in Humen Geography* 33: 840–848.

Crampton, J.W., and J. Krygier. 2001. An introduction to critical cartography. *ACME: An International E-Journal for Critical Geographies* 4: 11–33.

Crossman, N.D., B. Burkhard, S. Nedkov, L. Willemen, K. Petz, I. Palomo, E.G. Drakou, et al. 2013. A blueprint for mapping and modelling ecosystem services. *Ecosystem Services* 4: 4–14.

Daily, G.C., and P.A. Matson. 2008. From theory to implementation. *Proceedings of the National Academy of Sciences* 105: 9455–9456.

Davies, Z.G., J.L. Edmondson, A. Heinemeyer, J.R. Leake, and K.J. Gaston. 2011. Mapping an urban ecosystem service: Quantifying above-ground carbon storage at a city-wide scale: Urban above-ground carbon storage. *Journal of Applied Ecology* 48: 1125–1134.

Del Casino, V.J., and S.P. Hanna. 2006. Beyond the "binaries": A methodological intervention for interrogating maps as representational practices. *ACME: An International E-Journal for Critical Geographies* 4: 34–56.

Dominati, E., M. Patterson, and A. Mackay. 2010. A framework for classifying and quantifying the natural capital and ecosystem services of soils. *Ecological Economics* 69: 1858–1868.

Dufour, S., X. Arnauld de Sartre, M. Castro Larrañaga, S. le Clec'h, and J. Oszwald. 2014. Cartographie, services écosystémiques et gestion environnementale: entre neutralité technicienne et outil d'empowerment. In *Political Ecology Des Services Écosystémiques*, ed. X. Arnauld de Sartre, M. Castro, S. Dufour, and J. Oszwald, 225–246. Bruxelles: PIE Peter Lang.

Dymond, J.R., A.-G.E. Ausseil, J.C. Ekanayake, and M.U.F. Kirschbaum. 2012. Tradeoffs between soil, water, and carbon—A national scale analysis from New Zealand. *Journal of Environmental Management* 95: 124–131.

Eigenbrod, F., P.R. Armsworth, B.J. Anderson, A. Heinemeyer, S. Gillings, D.B. Roy, C.D. Thomas, and K.J. Gaston. 2010. The impact of proxy-based methods on mapping the distribution of ecosystem services. *Journal of Applied Ecology* 47: 377–385.

Ernstson, H., and S. Sörlin. 2013. Ecosystem services as technology of globalization: On articulating values in urban nature. *Ecological Economics* 86: 274–284.

European Commission, Directorate-General for the Environment. 2013. *Mapping and assessment of ecosystems and their services an analytical framework for ecosystem assessments under action 5 of the EU biodiversity strategy to 2020: Discussion paper—Final, April 2013*. Luxembourg: Publications Office.

Fairhead, J., and M. Leach. 1996. *Misreading the African landscape: Society and ecology in a forest-savanna mosaic, African studies series*. Cambridge and New York: Cambridge University Press.

Gomez-Baggethun, E., and M. Ruiz-Perez. 2011. Economic valuation and the commodification of ecosystem services. *Progress in Physical Geography* 35: 613–628.

Gould, P., and R. White. 1974. *Mental maps*. Harmondsworth: Penguin.

Grimaldi, M., J. Oszwald, S. Dolédec, M. del P. Hurtado, I. de Souza Miranda, X. Arnauld de Sartre, W.S. de Assis, et al. 2014. Ecosystem services of regulation and support in Amazonian pioneer fronts: Searching for landscape drivers. *Landscape Ecology* 29: 311–328.

Harley, J.B. 2001. *The new nature of maps: Essays in the history of cartography*. Baltimore, MD: Johns Hopkins University Press ed.

Harris, L.M., and H.D. Hazen. 2006. Power of maps: (Counter) mapping for conservation. *ACME: An International E-Journal for Critical Geographies* 4: 99–130.

Harwell, E. 2000. Remote sensibilities: Discourses of technology and the making of Indonesia's natural disaster. *Development and Change* 31: 307–340.

Hauck, J., C. Görg, R. Varjopuro, O. Ratamäki, J. Maes, H. Wittmer, and K. Jax. 2013. "Maps have an air of authority": Potential benefits and challenges of ecosystem service maps at different levels of decision making. *Ecosystem Services* 4: 25–32.

Hausermann, H. 2012. From polygons to politics: Everyday practice and environmental governance in Veracruz, Mexico. *Geoforum* 43: 1002–1013.

Hewson, J., M. Steininger, and S. Pesmajoglou. 2013. *REDD+ Measurement, Reporting and Verification (MRV) Manual*. Washington, DC: USAID-supported Forest Carbon, Markets and Communities Program.

Kandziora, M., B. Burkhard, and F. Müller. 2013. Mapping provisioning ecosystem services at the local scale using data of varying spatial and temporal resolution. *Ecosystem Services* 4: 47–59.

Kareiva, P.M., ed. 2011. *Natural capital: Theory & practice of mapping ecosystem services*. New York: Oxford University Press.

Karsenty, A., and D. Ezzine de Blas. 2014. Du mésusage des métaphores. Les paiements pour services environnementaux sont-ils des instruments de marchandisation de la nature? In *L'instrumentation de L'action Publique*, ed. Ch. Halpern, P. Lascoumes, and P. Le Galès, 161–189. Paris: Presses de Sciences Po.

Kronenberg, J. 2015. Betting against human ingenuity: The perils of the economic valuation of nature's services. *BioScience* 65: 1096–1099.

Lamarque, P., F. Quétier, and S. Lavorel. 2011. The diversity of the ecosystem services concept and its implications for their assessment and management. *Comptes Rendus Biologies* 334: 441–449.

Lascoumes, P., and P. Le Galès, eds. 2004. *Gouverner par les instruments, Gouvernances*. Paris: Presses de la Fondation nationale des sciences politiques.

Lave, R., M.W. Wilson, E.S. Barron, C. Biermann, M.A. Carey, C.S. Duvall, L. Johnson, et al. 2014. Intervention: Critical Physical Geography: Critical Physical Geography. *The Canadian Geographer/Le Géographe canadien* 58: 1–10.

Lavelle, P., T. Decaëns, M. Aubert, S. Barot, M. Blouin, F. Bureau, P. Margerie, P. Mora, and J.-P. Rossi. 2006. Soil invertebrates and ecosystem services. *European Journal of Soil Biology* 42: S3–S15.

Lavorel, S., A. Arneth, A. Bayer, A. Bondeau, S. Lautenbach, N. Marba, A. Ruiz, et al. 2014. Transferable geo-referenced metrics and GIS based quantification functions—Pathways to the incorporation of biodiversity into ecosystem service biophysical assessment. Operational Potential of Ecosystem Research Applications. Unpublished report, available at http://operas-project.eu/.

le Clec'h, S., S. Dufour, J. Oszwald, M. Grimaldi, and N. Jégou. 2014. Spatialiser des services écosystémiques, un enjeu méthodologique et plus encore. In *Political Ecology Des Services Écosystémiques*, ed. X. Arnauld de Sartre, M. Castro, S. Dufour, and J. Oszwald, 205–224. Bruxelles: PIE Peter Lang.

Maris, V. 2014. *Nature à vendre: les limites des services écosystémiques*. Paris: Quae éditions.

Milanesi, J. 2010. Éthique et évaluation monétaire de l'environnement: la nature est-elle soluble dans l'utilité? VertigO 10, https://doi.org/10.4000/vertigo.10050.

Millennium Ecosystem Assessment (Program). 2005. *Ecosystems and human well-being: Synthesis*. Washington, DC: Island Press.

Naidoo, R., A. Balmford, R. Costanza, B. Fisher, R.E. Green, B. Lehner, T.R. Malcolm, and T.H. Ricketts. 2008. Global mapping of ecosystem services and conservation priorities. *Proceedings of the National Academy of Sciences* 105: 9495–9500.

National Research Council (U.S.). 2005. *Valuing ecosystem services: Toward better environmental decision-making*. Washington, DC: National Academies Press.

Nemec, K.T., and C. Raudsepp-Hearne. 2012. The use of geographic information systems to map and assess ecosystem services. *Biodiversity and Conservation* 22: 1–15.

Norgaard, R.B. 2010. Ecosystem services: From eye-opening metaphor to complexity blinder. *Ecological Economics* 69: 1219–1227.

Noucher, M. 2013. Introduction. *L'Information géographique* 77: 6.

Pagella, T.F., and F.L. Sinclair. 2014. Development and use of a typology of mapping tools to assess their fitness for supporting management of ecosystem service provision. *Landscape Ecology* 29: 383–399.

Peluso, N.L. 1995. Whose woods are these? Counter-mapping forest territories in Kalimantan, Indonesia. *Antipode* 27: 383–406.

Pickles, J. 1994. *Ground truth: The social implications of geographic information systems.* New York: The Guilford Press.

———. 2004. *A history of spaces: Cartographic reason, mapping, and the geo-coded world.* London and New York: Routledge.

Portela, R., and I. Rademacher. 2001. A dynamic model of patterns of deforestation and their effect on the ability of the Brazilian Amazonia to provide ecosystem services. *Ecological Modelling* 143: 116–146.

Potschin, M., and R. Haines-Young. 2011. Introduction to the special issue: Ecosystem services. *Progress in Physical Geography* 35: 571–574.

Primmer, E., and E. Furman. 2012. Operationalising ecosystem service approaches for governance: Do measuring, mapping and valuing integrate sector-specific knowledge systems? *Ecosystem Services* 1: 85–92.

Rajão, R. 2013. Representations and discourses: The role of local accounts and remote sensing in the formulation of Amazonia's environmental policy. *Environmental Science & Policy* 30: 60–71.

Reid, W.V., and Millennium Ecosystem Assessment (Program). 2006. *Bridging scales and knowledge systems: Concepts and applications in ecosystem assessment, A contribution to the Millennium Ecosystem Assessment.* Washington, DC: Island Press.

Ruckelshaus, M., E. McKenzie, H. Tallis, A. Guerry, G. Daily, P. Kareiva, S. Polasky, et al. 2013. Notes from the field: Lessons learned from using ecosystem service approaches to inform real-world decisions. Ecological Economics. https://doi.org/10.1016/j.ecolecon.2013.07.009.

Tadaki, M., W. Allen, and J. Sinner. 2015a. Revealing ecological processes or imposing social rationalities? The politics of bounding and measuring ecosystem services. *Ecological Economics* 118: 168–176.

Tadaki, M., G. Brierley, M. Dickson, R. Le Heron, and J. Salmond. 2015b. Cultivating critical practices in physical geography: Cultivating critical practices in physical geography. *The Geographical Journal* 181: 160–171.

Unwin, G.L., and P.E. Kriedemann. 2000. Principles and processes of carbon sequestration by trees, Technical paper/Forest Research and Development Division. Sidney: State Forests of New South Wales.

Waage, S., K. Armstrong, and L. Hwang. 2011. New business decision-making aids in an era of complexity, scrutiny, and uncertainty. Tools for identifying, assessing, and valuing ecosystem services. BSR's Ecosystem Services, Tools & Markets Working Group. Unpublished report, available at https://www.bsr.org.

Willemen, L., B. Burkhard, N. Crossman, E.G. Drakou, and I. Palomo. 2015. Editorial: Best practices for mapping ecosystem services. *Ecosystem Services* 13: 1–5.

Wolff, S., C.E.J. Schulp, and P.H. Verburg. 2015. Mapping ecosystem services demand: A review of current research and future perspectives. *Ecological Indicators* 55: 159–117.

Wood, D. 1992. *The power of maps.* New York/London: The Guilford Press.

Wood, D., and J. Fels. 1986. Designs on signs/Myths and meaning in maps. *Cartographica* 24: 54–103.

Wood, D., and J. Krygier. 2009. Maps and protest. In *International encyclopedia of human geography*, ed. R. Kitchin and N. Thrift, 436–441. Oxford: Elsevier.

Zhu, X., S. Pfueller, P. Whitelaw, and C. Winter. 2010. Spatial differentiation of landscape values in the Murray river region of Victoria, Australia. *Environmental Management* 45: 896–911.

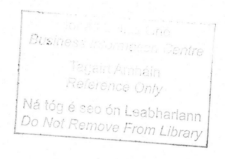

14

Beyond "the Mosquito People": The Challenges of Engaging Community for Environmental Justice in Infested Urban Spaces

Dawn Biehler, Joel Baker, John-Henry Pitas, Yinka Bode-George, Rebecca Jordan, Amanda E. Sorensen, Sacoby Wilson, Heather Goodman, Megan Saunders, Danielle Bodner, Paul T. Leisnham, and Shannon LaDeau

---

D. Biehler (✉) • J. Baker • J.-H. Pitas
Department of Geography and Environmental Systems, University of Maryland, Baltimore Country, Baltimore, MD, USA

Y. Bode-George
CommonHealth Action, Washington, DC, USA

R. Jordan
Department of Human Ecology, Rutgers University, Brunswick, NJ, USA

A.E. Sorensen
School of Natural Resources, University of Nebraska-Lincoln, Lincoln, NE, USA

S. Wilson
Applied Environmental Health, University of Maryland College Park, College Park, MD, USA

H. Goodman
Cary Institute of Ecosystem Studies, Center for Urban Environmental Research and Education, Baltimore, MD, USA

M. Saunders • D. Bodner • P. T. Leisnham
Department of Environmental Science and Technology, University of Maryland, College Park, MD, USA

S. LaDeau
Cary Institute of Ecosystem Studies, Millbrook, NY, USA

© The Author(s) 2018
R. Lave et al. (eds.), *The Palgrave Handbook of Critical Physical Geography*,
https://doi.org/10.1007/978-3-319-71461-5_14

One hot Thursday morning, in June 2012, residents and passersby in the Franklin Square neighborhood of southwest Baltimore, Maryland, witnessed a curious but not entirely unfamiliar sight. Eight people wearing matching gray t-shirts and carrying clipboards and caddies of equipment fanned out on a city block, some of them combing through backyards and alleyways, others knocking on front doors and chatting with residents on stoops. In this mostly black neighborhood, it was notable that seven of the visitors appeared white, and one Latino. "Are you from the city?" passersby asked. "What are you looking for?" Members of the group answered, "we're from the University of Maryland and the Baltimore Ecosystem Study. We're looking for mosquitoes."

Residents were accustomed to occasional visits from "the city"—municipal government workers—as well as from other Baltimore Ecosystem Study scientists, and throngs of volunteers planting trees. Soon, crews scrutinizing trash piles, tires, and flowerpots for mosquito larvae also became a regular sight, appearing throughout the summer in Franklin Square and adjacent neighborhoods. The researchers also attended community meetings and block parties, organized public meetings of their own, and engaged youth at summer camps. Their research prospectus stated their aim to advance environmental justice in communities long deprived of municipal services and investment. The socio-natural web that interested the researchers centered on mosquitoes but entangled more than these biting insects that, as of 2012, seemed to many residents quite disconnected from more pressing concerns in their neighborhoods. It is unsurprising, though, that many residents called this group "the mosquito people," given the prominence of their mosquito-sampling activities. Would this attention to mosquitoes brand the researchers as too narrowly focused, or even irrelevant?

In this chapter we, "the mosquito people," grapple with the challenges of engaging multiple modes of knowledge production in urban environmental science—not only that of ecologists and political ecologists but also of community members with their own social and environmental justice agendas. Physical and social scientists who study urban environments have already overcome prejudices in their disciplines against environments that are not "pristine" (see Urban, this volume). But choosing a city as one's research site does not automatically make research "critical" and supportive of environmental justice.

Environmental science often appears to isolate specific, non-human objects, focusing attention away from entangled networks of things and processes that concern justice activists. This chapter examines how a Critical Physical Geography approach can inform engagement with city residents while helping to supplant narrow and apolitical research approaches.

We begin by briefly examining the ways unwanted animals—the creatures we call "pests"—have in recent history been the focus of reductionist, apolitical science, which informs reductionist, apolitical control practices. We then discuss key recent insights in participation studies, drawing upon scholarship on citizen-science and environmental justice. The bulk of the chapter narrates how the Baltimore Mosquito Study kept its focus on mosquitoes while also striving to embrace residents' concerns for a wider range of neighborhood problems. We conclude by pointing to some key considerations for critical physical geographers as we strive to support social and environmental justice.

The Perils and Necessities of Focusing on the Pest

The Baltimore Mosquito Study is but one of the latest scientific and health-related endeavors to grapple with the place of unwanted animals in broader environmental and social issues. Prior to the late nineteenth century, public health authorities attributed disease primarily to broad environmental quality and, accordingly, aimed to create and improve sanitation systems to support health, sometimes also supporting social justice (Platt 2005; Taylor 2009). As the new science of medical entomology revealed that some diseases were transmitted by insects, authorities gained new insights into which specific parts of the environment could affect the spread of infections such as malaria or yellow fever (Patterson 2009). At the origins of medical entomology in the late nineteenth and early twentieth centuries, some public health agencies in the USA used new discoveries about insects' vector role to justify intensified attention to environmental sanitation (Rogers 1989). Malaria control, for example, often entailed application of arsenic-based insecticides as well as management of the water bodies where mosquitoes breed, and shoring up of houses to prevent mosquitoes from entering through open windows or crevices. In the 1930s, the Tennessee Valley Authority also foregrounded issues of poverty and underdevelopment in its efforts to reduce malaria infection in the southeastern USA (Humphreys 2001). Although socio-environmental holism helped inform this development effort, a more techno-scientific approach eventually dominated (Carter 2014). Thus the science and practice of vector management have a long tradition of integrating social and ecological knowledge, though often in a way that stresses technical intervention over systematic social and political change.

New pesticide technologies that became available after World War II enabled public health authorities and other pesticide users to downplay physical environmental conditions, at least for a time, while some pesticide users even saw dichlorodiphenyltrichloroethane (DDT) and similar chemicals as a way of overcoming social problems. Synthetic pesticide critics observed that DDT seemed so effective in the early years of civilian applications that many users abandoned previous holistic efforts to manage conditions that sustained unwanted animals. Users from housewives to farmers to landscapers sought an easy chemical kill rather than tending to sanitation or caring for crop health in ways that would limit animals' food or opportunities to reproduce (Carson 1962). Critics argued that reliance on synthetic pesticides was based in reductionist, mechanistic understandings of nature (Carson 1962). Still, many programs, from Peronist interventions in rural Argentina to the Black Panther Party's community services, saw the new pesticides as routes to better living conditions and development in poor regions and neighborhoods, even while bringing serious environmental impacts (Carter 2009; Shaw et al. 2010; Biehler 2013). Some projects attempted rudimentary forms of public participation to enroll community members and engage their knowledge about unwanted animals (Biehler 2013). Meanwhile, other programs simply prescribed standard pesticide application regimes without regard to local environmental *or* social conditions and against the resistance of targeted communities (Webb 2011; Biehler 2013).

Rachel Carson's 1962 book *Silent Spring* galvanized opposition to synthetic pesticides in the emerging mainstream (white, middle-class) American environmental movement and inspired ecologists to re-integrate knowledge of the physical environment into pest management. Ecologists and sympathetic entomologists advocated integrated pest management (IPM), a framework for *integrating* complex ecosystem relationships into decisions about whether to, for example, fortify human structures, release predatory insects, improve crop health, or, as the very last resort, apply judiciously small amounts of chemical pesticides. Advocates of IPM drew lessons from systems theory, emphasizing the need to understand feedbacks among multiple factors impinging on the ecology of pests, including both human systems and physical environment factors (Flint and Van Den Bosch 1981). Yet as Lave et al. point out in the introduction to this volume, when scientists imagine "factors" or "drivers" in a multifaceted system, they do not adequately account for the deep imbrication of human society and politics within nature. Thus IPM did not guarantee a truly *critical* approach to managing unwanted animals (Biehler 2013). Pest control in urban neighborhoods has seldom addressed issues of broad environmental injustice, that is, the systemic problems of racism and disinvestment that create ample niches for the disease vectors and other creatures that trouble marginalized communities (Biehler 2013). Mosquito-control programs that only call for residents to tip standing water containers similarly

miss the reasons why some neighborhoods have so many more container habitats than others (Tedesco et al. 2010; Unlu et al. 2013; LaDeau et al. 2013).
Critical Physical Geography can intervene by placing unwanted animals
within the context of processes—such as racial segregation—that are inextricably political and ecological.

While we urge greater attention to the complex socio-natural web that
sustains mosquitoes, we must not lose sight of mosquito ecologies. To put this
another way: while we problematize the ways that modern pest control methods have treated unwanted animals as distinct from the socio-natural webs in
which they are entangled, we must also recognize that animals possess their
own ecological and biological conditions of life (Nash 2006) and are not *only*
epiphenomena of the environment, racist policies, or capitalism. The arrival
of Zika virus in Maryland in late spring of 2016 via mobile human bodies
reminds us of the need for a framework that integrates knowledge of mosquito ecology *and* of urban political ecology and environmental justice (Abara
et al. 2012). The conditions that sustain the tiger mosquito, *Aedes albopictus*,
in west Baltimore are problems in themselves, and mosquitoes *also* bring distinct material problems of nuisance biting and disease potential to the city.
We need to *both* understand those material particularities of unwanted animals—where, when, and how quickly they breed; what hosts do they feed on
when and where—*and* connect them with the overarching process of disinvestment in west Baltimore. Thus the Baltimore Mosquito Study brought
together the specifics of mosquitoes and the political concerns of environmental justice to advance a critical approach to pest management.

Disparate Approaches to Participation

"The mosquito people" aimed to engage community members in the production of knowledge about the socio-natural web in which they lived. Public
participation is the subject of several growing bodies of scholarship in different disciplines, with disparate goals, values, and epistemologies. Baltimore
Mosquito Study team members themselves brought to the project different
scholarly and political perspectives on participation. A full review of the "participatory turn" (Eden and Bear 2012) is beyond the scope of this chapter, but
we will touch upon a few perspectives that inform our work and that deserve
consideration for scholars working toward inclusive processes of knowledge
production.

One lab that is part of the Baltimore Mosquito Study includes scholars who
study citizen-science; the principal investigator in this lab, Rebecca Jordan,

has in past projects engaged mostly privileged, highly educated environmentalists in gathering data about topics such as invasive plant species (Jordan et al. 2011). Among other things, her research measures participants' attitudes and sense of efficacy before and after engaging in data collection and interventions in environmental problems, while also investigating how useful citizen-science data is for testing hypotheses or monitoring environmental change. Some citizen-science scholars have shifted their attention away from citizens' contributions of data to scholarly study and more toward the role science activities might play in invigorating local cultures of environmental stewardship (Krasny et al. 2014). As we will discuss more later, this vein of citizen-science scholarship informed our own learning and activities in west and southwest Baltimore.

Another principal investigator, Sacoby Wilson, brought experience with participation from his environmental justice research. Scholarship in environmental justice is itself diverse; some of this literature relies on quantitative measures to gauge disparate impacts (distributive justice), but our team members mixed quantitative and qualitative methods to understand the processes that lead to these injustices and support communities' capacity to address them (Holifield 2001). Furthermore, Wilson et al. (2010) have criticized "extractive" research for deriving value from resources found in a location without benefitting the community itself. We encountered this very concern at a community meeting early on in the project, when residents accused previous researchers of only "taking their data." According to Wilson et al. (2010), so-called community-based participatory research (CBPR) does not sufficiently engage and benefit communities. Instead, they developed a community-owned and -managed research model (COMR), which they argue has the potential to sustain community reinvestment and job creation. The Baltimore Mosquito Study did not fulfill the ideal of COMR as Wilson et al. (2010) had in other settings, but Wilson's emphasis on qualitative narratives and process shaped our community engagement activities in important ways. Furthermore, the team also built feedback and support systems between community and researchers, even if this did not result in *ownership* of the research by local institutions.

Critical social scientists from several disciplines have also raised questions about the ethics of public participation in science and environmental decision-making. Some scholars argue that the neoliberal state may use participation to co-opt marginalized communities, appearing to include them while actually defining the terms of participation in ways that enroll layfolk in perpetuating existing power structures and practices of self-regulation (Holifield 2004; Ottinger 2013). Indeed, scientists could inadvertently extend dynamics of environmentality—whereby governments and quasi-governments enlist citizens in self-regulation of environmental behavior (Agarwal 2005)—by asking residents to dedicate their time to an activity that monitors populations and bodies in relation to their environmental impacts but does not turn their attention to more powerful entities such as investors, property owners, trash-hauling companies, and the state, which may be responsible for more environmental harm.

At the same time, the ecologies of unwanted animals often blur the boundaries of public and private space (Biehler 2013), and residents of disinvested communities have daily contact with and knowledge of mosquito habitats. Their knowledge and experience in managing urban space is crucial to their own comfort and environmental health. Citizen engagement in environmental issues can help invigorate and inform political engagement at a larger scale (Bartlett-Healey et al. 2014). Thus critical physical geographers have another thin line to tread when it comes to participation—we want to avoid deepening the neoliberal politics of personal responsibility while enlivening community knowledge and advocacy for broader-scale social change.

The Story of the Baltimore Mosquito Study

The Baltimore Mosquito Study grew out of a smaller study of mosquito ecologies and perceptions of mosquitoes in Washington, DC (Dowling et al. 2013b). Based on the earlier research, the principal investigators—a disease ecologist specializing in urban ecology and statistical modeling, an ecologist specializing in mosquito management, and an urban political ecologist/historical geographer—decided to pursue a larger study asking whether active resident involvement could result in improved mosquito management results at a community-wide scale. Two additional principal investigators joined the team— a scholar of environmental justice and environmental health and an evolutionary biologist who studies environmental education and citizen-science—along with a community partner, Baltimore's Parks and People Foundation.

Some previous studies of community mosquito management had tested the outcomes of providing educational materials to residents (Bartlett-Healey et al. 2011; Dowling et al. 2013a). The expanded team now hoped to engage community members in citizen-science, youth activities, civic actions, and public advocacy to address hazards that could breed mosquitoes and threaten residents with other insults to health and well-being. The team would engage residents in producing knowledge about mosquitoes and the socio-natural web to which insects as well as they themselves belong. Also, having examined mosquitoes in neighborhoods across a socio-economic spectrum in a city with a thriving housing market—Washington, DC—the researchers now hoped to learn whether a city that had suffered considerable population loss might feature different conditions for mosquito breeding and human exposure to mosquito-borne disease. How might mosquito ecologies be interwoven with social processes such as urban population loss, uneven development, and public and private disinvestment? And could active resident involvement in knowledge production about health hazards reshape the socio-natural environment of disinvested neighborhoods?

Site Selection: Uneven Development

The research team chose to study four neighborhoods in west and southwest Baltimore that varied in the degree to which they had been affected by disinvestment and population loss. Some team members framed this variation as a socio-economic spectrum, while others looked to neighborhoods' histories and housing conditions to examine the ways uneven development and legacies of racial discrimination shaped neighborhood ecologies. Researchers could see the privilege enjoyed by one neighborhood that has received robust investment in municipal services and infrastructure—Bolton Hill—alongside three others affected by dramatic legacies of racial redlining starting in the 1930s (Pietila 2010). In the early 1960s city planners targeted parts of two of these neighborhoods—Franklin Square and Harlem Park—for intensive demolitions under the so-called urban renewal program (Williams 2013). One of these—Harlem Park—had also been cut off from neighborhoods to the south by highway planning in the 1960s. The road project, now ruefully dubbed the "highway to nowhere," also entailed government seizure of properties, which worsened housing abandonment and population loss (Gioielli 2014). Meanwhile, the neighborhood furthest to the south—Union Square—has seen some reinvestment as gentrification has crept north and west from Baltimore's Inner Harbor. Thus Harlem Park, Franklin Square, Union Square, and Bolton Hill showed the ways some neighborhoods benefited and others lost out in uneven development processes (Fig. 14.1).

Harlem Park suffered repeated rounds of disinvestment and demolition, leaving it with the highest density of abandoned buildings and vacant lots and the lowest income levels. It also was home to the highest percentage of African-American residents among the study neighborhoods—both in the present day and historically, and historical research points to race-based discrimination in housing and development as a major reason for this area's decline. Franklin

Square has slightly lower rates of abandonment and vacancy and slightly higher income levels, with Union Square's income levels markedly higher than that. Bolton Hill differed sharply from the other neighborhoods, with rates of abandonment and vacancy near zero, a higher income level supported by nearby educational institutions where many residents work, and a much higher percentage of white residents (Baltimore Neighborhood Indicators

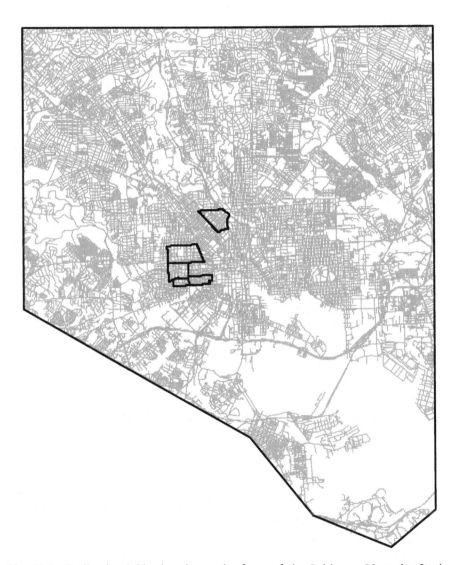

Fig. 14.1 Outlined neighborhoods are the focus of the Baltimore Mosquito Study. From North to South: Bolton Hill, Harlem Park, Franklin Square, and Union Square-Hollins Market

Table 14.1 Demographics and housing conditions in Baltimore Mosquito Study focal neighborhoods

	Median household income (avg household size)	Percent African-American	Life expectancy	Percentage of properties that are vacant and abandoned
Sandtown/Harlem Park	24,822 (2.6)	96.6	69.7	34.3
Franklin Square/ Union Square (Southwest)	25,199 (2.8)	75.8	68.3	27.1
Bolton Hill	36,070 (1.6)	32.1	76.0	3.6
Baltimore overall	41,385 (2.4)	63.8	73.5	8.0

The Baltimore Neighborhood Indicators Alliances groups Franklin Square and Union Square together
Source: Baltimore Neighborhood Indicators Alliance 2014

Alliance 2014). Urban environmental research can explain deep, systematic processes that shape and are shaped by physical environmental change if it examines activities such as race-based disinvestment and gentrification (Jakle and Wilson 1992). The research team was interested in how uneven development shaped habitats for mosquitoes, how disparate historical geographies might support or suppress community engagement in environmental management, and the degree to which municipal agencies served the communities (Table 14.1).

These neighborhoods were also a potentially receptive location for community engagement based on contacts already established by the Baltimore Ecosystem Study (BES), with which some researchers were affiliated, though the team planned engagement activities that BES had not attempted before. BES is one of two *urban* long-term ecological research sites (LTER) in the USA funded by the National Science Foundation. BES has strong relationships with local environmental organizations and schools and includes research linking inequality and urban ecology (Grove et al. 2015) though it is less active with social justice groups. BES supports community education and greening activities, but "the mosquito people" planned to add more qualitative and pluralistic knowledge production practices to BES's research profile.

Mosquitoes at the Center

Building on an existing adult mosquito-trapping program, we began larval sampling and door-to-door surveys of residents in one neighborhood in 2012. In unoccupied parcels, our field crew sampled all the containers we could safely reach, often clambering over piles of construction waste and wading through waist-high vegetation that grew in the backyards of abandoned buildings. We knocked on doors of occupied residences to seek permission to sample containers on each property for larvae. No one denied us access, and most were curious about the project. These visits often led to free-form conversations about environmental conditions, especially neglect by municipal services such as trash collection. The team's research specialist Heather Goodman was especially effective and compassionate in enlisting community members' support. People who were homeless or lacked stable housing became advocates and lookouts for the project, making sure no one tampered with adult mosquito traps stationed throughout the neighborhood.

We asked each resident who gave us permission to sample their parcel to complete a brief "KAP" survey, a common public health tool whose purpose was to understand "knowledge, attitudes, and practices" (KAP) surrounding mosquitoes (Tuiten et al. 2009). The original format of our KAP survey contributed to our reputation as "the mosquito people," leading with questions on mosquitoes rather than first inquiring about general environmental issues in the neighborhood. The survey asked about whether mosquitoes bothered the respondent, where they thought mosquitoes came from, whether they tried to control mosquitoes, who they thought should be responsible for mosquito control, and whether they were concerned about diseases spread by mosquitoes. Many respondents in both 2012 and 2013 said that mosquitoes did trouble them, and that they stayed indoors or avoided infested locations. But everyone ranked other creatures—rats, bedbugs, or cockroaches—as more troubling than mosquitoes. At the end of the survey we asked respondents what they considered the most important environmental problems in their neighborhoods; we later moved this question to the beginning to set an open-ended tone. Residents listed trash, abandoned buildings, vacant lots, trees and overgrown vegetation, drug dealers, crime, and rats as important environmental issues, but few mentioned mosquitoes. These other pressing priorities signaled the need to de-center the mosquito in our communications with residents. Furthermore, the very epistemology of individual surveys assumes that environmental knowledge is constructed individually rather than relationally. The surveys provided important information about how large numbers of residents experienced mosquitoes in the environment, and they helped us open up conversations with residents, but we would need to complement this research method to more fully explore the socio-natural webs that humans and mosquitoes shared.

KAP responses also forced us to grapple with environmental problems previously unknown to some team members. For example, many residents reported that mosquitoes came from "overgrown" trees and vegetation on vacant lots, backyards of abandoned buildings, and unmowed lawns. Suspicion of unmanaged vegetation was a common theme documented systematically (Williams 2013) and anecdotally by other researchers in Baltimore. At one community meeting we attended, for instance, residents responded to proposals by design students for new tree-planting by demanding whether resources allocated for new trees could be redirected to better maintain existing plantings. Many advocates for "urban greening" see new trees as unproblematically positive additions to urban neighborhoods (Pincetl et al. 2013), but for residents who cannot afford to hire an arborist, in areas underserved by municipal environmental services, trees and other vegetation represent a hazard that can fall on people, roofs, and cars that attract trash-dumping and other, worse crimes—and also breed mosquitoes. Anti-racist geographers such as Carolyn Finney have pointed to even more sinister meanings of trees in African-American culture (Finney 2014). Residents' concerns about unmanaged vegetation and other issues have complicated our understanding of mosquitoes in the local eco-social web, and have also helped encourage new research efforts, such as studying vegetation's role in mosquito ecology (Fig. 14.2).

Fig. 14.2 Unmanaged vegetation in the backyard of an abandoned house. Many residents attributed mosquito infestation to landscapes like these, and indeed, in late summer, our sampling crews experienced constant mosquito biting in areas with tall vegetation

In addition to KAP responses, we planned citizen-science studies that would also provide avenues for residents' knowledge to shape the study and inform our results. With both critiques (Wilson et al. 2010) and support of citizen-science in mind, the research team planned to channel participants' enthusiasm for studying mosquitoes into meaningful civic ecology projects such as a tire drive to remove a major source of standing water. In summer 2013, four residents completed a series of observations and returned data and mosquito samples to the team members leading this portion of the project. This small number itself became feedback that informed the future trajectory of the project. Part of the research question was whether it would be possible to engage sufficient numbers of residents and collect accurate enough data to expand mosquito monitoring through citizen-science. Neither the number of participants nor the accuracy of their data supported the prospect for citizen-based mosquito monitoring, so we would need a different approach to mosquito ecology that encouraged resident involvement in mosquito management even if they didn't produce data ready for analysis and publication.

While few adult citizen-scientists volunteered, youth programs found mosquito ecology a compelling topic. Working with a youth organization director, one of the principal investigators and her undergraduates developed an afterschool and camp curriculum that integrated mosquito ecology with civic education. Youth wrote poems about nature in their neighborhoods, discussed the history of vacant lots and envisioned future uses for them, and prepared a presentation that one of them delivered at a local community association meeting. Children came away more interested in both mosquitoes and issues of inequitable sanitation and neglect of neighborhood infrastructure.

We also invited residents to narrate their general concerns about environmental issues in a more relational setting than the individual survey. Fifty-seven survey respondents and youth citizen-scientists participated in longer group sessions, and the group facilitators took the survey responses as a cue to invite participants to elaborate on topics beyond the mosquito. After these focus groups and a year of participant observation in community association meetings and youth activities, we identified several themes and drew connections among them as suggested by residents' narratives, as shown in Fig. 14.3. Group interviews allowed residents to construct a story about how neighborhood change, particularly the loss of neighborhood leaders, the abandonment of buildings, neglect by government services, and the decline of quality public education, had detracted from health, well-being, and civic activities. We came to see residents' knowledge as a lay ecology (Eden and Bear 2011), which allowed us to better communicate the place of mosquitoes in the socionatural system of disinvested neighborhoods in southwest and west Baltimore.

Expanding Resident Engagement and De-centering the Mosquito

In early 2014, with two summers of mosquito ecology data and one summer of citizen-science and focus-group activities to build upon, the Baltimore Mosquito Study further extended our efforts to engage community members. The mosquito ecology data and residents' narratives together told stories about historic race-based disinvestment, continuing government neglect, and neighborhood change that created abundant habitats for breeding mosquitoes as well as other unwanted creatures. The neighborhoods with the highest densities of abandoned buildings and vacant lots had the highest numbers of standing water containers holding larval mosquitoes. Many of these containers were inside abandoned buildings too dangerous for our team members to enter. Many others were items ranging from tires to broken toilets to Tupperware containers strewn about in the backyards of abandoned buildings, and many of these were dumped there illegally by trash-hauling companies that visited under cover of night, avoiding transfer station tipping fees. Disinvested neighborhoods also suffered other waste disposal problems that left large numbers of containers for mosquito breeding, for example, trash can theft was rampant, and residents found municipal collections schedules inadequate, leading many to leave trash bags on the curb where they could attract rats and collect puddles. We trapped three times as many adult *Aedes albopictus* in these neighborhoods as well, and at the scale of the block, this species of

Fig. 14.3 Themes brought up by 57 adults and youth in 2013 in group discussions, and connections that participants made among neighborhood socio-environmental issues

mosquito was highly correlated with the presence of abandoned buildings. We organized a community meeting to share these results, seek community feedback, and develop plans for civic involvement (Figs. 14.4 and 14.5).

Residents were particularly troubled that the most serious neighborhood environmental hazards—namely, abandoned buildings and trash—that were the legacy of past injustices also led to mosquito infestation. Most had considered mosquitoes a nuisance but only a minor threat, but now some participants seemed to add an additional entity to their ecological understanding of injustice. (We were careful not to exaggerate the risk of disease from mosquitoes; as of 2014, West Nile Virus was a periodic and rather small risk, and there was also a small chance that dengue or chikungunya could be locally transmitted as happened in a few other sites outside these diseases' usual range.) This community meeting was invigorating for the researchers and seemed to affirm that residents grasped the connections among mosquitoes and other environmental injustices. We also hired residents from each neighborhood to help us communicate with their neighbors, paying them a small monthly stipend. We hoped that the community meeting and the new community liaisons would foster discussion of actions that might use our findings for advocacy with city government.

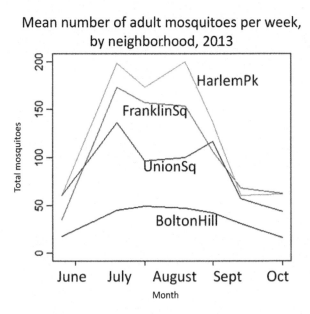

Fig. 14.4 Adult mosquito infestation in summer 2013. Infestation levels corresponded with the density of abandoned buildings and vacant lots

Fig. 14.5 Illegal dump in the backyards of several adjacent abandoned houses

We also introduced PhotoVoice to residents as a means of creative and visual modes of storytelling; participation numbers were small, but a dedicated core group of community members used photography to document numerous environmental hazards. It is hard to capture mosquitoes photographically, but the residents' expansive notion of ecological webs made mosquitoes legible in some scenes that stressed residents out, such as lots with tall vegetation and trash—and residents also portrayed a few bright spots, such as gardens planted by neighbors in vacant lots. The results of our PhotoVoice activities provided a vehicle for publicizing residents' perspectives at both neighborhood and citywide art events.

After low participation in the first summer of citizen-science, the team members leading this portion of the project determined that citizen-science was unlikely to yield results suitable for testing hypotheses about mosquitoes.

In subsequent years, the citizen-science protocol was shifted to focus on monitoring, and with 37 individuals contributing complete data over two years (2014–2015), we found that citizen-scientists' results generally did track the results of systematic monitoring by the research team (Jordan et al. 2017). During this time, we also shifted our emphasis to using mosquito studies to inspire interest in civic ecology. Advocates for civic ecology define it as "community-based, environmental stewardship actions taken to enhance green infrastructure, ecosystem services, and human well-being in cities and other human-dominated landscapes" (Krasny et al. 2014) and argue that it can invigorate grassroots efforts to build resilient ecosystem services in distressed communities (Krasny and Tidball 2015). In this case, the citizen-science scholars in the team found that residents who collected data about mosquitoes gained hope and a sense of efficacy in their ability to manage the environment, affects that might make them more likely to undertake civic ecology practices (Jordan et al. 2016).

Parallels with other citizen environmental activities point to promises and pitfalls of citizen-science and civic ecology. Medical anthropologist Alex Nading has documented the ways contemporary women health workers in Nicaragua become invigorated with the "politics of life itself" while searching out mosquito-breeding locations and educating neighbors about how to manage these insects (Nading 2012). These health workers exemplify the greatest promise of lay ecological knowledge, but citizen-science may also heighten dynamics of biopower—the control of populations by government and quasi-government entities—and reinforce the idea that citizens' greatest environmentalist focus should be on self-regulating their individual impacts. The civic ecology movement also recalls the early-twentieth-century "civic biology" movement, which sought to engage urban citizens in management of their environments and bodies. For example, urban high schools trained students to monitor flies—a supposed disease vector—and devise control methods (Biehler 2013). While self-regulation may seem a positive outcome, it may also distract attention from larger structures more responsible for environmental degradation. In the early twentieth century, failures of government code enforcement and sanitation contributed more to the production of flies—and more serious health threats—than did behaviors that health authorities demanded of (low-income, socially marginalized) individuals (Biehler 2010). It is all too easy for health and environmental authorities—not to mention an unsympathetic public—to take data like that which we have generated and use it to support a view that residents are themselves primarily responsible for poor quality of life and are in need of further discipline (Wilson 2010). But in neighborhoods such as Franklin Square today, especially in light of devastating population losses, residents lack the resources to extensively manage mosquitoes and other environmental hazards. It is extremely costly to remove illegal dumps or, even more troubling and expensive, buildings that have decayed as a result of disinvestment begun decades ago. Upstream mosquito management would require investment at a grander scale in reducing the devastating environmental hazards that breed mosquitoes—city, state, and federal investment in vacant building rehabilitation or demolition and enforcement of health and housing codes. While often better-resourced than community organizations, scientists still lack the funds to make up for decades of disinvestment in neighborhoods of color. What we can do is ensure that we send our data out into the world embedded in narratives by the people who live with the environmental conditions we measure, so that they help tell the story of neighborhood change and demand appropriate action from authorities.

As we strived to connect mosquitoes with the broader web of social and ecological injustice in west and southwest Baltimore, residents of these areas did take political action against injustice, articulating their own discourse of social-environmental holism. Freddie Gray, a young man who was harassed by law enforcement and in 2015 died after police gave him a "rough ride" and denied him medical attention, grew up within blocks of our study neighborhoods and was gravely affected by another environmental issue, lead poisoning. Some residents seemed unable to reconcile participation in mosquito monitoring with demands for bigger changes that, sparked by Gray's death, invigorated the Baltimore Uprising of Spring 2015. The research team extended offers of support to active community associations during and after the Uprising while also continuing the study. One community representative declared that summer that "mosquitoes ain't a big issue for people when there have been 30 homicides this month". Others seemed to lose interest in frustration after an entire year of failed attempts to get the city to clean up garbage piles that bred rats and mosquitoes. While some worried about West Nile Virus, most saw mosquitoes as a mere nuisance, not an inspiration to action or community engagement, regardless of how connected mosquitoes were with government neglect.

Conclusions

We write this as we end our fifth season of monitoring mosquitoes and talking with residents about their environment in west and southwest Baltimore, and as we wonder whether mosquitoes will transmit Zika in Baltimore in seasons to come. One of the most active community participants, who had been vocal in meetings and as a PhotoVoice contributor but not in the citizen-science program, said as Zika made headlines in late spring of 2016, "I've been wondering all along why you were so focused on mosquitoes. Now I understand why". Of course, the team had no way of knowing when we started that a mosquito-borne disease not yet prevalent in the Americas would arrive in Baltimore amid the project. The fact that this participant became active in spite of her lack of interest in mosquitoes confirms her passion for the other, broader environmental issues we addressed—and it might suggest also that we were effective in connecting mosquitoes to other issues at least for some residents. But her comment also reiterates the oddness, from residents' point of view, of our focus on mosquitoes after they had for decades watched so many other aspects of the physical and social environment going awry. Seeing such oddness reflected back at us, we feel moved to offer four considerations for critical physical geographers, particularly those working with marginalized communities.

First, a lack of interest in a particular environmental topic among residents must not be taken as a failure on their part. As critical physical geographers call for engagement of diverse publics in the production of environmental knowledge, we may find that the environmental topics most interesting to us are much less so to communities whose knowledge we wish to learn about and amplify. The Baltimore Mosquito Study did not abandon its focus on mosquitoes when residents told us that larger problems overshadowed these insects' importance, and neither did we dismiss residents as shortsighted or apathetic—a common stance that perpetuates negative stereotypes about people of color and environmentalism (Glave 2010; Finney 2014). Validating community members' own concerns is itself part of environmental justice. Critical physical geographers must honor socio-natural knowledge and politics that have grown through historical geographies and cultures different from those that have informed most environmental science.

Second, research into these complex historical geographies is vital for revealing marginalized communities' environmental knowledge, and also for guiding better research on the material, physical processes that constitute the environment. The life cycle of a mosquito lasts but a few weeks, but the processes that create mosquito-breeding habitat have been building for decades. Oral history narrators from the community have helped inform the Baltimore Mosquito Study's understanding of neighborhood history, as have archival sources and existing historiography of redlining, urban renewal, and sanitation problems. We strived to listen to residents' stories, to better understand the historical geographies of the neighborhoods we studied, and to iteratively alter our research and engagement plans to reflect what we learned. We can become (incrementally) better scientists and explain environmental processes better, and in a way that informs more meaningful interventions, if we develop our historical understanding of deep, unaddressed sources of injustice and environmental degradation. Such endeavors will always be challenging, however, because they force researchers to learn multiple methods, and to balance seemingly disparate research and political goals, such as monitoring mosquitoes *and* promoting environmental justice.

Third, reiterating a central theme of this chapter, knowledge of history and what we often call "context" can allow us to cast a broader net when defining the environmental issues we study—just as many communities of color do as they frame environmental justice problems. Past pest control activities, with a few exceptions, have failed to integrate social and ecological knowledge and have tended to focus on "the pest." But mosquito infestation is a consequence of a broader series of socio-natural processes that deserve at least as much attention as mosquitoes themselves. Critical Physical Geography can particularly invigorate studies of unwanted animals by imagining a wider network of things that matter for pest ecologies. But this consideration applies also to other research topics, from urban forests to water quality, where we might be tempted to narrow our sights and let politics and history fall away. De-centering the mosquito or the tree or the water does not mean eliminating all reference to this entity. Rather, we might think of our research as a network of nodes across which we shift our attention strategically in order to best understand and communicate the socio-natural dynamics of the communities where we work.

Fourth, critical physical geographers must seek iterative feedback about our own participatory activities in order to encourage meaningful political action and avoid extending the dynamics of environmentality, neoliberal biopower, and apolitical ecologies through public participation. This means focusing not just on individual data collection but also on narratives and on making sure any data collection that communities perform—be it photographic, historical, or scientific—helps contribute to shared, if shifting, narratives. Scientists often say that numbers in our spreadsheets "tell a story," and indeed, we interpret them to connect, say, the prevalence of abandoned buildings with the prevalence of mosquitoes. But we can enliven these stories and our relationships with communities if we embrace residents' explanations of what has happened to the places they live. Data can support political action, but when social and environmental harms have become entrenched over decades, it can take narrative to inspire connection and action for lasting ecological change.

Acknowledgments This research was supported by the National Science Foundation Coupled Natural Human Systems Program (lead PI, LaDeau; DEB 1211797). Field and data logistics were further supported by the Baltimore Ecosystem Study (NSF-LTER DEB 1027188). Thanks especially to the residents of west and southwest Baltimore who have told us their stories.

References

Abara, W., S. Wilson, and K. Burwell. 2012. Environmental justice and infectious disease: Gaps, issues, and research needs. *Environmental Justice* 5: 8–20.

Agarwal, A. 2005. *Environmentality: Technologies of government and the making of subjects*. Durham, NC: Duke University Press.

Baltimore Neighborhood Indicators Alliance. 2014. Community profiles. http://bniajfi.org/vital_signs/cprofiles/. Accessed 20 Feb 2016.

Bartlett-Healey, K., G. Hamilton, S. Healey, T. Crepeau, I. Unlu, A. Farajollahi, D. Fonseca, R. Gaugler, G. Clark, and D. Strickman. 2011. Source reduction behavior as an independent measurement of the impact of a public health education campaign in an integrated vector management program for the asian tiger mosquito. *International Journal of Research in Public Health* 8: 1358–1367.

Bartlett-Healey, K., G. Hamilton, T. Crepeau, and D. Fonseca. 2014. Integrating the public in mosquito management: Active education by community peers can lead to significant reduction in peridomestic container mosquito habitats. *PLoS One* 9: e108504.

Biehler, D. 2010. Flies, manure, and window screens: Medical entomology and environmental reform in early-twentieth-century US cities. *Journal of Historical Geography* 36: 68–78.

———. 2013. *Pests in the city: Flies, bedbugs, cockroaches, and rats*. Seattle: University of Washington Press.

Carson, R. 1962. *Silent spring*. Boston: Houghton Mifflin.

Carter, E. 2009. God bless General Peron': DDT and the endgame of malaria eradication in Argentina in the 1940s. *Journal of the History of Medicine and Allied Sciences* 64: 78–122.

———. 2014. Malaria control in the Tennessee Valley Authority: Health, ecology, and metanarratives of development. *Journal of Historical Geography* 43: 111–127.

Dowling, Z., P. Armbruster, S. LaDeau, M. DeCotiis, J. Mottley, and P. Leisnham. 2013a. Linking mosquito infestation to resident socioeconomic status, knowledge, and source reduction practices in suburban Washington, DC. *Ecohealth* 10: 36–47.

Dowling, Z., S. LaDeau, P. Armbruster, D. Biehler, and P. Leisnham. 2013b. Socioeconomic status affects mosquito (Diptera: Culicidae) larval habitat type availability and infestation level. *Journal of Medical Entomology* 50: 764–772.

Eden, S., and C. Bear. 2011. Models of equilibrium, natural agency and environmental change: Lay ecologies in UK recreational angling. *Transactions of the Institute of British Geographers* 36: 393–407.

———. 2012. The good, the bad, and the hands-on: Constructs of public participation, anglers, and lay management of water environments. *Environment and Planning A* 44: 1200–1218.

Finney, C. 2014. *White spaces, black faces: Reimagining the relationship of African-Americans with the great outdoors.* Chapel Hill: UNC Press.

Flint, M., and R. Van Den Bosch. 1981. *Introduction to integrated pest management.* New York: Plenum Press.

Gioielli, R. 2014. *Environmental activism and the urban crisis: Baltimore, St. Louis, Chicago.* Philadelphia: Temple University Press.

Glave, D. 2010. *Rooted in the earth: Reclaiming the African American Environmental Heritage.* Chicago: Chicago Review Press.

Grove, J.M., M. Cadenasso, S.T.A. Pickett, G. Machlis, and W. Burch. 2015. *The Baltimore School of Urban Ecology: Space, scale, and time for the study of cities.* New Haven: Yale University Press.

Holifield, R. 2001. Defining environmental justice and environmental racism. *Urban Geography* 22: 78–90.

———. 2004. Neoliberalism and environmental justice in the United States Environmental Protection Agency: Translating policy into managerial practice in hazardous waste remediation. *Geoforum* 35 (3): 285–297.

Humphreys, M. 2001. *Malaria: Poverty, race, and public health in the United States.* Baltimore: Johns Hopkins University Press.

Jakle, J., and D. Wilson. 1992. *Derelict landscapes: The wasting of America's built environment.* New York: Rowman and Littlefield.

Jordan, R.C., S.A. Gray, D.V. Howe, W.R. Brooks, and J.G. Ehrenfeld. 2011. Knowledge gain and behavioral change in citizen science programs. *Conservation Biology* 25: 1148–1154.

Jordan, R.C., S.A. Gray, A.E. Sorensen, G. Newman, D. Mellor, C. Hmelo-Silver, S. LaDeau, D. Biehler, and A. Crall. 2016. Studying citizen science through adaptive management and learning feedbacks as mechanisms for improving conservation. *Conservation Biology* 30: 487–495.

Jordan, R.C., A.E. Sorensen, and S.L. LaDeau. 2017. Citizen science as a tool for mosquito control. *Journal of American Mosquito Control Association* 33: 241–245.

Krasny, M., and K. Tidball. 2015. *Civic ecology: Adaptation and transformation from the ground up.* Cambridge: MIT Press.

Krasny, M., A. Russ, K. Tidball, and T. Elmqvist. 2014. Civic ecology practices: Participatory approaches to generating and measuring ecosystem services in cities. *Ecosystem Services* 7: 177–186.

LaDeau, S., P. Leisnham, D. Biehler, and D. Bodner. 2013. Higher mosquito production in low-income neighborhoods of Baltimore and Washington DC: Understanding the (re)emergence of mosquito-borne disease risk in temperate cities. *International Journal of Environmental Research and Public Health* 10: 1505–1526.

Nading, A. 2012. Dengue mosquitoes are single mothers: Biopolitics meets ecological aesthetics in Nicaraguan community health work. *Cultural anthropology* 27: 572–596.

Nash, L. 2006. *Inescapable ecologies: A history of environment, disease, and knowledge.* Berkeley: University of California Press.

Ottinger, G. 2013. *Refining expertise: How responsible engineers subvert environmental justice challenges.* New York: NYU Press.

Patterson, G. 2009. *The mosquito crusades: A history of the American anti-mosquito movement from the Reed Commission to the first Earth Day.* New Brunswick: Rutgers University Press.

Pietila, A. 2010. *Not in my neighborhood: How bigotry shaped a great American city.* Chicago: Ivan Dee Press.

Pincetl, S., T. Gillespie, D. Pataki, S. Saatchi, and J. Saphores. 2013. Urban tree planting programs, function or fashion? Los Angeles and urban tree-planting campaigns. *GeoJournal* 78: 475–493.

Platt, H. 2005. *Shock cities: The environmental transformation and reform of Manchester and Chicago.* Chicago: University of Chicago Press.

Rogers, N. 1989. Germs with legs: Flies, disease, and the New Public Health. *Bulletin of the History of Medicine* 63: 599–617.

Shaw, I., P. Robbins, and J.P. Jones III. 2010. A bug's life and spatial ontologies of mosquito management. *Annals of the Association of American Geographers* 100: 373–392.

Taylor, D. 2009. *The environment and the people in American cities, 1600–1900.* Durham: Duke University Press.

Tedesco, C., M. Ruiz, and S. McLafferty. 2010. Mosquito politics: Local vector control policies and the spread of West Nile Virus in the Chicago region. *Health & Place* 16: 1188–1195.

Tuiten, W., D. Koenraadt, K. McComas, and L. Harrington. 2009. The effect of West Nile virus perceptions and knowledge on protective behavior and mosquito breeding in residential yards in upstate New York. *EcoHealth* 6: 42–51.

Unlu, I., A. Faraji, D. Strickman, and D. Fonseca. 2013. Crouching tiger, hidden trouble: Urban sources of Aedes albopictus (Diptera: Culicidae) refractory to source-reduction. *PLoS One* 8: e77999.

Webb, J. 2011. The first large-scale use of synthetic insecticide for malaria control in tropical Africa: Lessons from Liberia, 1945–1962. *Journal of the History of Medicine and Allied Sciences* 66: 30.

Williams, Y. 2013. The socio-ecological system of vacant lot management for southwestern Baltimore neighborhoods. Unpublished PhD Dissertation, University of Maryland, Baltimore County.

Wilson, S. 2010. Environmental justice: A review of history, research, and public health issues. *Journal of Public Management and Social Policy* 16: 19–50.

Wilson, S., C. Heaney, and O. Wilson. 2010. Governance structures and the lack of basic amenities: Can community engagement be effectively used to address environmental injustice in underserved black communities? *Environmental Justice* 3: 125–133.

15

Circulating Wildlife: Capturing the Complexity of Wildlife Movements in the Tarangire Ecosystem in Northern Tanzania from a Mixed Method, Multiply Situated Perspective

Mara J. Goldman

Introduction

Wildlife conservation is inherently political—linked to questions of property rights and uneven distribution of costs and benefits. Throughout sub-Saharan Africa, wildlife conservation efforts began with European imperialism and continue today through ongoing efforts of international conservation agencies and national governments to protect a 'global heritage'. Such efforts have almost always involved large-scale evictions of local people for the creation of protected spaces for wildlife, heavily armed surveillance of local communities to prevent poaching, and the promotion of scientific knowledge production about wildlife populations over indigenous ways of knowing and living with wildlife. This story has been well covered elsewhere (Adams and Hutton 2007; Adams and Mulligan 2003; Bockington et al. 2006; Duffy 2014; Goldman 2003). Often less addressed is the role that conservation science plays in enabling conservation practices which promote the separation of wildlife populations from humans and define how we know about wildlife in Africa (and elsewhere) (Goldman 2007).

A large part of wildlife conservation science is concerned with tracking wildlife movements and population dynamics, the results of which are often used for political ends and to justify particular policies As Adams and McShane

M. J. Goldman (✉)
Department of Geography, University of Colorado-Boulder, Boulder, CO, USA

© The Author(s) 2018
R. Lave et al. (eds.), *The Palgrave Handbook of Critical Physical Geography*,
https://doi.org/10.1007/978-3-319-71461-5_15

(1992: 71) note, '[a] surefire way to excite donors is to trot out today's survey which demonstrates the latest grave development, while surveys showing more animals then expected are generally suppressed. The animal census has thus become an indispensable tool of park management'. Census information can readily be translated into promotional literature for wildlife conservation efforts. Whether documenting long-distance migratory patterns or a decline in wildlife numbers, large-scale data attracts attention; it is also a necessary component of wildlife management (Norton-Griffiths 1978). However, large-scale census counts are only one way of knowing animal populations, and often need to be used together with small-scale monitoring for efficient planning, particularly in combined human-wildlife systems, with high levels of seasonal and annual variability in mobility patterns (Reid, Rainy et al. 2003).

Counting animals may seem like a straightforward enough scientific undertaking. Yet locating wildlife in time and space is not easy, especially when populations are migratory, the area of concern is large, and forage resources vary with rainfall patterns (Gereta et al. 2004). The most common way of monitoring wildlife populations over large areas has been sample and total counts via regular aerial census surveys: the Systematic Reconnaissance Flight (SRF) (Norton-Griffiths 1978). In Tanzania, the SRF technique was initiated in Serengeti National Park in the 1950s by the Frankfurt Zoological Society to document migratory patterns that came to define the boundaries of the park and larger ecosystem (FZS 2000). This technique has since been regularly employed by the Tanzanian Wildlife Research Institute (TAWIRI) to conduct wet and dry season surveys of the Tarangire-Manyara Ecosystem (TME). The TME is approximately 35,000 km², including Tarangire and Lake Manyara National Parks and surrounding village land dominated by Maasai pastoralists (TWCM 1991). The focus of this chapter is on the western part of the TME between the two national parks, where wildlife data collection efforts have focused on documenting wildlife movements so as to introduce new conservation efforts (i.e. conservation corridors, see Goldman 2009) and support existing ones (the Manyara Ranch community-based conservation area, see Goldman 2011) (Fig. 15.1).

When I set out to conduct research in the TME in 2002, I was aware of the history of conservation politics in the area, that rested on the eviction of Maasai people from areas deemed 'natural' and worth saving inside national parks (Anderson and Grove 1987; Brockington 2002; Collet 1987), and a simultaneous rejection of Maasai ways of knowing and being with this 'nature' (Gardner 2016; Goldman 2003). I was interested in understanding how these politics played out on the ground, and in challenging them though in-depth ethnographic and ecological fieldwork. In line with one of the main tenets of CPG,

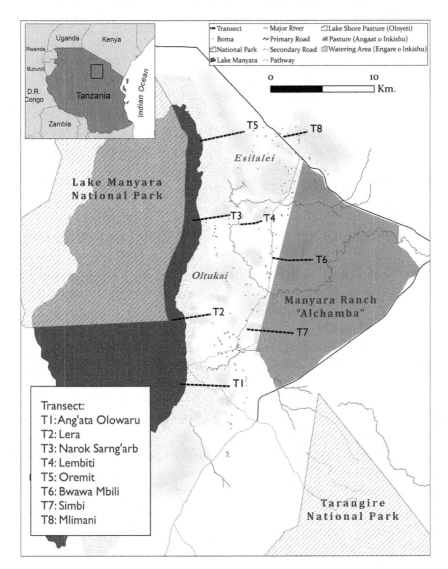

Fig. 15.1 Study area with transect placement

I was interested in challenging where 'nature' is, and how we come to know it—by looking at different ways of knowing how wildlife (i.e. 'nature') used the 'human' spaces of Maasai village lands adjacent to national parks and the new community-based conservation area, the Manyara Ranch, where livestock use was allowed on a limited basis (Goldman 2011). My research resulted in data that highlights the regular use of village land by wildlife, an important finding that challenges assumed nature-society divides promoted by conservation programs. Yet my goal was much broader: to illustrate the value of putting multiple ways of knowing wildlife into conversation (from conservation biology, ecology, Maasai, geography, history, etc.) in order to reveal the importance of scale (spatial and temporal) and uncover the complexities involved in measuring and predicting wildlife movements and resource use patterns. In this chapter I focus on what these different ways of knowing yield but also on how similarities in seemingly contradictory findings can be uncovered once data are closely analyzed, disaggregated, and compared. I also highlight the politics involved in reconciling knowledge conflicts and in really *knowing* where wildlife are in village lands (i.e. human/social space).

Different Ways of Knowing and Tracking Wildlife

Different actors living and working in the TME employ different knowledge about wildlife population trends and movement patterns at different geographic scales, relying on different methodological tools to acquire and communicate that knowledge—each with its own series of biases, limitations, and strengths. The various voices can be generalized into two groups: conservation scientists (ecologists, conservation practitioners, wildlife biologists) and Maasai (men, women, youth, elders, game scouts, village leaders, herders). The lines between science and conservation practice in the TME are blurred and often non-distinct, with researchers actively involved in conservation practice and practitioners engaged in research and drawing from published work. While all Maasai and all scientists in the study area are not always in agreement with each other, there are general statements that can be made regarding methodological approaches used by each group to locate wildlife in time and space. I conducted what can be thought of as a third 'hybrid' approach together with Maasai assistants using ecological methods and Maasai knowledge.[1]

Scientists Tracking Wildlife

In the early 1960s, a complete ground survey was conducted of wildlife in and around Tarangire National Park (TNP, a game reserve at the time) (Lamprey 1963, 1964), providing baseline data on large mammal densities and migratory movements beyond the park. The original maps produced by Lamprey continue to be used today as conservation efforts focus on protecting fairly regular 'migratory pathways as conservation corridors' (Goldman 2009).

Since that time data has continued to be collected through regular SRF counts (by Frankfurt Zoological Society) and with different techniques at smaller scales by individual researchers (Kahurananga 1981; Kahurananga and Sikiluwasha 1997; Gamassa 1989; Hassen 2000; Mmari 1989; Oikos 2002).

There has been much debate over the problems associated with different types of animal census methods, including the accuracy of sample versus total counts, aerial versus ground surveys, and the visual obstructions related to different ecotypes (e.g. forested areas having less visibility) (Caughley and Sinclair 1994; Hugo 2001). All of these methods provide snapshot data of a particular point in time and space, 'an instantaneous look into the larger, more complex trend of animal movement' (Foley 2002: 6). The most common technique is the SRF sample count, which is less expensive than total counts, but subject to high standard errors. The procedure involves counting animals along evenly spaced parallel transects flown over a defined area and then utilizing a model to calculate distribution and abundance estimates for the remainder of the area (Norton-Griffiths 1978). The major assumption of an SRF is that the density and distribution of animals is constant between transects. For animals that tend to cluster, this can lead to gross overestimates.[2] For this reason SRF data is less useful to discuss population trends than to delineate animal distribution patterns, and identify important areas for wildlife at certain times, where smaller-scale data can then be collected. Ideally, large-scale SRF distribution maps are combined with smaller-scale data and focused studies of wildlife movements in particular places to link up changes in land use on the ground with large-scale population changes monitored from above (Msoffe et al. 2007).

In late 1990s and early 2000s, in response to growing concerns by wildlife researchers in northern Tanzania over threats to wildlife migratory pathways and the increasing isolation of Tarangire National Park (Borner 1985), researchers from the Italian Istituto Oikos and the Tanzanian National Parks Authority (TANAPA) initiated the Tarangire-Manyara Conservation Project (TMCP), to produce multi-scale data of wildlife in the TME. They combined a series of sample and total counts and averaged across months and years to display levels of abundance of particular species by season. They also obtained exact locations through GPS radio tracking of a small sample of animals (12 wildebeest, 13 zebra, and seven elephants) to calculate home range, migratory pathways, and wet and dry season resource use. The combined results of these different methods produced distribution maps for the entire TME. The project also incorporated road (driving) transects to collect data at the smaller scale of (1) Tarangire National Park, (2) the Tarangire-Manyara Corridor area, and (3) the Manyara Ranch. This type of mixed methods approach is beneficial in providing different levels of detail and different approaches to 'seeing' animals that reflect different biases (i.e. driving vs. flying; total counts vs. SRFs).

The TMCP project was ambitious and impressive but also incorporated risks in aggregating data from different sources across years and reflected specific spatial and temporal limitations. Spatially, only areas covered by roads were surveyed. Temporally, transects were conducted during representative 'wet' and 'dry' season months: May (W), October (D), December (D), and March (W). Conducting transects the same month every year allows for cross-year comparison. However, averaging data across months erases much of the variability of wildlife resource use related to rainfall patterns, which vary markedly across months and years. Scientists working in the field realize that seasonal variability changes across years but they are often constrained by the time frame and finances of a specific grant or project and unable to collect more detailed data over a longer time frame. Many researchers thus also rely on 'non-scientific' methods—making note of wildlife sightings outside of official sampling segments. Word of mouth is used between colleagues and trusted friends, who might have seen migratory animals while in the field. Such information is not usually recognized as 'data' and either excluded from a scientific article or presented as 'anecdotal' information. Yet such information highlights the ways in which wildlife are living, acting beings and not just data that can be relied to 'show up' on schedule (Watson and Huntington 2008; Whatmore and Thorne 1994). GPS radio collars allow a higher degree of precision in tracking the movements of particular animals. The limitation is the small sample size and the assumption that one animal represents a larger population. Both word of mouth and GPS data are useful in determining when and where to conduct more detailed (and rigorous) animal censuses, to better track the complexity of wildlife movements across time and space.

Maasai Tracking Wildlife

Maasai living and working in TME also have an interest in knowing the whereabouts of wildlife, but in contrast to the conservation community, they are concerned with smaller-scale movement patterns in the places where they live or are likely to visit with their livestock. While Maasai do not hunt wildlife for meat, they have an interest in avoiding and/or following particular species at particular points in time. Elephants, for instance, raid crops and present a danger to people. Yet elephants can also 'open up the bush' and create a good grazing environment for livestock. Newborn wildebeest calves can spread a deadly disease, Malignant Catarrhal Fever (MCF), to cattle. Close observation of wildebeest movements during calving time is important to avoid the loss of cattle to disease. Yet at other times, close observation of the movement of wildebeest, known to follow rainfall and the growth of new grass, can be an important husbandry strategy. The behavior and whereabouts of predators, such as lions, hyenas, and leopards, are monitored for protection purposes. Lions are the only animals regularly hunted, to control their predation on livestock, and for the young men of the 'warrior' age-set (*ilmurran*) to fulfill their role in protecting Maasai society (people and livestock) (Goldman et al. 2013).[3]

Maasai knowledge of wildlife is related to non-utilitarian purposes as well. Open curiosity and playfulness is displayed by herd boys, who spend countless hours in the pasture where they observe wildlife, sometimes chase them, and often inspect their dung to figure out what they have foraged on. Women will often watch animals they encounter while out looking for water, fuel wood, and medicinal plants, or farming their small fields—sometimes to chase the animal away (from the fields), other times just to observe and admire them (de Pinho et al. 2014). These activities combined with stories from their elders teach Maasai men and women about different wildlife species. More recently, with increased crop cultivation, there is an increased interest in certain animals recognized as 'pests' on farms, such as zebra, eland, porcupine, and monkeys.

Maasai knowledge of wildlife relies on three general sources—direct observations, the transfer of information among themselves, and referral to 'trusted sources' (elders). Direct observations regarding wildlife movements come from living and herding around wildlife—going to different locations for grazing, water, firewood, medicinals, meetings, or visiting friends, relatives, or in-laws. Observations include direct sightings, signs of wildlife presence or recent passage (dung, footprints, tree or crop damage by elephant, heavy grazing by wildebeest), and the clearly visible paths created by wildebeest, zebra, and elephants as they move from place to place. Since wildlife and cattle are often in search of the same resources, herders are likely to see wildlife while out with their cattle. During interviews, Maasai named 25 different species of animals seen while herding. Where Maasai travel and what they see while there results in a particular partial perspective—or sampling—of wildlife.

When Maasai cannot be certain where wildlife are based on their own observations, they, like scientists and practitioners, rely on other sources: the stored knowledge from what the elders have taught them, from their own experience and observations over the years, and from recent news brought through extensive social networks. The words, knowledge, and experience of the elders act as an oral literature and constant source of consultation for the new generation.

This oral literature is comparable to early scientific articles, which are continuously called upon by scientists and conservationists trying to make sense of current wildlife patterns in the TME.[4] However, the words of the elders do not always reflect recent changes in movement patterns. Here word of mouth is an important methodological tool. From someone passing through, or a local resident returning from an excursion, people know where it is raining and where it is not, and they can often find out the latest whereabouts wildlife from someone who saw them directly. In the absence of direct observations, Maasai will deduce the likely location of particular animals based on other information received (i.e. recent rainfall and/or grass growth, the location of water or drying up of a dam). This is particularly important during wildebeest calving time when knowledge of even the passing of the animals while birthing is essential. The speed at which such knowledge transfers through Maasai communities, even more so with cellphones today, makes this a remarkably effective mechanism of tracking a moving target (Goldman 2007).

While ecologists tend to generalize wildlife movements by broad-scale 'wet' and 'dry' season descriptors, Maasai generalize wildlife movements along a finer-scale seasonal breakdown: (1) *Orkisirata*, the early rains (October/November to December); (2) *Oladalo*, the hot month in between the short and long rains (mid-January to mid-February); (3) *Engakwai*, the heavy rains (March to May); (4) *Koromare*, the end of the rainy season/beginning of the dry season (mid-May to June); and (5) *Alamei*, the dry season (July to October). This provides a finer resolution for data capture, while reflecting the variability of rainfall in the area, which does not always correspond to discrete months or 'wet' and 'dry' season aggregations (Goldman 2006; Goldman forthcoming).

A Hybrid Approach: A Local Enactment of Western Science Combined with Maasai Knowledge

My own fieldwork involved walking transects together with two Maasai assistants, and can thus be seen as a local enactment of western scientific practice combined with Maasai knowledge and skills. The advantages of walking transects are that they are inexpensive and fairly easy to conduct, can cover areas inaccessible by car, reduce the disturbance and flight of wildlife that can occur with a vehicle, can be useful for obtaining data on seasonal distribution patterns within different vegetation types (Norton-Griffiths 1978), and can easily be conducted by community members. The disadvantage is that the area covered is considerably smaller than that which can be covered in driving ground and aerial surveys. Walking transects are often sample counts, with systematic placement of transects at set distances across an area, strategically to monitor use of a particular habitat or migration route.

Our transects were devised to provide sample counts for different representative vegetative zones within the village utilized by both wildlife and livestock: short grass plains (T1, *Ang'ata Olowaru*), wooded grass savanna (T2, *Lera*), tall grass (T3, *Narok Sarng'arb*), mixed wooded/grass savanna (T4, *Lembiti*), scrub savanna (T5, *Oremit*), and mountain/wooded savanna (T8, *Mlimani*); see Fig. 15.1. Two additional transects (T6, *Bwawambili*, wooded savanna; T7, *Simbi*, grassland) were placed to enter the Manyara Ranch from village lands in areas used by villagers (legally and illegally). Transects were designed in consultation with Maasai over several months, and through observations of the area, interviews, and walking tours. Data collection techniques reflect those utilized by Lamprey (1964), Gamasa (1989), and Oikos (2002). Since the TMCP road transects described above were ongoing in the area, an effort was made to use comparable methods. Two Maasai men (Mungai Well and Kisiongo Makaa) worked with me for two years, and then without me for another two years, where they combined their knowledge of wildlife and their astute observation skills with standardized scientific methods.

Transects were walked in the early morning hours, every other month the first year (2003) and every month thereafter (2004–5), to capture more detailed seasonal change. Wildlife sightings on both sides of the transect line were recorded, the number of animals estimated, their activity noted, and their perpendicular distance to the line measured. A compass was used to direct movement along the transect, a GPS to mark all live sightings and record the distance walked, and binoculars and a counter to assist in counting animals. We devised a hybrid technique to measure distance of animals from the line (needed for analysis purposes). Maasai often calculate distance in time, such as before the cattle are all milked, by the time the cattle reach the pasture, and so on. In Western science, distance is measured in space. While walking transects, estimating the spatial distance of animals from the transect line was arguably less reliable than Maasai estimations based on time.[5] We used a GPS to measure the distance from our location on the transect to a nearby object (e.g. a tree at 'x' meters away), and then asked how far an animal sighting was in relation to this known distance, three times as far? Half as far? As such, we relied on a degree of certainty with the GPS measurement and Maasai ability to calculate distance in time (it would take us twice as long to reach those animals as it did this tree).

We also recorded other signs of wildlife presence on transects—footprints and dung deposited within the past couple of days directly on the transect line. Recording these 'marks' provided a crosscheck of the sightings data as well as additional information on the distribution patterns of wildlife and livestock not seen. This is particularly important regarding livestock, which were likely undercounted because the transects were conducted before they went out to pasture, and particular wildlife that are readily frightened. Both wildlife and livestock density calculations correlate positively with markings/ km at a significance level of 0.01 ($r = 0.28$ for livestock and 0.30 for wildlife). We also recorded observations on transects of pathways (wildlife, livestock, people, and cars), standing water, change in vegetation (structure, general composition, general 'greenness', and degree of defoliation), animal carcasses, farms, poaching tracks, and grazing reserves.

Data were analyzed to present wildlife densities within a particular transect area using a modified variable fixed width method (Norton-Griffiths 1978), with widths devised based on field of vision estimates for each transect, drawing from familiarity with the area and trends in the distance data collected after a two-year period. This method is biased toward large mammals and animals in groups (zebra, wildebeest, and cattle) while undercounting smaller and less visible animals (dikdik, warthog, gazelles, and goats, depending on the vegetation density). My interest was in determining the relative abundance of different animal species in different vegetation zones. We were particularly interested in wildebeest and zebra which (1) represent the highest wildlife biomass in the TME ecosystem, (2) reflect great seasonal variation in resource use across the TME, (3) are known to overlap with cattle in resource use patterns, and (4) are important tourist attractions. These are also the animals that most conservation science focuses on as well; thus our data were comparable. Density calculations are transect specific and speak less to animal presence across the study area. For discussions of wildlife across the entire area, we utilized a Kilometric Index of Abundance (KIA), as used by TMCP. This entails calculating how many animals were seen over the total number of kilometers covered, across transects, in a given month (an average of 25.5 km). While transects were not placed systematically to cover the whole study area, neither were those used by TMCP, and ours are likely more representative since they were conducted off the road, inside grazing areas.

The transect methodology provides reliable and comparable data across time but only a snapshot view. For this reason, additional 'opportunistic' sightings of wildlife (by myself, my assistants, or key interlocutors) were recorded with precise location and timing noted (GPSed if possible). See Fig. 15.2. These 'opportunistic' data represents an attempt to document (and thus make somewhat 'official') the unofficial knowledge about wildlife discussed above used by Maasai and scientists.

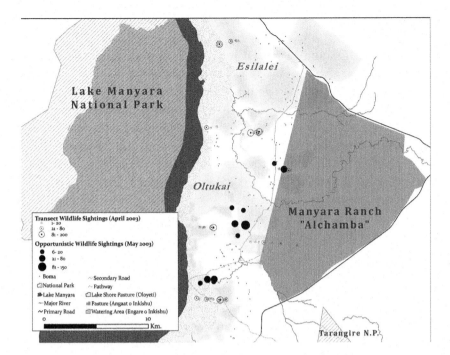

Fig. 15.2 Wildlife sightings

Where Are Wildlife? It Depends on Where and How One Is Looking

Different ways of seeing and knowing wildlife produce different information about where wildlife are at different points in time. I am not particularly interested in which way of seeing is 'better' or more accurate, nor do I think that is a useful inquiry. Yet it is helpful to explore disparities and similarities in findings, to expose the limitations of any one perspective, as well as the value of building on difference to understand complex nature-society relations. Paying close attention to multiple methods and ways of seeing can expose important differences, but also valuable areas of overlap and agreement, and highlight the value of bringing different knowledges into conversation.

Sampling and Aggregation

When making any generalized statements regarding the whereabouts of wildlife, or comparing the findings of scientific studies with what Maasai say regarding wildlife numbers, the specific ways in which these schemes work and how they compare to Maasai ways of seeing matter. For instance, TMCP used sample months to represent 'dry' and 'wet' seasons: May (W), October (D), December (D), and March (W). But according to Maasai, October and December are not always or even usually the dry seasons but more often represent a different season altogether—*Orkisirata*, the coming of the short rains—a season that has profound effects on wildlife movements. Rainfall patterns change from year to year, meaning that October is sometimes dry and sometimes receives rain, but this is not the same as it being a 'rainy' season—for the rain will be new, resulting in small pools of water available for migrating wildlife, but no new grass growth, and great visibility (i.e. of wildlife on transects) because of reduced vegetation. Sometimes October is dry, sometimes it is wet. Sometimes it brings larger numbers of wildebeest, sometimes the wildlife stay away. Our transect data went by month, and used Maasai seasons to refer to the month, but this changed every year. *Orkisirata* can come any time from October to December. It is not *Orkisirata* in October if it does not rain. This flexibility is important in making any generalized statements about 'wildebeest coming to the village in the dry season'.

Yet this variability is often collapsed in scientific studies that aggregate data across time and space. For instance, according to the TMCP report, 'the abundance of <u>zebra</u> in the area is higher during the wet season (March–June) compared to the dry season (October-December). Yet, "<u>wildebeest</u> abundance shows the opposite trend, being higher during the dry season (October) and almost zero in all other periods' (Oikos 2002: 28, emphasis in original). The report then averages across all transects, to show that cattle abundance is higher than wildlife in the area. However, when broken down by month and by transect, there are times (like October) and places, like on the transect that runs through Oltukai village to the lake, where wildebeest are more abundant then cattle (Oikos 2002), and that is important for showing wildlife use of village lands (sometimes outnumbering livestock!). It also reflects what Maasai say—during *Orkisirata* (often October) the wildebeest come to village land following the fresh rain. People spoke in particular about wildebeest liking an area by the edge of lake Manyara, called *Oloyeti*, the Maasai name for the *Sporobolus spicatus* grass that sprouts quickly as the lake retreats and new rain falls. Maasai also refer to this area is *Naoong*, the sound that wildebeest make when they occupy the area with their small calves during calving time. When the TMCP data is broken down by month and then addressed as if it were a Maasai season, there is more overlap with what Maasai say about wildlife presence than disagreement, though information on rainfall is also helpful. For this reason, in our data, we named the months differently depending on the rainfall (October was sometimes 'dry season' and sometimes *Orkisirata*).

Sampling is also spatial, with transects often assumed to 'represent' parts of a larger area and averages across transects used to discuss wildlife trends. This can be problematic when an area of concern is not uniform in vegetation coverage, soil type, rainfall, or other potential draws or obstacles for wildlife (roads, settlements, livestock). Disaggregating data to look at differences across transects may expose patterns of wildlife presence that compare differently with Maasai ways of seeing wildlife than aggregate trends do, often aligning more for specific species and places. For both Maasai knowledge and transect data represent different snapshot views, but often of different places and at different times.

TMCP uses an index to talk about the presence of animals across transects (1 = on all transects). They found zebra and ostrich present on all the road transects (index of presence = 1). Other common species included Thompson's gazelle, giraffe, and impala. Zebra were the most abundant wildlife species with a maximum relative abundance of 6.6 animals/km. Wildebeest, while less common (index of presence = 0.5), also had a high relative abundance (four animals/km in October 1999). Maasai reports of animals seen out while herding were similar, with some variation: the most common animals reported across seasons were gazelle (mostly Thompson but also Grant's),[6] zebra, wildebeest, ostrich, and giraffe. These were among the most frequently sighted species on transects as well, with some variation. For instance, on transects, impala were more frequently observed than wildebeest (0.8 vs. 0.5 index of presence), but this reflects their presence in only two transects (T5, *Oremit*; T6, *Bwawambili*). Maasai only frequent these areas at certain times of the year and are therefore less likely to see these animals, and these areas were not included in the TMCP sampling. On the other hand, Maasai spoke of commonly seeing reedbuck and warthog while herding, two species which were not well represented in the transect data because of their small size and elusive behavior.[7] There were also animals listed in interviews that were never observed on transects (gerenuk, hartebeest, fox, hyena, monkey, porcupine, mongoose, and rabbit). Of these rabbit, mongoose, and hyena were identified through markings on transects, which allowed us to record the presence of animals that are difficult to see. The others are likely seen by Maasai in areas which transects did not cover (i.e. monkeys and hartebeest by the river area). Observations made on the transects, and by Maasai while herding, reflect different biases. Both are snapshots in time of particular places. From a scientific perspective Maasai observations as a sampling frame are not consistent from one window to the next and are temporally heterogeneous and spatially biased. However, if wildlife and livestock resource use patterns consistently overlap in time and space, then Maasai observations can be viewed as a more targeted sampling strategy. Additionally, since most herders tend to frequent the same areas over the years, their observations can also be discussed in terms of longitudinal trends.

Clearly, where transects are placed and when they are walked/driven matter. For example, *Oremit* is an area (T5) that gets rain first during *Orkisirata*; the animals come near the lake and are also more visible at this time because of the preceding dry season leading to reduced vegetation and more visibility. TMCP did not sample this area and we did, leading to differences in our findings. Only some Maasai use this area for grazing at certain times, meaning not all were able to report on wildlife found here at certain times of the year. The area all along the lake is important for wildlife, according to Maasai reports and our own transect data, but it is very difficult to get to when it is raining—particularly in a car. Most places that are difficult if not impossible to drive through during the rains because of mud are favored by animals but rarely sampled on the ground.

'Other/Anecdotal' Data

Sampling is just one way of making complex realities legible. The other is by limiting what counts as 'data'. Samples assume representation and continuity. But what happens when the rain patterns change? For instance, what happens when October is not the dry season this year though it was last year? Or when the rains come so late that even March is not the wet season? What happens when residence and farming patterns change or poaching occurs? All these factors affect wildlife numbers and movement patterns and yet are not often captured in wildlife census measurements. Much of this data is considered anecdotal and put in footnotes or ignored. But if listened to carefully, such data can tell us a lot about changing wildlife patterns. For instance, when Maasai talk about the whereabouts of wildlife they often talk about changes in dam levels, farming, closing of pathways, and grass growth. While there might be 'regular' patterns, such as wildebeest being by the lake in the wet season, even these patterns vary. As one man explained, 'if it rains a lot during *Engakwai* [wet season] the wildlife leave because there is too much water and mud, they go to more mountainous areas, especially wildebeest. If the rains are not very big, they can stay in *Lera* [a slightly wooded area up a bit from the lake]'. There was much discussion about the lack of wildebeest by the lake in recent years, and it was attributed to changes made to the water flow, diverting it into the park and thus changing wildlife numbers. This sort of information could help to clarify disparities or changes in scientific monitoring results, or point to the need for different types of data collection. But this demands taking such knowledge seriously and building collaborative dialogues between conservationists and community members. This leads to the final issue: politics.

The Politics

I began this chapter discussing the politics of wildlife census counts—used to promote or support conservation interventions. Politics is present in every aspect of counting wildlife. Conservationists often draw on scientific data from 1963 to the present to talk to Maasai communities about declining wildlife numbers. Sometimes Maasai agree with these assessments and sometimes they do not. For many Maasai, wildlife numbers seem to be increasing. They farm more now, so more wildlife are seen as 'pests' and noticed than before. With increased land fragmentation and loss (to national parks and farms), more people now occupy smaller spaces, and more wildlife seem to be around and closer to people than before. Many people complained about wildlife from the nearby conservation area coming to their settlements at night for protection against predators, something not often picked up in scientific sampling schemes (but see Reid et al. 2003). So for many Maasai, wildlife numbers *have increased*. There are also differences in where wildlife are at different times, and because these differences are not measured scientifically, conservation practitioners and scientists often discount Maasai knowledge.

Part of our transect work was an attempt to show the continual presence of wildlife in village land, and thereby challenge conservation practices that demand wildlife need to be separated from people. But there are also potential political implications of this work. While our intention was to say, 'hey look wildlife are doing fine in village lands', conservationists often respond to such data with suggestions for increased restrictions on land use to protect such land from farming, habitation, and sometimes herding. I was living in a Maasai village, and concerned about their well-being, so these politics needed to be carefully considered. As a result, certain areas were excluded from our sampling scheme because they were too important for herding, farming, and housing—even if they would have exposed high wildlife presence. These areas were small, but their exclusion matters, because it reflects the ongoing antagonism preventing better communication and collaboration between communities and conservation practitioners. Perhaps an increased awareness of how wildlife can and do co-exist with people in this landscape would minimize the risks Maasai (and many others) face, but that is not something we were willing to gamble on. So our methods were limited by the politics of conservation.

Conclusion

In the end, what did our research do? What did we succeed at showing? And why is this Critical Physical Geography? Our data showed high numbers of wildlife in Maasai village land throughout the year, and the regular patterns of wildlife following rainfall and vegetation growth. But we also showed that Maasai knowledge about wildlife is important for understanding wildlife presence in more than an anecdotal way, in a way that can be brought into dialogue with scientific ways of knowing wildlife and thus impact how conservationists too know wildlife. It shows that there are ways to expose complexity and incorporate multiplicity, while still discussing trends that can be helpful for people—Maasai residents and conservationists.

For example, there are different places that the wildebeest and other animals frequent at different times of the year. Different people have different knowledge about that. Maasai out herding have information that is not covered through transects. There were many agreements between what the elders say and our transect data showed, and other times these two sources of knowledge diverged. Where the transects were placed clearly results in what gets seen, just as where an individual lives and where s/he goes for grazing or water determines what s/he sees. Animals move to follow rainfall, which changes every year. This seems to be particularly true for wildebeest but also, to a lesser extent, zebra and gazelle. However, wildebeest may also be reacting to rainfall at a larger scale than we were able to measure. If it rains in other places that the animals prefer, they may not come to this particular village where I was working. For this reason it would be useful to look at larger-scale data of the TME, such as through the SRF counts, and the work of TMCP. Perhaps more collaboration between scientists and local people could help determine where wildlife are when and how this is changing. This would demand taking multiple ways of knowing and seeing seriously.

To do all this demands a real engagement with ecology (counting animals, closely analyzing existing data), grounded long-term ethnography in a particular place, trusted relations with local community members and conservationists, and a critical awareness of the politics involved—in terms of what knowledge counts, and what the on-the-ground implications may be for producing (and sharing) certain information. It demands simultaneously 'doing' science, critically investigating scientific knowledge production, and other ways of knowing. And it involves a constant back and forth between the politics of knowledge production, application, and circulation (Goldman et al. 2011), with an eye for producing policy-relevant findings.

Notes

1. Other similar forms of 'hybrid' research with pastoralists on wildlife ecology include the work of Reid et al. (2003) and Turner and Hiernaux (2002).
2. In the 2001 wet season, elephants were counted in the TME using SRF and total count methods in rapid succession. Total count results reported 1523 elephant, while SRF figures reported 5325 animals (Foley 2005: 15).
3. Maasai society is socially organized around corporate 'age-sets', where young men are initiated into a particular age-set at circumcision and then become *ilmurran*. The *ilmurran* have historically been responsible for protecting people and livestock, but also going to war, whereas the elders are more responsible for governance, though this is rapidly changing across Maasailand. See Goldman, M. J., Roque de Pinho, J. & Perry, J. 2010. Maintaining complex relations with large cats: Maasai and lions in Kenya and Tanzania. *Human Dimensions of Wildlife*, 15, 332–346, Hodgson, D. L. 2001. *Once Intrepid Warrior: Gender, Ethnicity and the Cultural Politics of Maasai Development*, Bloomington, Indiana University Press.
4. Researchers working in the TME, for instance, almost always use Lamprey (1963, 1964) as a standard for the way things used to be and a model of sound counting techniques.
5. There are tools to measure distances that utilize laser technology and are very expensive. It is also questionable how good they are at detecting moving targets. The other type is based on light, and works much like an old-fashioned viewfinder, but was very inaccurate in the hot, bright sun.
6. In Maa, the Maasai language, both gazelles are referred to as *enkoilii*. Specific names exist for the Thompson's (*eminimin*) and Grant's (*Olwargas*), but it is common for people to say *enkoilii* without specifying which unless asked. Most often reference is to Thompson's gazelle which is the most common in the village year round.
7. Both animals had only a 0.3 index of presence, seen on only 30% of the transect periods. While walking transects are better in capturing such species, they are still limited.

References

Adams, W.M., and J. Hutton. 2007. People, parks and poverty: Political ecology and biodiversity conservation. *Conservation and Society* 5: 147–183.

Adams, W.M., and M. Mulligan, eds. 2003. *Decolonizing nature: Strategies for conservation in a post-colonial era*. Sterling, VA: Earthscan.

Anderson, D., and R. Grove, eds. 1987. *Conservation in Africa: People, polices and practice*. Cambridge: Cambridge University Press.

Bockington, D., J. Igoe, and K. Schmidt-Soltau. 2006. Conservation, human rights, and poverty reduction. *Conservation Biology* 20: 250–252.

Borner, M. 1985. The increasing isolation of Tarangire National Park. *Oryx* 19: 91–96.

Brockington, D. 2002. *Fortress conservation: The preservation of the Mkomazi Game Reserve, Tanzania.* Bloomington: Indiana University Press.

Caughley, G., and A.R.E. Sinclair. 1994. *Wildlife ecology and management.* Cambridge, MA: Blackwell Science.

Collet, D. 1987. Pastoralists and wildlife: Image and reality in Kenya Maasailand. In *Conservation in Africa: People, policies, and practice*, ed. D. Anderson and R. Grove. Cambridge, UK: Cambridge University Press.

Duffy, R. 2014. Waging a war to save biodiversity: The rise of militarized conservation. *International Affairs* 90: 819–834.

Foley, L.S. 2002. *The influence of environmental factors and human activity on elephant distribution in Tarangire National Park, Tanzania.* Master of Science, International Institute for Geo-information Science and Earth Observation.

Gamassa, D.M. 1989. *Land use conflicts in arid areas: A demographical and ecological case study: The Kwakuchinja wildlife corridor in northern Tanzania.* MSc, Agricultural University of Norway.

Gardner, B. 2016. *Selling the Serengeti: The cultural politics of Safari Tourism.* Athens, GA: University of Georgia Press.

Gereta, E., G.E. Ole Meing'ataki, S. Mduma, and E. Wolanski. 2004. The role of wetlands in wildlife migration in the Tarangire ecosystem, Tanzania. *Wetlands Ecology and Management* 12: 285.

Goldman, M. 2003. Partitioned nature, privileged knowledge: Community based conservation in Tanzania. *Development and Change* 34: 833–862.

———. 2007. Tracking wildebeest, locating knowledge: Maasai and conservation biology understandings of wildebeest behavior in Northern Tanzania. *Environment and Planning D: Society and Space* 25: 307–331.

———. 2009. Constructing connectivity? conservation corridors and conservation politics in East African Rangelands. *Annals of the Association of American Geographers* 99: 335–359.

Goldman, M.J. 2006. *Sharing Pastures, Building Dialogues: Maasai and Wildlife Conservation in Northern Tanzania.* PhD, University of Wisconsin-Madison.

———. 2011. Strangers in their own land: Maasai and wildlife conservation in Northern Tanzania. *Conservation and Society* 9: 65–79.

———. forthcoming. *Enacting Nature, Performing Conservation: Maasai, Conservation Scientist, and Wildlife in Northern Tanzania.* Book manuscript in review at University of Arizona Press.

Goldman, M.J., J. Roque de Pinho, and J. Perry. 2010. Maintaining complex relations with large cats: Maasai and lions in Kenya and Tanzania. *Human Dimensions of Wildlife* 15: 332–346.

Goldman, M.J., P. Nadasdy, and M.D. Turner, eds. 2011. *Knowing nature: Conversations at the intersection of political ecology and science studies.* Chicago: University of Chicago Press.

Goldman, M.J., J. Roque de Pinho, and J. Perry. 2013. Beyond ritual and economics: Maasai lion hunting and conservation politics. *Oryx* 47: 490–500.

Hassen, S.N. 2000. Conservation status of the Kwakuchinja Wildlife corridor in the last two decades. *Kakauona* 16: 12–15.

Hodgson, D.L. 2001. *Once intrepid warrior: Gender, ethnicity and the cultural politics of Maasai Development.* Bloomington: Indiana University Press.

Hugo, J. 2001. *Estimating abundance of African wildlife: An aid to adaptive management.* Boston: Kluwer Academic Publishers.

Kahurananga, J. 1981. Population estimates, densities and biomass of large herbivores in the Simanjiro Plains, Northern Tanzania. *African Journal of Ecology* 19: 225–238.

Kahurananga, J., and F. Sikiluwasha. 1997. The migration of zebra and wildebeest between Tarangire National Park and Simanjiro Plains, northern Tanzania, in 1972 and recent trends. *African Journal of Ecology* 35: 179–185.

Lamprey, H.F. 1963. Ecological separation of the large mammal species in the Tarangire Game Reserve, Tanganyika. *East African Wildlife* 1: 63–92.

———. 1964. Estimation of the large mammal densities, biomass and energy exchange in the Tarangire Game Reserve and the Maasai Steppe in Tanganyika. *East African Wildlife* 2: 1–46.

Mmari, A.R. 1989. *The Lake-Manyara-Tarangire-Simanjiro complex: A case study of increasing isolation of national parks with emphasis on the Kwakuchinja wildlife corridor.* MSc, Agricultural University of Norway.

Msoffe, F., F.A. Mturi, V. Galanti, W. Tosi, L.A. Wauters, and G. Tosi. 2007. Comparing data of different survey methods for sustainable wildlife management in hunting areas: The case of Tarangire–Manyara ecosystem, northern Tanzania. *European Journal of Wildlife Research* 53: 112–124.

Norton-Griffiths, M. 1978. *Counting animals.* Nairobi: African Ecological Monitoring Programme.

OIKOS. 2002. TMCP (Tarangire-Manyara Conservation Project): Wildlife monitoring in the Tarangire-Manyara Area. Prepared for WWF/TPO.

de Pinho, J.R., C. Grilo, R.B. Boone, K.A. Galvin, and J.G. Snodgrass. 2014. Influence of aesthetic appreciation of wildlife species on attitudes towards their conservation in Kenyan agropastoralist communities. *PloS One* 9: e88842.

Reid, R.S., M. Rainy, J. Oguto, R.L. Kruska, M. Mccartney, M. Nyabenge, K. Kimani, et al. 2003. People, wildlife and livestock in the Mara ecosystem: The Mara count 2002. *Report, Mara Count 2002.* Nairobi: International Livestock Research Institute.

TWCM. 1991. *Wildlife Census, Tarangire, 1990.* P.O. Box 3134. Arusha, TZ: Frankfurt Zoological Society.

Watson, A., and O.H. Huntington. 2008. They're here—I can feel them: The epistemic spaces of indigenous and western knowledges. *Social and Cultural Geography* 9: 257–281.

Whatmore, S., and L. Thorne. 1994. Elephants on the move: Spatial formations of wildlife exchange. *Environment and Planning D: Society and Space* 18: 127–284.

16

Race, Nature, Nation, and Property in the Origins of Range Science

Nathan F. Sayre

Introduction

United States Forest Service photograph 381785 (Fig. 16.1), taken in June 1939 by one W. H. Shaffer and filed under the heading "Handling Stock, New Mexico, Carson N[ational] F[orest]," bears the innocuous caption: "Sheep herder Damacio Lopez talking with Forest Ranger R. L. Grounds." This apparently banal and innocent encounter, dutifully recorded in black and white, is arresting and troubling for anyone familiar with the history and politics of national forests in northern New Mexico (Kosek 2006). Manifold signs of unequal power can be found encoded in the bodies, clothing, postures, and expressions of the two men. Lopez is clearly *not* talking but *listening* while Grounds speaks, a pen in his right hand, gesturing as though to emphasize a point and casting his gaze into the distance over Lopez's head. The sheepherder, by contrast, holds his hands together at the waist, patient and impassive, looking up into the ranger's face. Grounds does not appear angry or argumentative, merely didactic. Tall grasses shroud both men's feet from view, which in June in northern New Mexico suggests either good range conditions, a wet previous year, or both. Yet it is hard not to sense that Grounds is lecturing to Lopez, and that he feels authorized to do so both by his office and by his knowledge of how rangelands work; a notepad, open in his left hand, signals the nature of his expertise. Lopez neither smiles nor frowns, his

N. F. Sayre (✉)
Department of Geography, University of California-Berkeley, Berkeley, CA, USA

© The Author(s) 2018
R. Lave et al. (eds.), *The Palgrave Handbook of Critical Physical Geography*,
https://doi.org/10.1007/978-3-319-71461-5_16

Fig. 16.1 This photograph was taken in the Carson National Forest in northern New Mexico in June 1939. The Forest Service captioned it "Sheep herder Damacio Lopez talking with Forest Ranger R.L. Grounds." By W.H. Shaffer. US Forest Service photograph 381785, National Archives and Records Administration, College Park, Maryland

lips pursed or just barely parted, while his eyes fix on Grounds's visage with a mixture of bemusement, forbearance, and anger, like a parent who sees through a child's story. He looks as though he has heard such talks before and has learned the wisdom of pretending to listen respectfully, even if it requires disciplined self-restraint. Lopez had probably been herding sheep in those mountains since before Grounds was born, as his ancestors had done for some 200 years before that, and he very likely disputed the legitimacy of the Forest Service's possession of the land around them. He may have lacked written notes—indeed, he may have been illiterate—but chances are he knew a great deal about sheep, grazing, and local plants. In this chapter, I hope to explain the dispositions visible in photograph 381785, in particular, the insouciant hubris of Grounds, by examining the origins of range science, from which he derived his authority as an expert.

It is well established that access to and use of natural resources in the United States have been systematically skewed along racial and ethnic lines since European settlement, and that the resulting patterns of inequality and discrimination persist into the present (Schelas 2002). But the roles of biophysical scientists and their knowledge in this history are obscure. Unlike social scientists who sought evidence of racial difference through methods such as anthropometry, natural scientists studied plants, animals, soils, climate, and so forth—everything *but* people. Accordingly, the knowledge range scientists produced was ostensibly about "nature," not society. One premise of Critical Physical Geography (CPG) is that science and politics are never so cleanly divided. Christine Biermann's analysis of the American chestnut blight, for example, demonstrates that "Progressive Era conservation melded together a nostalgic desire for cultural, racial, and environmental purity with a future-oriented project aimed at securing resources into an unknowable and dynamic future…[N]ature and the environment functioned as key sites through which ideas about race and nation circulated" (Biermann 2016, 213). Much the same could be said of the (concurrent) history of range science, as we will see, and this case provides a more specific account of how scientific practices and discourses articulated with race, gender, nation, and nature. Norms of experimental control and data collection ratified the exclusion of herders from research and ultimately rationalized the demise of herding. Additionally, Diana Davis's (2007) pioneering work has illuminated the colonial origins and influences of scientific theories about livestock grazing in arid environments. In the internal, settler-colonial context of the western United States, range science not only echoed European forestry's prejudice against mobile pastoralists, but it also generated a putatively scientific basis for an alternative form of rangeland livestock production, one that was compatible with capitalist norms of property and governance: ranching (Ingold 1980). But by omitting any discussion of these norms, range science euphemized and legitimized the domination of Anglo livestock owners over western rangelands.

Botanical Nationalism

Complex issues of racial, ethnic, and national identity were unavoidable in the settlement of western rangelands. Spanish missionaries, soldiers, and settlers had preceded US citizens by as much as three and a half centuries in parts of what are now California, Arizona, New Mexico, and Texas—areas wrested by force from the sovereign nation of Mexico in 1848. Conquest and settlement of Texas was further inflected through the politics of slavery, as Southerners predominated among early westward migrants there (Perkins 1992). Contrary to Hollywood's later representations, African-Americans were numerous among early cowboys, so too were Mexicans and Native Americans (Jordan 1993). The core practices of range livestock production, moreover, descended not from Great Britain or New England but from Iberia and North Africa (Sluyter 2012). Throughout the west, of course, imperial expansion also involved violent dispossession of Native Americans, almost always rationalized through narratives of racial difference and civilizational "progress" (Brechin 1996, DeLuca and Demo 2001). Rangelands were where native tribes succeeded the longest in resisting US conquest: the Comanche and others in Texas until 1875, the Sioux in the northern Great Plains until 1881, and the Apache in Arizona and New Mexico until 1886. But most scientists engaged these landscapes only after "pacification" was complete, and how their activities might relate to the practices and ideologies of Manifest Destiny was neither preordained nor obvious.

The first biophysical scientists to study western rangelands were botanists sent by the US Department of Agriculture to collect and describe the plants found there.[1] One of the earliest and most famous, George Vasey, unabashedly framed his inquiries in terms of maximizing production in aid of settling the nation's "extensive territory." "Every thoughtful farmer realizes the importance of the production on his land of a good supply of grass for pasturage and hay," Vasey wrote in the opening lines of an early report. "He, who can produce the greatest yield on a given number of acres, will be the most successful man [sic]" (Vasey 1884, 5). Reporting on the grasses of Kansas, Nebraska, and Colorado, he articulated the widespread view that cultivation was invariably more productive than unaided nature:

Man has learned to select those plants, grains, and grasses which are best adapted to his wants, and to grow them to the exclusion of others. This is the essence of agriculture. Nature shows her willingness even here to respond to the ameliorating influences of cultivation. No sooner is the ground plowed, and corn, sorghum, or millet planted, than a crop many times as heavy as that of the native soil is at once produced. (Vasey 1886, 10)

Vasey's patriarchal, post-lapsarian formulation was typical of his time and station: through the toil of cultivation, "man" improved a fecund, feminized, incomplete "nature" (Merchant 1980). And although *human* races appeared nowhere in Vasey's reports, races of *plants*—species, subspecies, and varieties—were very much at issue, arranged in a hierarchy which considered cultivated plants intrinsically superior. By an implicit transitive principle, the greater the improvements achieved, the greater the (type of) "man" and the more "advanced" the civilization.

Geographical provenance served as a second axis of difference, smuggling political divisions into botanical classification as either "native" or "foreign." Imported crop species were widely considered superior, even necessary, for successful Anglo-American settlement: they were the ones that settlers recognized and knew how to grow, and also the ones for which markets already existed. As Alfred Crosby (1986) has shown, exotic plants underwrote colonization of the Americas, aided by livestock and microorganisms that likewise hailed from overseas. One of the USDA's core activities was testing plants from around the world in greenhouses and experimental gardens for distribution to farmers across the nation (Kloppenburg 1988). Turkestan alfalfa, to take one example, was studied in 45 states as well as the District of Columbia and Indian Territory, with positive results reported in 237 of 466 trials (Kennedy 1900, 4). Old World crops, like Old World peoples, were considered naturally superior to their New World counterparts.

As Patricia Seed has shown, cultivation had long been understood in English language and law as the basis of private property. Planting a garden signified and performed possession; in the New World, replacing "wild" vegetation with "fruits and vegetables not indigenous to the country" was not only a way to provide food; it also "justified English title to the Americas" (Seed 1995, 29, 35). Western rangelands posed a challenge to this practice and its underlying prejudices, however. Without irrigation, farming west of the 100th meridian was impossibly unreliable, with catastrophic consequences during periodic severe droughts (Worster 1979). Moreover, bison and perennial grasses "formed a tight partnership" in the Great Plains, "fending off the

entry of any great number of exotic plants and animals," and thereby constituting "the most mysterious exception to the success story of Old World
weeds in the Neo-Europes" (Crosby 1986, 290).

After the extermination of the bison opened an ecological niche for cattle,
native plants soon proved superior to imported species for livestock production. This discovery, in turn, enabled nationalist sentiments to find oblique
expression in the otherwise clinical prognostications of government scientists.
H. L. Bentley, hired by the USDA's new Division of Agrostology[2] in the late
1890s, catalogued the numerous failures and frustrations of livestock producers in central Texas who had tried to use "forage plants not native to the country" to increase the productivity of their rangelands. "What stockmen need,"
Bentley insisted, "are hay meadows of *native* grasses that have shown in past
years all the best qualities of the best hay grasses elsewhere, and that do not
require any experimental work to determine their adaptability and general
value" (Bentley 1898, 18, emphasis in original). The Division of Agrostology's
first circular lamented that "Nearly all of our cultivated grasses and clovers are
of foreign origin," whereas few native species had yet been studied for possible
cultivation. "We want to know what plant will provide the greatest amount of
the most nutritious forage in the shortest season at the least expense to the
farmer…In short, we want the best, and we believe the best can be grown on
American soil from native species" (Smith 1895, 2–3). Similarly, Bentley
(1898, 18) asked, "why should stockmen look to foreign countries or even to
other sections of Texas for grass seeds and hay?…. Let us take care of what we
have and develop them. They are here now. They are here because the soil and
climatic conditions are favorable. About the only question we have to determine is, Which of these are best for hay and which for grazing purposes?"

Ideologically, western rangelands were both threatening and promising.
On the one hand, they challenged the transitive principle that valorized people in proportion to their power to remake nature. On the other hand, they
afforded the possibility of endowing settlers with their own "native" status,
grounded in the apparently natural fit between their imported livestock and
the indigenous environment. "[N]ative grasses are by far the best for home
use; they are suited to the climate and the climate is suited to them" (Bentley
1898, 21, emphasis in original). In retrospect, resolving this tension was made
easier, socially and biophysically, by the fact that Native North Americans had
not domesticated any grazing animals and therefore could not be seen as ante-
cedents or rivals. To simplify: if cattle could occupy the niche previously held
by bison, then the replacement of native tribes by Euro-American ranchers
could be constructed as *both* natural *and* a step forward in the march of civilization. Early range scientists such as Bentley did not make any such claims

explicitly—they did not necessarily *think* them either—but they didn't really need to: the larger discursive field tacitly did it for them.

Blooded Stock

While botanists and early range scientists debated the merits of native versus foreign grasses, experts in the USDA's Bureau of Animal Industry, in concert with agricultural extension agents throughout the country, endeavored to convince farmers and ranchers to pay more attention to the breeding of their livestock. Especially in the west, most herds' lineages were indistinct and effectively unknown. Most cattle (including the famous Texas Longhorns) were descendants of animals imported from Spain long before, as were the sheep in the southwest. Some herds hailed from various other places and times, but in nearly all cases, they were deemed "unimproved" or "scrub" animals, the products of males and females breeding more or less by chance under whatever environmental and other conditions they happened to inhabit. Such mixing was viewed as the animal equivalent of miscegenation, and the "Mexican" breeds were not proper "breeds" at all but mongrels of little value—market value, that is—to their owners or to the nation. What was needed, according to the government experts, was "blooded" or "purebred" animals, especially bulls and rams, who would "improve" the nation's herds through carefully controlled crossbreeding.

The preferred cattle breeds—Shorthorns, Herefords, and Angus—all came from Great Britain, where pioneering breeders such as Robert Bakewell had created them over many generations, "fixing" specific characteristics or traits by mating like to like (inbreeding) and culling non-conforming individuals (Trow-Smith 1959). As Harriet Ritvo (1987) has shown, blooded stock not only reflected the British upper classes' obsession with human bloodlines but also legitimized their dominance by cloaking it in the guise of economic efficiency for national prosperity. In the United States, government scientists such as Charles Curtiss (1898), Director of the Iowa Agricultural Experiment Station, employed the same logic to appeal to ordinary citizen-farmers. Controlled breeding could indeed instill desired traits over the course of generations, but as a strategy for increasing profits, the argument was circular: "superior" breeds were those that middlemen and consumers would pay more for, and they would pay more because they had come to believe those breeds' traits were superior. As in Britain, then—though without (human) aristocrats—blooded stock euphemized market power as a function of breeding and descent (Sayre 2002). Judging from the persistence of government efforts (e.g., Pickrell 1925), however, many Western livestock owners ignored the experts' admonitions, perhaps because they could not afford the high-priced animals or the fencing that controlled breeding required.

Issues surrounding the breeding of sheep were more complex, in part because the most highly valued breeds were the Merino and the Rambouillet, which originated in the royal flocks of Spain and France, respectively. These noble, Latin roots necessitated more intricate ideological work, but the outcome was the same as in the case of cattle. In 1892, the Bureau of Animal Industry published a comprehensive, two-volume *Special Report on the History and Present Condition of the Sheep Industry of the United States,* in which the racial homology of humans and livestock was conspicuous. The section on California, for example, opened with the observation that the first domestic sheep there were "of very low grade, both as to wool and mutton products… These sheep were of nearly all colors, indicative of carelessness in breeding for many generations." The very next paragraph stated that the gold rush had triggered the transition between "California under the rule of the Latin race and its Saxon successors" (Carman et al. 1892, 947), and that since that time steady improvements in the state's flocks had occurred. Mexicans had driven more than half a million sheep into the state from New Mexico in the 1850s, while Anglo-Americans had imported "a much better class of sheep" from the eastern United States, with breeds carefully chosen "with reference to improving the Mexican sheep" (Carman et al. 1892, 952). The report praised "the successful crossing of the Merino with the British breeds" in the counties north of San Francisco (Carman et al. 1892, 961). For Oregon, the report linked individual Anglo sheep owners with specific "improved" breeds: Hiram Smith had brought "pure-blood Merino rams from Ohio" in 1851, for example, while Martin Jesse's Australian Merinos, imported in 1858, "were certified as being pure descendants of Spanish Merino flocks of King George III of England" (Carman et al. 1892, 977). Other men imported British breeds: New Oxfordshires, Hampshire Downs, Southdowns, and New Leicesters. As one of the report's co-authors, John Minto (himself a prominent sheep breeder in Oregon [Rakestraw 1958]), wrote ten years later, the combination of these various pure breeds, in a favorable climate, had made Oregon "the greatest Merino breeding station in the world at present" (Minto 1902, 242). The "American improved Spanish Merino" (Carman et al. 1892, 982) was constructed as the best breed of all, combining British and Spanish virtues, while erasing any aristocratic or Latin residues. As with cattle, then, sheep breeding could imbue American settlers with a kind of nativeness or autochthony via interlinked ideas about animals, race, and nature. At least it could under certain conditions.

Anti-herder Chauvinism

Not surprisingly, opinions about the people involved in livestock production were comingled with opinions about the breeds of their livestock, and this was especially true in relation to sheep. Unlike cattle, sheep required constant human oversight to protect them from predators and to lead them to water and fresh pasture. "Domestication had given sheep…too great an inventiveness in finding ways to die. Although hardy and able to withstand drought, sheep could not live without shepherds" (White 1994, 239). "The occupation of a herder," Minto (1902, 238) wrote, "is that of a protector…a good herder has his flock within his sight every waking hour." Herders' work was tedious, lonely, and sometimes dangerous, and they endured both low wages and lowly status. Many hailed from Asia, Mexico, France, Spain or Portugal, and herding neither required nor instilled proficiency in English. The Bureau of Animal Industry's 1892 report complained that in southern California,

> a large proportion of the sheep and wool industry is in the hands of a foreign element called Basques. As a class they are described by their intelligent countrymen as 'very ignorant about anything except their special calling, few of them being able to read or write in their own language'… As a class, they occupy the same relation to English-speaking men engaged in sheep-raising that Chinese laborers held to white laborers on this coast before they were excluded by law. (Carman et al. 1892, 972–973)

The Basques were not the only sheep raisers to experience prejudice—Chinese, Mexicans, and Mormons were among the others (Perkins 1992). But the Basque case makes clear that such persecution was not simply ethno-racial; citizenship and nationalism were also involved. "A large proportion of the herders are of foreign birth" (Carman et al. 1892, 985). Reporting ten years later on the Northern Great Basin, the Bureau of Plant Industry's David Griffiths referred to "alien sheep interests. It is said that a very large proportion of the sheep in the region belong to Basques, who own no land, and who in many cases are not citizens" (Griffiths 1902, 23). In a country of immigrants, many of whom had only recently arrived, those willing to become US citizens were much less objectionable than "foreigners in the range country (generally with sheep), who are there only to gather wealth and go away with it. This evil is not as great in Oregon as it was ten years ago, and not nearly as bad as it is now in southern California" (Carman et al. 1892, 983).

How, then, to discern and shape the intentions of people who might or might not settle down permanently and help build the new nation? The answer lay in property and land and therefore also in class. The Bureau of Animal Industry report made the link explicitly: "The peace and permanency of wool-growing, as the pursuit to which these lands are best adapted, requires that means to secure private control [of land] should be adopted as soon as possible," whether by lease, sale, or outright grant and "in such quantities as would enable a man of average industry to support a family from their use" (Carman et al. 1892, 983). Sheep could be markers of poverty and landlessness, especially by association with the herders who tended them day in and day out. But sheep could also be a ladder out of poverty and wage labor altogether. The costs of entry were low; H. A. Heath, another co-author of the 1892 report, estimated that "one-third the capital required to stock up with cattle is sufficient to start with sheep" (quoted in McGregor 1982, 28). "The very poorest men may, and often do, enter the business with their labor only, by undertaking to care for a flock purchased by the capital of others" (Carman et al. 1892, 982). Meanwhile, the rate of return was high, not so much from the sale of mutton but from wool: a non-perishable, readily transportable product in high demand, one which did not require the slaughter of any animals. A healthy, well-tended flock could more than double in number in a single year, twins being quite common, and part of a herder's wages was typically paid in lambs from the flock in his care. If things went well, a herder could amass a herd of his own in just a few years. In short, sheep could be a uniquely rapid means of socio-economic mobility. When cattle prices collapsed in the 1880s and 1890s, range sheep production boomed, with some "fifteen million sheep trailed eastward from the Pacific Slope during the last third of the nineteenth century, a movement of livestock five times larger than the fabled 'long drives' of cattle north from Texas after the Civil War" (McGregor 1982, 28).

Range livestock production was thus a means both of "naturalizing" European animals and of transforming immigrant people from wage laborers into capitalists; if they also chose to become naturalized Americans, they could metamorphose from larcenous foreigners into productive citizens. Published opinions about such mobility displayed a striking double standard, however. A herder who attained sheep owner status might be praised for diligence and hard work: "The best herders are…Americans, and generally have aspirations to become flock-owners or something else they prefer. The wealthiest men now in the business are Americans, many of whom started as herders for themselves or others" (Carman et al. 1892, 985). But Basque herders who did the same thing were condemned for their success, described as "suspicious, secretive," and clannish, "active, economical, and expert rivals for public range" (Carman et al. 1892, 972–974)—in other words, threatening.

Meanwhile, it was common to accuse other non-Anglo herders of "pure laziness" (Wooton 1908, 27), especially if they were Mexican or didn't speak English. In at least some times and places, meanwhile, the double standard was enforced by outright violence: "According to one estimate, between 1893 and 1903 in Wyoming alone, over twenty sheepherders and sheepmen were murdered, and at least five times that number maliciously wounded, for no other apparent reason than their association with the ovine species of livestock" (Perkins 1992, 1). Roughly 100,000 sheep were killed on the range in Wyoming in the 1890s and 1900s, and many more killings occurred in the wider western United States.

The precise motivations behind these attacks are unknown. The legendary "range wars" between cattlemen and sheepmen, cowboys and sheepherders, have long been prone to Hollywood hyperbole and pendulum-like revisionism among historians. No matter how extensive (or not) these conflicts were, the question remains unsettled: why were the two kinds of livestock considered incompatible? In a remarkable (and remarkably overlooked) article, John Perkins (1992) argued that anti-sheep violence and prejudice in the west was rooted in the attitudes of Southerners who migrated to Texas, adopted Spanish and Mexican range cattle production practices there, and then moved north and west as the harbingers of White American settlement. If they were racists with respect to non-White people, they were biased against sheep for political-economic reasons. Before the Civil War, northern and southern states had feuded in Congress over tariffs on wool, with the North in favor of levies to protect domestic production—which was concentrated in New England and supplied factories there—and the South opposed out of fear that Britain would retaliate with tariffs on cotton. An 1828 wool tariff "caused many in the South to question the benefits of remaining in the Union…It would be no exaggeration to say that for the typical Southerner, the sheep came to represent the detested 'Yankee' and the threat he posed to the existence of the South" (Perkins 1992, 9). The notion that sheep were uniquely potent agents of land degradation—on which more in a moment—may also have its origins in Southern sensibilities, as plantation owners sought to deflect blame for "soil exhaustion" away from cotton cultivation (Perkins 1992, 8).

The evidence presented above—linking botanical nationalism, blooded stock, and anti-herder chauvinism from Texas to Oregon—amounts to the exception that proves the rule for Perkins's thesis. Hispanic, Asian, Basque, and Mormon sheep producers experienced widespread prejudice and extra-legal persecution, in which multiple axes of difference—ethnicity, language, religion, gender, nationality, property, and class—were alloyed in various combinations with racial categories and hierarchies. White, Protestant,

English-speaking, male, propertied, capitalist American citizens could and did succeed in the sheep business, especially in the Pacific Northwest, because they defined and occupied the niche where the privileged pole of all these axes intersected and reinforced each other. "In the end Americans made the best success, both as herders and flock masters. Not rarely a young man starting as herder ended as a wealthy sheep and land owning banker" (Minto 1902, 230). On western rangelands as a whole, however, sheep production would nonetheless be subordinate to cattle production, with sheep numbers plummeting and sheepherders effectively disappearing by the Second World War. In these developments, the Southerners' prejudices were transmuted into a suite of scientific knowledge claims that erased or euphemized racial issues.

Eliminating Herders Scientifically

The first scientific research on western range livestock production took place in central Texas in the late 1890s. Scientists in the USDA's Division of Agrostology (founded in 1895 and merged into the new Bureau of Plant Industry in 1902) were instructed to describe and measure the major forage plants, estimate the damage done by widespread overgrazing in the preceding decade, and determine how many livestock the range could support on an ongoing basis—the so-called carrying capacity of the land. In all this work, the livestock in question was cattle (Sayre 2017). But the most influential early range science experiment—so influential that it is sometimes erroneously described as the first—involved sheep, and it took place in eastern Oregon in 1907–1909, co-sponsored by the United States Forest Service and conceived by its founding chief, Gifford Pinchot. The goal of the Coyote-Proof Pasture Experiment was to demonstrate the advantages of grazing sheep in the absence of predators. Ironically, the experiment sought to reverse one of the very traits that had been selected for in the development of the Merino and Rambouillet breeds: the instinct to bunch and move together so that a herder could tend to them on unfenced rangelands. This "flocking instinct" was necessary for long seasonal migrations or transhumance, and it was as much a part of *being* Merino and Rambouillet as "large fleeces of fine-grade wool that commanded a premium in the marketplace" (McGregor 1982, 30). It was also why "if one sheep bolted the rest would follow," potentially endangering the entire flock and making a quality herder so "vitally important" (McGregor 1982, 84–85).

I have examined the Coyote-Proof Pasture Experiment in detail elsewhere and shown that its ulterior motive was to reduce or eliminate the need for herders (Sayre 2015, 2017). The experiment's "success" was measured and proclaimed in terms of the economic rate of return to constructing the necessary fence, understood as a capital investment that could be recouped through increased wool production, reduced sheep mortality, and reduced labor costs. But the flaws in the analysis were numerous and severe: the actual cost of the fence was arbitrarily reduced by nearly half, while the cost of a hunter, who was hired to patrol the fence and kill any predators he encountered, was omitted entirely; the ecological data from inside and outside of the fence were confounded and non-comparable. It is difficult to avoid the conclusion that the "successful" outcome was a foregone conclusion. Here I ask: can this be attributed to issues of race?

On first glance, evidence of racial bias in range science is vanishingly slim. The reports, scientific publications, and internal memos and notes of government range scientists contain no overt racism, nor any pejorative or derogatory language that might suggest racist attitudes. The scientist who oversaw the Coyote-Proof Pasture Experiment, Pinchot's close friend Frederick Coville, was if anything an advocate for sheepherders. At the request of the secretary of agriculture, Coville conducted extensive fieldwork on sheep grazing in the Cascade Forest Reserve in 1897 in the wake of a controversial National Academy of Sciences report that singled out "nomadic sheep husbandry" as a singular menace to the nation's forests: the sheep themselves were "hoofed locusts"[3] who ate grasses, shrubs and trees alike "to the ground"; their owners were "foreigners, who are temporary residents of this country"; and the herders were "dreaded and despised" in "every Western State and Territory" for setting fires that destroyed forests and unleashed erosion on downstream communities (National Academy of Sciences 1897, 18–19). Coville's report—"a model of fairness and thoroughness" (Rakestraw 1958, 377)—bluntly rejected all these accusations: "All the sheep owners in eastern Oregon appear to be American citizens," he wrote, and many "are prominent influential citizens of the highest character."

A popular impression seems to prevail that sheep herders in Oregon, as elsewhere, represent a comparatively low class of humanity. This impression as applied to the majority of sheep herders many years ago was perhaps correct. At the present time, however, many bright and reputable young men have undertaken sheep herding in default of opportunities for more desirable work, and as a whole they probably average as well in character as the men engaged in any other branch of agricultural industry. (Coville 1898, 12)

The association of herders with forest fires was erroneous or exaggerated, Coville wrote, and the effects of sheep on vegetation and watersheds were modest and manageable. Echoing the Bureau of Animal Industry's 1892 report, he called for a system of leases or permits that would secure individual sheep owners' access to public rangelands.

For the scientists, the effects of herding took priority over the ethno-racial makeup of herders. In his official reports, the scientist who conducted the Coyote-Proof Pasture Experiment, James Jardine, showed no sign of *racial* prejudice against sheepherders. He did complain about herders' *skills*, however, in ways that invoked their nationalities and language skills. Damage to vegetation by sheep could be attributed, he wrote, to "poor herders, who were of French descent, unable to speak English" (Jardine 1909, 32). The specious implication suggests some sort of prejudice—neither language nor nationality has any necessary relation to herding ability—but quality herders were in high demand and hard to find. "Sheepmen of the Columbia Plateau, as well as woolgrowers in other western range areas, frequently complained about the shortage of good herders. Sloppy herding could be costly" (McGregor 1982, 85). One of the largest range sheep operations in the region, the McGregor Land and Livestock Company, relied for decades on French immigrant herders, whose skills were invaluable to the firm's success (McGregor 1982, 84).

If sheep owners hoped that fencing and predator control could reduce their labor costs, scientists hoped the same innovations could improve their experimental controls. Jardine, like Coville before him, recognized that herding could be done well or poorly, and that the differences would manifest both in the performance of the sheep—in weight gain, wool clip, survival, and mortality—and in the effects on vegetation. In the Coyote-Proof Pasture Experiment, they compared the sheep inside the pasture to flocks of sheep on the adjacent open range in terms of these variables. But the scientists could not control the herders of those sheep, let alone quantify and measure their skillfulness. Herders were a variable that threatened the "scientific" rigor of range science, *even if they were highly skilled*. In subsequent decades, the Forest Service and other government agencies spent millions of dollars building fences, exterminating predators, and thereby rendering herders obsolete throughout the West (Sayre 2017). Concurrently, the sheepherder/owner who lacked landed property—known at the time as "the coyote sheepman"—was driven out of business by state and federal laws. In the words of the *National Wool Grower* in 1907: "Like the coyote he was a vagrant and his extinction will not be regretted. Eventually he had to go and the industry will hereafter be on a permanent basis" (quoted in McGregor 1982, 110).

Conclusion

Range science transmuted the private economic risk that bad herders posed for sheep owners into a condemnation of herding itself as a public threat to the nation's rangelands, forests, and watersheds. In the process, range scientists simultaneously embraced and euphemized a complex set of interconnected ideas about plants and animals, land and property, efficiency and national progress. "Even with the best herders it is impossible to handle large bands of sheep with the same grazing efficiency as is secured in the fenced pastures of the eastern United States, and when one considers the large percentage of herders who are not skilled or who have a greater regard for their own comfort than for the interests of the owner of the sheep or for the permanent welfare of the range, the aggregate waste can be regarded in no other light than as a matter of serious public concern" (Jardine 1908, 5–6). Thus were issues of race and class both "etched and elided" (DeLuca and Demo 2001, 542) in (to) range science: etched because herders were wage laborers and often non-Anglo Saxon, widely "othered" and sometimes violently persecuted; elided because the ostensibly ecological evaluation of grazing that range scientists developed erased human agency and its attendant political issues altogether.

A CPG of range science, one that "investigate[s] material landscapes, social dynamics, and knowledge politics together, as they co-constitute each other" (Lave et al., this volume), has potentially far-reaching implications for how rangelands are managed today. The ideal to which early range science and government policy aspired was one in which fences replaced herders, predators were exterminated, and livestock achieved a "natural" balance with forage growth. Just over a century later, this vision is viewed as fundamentally mistaken. Fences fragment rangelands and impede the mobility of livestock and wildlife (Galvin et al. 2008); predators may play outsized, "keystone" roles in ecosystems (Marris 2014); static carrying capacities cannot be determined for many rangelands (Behnke, Scoones and Kervin 1993); the "balance of nature" is a chimera (Wu and Loucks 1995). In light of these new—or rather, recently rediscovered—eco-social insights, there is every reason to conclude that *good* herders may be the most economical and ecologically sustainable means of producing livestock—cattle as well as sheep—on western rangelands (Meuret and Provenza 2014).

Acknowledgments In writing this chapter, I benefited from generous and constructive feedback from many people, including Rebecca Lave, Christine Biermann, Diana Davis; participants in a seminar discussion hosted by University of California-

Berkeley's Social Science Matrix, including Bill Hanks, Carla Hesse, Marion Fourcade, Terry Regier, Lynsay Skiba, and Istvan Rev; and a number of graduate students with whom I am fortunate to work, including Christopher Lesser, Julia Sizek, Robert Parks, and Mike Simpson. The usual disclaimers apply.

Notes

1. By this, I mean the first to study rangelands as such, rather than incidental to other kinds of research and exploration, such as geological investigations and surveys for railroad routes.
2. "Agrostology," from the Greek root *agrōstis*, is the botanical study of grasses. As the name of a government agency, it seems never to have caught on, requiring a parenthetical definition "(Grass and Forage Plant Investigations)" in official publications.
3. The phrase appeared in quotation marks in the report, presumably alluding without attribution to John Muir. Muir was not a member of the Academy's National Forestry Committee, which wrote the report, but he was closely and publicly associated with it. Five years later, John Minto (1902, 233, emphasis in original) pithily wrote, "The epithets used [to disparage sheep] are the *worn coin* of the half insane but charming Carlylian writer on mountains and forests, John Muir."

References

Behnke, Roy H., Ian Scoones, and Carol Kerven, eds. 1993. *Range ecology at disequilibrium: New models of natural variability and pastoral adaptation in African Savannas.* London: Overseas Development Institute.

Bentley, H.L. 1898. *A report upon the grasses and forage plants of Central Texas.* USDA Division of Agrostology Bulletin 10. Washington, DC: Government Printing Office.

Biermann, Christine. 2016. Securing forests from the scourge of chestnut blight: The biopolitics of nature and nation. *Geoforum* 75: 210–219.

Brechin, Gray. 1996. Conserving the race: Natural aristocracies, eugenics, and the U.S. Conservation Movement. *Antipode* 28: 229–245.

Carman, Ezra A., H.A. Heath, and John Minto. 1892. *Special report on the history and present condition of the sheep industry of the United States.* Washington, DC: Government Printing Office.

Coville, Frederick Vernon. 1898. *Forest growth and sheep grazing in the Cascade Mountains of Oregon.* Bulletin no. 15. Washington, DC: USDA Division of Forestry.

Crosby, Alfred W. 1986. *Ecological imperialism: The biological expansion of Europe, 900–1900.* Cambridge and New York: Cambridge University Press.

Curtiss, Charles F. 1898. *Some essentials of beef production.* USDA Farmer's Bulletin No. 71. Washington, DC: Government Printing Office.

Davis, Diana K. 2007. *Resurrecting the granary of Rome: Environmental history and French Colonial Expansion in North Africa.* Athens, OH: Ohio University Press.

DeLuca, Kevin, and Anne Demo. 2001. Imagining nature and erasing class and race: Carleton Watkins, John Muir, and the construction of wilderness. *Environmental History* 6: 541–560.

Galvin, Kathleen A., Robin S. Reid, R.H. Behnke, and N. Thompson Hobbs, eds. 2008. *Fragmentation in Semi-Arid and arid landscapes: Consequences for human and natural systems.* Dordrecht: Springer.

Griffiths, David. 1902. *Forage conditions on the Northern Border of the Great Basin.* Bureau of Plant Industry Bulletin No. 15. Washington, DC: Government Printing Office.

Ingold, Tim. 1980. *Hunters, pastoralists, and ranchers: Reindeer economies and their transformations.* Cambridge: Cambridge University Press.

Jardine, James T. 1908. *Preliminary report on grazing experiments in a Coyote-Proof Pasture.* Circular no. 156. USDA Forest Service: Washington, DC.

———. 1909. *Coyote-Proof pasture experiment, 1908.* Circular no. 160. Washington, DC: USDA Forest Service.

Jordan, Terry G. 1993. *North American cattle ranching frontiers: Origins, diffusion, and differentiation.* Albuquerque: University of New Mexico Press.

Kennedy, P. Beveridge. 1900. *Turkestan Alfalfa.* Division of Agrostology Circular 25. Washington, DC: US Department of Agriculture.

Kloppenburg, Jack Ralph. 1988. *First the seed: The political economy of Plant Biotechnology, 1492–2000.* Cambridge: Cambridge University Press.

Kosek, Jake. 2006. *Understories: The political life of forests in Northern New Mexico.* Durham: Duke University Press.

Marris, Emma. 2014. Rethinking predators: Legend of the wolf. *Nature* 507: 158–160.

McGregor, Alexander Campbell. 1982. *Counting sheep: From open range to agribusiness on the Columbia Plateau.* Seattle and London: University of Washington Press.

Merchant, Carolyn. 1980. *The death of nature: Women, ecology, and the scientific revolution.* San Francisco: Harper and Row.

Meuret, Michel, and Fred Provenza, eds. 2014. *The art and science of shepherding: Tapping the wisdom of French Herders.* Trans. Bruce Inksetter and Melanie Guedenet. Austin, TX: Acres USA.

Minto, John. 1902. Sheep husbandry in Oregon. *The Quarterly of the Oregon Historical Society* 3: 219–247.

National Academy of Sciences. 1897. *Report of the committee appointed by the National Academy of Sciences upon the inauguration of a forest policy for the forested lands of the United States to the Secretary of the Interior, May 1, 1897.* Washington, DC: Government Printing Office.

Perkins, John. 1992. Up the trail from Dixie: Animosity toward sheep in the culture of the U.S. West. *Australasian Journal of American Studies* 11: 1–18.

Pickrell, Charles U. 1925. *The range bull*. University of Arizona College of Agriculture Extension Service Circular No. 51. Tucson: University of Arizona.

Rakestraw, Lawrence. 1958. Sheep grazing in the cascade range: John Minto vs. John Muir. *Pacific Historical Review* 27: 371–382.

Ritvo, Harriet. 1987. *The animal estate: The English and other creatures in the Victorian age*. Cambridge: Harvard University Press.

Sayre, Nathan F. 2002. *Ranching, endangered species, and urbanization in the Southwest: Species of capital*. Tucson: University of Arizona Press.

———. 2015. The coyote-proof pasture experiment: How fences replaced predators and labor on US Rangelands. *Progress in Physical Geography* 39: 576–593.

———. 2017. *The politics of scale: A history of rangeland science*. Chicago and London: University of Chicago Press.

Schelas, John. 2002. Race, ethnicity, and natural resources in the United States: A review. *Natural Resources Journal* 42: 723–763.

Seed, Patricia. 1995. *Ceremonies of possession in Europe's conquest of the new world, 1492–1640*. Cambridge: Cambridge University Press.

Sluyter, Andrew. 2012. *Black ranching frontiers: African cattle herders of the Atlantic World, 1500–1900*. New Haven: Yale University Press.

Smith, Jared G. 1895. *A note on experimental grass gardens*. USDA Division of Agrostology circular no. 1. Washington, DC: Government Printing Office.

Trow-Smith, Robert. 1959. *A history of British livestock husbandry, 1700–1900*. London: Routledge and Kegan Paul.

Vasey, George. 1884. *The agricultural grasses of the United States*. Washington, DC: Government Printing Office.

———. 1886. *Report of an investigation of the grasses of the Arid Districts of Kansas, Nebraska, and Colorado*. USDA Botanical Division Bulletin 1. Washington, DC: Government Printing Office.

White, Richard. 1994. Animals and enterprise. In *The Oxford history of the American West*, ed. Clyde A. Milner II, Carol A. O'Connor, and Martha A. Sandweiss, 237–273. New York and Oxford: Oxford University Press.

Wooton, E.O. 1908. *The range problem in New Mexico*. Bulletin no. 66. Albuquerque, NM: New Mexico College of Agriculture and Mechanic Arts Agricultural Experiment Station.

Worster, Donald. 1979. *Dust bowl: The Southern Plains in the 1930s*. Oxford: Oxford University Press.

Wu, Jianguo, and O.L. Loucks. 1995. From balance of nature to hierarchical patch dynamics. *Quarterly Review of Biology* 70: 439–466.

17

Coffee, Commerce, and Colombian National Soil Science (1929–1946)

Greta Marchesi

In 1940, Juan Pablo Duque, director of the Technical Department of the National Federation of Colombian Coffee Growers (*Federación Nacional de Cafeteros Colombianos* or FedeCafé) prepared a detailed presentation for the Commission on Plant Sanitation in the Department of Caldas announcing an unconventional shift in his organization's field research program.

Through the decade previous, local planters had been plagued by *la gotera*, or coffee rust, prompting a new technical campaign in 1938 to improve plant health. After two years of observations, FedeCafé technicians had concluded that the health of the industry lay not in the trees themselves but rather in the vitality of plantation ecosystems, and most specifically in the soil. Soil degradation rather than insects or disease, Duque attested to the planters who gathered to hear field results, was the central ecological injury from which all other coffee ailments sprung. Going forward, the Federation's research campaigns would be reoriented from plant canopies to roots and soils.

> The symptoms of decline that I encounter, each time more accentuated and with variable intensity through the whole of the country, have convinced me over the past three years that the gravest problem of our industry is not disease or infestations nor is it problems of cultivation. Rather, there is an *original sin* that obeys the topography of hillsides and patterns of rainfall in our coffee-growing region. The devastations of erosion follow a veritable chain of maladies that increase the vulnerability of plants to disease. (Duque 1940: 2615) (translation and all subsequent translations by author)

G. Marchesi (✉)
Department of Geography, Dartmouth College, Hanover, NH, USA

© The Author(s) 2018
R. Lave et al. (eds.), *The Palgrave Handbook of Critical Physical Geography*,
https://doi.org/10.1007/978-3-319-71461-5_17

Duque blamed nineteenth-century frontier cultivation practices for these degradations, citing techniques developed when nature was "wild and abundant" and land and labor were cheap. By the 1930s, the largely poor community of coffee growers had neither the opportunity nor the resources to follow an earlier generation of growers in opening new terrain. Rather, Colombian *cafeteros* would need to develop practices directed toward the long-term viability of existing plantations. The work of the Federation would be to provide all growers, regardless of the size of their operation, with the technical and educational support—and financial credit—to enable long-term investments in economically and biologically sustainable production. That sustainability, FedeCafé's agronomists had concluded, would be rooted in soil health.

Founded in 1927, FedeCafé represented predominantly small-holding Colombian coffee growers in a global market long dominated by Brazilian plantation agriculture (Koffman 1969). The Federation's work was framed around the unique physical and social relationship between coffee production and Colombia itself, both the nation and its landscape. Coffee exports represented nearly 66% of the total value of the nation's exports (Posada 1976: 114), but given, as Marco Palacios points out, that profits from the other dominant export sectors of gold, bananas, and oil were destined for foreign coffers, the significance of Colombian coffee exports to the national export economy was even greater. In an era of nationalist development around Latin America but a weak national government at home, Colombia's coffee growers' association assumed a pivotal governmental role in coffee-growing communities, assuming what might be considered a kind of *cafetero* hegemony (Palacios 1980) rooted in field-based research and development programs.

FedeCafé's constituency in this period was composed primarily of poor and working class farmers recently radicalized by a decade of populist mobilizations. Widespread poverty among *cafeteros*—the great majority of whom farmed plots smaller than 5.12 hectares— meant that the community had only a limited capacity to invest in chemical inputs and technical innovation (Safford and Palacios 2002: 267). Marginal conditions among *cafeteros* were compounded by plummeting coffee prices in the 1930s. In the past, the Colombian coffee industry had weathered such international crises by relying on intensified production among self-exploiting smallholders, raising production to offset declining prices per kilo. However, declining yields on established plantations in the 1920s and 1930s were evidence that such a policy on its own was insufficient to guaranteed *cafetero* welfare into the future. Further, given the 15-year life span of the coffee tree, frontier zone *fincas* planted during 1920s would face an additional crisis of yields in less than a decade. FedeCafé's central challenge was to support its membership of predominantly small-scale, poor farmers in sustainably maximizing yields without significant capital outlay. They turned in this pursuit to the resources already widely available to small-scale *cafeteros*: the inherent capacities of the land and its inhabitants.

If Critical Physical Geography (CPG) grounds "a critical attention to relations of social power" in "deep knowledge of … biophysical science and technology" (Lave et al. 2014), then the case of the Colombian Federation of Coffee Growers offers an opportunity to consider how science and technology have themselves been transformed by socially produced changes in the material world. Such an enterprise enacts CPG's commitment to the simultaneous study of landscapes, human communities, and the knowledge practices that bind them together in place. FedeCafé agronomists began their work from an assumption of the fundamental inextricability of social and environmental well-being across Colombia's coffee-growing landscapes. Their work was embedded not only in the economic constraints of local communities but also in the global political and economic structures that shaped international coffee markets. The field research and education program that emerged from these concurrent concerns were unapologetically staked to those understandings and to their specific social and political commitments.

Geographers, historians, and sociologists of science have effectively demonstrated how the global expansion of European and North American market interests spurred new bodies of scientific expertise. Scholars have shown how epistemological commensurabilities were created between far-flung environmental contexts via new universalized taxonomies and systems of measure (Koerner 2001; Livingstone and Withers 2005; Scott 1999) even as liberal capitalist states applied surveyed and mapped landscapes to extend private property law across those same spaces (Craib 2004; Banner 2005). However, much less work has been done exploring the ways that *reforms* to that global market sociality create their own scientific and technical forms, particularly in light of nature-society engagements. These questions are especially salient in the realm of agrarian production. As Franklin (2007) and Stassart and Whatmore (2003) have argued, agriculture breaks down key ontological distinctions between human and non-human bodies, connecting societies to environments across time and space in ways that are simultaneously political, discursive, and metabolic.

This chapter considers the florescence of FedeCafé soil management during the 1930s. It shows how FedeCafé scientists' implication in Colombia's central export sector generated a geographically specific field of scientific concerns closely bound to the social and economic conditions of commodity crop production and, by extension, to the project of Colombian nation-building. Finally, it shows the devolution of that mode of land management under the influence of North American experts, a reflection of shifting hemispheric politics and resource streams during the Cold War era.

Like other commodities in liberal capitalist society, agricultural products are conventionally abstracted from the social and ecologic relations of their production in the process of market exchange, what Polanyi (2001) has termed a "dis-embedding" from ecological context. I argue that Colombian coffee in this period offers an intriguing counter-example of market-based sociality consciously embedded in rather than abstracted from ecological conditions of production: effectively CPG in practice. Coffee implicated *cafeteros* in dynamic ecological and social systems, the particularity of which was affirmed by the national institutions, popular knowledge, and biological and social sciences orbiting Colombia's central export industry.

Cultivating Colombian Nationhood

As an industry controlled by domestic producers, coffee represented Colombian self-determination in a world market long tilted toward European and North American interests (Bergquist 1986: 258), producing a nationalist space that was, simultaneously, *cafetero* space. Colombians in the early twentieth century entered *cafetero* society through the cultivation, care, and harvest of a singular crop. Further, community commitment to one crop meant a similarly long-term commitment to the ecological conditions that would best support it. The high-quality, shade-grown coffee plants that distinguished the Colombian industry in this period yielded smaller annual harvests but had longer lifetimes than those grown in the absence of forest canopies (Guhl 2008). Investment in a coffee crop required that *cafeteros* care for particular organisms over many years, an intimate and sustained attention.

In some ways, FedeCafé's new focus on soils mirrored a growing international attention. Heightened concerns about agricultural soil reflected the reality of widespread land exhaustion in the wake of late nineteenth and early twentieth century commodity and credit booms. International market bubbles had encouraged both intensified and extensified production of commodity crops while frequently pushing local subsistence production onto land ill-suited for cultivation. When markets crashed, agrarian communities around the world found themselves facing degraded landscapes, dwindling yields, and few cash resources for improvements. In some contexts like Colombia and Mexico, local leaders devised innovative strategies to ameliorate conditions in the countryside, but many also looked abroad for examples of how to address soil degradation on unprecedented scales. The work of the

United States Soil Conservation Service (SCS) became the most prominent international model of the emerging science of soil conservation, disseminated by the circulation of written materials as well as technical missions of the agency's Chief, Hugh Hammond Bennett and other SCS technicians (Phillips 1999; Marchesi 2016).

In Latin America, the US State Department funded a technical cooperation program through the Office of Inter-American Affairs, recruiting cohorts of Latin American agronomists to train for field training in the United States and Puerto Rican conservation districts, translating SCS technical materials into Spanish, and sending US technicians abroad to advise new conservation programs in the countryside (Marchesi 2016). Despite the prevalence of the US' technical influence, these nationally scaled soil conservation programs also bore the imprint of local concerns. In Colombia, coffee sector scientists conceived their object of study in relation to the particular conditions of coffee cultivation. These included both material factors associated with coffee plants as well as social and economic factors effecting *cafetero* households. As a result, FedeCafé agronomists' attention to on-farm ecology as a function of soil health during this period deviated from US technicians' central concern with particle stability. Within a decade, however, international development efforts brought Colombian agricultural concerns more closely into alignment with their northern practices. Such efforts included but were not limited to the introduction of US-sponsored Green Revolution programs in the 1940s. These programs not only strengthened the role of the state and state-affiliated foreign agents in rural technical programs but also brought a new cohort of experts—with a new assemblage of interests and experience—to the countryside. While the coffee industry was less central to these programs than other farming sectors, it was still implicated in a sea change of changing technical practices (Havens and Flinn 1973).

FedeCafé and Colombian Science

FedeCafé policy and discourse emphasized the centrality of coffee to national development and self-determination. However, its rhetoric was in many ways simply an amplification of notions already at large among the Colombian public, notions that simultaneously reflected and legitimated FedeCafé's institutional project. Prominent Colombian intellectuals in the first half of the twentieth century linked social and biological natures, including race and gender, to coffee-country ecology. As the century progressed, Colombian politicians, scholars, and social activists refined and developed this mythos.

Coffee was practically at the economic center of national productivity; these intellectuals affirmed its discursive centrality to Colombian national development in the independence period.

The designation of "la Raza Antioqueño"—that is, the "race" of inhabitants of the coffee-growing frontiers of the state of Antioquia as well as Caldas, northern Tolima and Valle—was a nationally familiar truism (Parsons 1948). In 1934, historian Luis López de Mesa published the influential *De cómo se ha formado la nación colombiana* citing the geography of coffee-growing regions as the key formative agent of the Colombian "race," which partnered indigenous, African, and European bloodlines with the fertile soils, ample water supply, and temperate climate of the country's mountainous areas. The intrepid colonizers of the Colombian coffee-zones, he explained, mixed the biological potential of their own bodies and the land itself to clear forests and begin a new society of entrepreneurial smallholders who would drive a new, democratic, republican society (Bergquist 1986). Colombian coffee growing and coffee growers, then, would act as both a material and cultural anchor as the nation faced future uncertainties, securing Colombian self-determination through personal effort, innovation, and capital accumulation.

If the bodies of *cafeteros* were a vital element in Colombian coffee production, the body of the land was no less critical. Soil health came to be seen as foundational to coffee enterprise. Industry leaders were anxious to avoid the outbreaks of diseases that had crippled the industry in the past. Over the past decades, frontier colonization had involved the swidden plantation of subsistence crops for settlers and their families. Subsequent coffee plantings helped secure previously exposed soils. However, the accumulated loss of forest cover on steep hillsides meant that coffee farmers faced diminished shade cover and accelerated erosion in comparison with an earlier generation of growers (Guhl 2004: 126–127). In 1933, Clemente Lopez Lozano, Director of FedeCafé's new Central Experiment Station argued that the health of the soil was prior in importance to any technical advances not only for the coffee industry but for the nation:

> The defense of the soil is the defense of the economic base of the nation. We want to produce and increase production, improve agriculturally, but to achieve that goal [Colombians] believe that the base of industrial agricultural improvement that everyone so desires must be initiated by seed selection and hybrids, with genetics; we believe that the next step is the perfection of our work methods, sanitation, machinery, and then markets, etc. *But before all of this there is something even more essential, of greater importance, and that is the soil, whose defense and improvement are indispensable* (emphasis added). (Lopez Lozano 1933b)

The *Cafetero* Congress in December 1935 listed soil erosion as the industry's most pressing problem, (RCC 1938). *Revista Cafetera* admonished growers to "have the patriotic foresight not to bequeath to our children impoverished lands." Readers had only to glance at the journal's back cover in 1937 to learn that "THE WASHING AWAY OF SOILS IS THE GREATEST PROBLEM OF THE COLOMBIAN COFFEE INDUSTRY" (RCC 1937b).

For *cafetero* leaders, the future of the nation—its country and its people—rested on their capacity to protect and improve its soils. The science developed within that undertaking would also be the science of nation-building, even as that science bound Colombia more tightly within its international networks.

A mid-century primer produced by the Catholic Church for rural farmers echoed these links between soil and nation:

> This earth that you cultivate, to which you are bound by need and by love- like an infant at its mother's breast- is tired; it is exhausted; it is dying in your hands, victim of an unrelenting enemy called erosion. And you aren't alone in depending on this land; we all depend on it. The soil is the most precious material resource of the nation, and you, *amigo campesino*, are its steward and guardian. The subsistence of you and your family depends on the fertility of the soil you cultivate, as does the prosperity of your country and the security of the nation… You, *amigo campesino*, with your virile spirit, with your power and your constancy, must transform them into fecund reality… Once again you will be the defender of the Country. You have the blood of heroes. (Vargas Vanegas 1956)

While soil conservation was adopted as a broad national concern during these decades, it is relevant that the impetus and technical expertise for this work emanated from the coffee growers. When the national government moved to begin its own soil conservation agency in 1948, for example, it contracted with FedeCafé to train field agents working in other agricultural sectors (Ministerio de Agricultora de Colombia 1953).

Moreover, while early SCS innovators advocated different techniques under different soil and topographic techniques, the diversity and annual nature of US commodity crops meant that over time standardized technical prescriptions did not consider soil in relation to particular plant species but rather as a potential (if critical) medium for many different potential crops. FedeCafé commitment to coffee meant a similarly long-term commitment to the specific ecological conditions that would best support that crop. The high-quality, shade-grown coffee plants that distinguished the Colombian industry in this period from its competitors yielded smaller annual harvests but had longer lifetimes than those grown in the absence of shade trees, reaching a peak of productivity in its second decade of growth (Guhl 2008). Under these conditions, the time scale of FedeCafé ecological management extended to decades, significantly longer than other cultivators for whom changing crops presented less of a burden.

As such, the long-term success of the coffee industry also depended on a local set of dynamic connections: the relationships between coffee trees, soils, microbes, plants, and growers in the coffee zones. In Federation discourse, the generative relationship between all of those elements formed the living warp and weft of *cafetero* society more generally. FedeCafé agents took a two-pronged approach to facilitating those connections. First, the organization funded an active field research program to be conducted by a new National Center for Coffee Research (*Centro Nacional de Investigaciones de Café* or Cenicafé), establishing training centers in the departments of Caldas and Tolima with the purpose of studying and protecting coffee-growing soils (Lopez Lozano 1933a). Second, the Federation established diverse mechanisms for the dissemination of research among growers, including collaborations with farmers, public presentations, and the circulation of printed materials.

By 1945, the Technical Committee of the 15th National Congress of *Cafeteros* laid the institutional foundations for a nationally scaled "Campaign for Soil Defense and Restoration." With a three-fold mission of education, research, and extension, the Campaign extended to all 12 of Colombia's coffee-growing departments. The Federation based its Campaign in part on the work of the SCS, contracting with two US-trained Puerto Rican soil conservation technicians to guide its technical and organizational aspects. In 1946, after nearly 5 years of preparation, 20 trained FedeCafé soil conservation technicians, 70 field assistants, and an even greater number of trained laborers dispersed throughout the coffee country (Ministerio de Agricultura de Colombia 1953: 1–2).

FedeCafé researchers departed from their North American colleagues, however, in their assessment that the most important aspect of both soil fertility and stability was not chemistry or engineering but biology and interactions between living parts. As Showers (2006) has argued, US soil conservation practices were shaped by the challenges of farming annual crops on fragile frontier grasslands. As such, US soil management centered primarily on soil particle stability. In contrast, FedeCafé technicians focused on long-standing coffee plantings considered soil particles as just as one aspect of a diverse living pedosphere that included microorganisms, root structures, and other important organic matter. The introduction to a special soils issue of the FedeCafé journal was especially sanguine about soil biology's relevance to a host of national problems:

We would like to draw special attention to our investigations on soil biology, as our country is situated in a part of the world where the living section of the soil … offers nearly unlimited perspective, sufficient that a small human effort can provoke maximum activity on the part of the infinitely small universe of living beings, invisible collaborators of immense power in humanity's fight for new fields for sustaining its population.

In the course of a few years, research in the field of soil biology has yielded huge surprises and immeasurable benefits which may well hold the key to solving our great problems caused by deforestation, diminishing waterways, climate change and all the negative effects that follow the erratic and wasteful methods of natural resource use employed until now the man of the tropics. (RCC 1940)

A 1940 assessment gave similar primacy to soil ecology, placing healthy communities of microorganisms ahead of chemical amendments for encouraging plant growth. While fertilizers could be used to enhance the fertility of already healthy soil, field researchers concluded, chemical fertilizers could not substitute for biological action. FedeCafé's director of research cautioned that "before using [commercial fertilizers] it is necessary that soil have achieved its maximum natural fertility… The basis for achieving the best returns from a soil is the exploitation of its natural fertility" (Schaufelberger 1940).

These conclusions were based not just on formal field research but also on the past experience of hacienda owners on the older plantations of the central regions. In the 1910s and 1920s in Viotá, Cundinamarca, for example, soil exhaustion and diminishing yields had prompted local landowners to invest heavily in chemical fertilizers. While such efforts succeeded in raising yields from 0.75 to 1.7 pounds per bush, *cafeteros* of southwestern Cundinamarca subsequently found themselves facing an expensive battle with repeated outbreaks of crop disease (Jimenez 1989: 199). Though FedeCafé in this period did not object in principle (or in practice) to chemical amendments, those inputs on their own could not stand in for healthy soil biology.

FedeCafé researchers similarly argued that erosion control could not be based solely on structural intervention but rather required healthy soil biology. While most international anti-erosion work emphasized structural elements like terracing, embankments, and leguminous field barriers as the foremost anti-erosion practices, Duque's team of researchers expanded their scope of intervention to include microorganisms, arguing that the subterranean biota provided a fundamental structural integrity that preceded the macro-level supports of terraces and wind-breaks. "Paradoxically," Duque told the assembled growers, "man can most actively provoke plant response by operating on invisible rather than visible parts" (Duque 1940: 2620). In contrast, US soil conservationists saw biology as secondary to mechanics in anti-erosion work. For example, the USDA Yearbook also advocated the introduction of cover crops but explained their utility in terms of the physical protection they afforded soil particles against wind and water (Pieters and McKee 1938).

FedeCafé also emphasized the importance of biological diversity on the macro scale. Educational materials advocated the cultivation of shade trees in the denuded frontier zones for the dual purpose of protecting coffee plants as well as national soils in the coffee zones (RCC 1937a) and readers of the *Revista Cafetera* were frequently warned against the economic and ecological dangers of mono-cropping. If such an injunction coming from a national cooperative devoted to a single crop seemed paradoxical on its face, FedeCafé's position reinforced the realities of production on most Colombian coffee *fincas*. Historically, peasant farmers grew coffee trees as a source of cash alongside subsistence and off-season commodity crops; large plantations unwittingly ameliorated the dangers of mono-cropping via the subsistence garden plots of hired laborers and sharecroppers (Reinhardt 1988). In a strategy as much practical as political, FedeCafé recommendations upheld ongoing peasant and smallholder habits as the basis for healthy farm ecology and economy.

Cold War International Development and the Erosion of Domestic Expertise

Though the Federation would retain its democratic structure and commitment to collective representation in the decades that followed, the changing political and economic landscape of the Cold War era drove yet another reconceptualization of human and ecological subjects in coffee country. In 1946, FedeCafé President and third generation coffee *hacendado* Mariano Ospina Pérez was elected to the Colombian Presidency. While his political ascent confirmed the national power of the coffee industry, it also coincided with a broader shift in hemispheric relations. During the 1930s and early 1940s, coffee-growing elites had contained the demands of peasant farmers through policies that attended to the needs of small and medium-sized growers. As the Second World War drew to a close, changing international forces tipped the scale of influence away from the masses and toward a resurgent international capitalist elite (Jiménez 1995).

National-level action and discussion about national soils during the 1940s reflected the complex and often conflicting forces at work within Colombian society more broadly, channeling different aspirations for sovereignty, national resource use, and rural governance. At the close of the Second World War, the Colombian national government hosted a year-long mission from the United States "toward development of agricultural resources, particularly the production of crops that are complementary to those of the United States" (*Agriculture in the Americas* 1944a; see also Offner 2012).

Influenced in part by the increased traffic of US' technical resources during the 1940s supported by the Office of Inter-American Affairs, the interests of FedeCafé's technical department shifted during this period from on-farm ecology to a focus on soil's chemical and mechanical aspects. These changes accompanied a larger development push by the United States in Colombia that included the simultaneous modernization of both urban and rural environments (Hirschman 1963; Offner 2012). In 1947, FedeCafé joined forces with the Colombian Geographical Military Institute and University of California soil scientist, Earl Storie, to begin a comprehensive technical soil and mapping survey of the nation's agricultural lands (RCC 1947). Conducted by a team of agricultural engineers, Geographical Institute surveys were designed to determine where the nation might most productively undertake industrialized agricultural development, dividing soils between the categories of "totally mechanizable" and "impossible to mechanize or largely eroded," noting location, approximate spread, elevation, dominant soil types (according to international classificatory categories), chemical analyses, levels of erosion, principle crops, and yields of the same (*Agriculture in the Americas* 1944b; Ruiz and Garcia Espinel 1951).

Among the leadership of FedeCafé, also, earlier commitments to sustainable smallholder production were eclipsed by a new emphasis on expansion and productivity supported by increased chemical inputs and a move away from shade-grown coffee. Over the next decade, the introduction of higher-yield coffee varieties brought significant benefits to large and medium-sized growers who could afford the cost of new varietals and the chemical inputs their upkeep required. While new varietals increased wealth for many growers, they also intensified erosion in the coffee zones (Havens and Flinn 1973).

These new management regimes drove many of the poorest farmers from the ranks of independent producers. A study of 65 *cafetero* families in Támesis, Antioquia, found that less than a third of growers were able to adopt these new technologies; eight years later, the per acre yield was 1642 pesos and 632 pesos for adopters and non-adopters, respectively. Land ownership also became more concentrated, with adopters of new technologies increasing farm size from 18.86 to 33.13 acres and non-adopters decreasing average holdings from 7.97 to 6.42 acres. Indeed, nearly a quarter (23%) of non-adopters who were owners or renters of land in Támesis had become day laborers within the eight-year survey period (Havens and Flinn 1973).

Conclusion

The development of scientific knowledge alongside global market expansion is often associated with the erasure of locally specific interests and practices. Such expertise has also been frequently deployed as part of a broader conceptual dis-embedding of land from society as part of the expansion of liberal capitalist sociality. In the case of the Colombian coffee industry in the 1930s and 1940s, by contrast, the mutual dependence of coffee crops, small-scale coffee growers, and national economic sovereignty birthed a distinct body of knowledge linking the development of all three to the land itself.

This body of knowledge was an essential element of what this chapter has termed FedeCafé hegemony in Colombia's coffee-growing regions, a project that was simultaneously solicitous of and defensive against transnational capital flows. If FedeCafé authority was amplified by the absence of a strong central state in this period, the exit of foreign investment capital following the 1929 crash evacuated another powerful source of social influence, clearing space for new assertions of national sovereignty through productive independence. Via developments in social and ecological science, FedeCafé posited Colombian coffee growing and coffee growers as both a material and a cultural anchor for a destabilized nation. The ensuing transformations wrought by FedeCafé's territorial project leveled an epistemological and practical challenge to the forces that had left Colombians and the Colombian countryside vulnerable to the exploitative investments and economic vagaries of global commodity markets in the early decades of the twentieth century.

The distinctive scientific knowledge production associated with FedeCafé governance in this period is notable for its challenge to universalist models of modern cosmopolitan science. At the same time, the seismic economic and political shifts that enabled that project rendered it unstable in the face of subsequent transformations of the international landscape. During the Cold War era that followed, the nationalist bent of Depression-era FedeCafé science, and particularly its two-fold advocacy of smallholder production and soil ecology, gave way to a new commitment to modernization and industry-wide yield maximization. As such, this example is also instructive as to the vulnerability of such place-specific projects to the transnational forces to which they are a response.

For scholars seeking to implement critical physical geographic scholarship, FedeCafé agronomy offers key insights into CPG's three core tenets regarding the fundamental entanglements of societies, environments, and power. As a research program premised on the mutual constitution of social and environmental well-being, FedeCafé agronomy began from the assumption that human activity was constitutive of rather than disruptive to local ecologies. And conversely, they also assumed that market-based agriculture could not be scientifically abstracted from its broader ecological context. Thus, researchers charged with supporting coffee production concerned themselves with soil bacteria, forest diversity, and farmworker health and education alongside coffee plant structures. Second, FedeCafé agronomy in the 1930s and 1940s not only acknowledged the imbrication of social with ecological systems, it also conceived coffee growers and landscapes in ways that had real material implications for *cafetero* communities. For example, FedeCafé field research was premised on cash-poor farmers managing relatively small plots of land on which other kinds of subsistence activities might take place. As such, the interventions they recommended relied on small-scaled interventions coupled with careful, ongoing observations and relatively few off-farm inputs to support a particular kind of *cafetero* space; subsequent management schemes premised on different productive scales and farmer capacities emphasized alternative forms of expertise. Finally, these conceptual frames helped dictate who and what flourished in *cafetero* landscapes and, in turn, shaped the allocation of power and resources in the countryside. Indeed, *cafetero* researchers were far from alone in abstracting their research subjects according to particular visions of social and environmental good. Attending to embedded assumptions about people and places reveals the power of research questions to make ideas into realities. That observation is relevant both for critics of scientific interventions and those hoping to craft scientific interventions in support of social and environmental change. As CPG asserts, scholars seeking to understand place-based transformations must begin from the premise that not only our conclusions but also our investigations enact specific commitments in the worlds we investigate.

References

Agriculture in the Americas. 1944a. Names and news- agricultural mission to Colombia. 4 (6): 102.

———. 1944b. Land use study in Colombia. 4 (6): 102.

Bergquist, C. 1986. *Labor in Latin America: Comparative essays on Chile, Argentina, Venezuela, and Colombia.* Stanford, CA: Stanford University Press.

Banner, S. 2005. *How the Indians lost their land: Law and power on the frontier.* Cambridge, MA: Harvard University Press.

Craib, R. 2004. *Cartographic Mexico: A history of state fixations and fugitive landscapes.* Durham, NC: Duke University Press.

Duque, J. 1940. El mejoramiento de los cafetales por medio reconstrucción de suelos [Improvement of coffee plantings via soil rehabilitation]. *Revista Cafetera de Colombia* 8: 2614–2621.

Franklin, S. 2007. *Dolly mixtures: The remaking of genealogy.* Durham: Duke University Press.

Guhl, A. 2004. Coffee and landscape change in the Colombian countryside, 1970–2002. PhD Dissertation, University of Florida.

———. 2008. Coffee production intensification and landscape change in Colombia, 1970–2002. In *Land-change science in the tropics: Changing agricultural landscapes,* ed. A. Millington and W. Jepson, 93–116. New York: Springer.

Havens, E.A., and W. Flinn. 1973. *Green revolution technology and community development: The limits of action programs.* Madison, WI: The Land Tenure Center, University of Wisconsin.

Hirschman, A. 1963. *Journeys toward progress: Studies of economic policy-making in Latin America.* New York: The Twentieth Century Fund.

Jimenez, M. 1989. Travelling far in grandfather's car: The life cycle of central Colombian coffee estates: The case of viota, cundinamarca (1900–1930). *Hispanic American Historical Review* 69: 185–219.

Jiménez, M. 1995. The limits of planter hegemony in early twentieth century Colombia. In *Coffee, society, and power in Latin America,* ed. William Roseberry, Lowell Gudmundson, and Mario Samper Kutschbach, 262–293. Baltimore: The Johns Hopkins University Press.

Koerner, L. 2001. *Linnaeus: Nature and nation.* Cambridge, MA: Harvard University Press.

Koffman, B. 1969. The national federation of coffee growers of Colombia. PhD Dissertation, University of Virginia.

Lave, R., et al. 2014. Intervention: Critical Physical Geography. *The Canadian Geographer* 58 (1): 1–10.

Livingstone, D., and C. Withers. 2005. *Geography and revolution.* Chicago, IL: University of Chicago Press.

Lopez Lozano, C. 1933a. El problema de la erosion del suelo, Parte I [The problem of soil erosion, part I]. *Revista Cafetera de Colombia* 5: 1707–1710.

———. 1933b. El problema de la erosion del suelo, Parte II [The problem of soil erosion, part II]. *Revista Cafetera de Colombia* 5: 1755–1757.

Marchesi, G. 2017. The other green revolution: Land epistemologies and the mexican revolutionary state. *Antipode* 49 (4): 1060–1078.

Ministerio de Agricultura de Colombia. 1953. *Creacion y organizacion de Servicio Nacional de Conservacion de Suelos* [Creation and organization of the national soil conservation service]. Bogotá: Editorial MINIAGRICULTURA.

Offner, A. 2012. Anti-poverty programs, social conflict, and economic thought in Colombia and the United States. PhD Dissertation, Columbia University.

Palacios, M. 1980. *Coffee in Colombia 1850–1970: An economic, social and political history.* Cambridge: Cambridge University Press.

Parsons, J. 1948. Antioqueño colonization in Western Colombia as historical geography. PhD Dissertation, University of California, Berkeley.

Phillips, S. 1999. Lessons from the dust bowl: Dryland agriculture and soil erosion in the United States and South Africa, 1900–1950. *Environmental History* 4: 245–266.

Pieters, A., and R. McKee. 1938. The use of cover and green-manure crops. In *Soils and men: Yearbook of agriculture 1938*, 431–444. Washington, DC: United States Department of Agriculture.

Polanyi, K. 2001. *The great transformation: The political and economic origins of our time.* 2nd ed. New York: Beacon Press.

Posada, C. 1976. *La crisis del capitalismo mundial y la deflacion en Colombia, 1929–1933* [The global capitalist crisis and deflation in Colombia, 1929–1933]. Medellín: Universidad de Antioquia, Facultad de Ciencias Económicas, Centro de Investigaciones Económicas.

Reinhardt, N. 1988. *Our daily bread: The peasant question and family farming in the colombian andes.* Berkeley and Los Angeles: University of California Press.

Revista Cafetera de Colombia (RCC). 1937a. El problema del monocultivo en Colombia [The problem of monoculture in Colombia]. 6: 2155–2158.

———. 1937b. En deslave de los suelos es el magno problema de la industria cafetera Colombiana [Soil erosion is the greatest problem facing the Colombian coffee industry]. 6: 2200.

———. 1938. Apartes del informe del Jefe del Departamento Tecnico al VIII Congreso Nacional de Cafeteros [Presentation proceedings from the technical department head at the VIII national conference of cafeteros]. 7: 2246–2247.

———. 1940. Introducion [Introduction]. 8: 2584.

———. 1947. Plan para la Campaña de Defensa y Restauración de Suelos, adoptado por la Federación Nacional de Cafeteros de Colombia, aprobado por el XVI Congreso Cafetero [Plan for the campaign for the defense and restoration of soils, approved by the XVI coffeegrowers' Congress]. 8 (115): 3300–3305.

Ruiz, J., and A. Garcia Espinel. 1951. *Breve descripcion de las series de suelos dominantes en las principales regiones agricolas de Colombia* [Brief description of dominant soil types of the principal agricultural regions of Colombia]. Bogota: Ministerio de Hacienda y Credito Publico, Instituto Geografico de Colombia 'Agustin Codazzi', Seccion de Suelos.

Safford, F., and M. Palacios. 2002. *Colombia: Fragmented land, divided society.* Oxford: Oxford University Press.

Scott, J. 1999. *Seeing like a state: How certain schemes to improve the human condition have failed.* New Haven: Yale University Press.

Schaufelberger, P. 1940. Apuntes de pedologia: La fertilidad del suelo, su conservación y mejoramiento [Pedological notes: Soil fertility, conservation, and improvement]. *Revista Cafetera de Colombia* 8: 2604.

Showers, Kate. 2006. Soil erosion and conservation: An international history and cautionary tale. In *Footsteps in the soil: People and ideas in soil history*, ed. Benno P. Warkentin, 369–408. Elsevier Science.

Stassart, P., and S. Whatmore. 2003. Metabolising risk: Food scares and the un/remaking of Belgian beef. *Environment and Planning A* 35: 449–462.

Vargas Vanegas, C. 1956. *Defendamos Nuestro Suelo* [Defending our soil]. Segunda Edicion, Biblioteca Campesina (Accion cultural popular—Escuelas radiofonicas), Bogotá.

18

Who Values What Nature? Constructing Conservation Value with Fungi

Elizabeth S. Barron

Introduction

Conservation is an applied field with a specific goal: to protect and maintain nature. The politics of conservation are value-based negotiations: what should be conserved, for whom, how, and why? Typically, how people value nature is directly related to their knowledge of and interaction with it. Some are interested in conserving nature because of its *instrumental value*, as something to enjoy and benefit from (Chan et al. 2016). These benefits are often measured economically. By contrast, others argue that nature has *intrinsic value*, its own value independent of people, which is sometimes also articulated as ecological value. Finally, Bryan et al. (2011) suggest that value is created, defined, and changes over time through social practices relating to uses and "non-uses" of nature. They argue that "successful conservation depends not only on identifying ecological and economic priorities for specific areas, but also on how these priorities align with the *social values* assigned these areas by the community" (p. 173, emphasis added). In this chapter, I refer to these three forms of value as economic/instrumental, ecological/intrinsic, and social, respectively.

CPG suggests a distinct position on how to craft and practice new knowledges of nature. As the introduction to this volume suggests, if we have begun to see the world as eco-social, we must adapt how we study and learn to know the world in this new reality. This adaptation must include fundamental shifts

E. S. Barron (✉)
Department of Geography and Urban Planning,
University of Wisconsin - Oshkosh, Oshkosh, WI, USA

© The Author(s) 2018
R. Lave et al. (eds.), *The Palgrave Handbook of Critical Physical Geography*,
https://doi.org/10.1007/978-3-319-71461-5_18

in value where we allow ourselves to be "transformed by the world in which we find ourselves" (Gibson-Graham and Roelvink 2009: 322) in order to recognize the interdependence of intrinsic, instrumental, and social values of nature. This means going beyond separate claims that biodiversity is primarily a social construction based on instrumental values (Takacs 1996; Lorimer 2012), intrinsically valuable (Hamilton 2005), or that if instrumental and intrinsic values are combined with social values, the total value will finally result in conservation. Indeed, CPG necessitates that we root our social analysis in the material world around us *and* that we consider how our practices of conservation are shaped by the natural world as much as they shape it in return. In other words, it requires new ways to construct value in eco-social futures. The concept of econo-ecologies (Barron 2015) can be useful in this respect.

I have previously introduced the concept of econo-ecologies to "foreground everyday economic practices and choices into not only the social dimensions of natural resource use, but also the ecological dimensions of natural resources themselves" (Barron 2015: 174). Econo-ecologies focus on moments of engagement as forms of work that build and maintain interconnected values between people and nature. This process shows all values are constructed through engagement and therefore open to reinterpretation as eco-social. Although they do not use the term "eco-social," Gibson-Graham and Roelvink's (2009) work on learning to be affected in the Anthropocene sets the stage for econo-ecological conservation values by highlighting humans' interconnectedness with the biota: "recognizing earth others as not-other than ourselves; … acknowledging our co-constituted being as body-world" (p. 324). To ground this exploration of transforming conservation value, I focus on a group of organisms often overlooked in conservation and society: the fungi.

Biologists and policymakers rarely consider fungal conservation (Heilmann-Clausen et al. 2015), and it can be challenging to identify how fungi are valuable economically, ecologically, and socially. With perhaps one notable exception (matsutake), economically valuable fungi are not rare or threatened and most endangered fungi have little or no monetary value (The IUCN Red List of Threatened Species 2015). The vast biodiversity of fungi does not recommend the group broadly for conservation (Blackwell 2011), and for most people, fungi are actually undesirable pests or disease agents (Moore 2001), often targeted for extermination (e.g. *Cryphonectria parasitica* for causing chestnut blight).

Despite the awkward fit, there is a growing discourse around their protection, passionately promoted by fungal conservationists (Heilmann-Clausen et al. 2015; Pringle et al. 2011; Hawksworth 2003; Suryanarayanan et al. 2015). They cite the economic and sociocultural values of fungi for food and medicine and as principle sources of enzymes, antibiotics, and in biotechnology. They highlight the importance of fungi as food sources for other animals, major actors in ecosystem processes of decomposition, and in supporting the vast majority of flowering plants on Earth through mutualisms (Heilmann-Clausen et al. 2015). A CPG stance acknowledges these different values of fungi and pushes us to consider how they shape and are shaped through interaction grounded in the material realities of exchange. Econo-ecologies shifts the focus to human *and* biotic work, both clearly visible and co-constituted in these examples.

The chapter proceeds as follows: in the next section, after a short introduction on valuing nature, I explore how conservation values are normatively constructed vis-à-vis economy, ecology, and society. I include a review and critique of the relevant conservation literature that has adapted these values to fungi. Following this, I present the concept of econo-ecologies more fully in relation to fungal conservation. This analysis highlights the interdependency of physical and human systems at the very heart of CPG, results in a new imperative for fungal conservation, and provides an alternative for other conservationists.

Construction of Normative Conservation Values: Background and Challenges

The logic of the scientific method is premised on isolating and testing individual variables in order to identify sources of variability. This method suggests that isolating different forms of value from each other can help identify mechanisms for more effective conservation management. Below I explore economic, ecological, and social values in turn and pay special attention to how they have motivated fungal conservationists to action.

Economic Value

In public discourse, the all too common "environment vs. jobs" rhetoric reduces any discussion of value to an impossible (and problematic) choice between nature's need to exist and humans' need to work (Chan et al. 2016; Mansfield et al. 2015; Burke and Heynen 2014), suggesting that humans' primary relationship with nature is centered on the need to make money in a capitalist economy. This perspective assumes that what is good for the environment is bad for the economy and vice versa. On a deeper level, these effects are premised on one way of knowing the environment—as something separate from people and one way of knowing the economy—as a capitalist marketplace in which people are either workers or owners of businesses that produce goods and services. In this system, the only way to maintain jobs and protect the environment is through a process often called the neoliberalization of nature.

Neoliberalism, simply stated, is "an economic doctrine that favors free markets, the deregulation of national economies, decentralization and the privatization of previously state-owned enterprises (e.g. education, health care)" (Cloke et al. 2013: 608). The neoliberalization of nature refers to the extension of these ideas to natural resource management and conservation. This approach to conservation is based on the belief that market mechanisms are the best way to create value in nature, making things like clean water, clean air, or biodiversity valuable enough to protect them. In other words, laws and regulations to protect water, air, biodiversity, and so on are unnecessary because if capitalism is allowed to operate freely, conservation will happen.

Buscher et al. (2012) assert, "there has been a conflation of what is generally (and simplistically) referred to in conservation discourses as *economics* with the ideological assumptions of neoliberalism" (p. 5, emphasis in original). Neoliberal conservation, they contend, "shifts the focus from how nature is *used* in and through the expansion of capitalism" to "*conserved* in and through the expansion of capitalism" (p. 4, emphasis added). Nature conservation essentially becomes a growth market (Sullivan 2013). This ideological foundation for nature conservation appears self-evident in a world seemingly dominated by capitalism, making it seem quite logical for non-governmental organizations to work with capitalist enterprises to determine the economic value of nature.

A neoliberal conservation approach is evident in the work of organizations like Conservation International and international conservation bodies like the International Union for the Conservation of Nature (IUCN), who are taking on the work of conservation with significant underwriting from global and transnational corporations (Brockington 2009; MacDonald 2010). For fungal conservationists, the most widely recognized and authoritative conservation organization is the IUCN. Most fungal conservationists are professional scientists (biologists who study fungi are called mycologists). They see their scientific colleagues, working on organisms ranging from coral to tigers, shaping conservation discourse at the IUCN and have been working for decades to make inroads there (Barron 2011). Beyond their own professional societies and a few EU agencies, as recently as 2014, the IUCN was the only international conservation forum where mycologists were active. Those mycologists have in turn been very active at international, national, and local levels to educate other members of the mycological community on the IUCN specialist groups and the process of creating Red Lists (Dahlberg and Mueller 2011), the main mechanism through which the IUCN draws attention to threatened and endangered species (IUCN 2012).

While fungal conservationists have put their faith in the IUCN, the IUCN has in many ways put its faith in global capitalism to help save nature. MacDonald (2010) observed this firsthand at the World Conservation Congress of the IUCN in 2008, which he deemed "a site in the neoliberal restructuring of conservation ... in which the interests of capital accumulation receive an unparalleled degree of access and consideration in conservation planning and practice" (p. 271).

By focusing their efforts at the IUCN, mycologists' arguments about the economic value of fungal conservation have been enrolled into neoliberal conservation strategies where they must demonstrate the value of fungi in neoliberal terms. In this arena, the fate of fungi becomes linked to their ability to prove their worth in the new marketplaces of conservation. For rare, threatened, and endangered fungal species, this is a hard argument to make since only one (*Pleurotus nebrodensis*) of the 35 fungal species Red Listed at the IUCN has any commercial market (The IUCN Red List of Threatened Species 2015).

Ecological Value

Biodiversity is the cornerstone of building ecological value. Since the publication of the Millennium Ecosystem Assessment (MA) in 2005, the value of the environment has increasingly been discussed in terms of the many "services" and benefits it provides to humans (i.e. instrumental values). These services are broken down into four categories: provisioning (e.g. food, water), regulating (e.g. climate, decomposition), cultural (e.g. aesthetics, religious connections), and supporting (e.g. nutrient cycling, soil formation) (Millennium Ecosystem Assessment 2005). Importantly, as specified in a key figure in the MA (Fig. A on p. vi), biodiversity, or simply the diversity of life on Earth, is not considered an ecosystem service but rather is foundational to all services. This distinguishes biodiversity as intrinsically/ecologically valuable, outside of the service framework based on instrumental/economic valuation.

The concept of biodiversity emerged in the 1980s in political debate and was rapidly picked up and used by scientists to secure research funding to demonstrate the applied value of their work (Hamilton 2005; Ghilarov 1996). Biodiversity is an especially interesting concept in relation to ecological value, because it is both from abundance and scarcity that its scientific and public worth is generated (Stuart et al. 2010). One may simply recall the amazing images from the recent Planet Earth 2 trailer, from the massive flocks of penguins covering the Antarctic tundra to the lone cheetah on the African savanna, to recognize this odd paradox (BBC Earth 2016).

As the MA suggests, biodiversity is foundational to all Earth processes and therefore at the heart of conservation. This is also evidenced in the Convention on Biological Diversity (CBD), written at the United Nations Conference on the Environment and Development (the "Earth Summit") in Rio de Janeiro in 1992, which codified the connections among conservation, biodiversity, and development in international environmental law.

For ecologists, however, the definition and scope of biodiversity is more constrained. Simply put, it is difference at varying scales: genetically within species, among species (species richness, species diversity), and among ecosystems at all trophic levels (Hamilton 2005). Ecologists are interested in diversity because they interpret it as a predictor of community stability and hypothesize that higher numbers of species protect an ecosystem from various forms of disturbance (Hamilton 2005). The role of biodiversity in ecosystem function may make the case for conservation, but that is not the scientific goal in documenting it (Ghilarov 2000).

While ecologists and policymakers strive to protect overall biodiversity, emphasis is often placed on the special values of rare species, such as their rare genetics, their unique role in an ecosystem, or their value as an indicator species of some kind. Recognizing specific species as ecologically valuable because of their rarity, and also vulnerability, leads to their placement on lists: IUCN Red Lists (discussed in the previous section), lists of species recognized under the Endangered Species Act, and/or appendices in the Convention on International Trade in Endangered Species (CITES: an international treaty to ensure that trade does not threaten the survival of endangered species). Legal status generates ecological value and makes these species legible and eligible for attention, funding, and special treatment (Burke and Heynen 2014).

Studying relationships among species is the work of ecologists; studying individual species populations, like those placed on lists, is the work of biologists, mycologists in the case of fungi. Prior to 2014 there were only three species of fungi (one macro-fungus and two lichenized fungi) on the global Red List (Dahlberg and Mueller 2011); as of 2015, there are 35 (The IUCN Red List of Threatened Species). This >1100% increase represents a significant achievement for mycologists. By contrast, no macro- or micro-fungi are listed on the USA Endangered Species Act. Two lichenized fungi (*Cladonia perforate* and *Gymnoderma lineare*) are listed as endangered in the "non-flowering plants" category (a problematic listing by the US Fish & Wildlife Service [USFWS] since fungi are a different taxonomic group than plants) (US Fish & Wildlife Service 2016). As of March 2016, no fungi were listed in the appendices to CITES, not even the highly valuable caterpillar fungus (*Ophiocordyceps sinensis*), a rare fungus internationally traded as a medicinal product in traditional and Western medicine (Stewart 2014). Clearly, despite inroads at the IUCN, the conservation status of fungi in the USA (via the ESA) and internationally (via CITES) is low. Conservation mycologists argue that this is because people are unaware of the need for fungal conservation, despite the many ecological values of fungi (Heilmann-Clausen et al. 2015; Griffith 2012). I suggest that the problem also stems from challenges in fungal biology.

Biodiversity conservation policy rests strongly on the ability of biologists to measure and assess a number of characteristics about the diversity of life. The core concept in this formulation is the species concept, which is premised on the ability to identify, define, and differentiate living organisms from each other. Once identified, a suite of characteristics, including population size, abundance, range, habitat, and how these characteristics are changing over time, may be assessed. Species may be compared and contrasted using these metrics and valued accordingly. Species are literally the currency of ecological value in conservation.

Species are difficult to identify and assess in mycology and thus to value ecologically. The total estimated number of fungal species ranges from 250,000 to 5.1 million (Hawksworth 2001; Blackwell 2011). Because of changing species concepts and changes in fungal taxonomy, previously identified species are regularly reclassified. There is an abundance of examples: based on genetic analysis, the morphological species *Armillaria mellea*, a popular wild edible species in different parts of the world previously identified based on its structure, was broken up into 15 "genetic" species (Coetzee et al. 2000). The morphological species *Boletus regius*, a species listed under the UK Biodiversity Action Plan (Fleming 2001), was recently split into two species. The new species, *B. pseudoregius*, is now also listed as a priority species, simply because of its relationship to *B. regius* (Joint Nature Conservation Committee 2010). Beyond species identification, mycologists also regularly deal with high levels of uncertainty regarding several population biology metrics that form the foundation for the inventory and monitoring assessments that establish ecological value, such as identifying population size and location, mature individuals, and generation length (i.e. average lifespan) (Dahlberg and Mueller 2011).

Conservation mycologists, then, face many challenges in constructing ecological value for fungi. Fungal conservation is still in its infancy and thus is often discussed in relation to the entire kingdom because few people are conversant in individual fungal species of conservation interest. As a kingdom, fungi are wildly abundant and more diverse than the plant and animal kingdoms. Their biology, however, is complicated and not as well understood as flora and fauna. When making a case for fungal conservation, in fact, it is the value of fungi as drivers of many ecosystem services, including nutrient cycling, soil formation, decomposition, disease regulation, and waste mitigation, that gives fungi significant ecological value. However, these ecosystem services are considered separately from biodiversity. This suggests that in regard to ecological valuing, it is problematic to discuss the conservation of fungi solely in terms of biodiversity, since their ecological value is mostly derived in relation to non-biotic-based ecosystem services.

Social Value

The rise of the ecosystem services discourse highlights that for many people, the intrinsic values of nature are not sufficient to change behaviors on a large scale. Economic values represented through ecosystem services create some additional value, but the combination of instrumental and intrinsic values does not go far enough; both of these mechanisms are based on rational, logical valuations. They do not account for the social, emotional, and spiritual connections many people feel with the world around them, which at the individual and communal level are often the most valuable premises for conservation.

For the public, concepts of biodiversity often hinge on a few key species or special landmarks which are consistent with individuals' perspectives on health, balance, wellness, and other personal and social values (Fischer and Young 2007). Bennett (2016) showed that conservation programs were more successful and supported by local communities when people felt positively about them, rather than when they were based on objective scientific evidence. Studies like this one suggest that public perception, including peoples' feelings toward different organisms and landscapes, plays a more significant role in conservation than scientific data and ecological value. For example, visiting Cape Cod National Seashore every summer to breathe in and be renewed by the sea air or traveling to Yellowstone National Park to see wolves in the wild are iconic American experiences of high emotional value that result in stewardship of national parks and may produce "trickle-down" environmental stewardship at the local level.

Popular conservation outlets strongly rely on peoples' emotional and psychological connections to nature rather than the public's ability to process scientific evidence and reasoning (Fischer and Young 2007). In mainstream conservation culture, social values are grounded in what is beautiful, majestic, and invokes emotions. For the biota, "cute and cuddly" or "charismatic" species are at the top of the list. The logo and homepage of the World Wildlife Fund make this point very clearly. The panda logo is surrounded by panels featuring elephants, sea turtles, and a "donate" arrow pointing at a photo of a young boy at a candlelight vigil holding a handmade sign that reads "I love this planet!" (World Wildlife Fund 2016).

Fungal conservationists share an interest in building social value, but fungi tend to trigger associations with rotting food, infections, and poison; mycologists have little scope to invoke positive perceptions or warm and comforting moments. Instead, they emphasize the significant use values (yeast makes bread and beer possible), the major ecosystem functioning values (decomposition makes nutrients available for plants), and the novelty values of fungi (many are used to make natural dyes) (Moore 2001; Money 2012; Heilmann-Clausen et al. 2015). They host events ranging from "mycoblitzes" at national parks (a 24-hour citizen science event) to mycological forays sponsored by specialty grocers, to draw in the public and expose them to the unique and interesting world waiting just below their feet (Barron 2010).

There are also distinct subcultures of people who are deeply passionate about fungi and ascribe significant social value to them. Fungal bodies are turned into many kinds of art (mykoweb.com accessed 7/14/16), fungi have been worshipped as symbols of the Gods (Feinberg 2003), and there is a long history of association between specific fungi and magic, both for medicinal, spiritual, and recreational uses (Letcher 2006). Like with many subcultures, fungal enthusiasm has blossomed on the internet, where amateur and professional mycologists maintain active webpages (e.g. mushroomexpert.com), blogs, and discussion boards for everything from mushroom identification to club organization to tracking the fruiting of different species through mycological association websites and email list serves.

Within the academic community, there is increasing interest in assessing, quantifying, and incorporating social values into conservation planning and management. Ostrom (2007, 2009) outlines a social-ecological-systems framework where she examines human choices and their effects on ecological systems. Bryan et al. (2011) base their assessment of social values on research participants' connection to specific landscapes and specific ecosystem services that they value. Bennett (2016) explores the role of perception in the affectivity of conservation management. This research suggests greater understanding and inclusion of social concerns will improve and enhance wilderness protection and ecosystem management.

Chan et al. (2016) use the concept of relational values to explore a range of social values including forms of identity, stewardship, and responsibility toward nature. They emphasize the value of these human-nature relationships to individual and community well-being. They suggest that seeing conservation through relational values opens up possibilities for more collective negotiation, local knowledge traditions, and local practices. Leveraging social and place-based relationships in conservation, they argue, can then be extended to other places as we build and expand the scope of social relations.

The focus on relational values draws attention to often-neglected dimensions of environmental management but maintains a fundamental separation between humans/society and nature which itself has distinct effects on how nature and people are valued. When people and the environment are conceptualized separately, the underlying premise is a distinct, external nature with essential, intrinsic qualities like biodiversity and clean water. People use and appreciate these qualities as part of social relations, but nature remains outside the social. As a result, the value of nature also continues to exist outside of human and community well-being.

In eco-social futures, being "transformed by the world in which we find ourselves" (Gibson-Graham and Roelvink 2009: 322) means that conservation of nature becomes about self-recognition—that we must know nature differently because to know ourselves we must look to nature and to know nature we must look to ourselves. As organisms that exist within and all around us, fungi provide a unique group of organisms with which to begin this work.

Alternative Value Systems and How They Work for Fungi

The concept of econo-ecologies is based on two geographical literatures: social nature and diverse economies. Geographers (and others) have been working with the concept of social nature for some time now (Castree and Braun 2001; Puig de la Bellacasa 2010) and many take the conceptual (re)unification of nature and society as axiomatic in their writing on environmental management. For example, Burke and Heynen (2014) examine the "valuing and devaluing of natures, knowledges, and peoples" (p. 8) by linking three common systems of valuation (private, public, and household) to particular ways of knowing nature, which they categorize as (1) science based and market friendly, (2) publicly engaged, or (3) outside of science. They identify an expert-only neoliberal knowledge, which they argue dominates what they refer to as socioecological discourses, with negative consequences leading to and maintaining social inequalities. Like "much scholarship on environmental conflict [that] re-externalizes nature by treating it as an abstraction over which people struggle both materially and discursively" (Mansfield et al. 2015: 285), Burke and Heynen's argument about what is valued and what is devalued centers on socioeconomic practices and power relationships in environmental decision-making and therefore maintains, analytically and conceptually, a separation between humans and an external nature.

Mansfield et al. (2015) address the reunification of society and nature by "conceptualiz[ing] people and their needs, visions, and actions as internal to what nature is and does. [They] reject identifying groups of people that come into conflict over an externalized nature, instead considering the inherently political process through which particular social natures are fostered and contested" (p. 285). They enact their eco-social conceptualization by defining distinct forest types based on social groups, management actions, and ecologies. While specific biotic species are identified, they are never the defining drivers of the typology. Rather, forests are named according to differences in practices and accessibility based on land tenure. For example, instead of an "oak-hickory complex" or "pinion-juniper woodland" subjected to different social values and demands, Mansfield et al.'s forests are named "silvicultural forests" or "livelihood forests."

Although Burke and Heynen (2014) suggest their work is premised on a unified vision of nature and society, by comparing it with the work of Mansfield et al. (2015), it appears less so. Similarly, critical research on conservation is often purported to be about conservation (Neumann 2004) but does not always seem attendant to the biophysical material conditions of rapid widespread population declines and accelerated rates of species extinction (Bauer et al. 2016). Critical Physical Geography is, by design, intended to "go the extra step" that is visible in Mansfield et al.'s work, to construct new styles of research design, conceptualization, and writing to internalize nature's struggles as co-constitutive of human material and discursive struggles.

To construct values for effective conservation practices, ecological and social values must be recognized as interdependent. Ecosystem services and neoliberalization of nature are premised on capitalism, which does not allow for this type of interdependency. Econo-ecologies, by contrast, are premised on the diverse economy and thus provide an economic discourse through which values can be intertwined.

The diverse economy is an idea introduced by Gibson-Graham (1996) to critique the widespread belief that the economy is dominantly capitalist and to show the economy as diverse and always changing. She argues that the world cannot be reduced to a few key determinants (e.g. economic, ecological, social values) and instead can be understood "as shaped by multiple and interacting processes, only some of which we can apprehend. This approach helps us recognize the power and efficacy of things that might seem small and insignificant. It also means that we are open to the unexpected and the unknown" (Gibson-Graham and Community Economies Collective 2016). The diverse economy framework can thus provide theoretical tools for Critical Physical Geography to avoid reductionist analysis.

Econo-ecology, a concept developed based on diverse economic theory, is one such tool (Barron 2015). It encapsulates multiple forms of economic value grounded in social nature relationships. Conservation in this framework involves ongoing processes of learning and "becoming ethical subjects" through negotiations among humans, species, landscapes, that is, organisms and the places we all inhabit together. Conceptualizing those negotiations as moments of work, in which all involved are invested in maintaining and supporting well-being, foregrounds conservation values premised on ethical exchange and for long-term sustainability.

To somewhat mirror the structure above, I discuss fungi in an econo-ecological framework by moving from more economic to more ecological aspects. The entire section is premised in the valuing of social relationships inclusive of fungi. Unlike the sections above, these areas are intentionally not separated out in order to emphasize their interdependence.

In earlier work (Barron 2015), I suggest that considering society, economy, and ecology in isolation, as is often done in capitalist approaches to natural resource management, essentially pits these interests against each other and sets management up for failure. Econo-ecology is an alternative framework focused on engagement and exchange—that is, when value is created and materializes—as moments of ethical interaction among organisms. For example, the act of picking a mushroom in the forest includes an ethical choice: "For me to have this mushroom to eat/share/sell, I may decrease its reproductive success (mushrooms are the fruit of fungi). Is this OK? Yes, under these specific conditions…." Expanding on the economic value of wild edible fungi, I present a range of "transactions, labor practices, and enterprises found in wild product harvesting" (Ibid., p. 182).

Shifting the scale of engagement from conservation institutions to personal individual exchanges like the one above highlights economic relationships outside of capitalism. As discussed in the previous section, neoliberalism has colonized conservation spaces and discourse. Critical scholarship suggests that everything that happens within neoliberalized spaces, like the IUCN, specifically furthers neoliberalism (MacDonald 2010; Buscher et al. 2014). In other words, neoliberalized spaces are absolute. In econo-ecologies, we can observe economically driven conservation moments outside of these spaces by thinking about biodiversity as a scale. With a diverse economies perspective in mind, this enables us to observe that while perhaps capitalism has colonized conservation at certain scales and in certain spaces, there is much more to conservation. For fungi, the scale of engagement is critical precisely because they are not highly visible in large-scale conservation institutions and likely never will be. Rather, fungi's greatest value, their best chance at contributing to conservation, becomes clearer at different scales.

Species are the currency of biodiversity. Ecologists discuss biodiversity in terms of species richness and overall species diversity. These are intrinsic values. However, biodiversity can also be considered as difference across all trophic levels, meaning that it occurs on a scale and can be studied biogeographically. On a "biodiversity scale," distinct levels may be: genes (micro-level), species (meso-level), and ecosystems (macro-level). Less than 10% of fungal species have been discovered and described, suggesting that meso-level conservation strategies are not ideal for fungi.

Interpreting biodiversity as a scale conceptually "opens up" the "species currency market" beyond species, meaning fungal gene fragments found in soil have conservation value, and fungal functional groups have conservation value as drivers of ecosystem services. Thus, econo-ecological value can be found in the following examples: fundamental aspects of major biogeochemical cycles are influenced by fungi, such as the weathering of Earth's surfaces (Hoffland et al. 2004; Taylor et al. 2009) and decomposition in terrestrial (Hattenschwiler et al. 2005) and freshwater (Nilsson et al. 1992; Hackney et al. 2000) systems. Moreover, the diversity of fungi in a community impacts the diversity of plants in a community, as when mutualist fungi serve as positive drivers of plant diversity (van der Heijden et al. 1998; Dighton 2009) but also when emerging pathogens (chestnut blight or more recently sudden oak death) kill common species (Rizzo et al. 2002). Many of the specific species involved in these processes are not known, and as the discussion of fungal biodiversity above suggests, the number of functionally active species is likely much higher than currently thought. What is increasingly observed, however, is that fungal species do have distinct biogeographies (Griffith 2012) and that failing to recognize this could put dependent plant communities at risk. Ecosystem functioning is affected by biogeography because as species assemblages change, who is completing which activities (physical-chemical-biological processes of energy transformation and nutrient and matter cycling (Ghilarov 2000)) may shift.

Biogeographies are econo-ecological relationships because people have been intentionally and unintentionally moving organisms at great distances for millennia (Barron 2015). They move spores when they collect wild edible mushrooms in the woods of Oregon and ship them to Japan for consumer markets. They intentionally move mycelia when they inoculate trees with truffle spawn and distribute them for sale across the northwest of the USA. More often, people unintentionally move spores and mycelia when they cut down timber, package soil, ship horticultural plants, move food, drive cars, walk through woods, and so on. Based on recent research on the human gut microbiome, when people move, microbes move (Huffnagle and Noverr 2013).

Econo-ecology draws attention to the interconnectedness of microbial biodiversity, ecosystem services, and interpersonal relationships with and among humans, fungi, and other organisms. It shows the micro-level and macro-level of conservation as mediated by human actions and therefore as sites of ethical decision-making and practice. Spatially, the econo-ecology framework advocates for more place-based, context-dependent formulations of value attentive to experiential knowledge. This idea, that ecosystem functioning is co-constitutive with human choices, actions, and values about where to move and how to interact with plants and animals, is an eco-social stance on how to know nature.

Conclusion

The idea that fungi need the same form of conservation as "whales, primates, orchids and albatrosses" (Dahlberg and Mueller 2011: 149) may seem surprising to many, including other scientists and the public at large. Fungi are relatively unknown, not liked, and essentially *undervalued*. Not only is the conservation value of fungi not immediately obvious to most people, for some fungi are anathema to conservation; the idea that people should divert energy and resources to protecting a fungus is akin to the idea that we should protect mosquitoes or leeches. In the face of this widespread undervaluation, the fact that fungal conservation is being championed at all is itself notable.

The monetary values of different fungi, as direct and value-added commodities, go some way toward making the case for fungal conservation because it serves as a tool to educate scientists and the broader public, already steeped in neoliberal conservation, about beneficial "friendly" fungi. However, it is also limiting because it suggests that if fungal bodies or fungal labor cannot be converted into a material or service commodity, they are without worth.

In conservation, economic/instrumental, ecological/intrinsic, and social values are identified and rationalized as separate values with a shared goal: to protect nature from collapse while maintaining society (and nature) as we know it. In econo-ecology, these values are not separated. Instead, species and environments that we personally, spiritually, and emotionally value (social) are sites where we build well-being through moments of exchange and work (economic), leading to conservation of a co-constituted eco-social environment (ecological). Fungi become part of our exchange networks in a diverse economy and therefore must be maintained for our own survival.

For conservation, value is thus constructed through moments of exchange and is only as stable for as long as the moment or specific relationship exists. Change may not be frequent, but the possibility is always present. An approach to conservation that is based on adapting to ongoing change is important if we are to adapt and change conservation to include fungi. Most immediately, an econo-ecological perspective on fungal conservation might bring value to species linked to the need for new fungal-based medicines (Sliva 2003), those foundational to previously disregarded ecosystems now important to combat global change (Wieder 2014), or changing harvesting practices of matsutake in the USA based on their extirpation from Japan, where they are highly culturally valued (Hosford et al. 1997).

Value systems emphasizing specific forms of valuation are not effective for all organisms, and normative valuations can be detrimental to conservation overall because they undervalue organisms that are fundamental to the long-term success of conservation and neglect to recognize the interdependent nature of value. Consider, for example, the noble polypore (*Bridgeoporus nobillissimus*), a critically endangered fungus that only occurs in the Pacific northwest USA. The logging of old growth forests, changes in forest composition, and disturbance regimes have decreased its habitat (old growth Fir (*Abies*) forest), by over 90% in the last 100 years. The known sites of the species are protected, but the tree composition has shifted to a Douglas Fir (*Pseudotsuga*)-dominated forest (The IUCN Red List of Threatened Species 2015). The species is listed as critically endangered by the IUCN, but the Endangered Species Act regulates US management of endangered species, so at the federal level no action is mandated. There is not space here to discuss the complexities of forest management that resulted in this shift in noble polypore habitat. The salient point is despite its global endangered species status, what has happened and continues to happen to this fungus is mostly a series of side effects of the active management of trees.

Employing an econo-ecological perspective shows us that we live in an eco-social world in which our basic needs are coupled with those of other organisms and center on work: we work to eat, to reproduce, to move, and for joy. Other organisms also perform work for most of these same reasons. These are all fundamental to our survival as individuals and together. But for humans, work is more than that. It is not just about material survival or having the resources to meet one's basic needs. We work to maintain social and communal networks and physical health and security (Gibson-Graham et al. 2013). These are all enhanced in environments that are flourishing.

Econo-ecologies highlight relationships that cannot be easily quantified, categorized, or regulated but are real and worthy of nourishment and protection. Using econo-ecologies to reframe biodiversity conservation means reclaiming the concept of value as one that cannot be broken down into composite values and subsequently aggregated for better management. It means recognizing the diverse ways humans work with nature *as a part of it*. For example, ecotourism businesses provide paid employment and are dependent on intact and functioning ecosystems filled with clean air, clean water, and beautiful creatures. Alternative economic sector activities from fair trade coffee to organic cotton rely on humans' ability to work in and with coffee and cotton plants. Unpaid activities like berry and mushroom gathering generate important food products for families in many parts of the world and support close connections with the environment premised on the availability and safety of eating these wild foods.

An econo-ecological framework shifts the focus away from quantification of individual values, and is therefore less stable, but it can still be assessed: is this an exchange that benefits those making it now and does not harm the possibility of those in the future to make their own choices? This sustainability-based approach radically alters value, because exchanges that do not consider the present *and* the future are not valuable. For Critical Physical Geography, this means that value does not rest in the landscape, the biota, or the human systems but in their co-constitution.

References

Barron, E.S. 2010. Situated knowledge and fungal conservation: Morel mushroom management in the mid-Atlantic region of the United States. Ph.D. Dissertation, Rutgers University.

———. 2011. The emergence and coalescence of fungal conservation social networks in Europe and the USA. *Fungal Ecology* 4: 124–133.

———. 2015. Names matter: Interdisciplinary research on taxonomy and nomenclature for ecosystem management. *Progress in Physical Geography* 39 (5): 640–660.

Bauer, H., C. Packer, P.F. Funston, P. Henschel, and K. Nowell 2016. *Panthera leo. The IUCN red list of threatened species 2016: e.T15951A107265605.* https://doi.org/10.2305/IUCN.UK.2016-3.RLTS.T15951A107265605.en.   Downloaded on 04 January 2017. [Online]. [Accessed].

BBC Earth. 2016. *Planet earth II: Official extended trailer.* youtube.com

Bennett, N.J. 2016. Using perceptions as evidence to improve conservation and environmental management. *Conservation Biology* 30: 582–592.

Blackwell, M. 2011. The fungi: 1, 2, 3…5.1 million species? *American Journal of Botany* 98: 426–438.

Brockington, D. 2009. *Celebrity and the environment: Fame, wealth, and power in conservation.* London: ZED Books.

Bryan, B.A., C.M. Raymond, N.D. Crossman, and D. King. 2011. Comparing spatially explicit ecological and social values for natural areas to identify effective conservation strategies. *Conservation Biology* 25: 172–181.

Burke, B.J., and N. Heynen. 2014. Transforming participatory science into socioecological praxis: Valuing marginalized environmental knowledges in the face of the neoliberalization of nature and science. *Environment and Society: Advances in Research* 5: 7–27.

Buscher, B., W. Dressler, and R. Fletcher, eds. 2014. *Nature inc.: Environmental conservation in the neoliberal age.* Tuscon, AZ: The University of Arizona Press.

Buscher, B., S. Sullivan, K. Neves, J. Igoe, and D. Brockington. 2012. Towards a synthesized critique of neoliberal biodiversity conservation. *Capitalism Nature Socialism* 23: 4–30.

Castree, N., and B. Braun, eds. 2001. *Social nature: Theory, practice, and politics.* Malden, MA: Blackwell.

Chan, K.M., P. Balvanera, K. Benessaiah, M. Chapman, S. Díaz, E. Gómez-Baggethun, R. Gould, N. Hannahs, K. Jax, S. Klain, G.W. Luck, B. Martin-Lopez, B. Muraca, B. Norton, K. Ott, U. Pascual, T. Satterfield, M. Tadaki, J. Taggart, and N.J. Turner. 2016. Opinion: Why protect nature? Rethinking values and the environment. *Proceedings of the National Academy of Sciences* 113: 1462–1465.

Cloke, P., P. Crang, and M. Goodwin, eds. 2013. *Introducing human geographies.* London: Routledge.

Coetzee, M.P.A., B.D. Wingfield, T.C. Harrington, D. Dalevi, T.A. Coutinho, and M.J. Wingfield. 2000. Geographical diversity of armillaria mellea s.s. based on phylogenetic analysis. *Mycologia* 92: 105–113.

Dahlberg, A., and G.M. Mueller. 2011. Applying IUCN red-listing criteria for assessing and reporting on the conservation status of fungal species. *Fungal Ecology* 4: 147–162.

Dighton, J. 2009. Mycorrhizae. In *Encyclopedia of microbiology*, ed. M. Schaechter. Oxford: Elesevier.

Feinberg, B. 2003. *The devil's book of culture: History, mushrooms, and caves in Southern Mexico*. Austin: University of Texas Press.

Fischer, A., and J. Young. 2007. Understanding mental constructs of biodiversity: Implications for biodiversity management and conservation. *Biological Conservation* 136: 271–282.

Fleming, L.V. 2001. Fungi and the UK biodiversity action plan: The process explained. In *Fungal conservation: Issues and solutions*, ed. D. Moore, M.M. Nauta, S.E. Evans, and M. Rotheroe. Cambridge: Cambridge University Press.

Ghilarov, A. 1996. What does 'biodiversity' mean—Scientific problem or convenient myth? *Trends in Ecology & Evolution* 11: 304–306.

———. 2000. Ecosystem functioning and intrinsic value of biodiversity. *Oikos* 90: 408–412.

Gibson-Graham, J.K. 1996. *The end of capitalism (as we knew it): A feminist critique of political economy*. Blackwell: Malden, MA.

Gibson-Graham, J.K., J. Cameron, and S. Healy. 2013. *Take back the economy: An ethical guide for transforming our communities*. Minneapolis: University of Minnesota Press.

Gibson-Graham, J. K., and Community Economies Collective. 2016. Cultivating community economies: Tools for building a liveable world. A Contribution to the Next System Project Comparative Framework.

Gibson-Graham, J.K., and G. Roelvink. 2009. An economic ethics for the anthropocene. *Antipode* 41: 320–346.

Griffith, G. 2012. Do we need a global strategy for microbial conservation? *Trends in Ecology and Evolution* 27: 1–2.

Hackney, C., D. Padgett, and M. Posey. 2000. Fungal and bacterial contributions to the decomposition of cladium and typha leaves in nutrient enriched and nutrient poor areas of the everglades, with a note on ergosterol concentrations in everglades soils. *Mycological Research* 104 (Part 6): 666–670.

Hamilton, A. 2005. Species diversity or biodiversity? *Journal of Environmental Management* 75: 89–92.

Hattenschwiler, S., A. Tiunov, and S. Scheu. 2005. Biodiversity and litter decomposition in terrestrial ecosystems. *Annual Review of Ecology Evolution and Systematics* 36: 191–218.

Hawksworth, D.L. 2001. The magnitude of fungal diversity: The 1.5 million species estimate revisited. *Mycological Research* 105: 1422–1432.

———. 2003. Monitoring and safeguarding fungal resources worldwide: The need for an international collaborative mycoaction plan. *Fungal Diversity* 13: 29–45.

Heilmann-Clausen, J., E.S. Barron, L. Boddy, A. Dahlberg, G.W. Griffith, J. Nordén, O. Ovaskainen, C. Perini, B. Senn-Irlet, and P. Halme. 2015. A fungal perspective on conservation biology. *Conservation Biology* 29: 61–68.

Hoffland, E., T. Kuyper, h. Wallander, C. Plassard, A. Gorbushina, K. Haselwandter, S. Holmstrom, R. Landeweert, U. Lundstrom, A. Rosling, R. Sen, M. Smits, P. Van Hees, and N. Van Breemen. 2004. The role of fungi in weathering. *Frontiers in Ecology and the Environment* 2: 258–264.

Hosford, D., D. Pilz, R. Molina, and M.P. Amaranthus. 1997. *Ecology and management of the commercially harvested American matsutake mushroom*. Portland, OR: USDA Forest Service.

Huffnagle, G.B., and M.C. Noverr. 2013. The emerging world of the fungal microbiome. *Trends in Microbiology* 21: 334–341.

IUCN. 2012. The IUCN Red List of Threatened Species [Online]. http://www.iucnredlist.org. Accessed Downloaded 12 June 2012.

Joint Nature Conservation Committee. 2010. *UK priority species data collation*. UK Biodiversity Action Plan.

Letcher, A. 2006. *Shroom: A cultural history of the magic mushroom*. New York, NY: Harper Perennial.

Lorimer, J. 2012. Multinatural geographies for the anthropocene. *Progress in Human Geography* 36: 593–612.

Macdonald, K.I. 2010. Business, biodiversity and new 'fields' of conservation: The world conservation congress and the renegotiation of organisational order. *Conservation and Society* 8: 256–275.

Mansfield, B., C. Biermann, K. Mcsweeney, J. Law, C. Gallemore, L. Horner, and D. Munroe. 2015. Environmental politics after nature: Conflicting socioecological futures. *Annals of the Association of American Geographers* 105: 284–293.

Millennium Ecosystem Assessment. 2005. *Ecosystems and human well-being: Synthesis*. Washington, DC: Island Press.

Money, N. 2012. *Mushroom*. New York, NY: Oxford.

Moore, D. 2001. *Slayers, saviors, servants and sex: An expose of kingdom fungi*. New York: Springer.

Neumann, R.P. 2004. Nature-state-territory: Toward a critical theorization of conservation enclosures. In *Liberation ecologies: Environment, development, social movements*, ed. R. Peet and M. Watts, 2nd ed. New York: Routledge.

Nilsson, M., E. Baath, and B. Soderstrom. 1992. The microfungal communities of a mixed mire in Northern Sweden. *Canadian Journal of Botany* 70: 272–276.

Ostrom, E. 2007. A diagnostic approach for going beyond panaceas. *PNAS* 104: 15181–15187.

———. 2009. A general framework for analyzing sustainability of social-ecological systems. *Science* 325: 419–422.

Pringle, A., E.S. Barron, K. Sartor, and J. Wares. 2011. Fungi and the anthropocene: Biodiversity discovery in an epoch of loss. *Fungal Ecology* 4: 121–123.

Puig de la Bellacasa, M. 2010. Ethical doings in naturecultures. *Ethics, Place and Environment* 13: 151–169.

Rizzo, D., M. Garbelotto, J. Davidson, G. Slaughter, and S. Koike. 2002. Phytophthora ramorum as the cause of extensive mortality of Quercus spp. and Lithocarpus densiflorus in California. *Plant Disease* 86: 205–214.

Sliva, D. 2003. Ganoderma lucidum (Reishi) in cancer treatment. *Integrative Cancer Therapies* 2: 358–364.

Stewart, M.O. 2014. *The rise and governance of 'himalayan gold': Transformations in the caterpillar fungus commons in tibetan yunnan, china.* Ph.D., University of Colorado Boulder.

Stuart, S.N., E.O. Wilson, J. Mcneely, R.A. Mittermeier, and J.P. Rodriguez. 2010. The barometer of life. *Science* 328: 177.

Sullivan, S. 2013. Banking nature? The spectacular financialisation of environmental conservation. *Antipode* 45: 198–217.

Suryanarayanan, T.S., V. Gopalan, D. Sahal, and K. Sanyal. 2015. Establishing a national fungal genetic resource to enhance the bioeconomy. *Current Science* 109: 1033–1037.

Takacs, D. 1996. *The idea of biodiversity.* Baltimore, MD: The Johns Hopkins University Press.

Taylor, L.L., J.R. Leake, J. Quirk, K. Hardy, S.A. Banwart, and D.J. Beerling. 2009. Biological weathering and the long-term carbon cycle: Integrating mycorrhizal evolution and function into the current paradigm. *Geobiology* 7: 171–191.

The IUCN Red List of Threatened Species. 2015. *Version 2015-4.* [Online]. www.iucnredlist.org. Accessed Downloaded 28 June 2016.

U.S. Fish & Wildlife Service. 2016. *Endangered species* [Online]. https://www.fws.gov/endangered/species/us-species.html. Accessed 28 June 2016.

van der Heijden, M., J. Klironomos, M. Ursic, P. Moutoglis, R. Streitwolf-Engel, T. Boller, A. Wiemken, and I. Sanders. 1998. Mycorrhizal fungal diversity determines plant biodiversity, ecosystem variability and productivity. *Nature* 396: 69–72.

Wieder, W. 2014. Soil carbon: Microbes, roots and global carbon. *Nature Climate Change* 4: 1052–1053.

World Wildlife Fund. 2016. *Homepage* [Online]. www.worldwildlife.org. Accessed 14 July 2016.

19

Soils in Ecosocial Context: Soil pH and Social Relations of Power in a Northern Drava Floodplain Agricultural Area

Salvatore Engel-Di Mauro

Soils, Social Relations, and Critical Physical Geography

With roughly 40% of the Earth's land surface under cropland and pasture alone (Ramankutty et al. 2008), the importance of studying the effects of human activities on soils can scarcely be over-emphasized. Much light has been shed by laboratory or long-term field experiments, modeling, and chronosequences on the form, extent, and lasting effects of human impacts (Richter 2007), but livelihood-based soil use and its social determinants are often omitted. In doing so, research on human-induced changes in soils lacks sufficient contextualization (Phillips 2001) and fails to address "distal social processes mediating proximal soil disturbance" (McClintock 2015: 70). The result is a skewed explanatory framework which risks reinforcing prevailing unsupported assumptions about society and their relationships to soils (see Kiage 2013; Scoones 2001). Conversely, research explicitly addressing social processes, especially power relations and the politics of knowledge, tend to turn soils (and other biophysical processes) into an analytical backdrop (e.g., Bell and Roberts 1991; Blaikie 1985).

Critical Physical Geography (CPG) offers the possibility of overcoming such explanatory inadequacies. While encompassing potentially the breadth of physical geography, CPG includes different perspectives on the meaning of

S. Engel-Di Mauro (✉)
Department of Geography, SUNY New Paltz, New Paltz, NY, USA

© The Author(s) 2018
R. Lave et al. (eds.), *The Palgrave Handbook of Critical Physical Geography*,
https://doi.org/10.1007/978-3-319-71461-5_19

critique. To Lave et al. (2014), it implies extending to physical geography what are diverse, and sometimes contrasting theories from critical human geography. More recently, Tadaki et al. (2015) contend that CPG entails cultivating a critical disposition, aware of and acting on the intrinsically political character of practicing environmental science. This overlaps with that "critical" side of human geography concerned primarily with structural social inequalities and the context and effects of physical geography knowledge (Lave 2015; Lave et al., this volume), but eventually the divergent political projects represented in human geography will have to be confronted within CPG as well. Be that as it may, the scope of this chapter is largely confined to studying social relations of power rather than issues of knowledge production and physical geography practices.

Soils may have garnered little attention in CPG so far, but its extension to explaining human-influenced soil dynamics shows much promise. In a study of urban soil lead (Pb) pollution in the Oakland area (California, USA), McClintock (2015) shows how racial capitalist urbanization history must also be considered to explain pollution sources and their highly uneven racialized distribution and effects. As far as the author is aware, this constitutes the sole existing work explicitly analyzing soils through an explicitly CPG lens. This chapter elaborates on such work, as well as the author's previous research, to explore how soil pH is shaped by intrinsic soil properties, wider environmental processes, farming practices, as well as social relations of power. Otherwise put, the objective is to investigate not only environmental but also social factors that constrain or enable soil-modifying activities. There are three kinds of contributions made thereby. One is to extend the breadth of CPG to understudied, or thus far missing areas of research (pedochemistry and soils generally), gender and class relations, and agriculture in a formerly state-socialist context. Second, the case study calls attention to subtle instances of environmental change that are still important in explaining general dynamics, like soil development. Finally, this work illustrates how human-induced changes in soil characteristics can contribute to reinforcing social inequalities.

This chapter consists in fusing what continues largely to be separated yet should be viewed as inextricable, the biophysical and social processes affecting soil pH. It is an extension of prior research that could only be institutionally legible if published in separate academic compartments, one physical (Engel-Di Mauro 2003) and the other human (Engel-Di Mauro 2006a). Unlike much earlier attempts at combining ecology (or biology) with social theory, such as ecofeminism, political ecology, and eco-Marxism, CPG is based on studying and explaining biophysical phenomena while accounting for (not explaining, it should be underlined) the social contexts wherein human impacts as well as scientists are embedded (Lave et al. 2014: 3). Thus, CPG not only helps reveal explanatory processes unaccounted for in society-focused theories but also helps unite what remains fragmented within geography. To illustrate this kind of CPG contribution, the chapter first includes a discussion of soil pH and acidification processes and existing explanations. This is followed by a description of the study area, the northern Drava floodplain (SW Hungary), and an abbreviated review of methods. Results are subsequently presented with an ensuing discussion where salient relations of power affecting soil use and human-impact outcomes are identified. The conclusion includes issues for further investigation and ideas toward socially reflexive soils geography.

Soil Acidification and Prevalent Explanations

Pedochemists have long recognized pH as governing nutrient cycling and availability, soil ecosystem composition, and trace element mobility, among other processes (Sparks 2003). It is associated with acid and base additions and losses (Conyers et al. 1991; Helyar and Porter 1989; van Breemen et al. 1983). These mainly occur in soil solution (water between solid particles) and on exchange sites (colloid surfaces, often clay and organic matter). Where annual precipitation exceeds evapotranspiration and alkali inputs are negligible (as in the Drava Floodplain), soils tend to acidify with or without human impact. Net H^+ additions can result from rainfall; organism-led C, N, and S cycling; and breakdown of many forms of soil organic matter (SOM) in neutral to alkaline soils. Sources of acidity have varying effects over diverse scales, from within a meter over days (e.g., acids released by roots) to hundreds and thousands of hectares over decades (e.g., acid rain) and centuries (e.g., precipitation and mineral weathering). Acidity (H^+ input) is buffered (neutralized) by high levels of reactive clays, SOM (at $pH > 5.5$), base cations from water-table fluctuations, and preexisting alkaline substance from parent material (Prasad and Powers 1997; Weaver et al. 2004). Reactive clays (e.g., smectites) and SOM (over the short term) draw acid cations to their surfaces and exchange them with other cations (other poorly crystalline and amorphous minerals are also involved but to a minor degree in the soils considered here). Preexisting and introduced alkaline substances may also neutralize acids. Measurements of cation exchange capacity (CEC, a soil's ability to hold on to cations) and exchangeable acidity (EA, the sum of H^+ and Al^{3+} ions) are ways to estimate these factors' combined effects (Chadwick and Chorover 2001; Sumner and Noble 2003).

Human-induced acidification is mainly associated with acid deposition (industrial sources) and fertilizer N application (especially with NH_4^+–N nitrification), C cycle disturbance (SOM decline), and base cation removals from intensive agriculture and pasture management (Sumner and Noble 2003). The process can be mitigated or reversed by liming and adding alkali-rich manure (Porter et al. 1995; Richter and Markewitz 2001). Persistent, long-term acidity affects most organisms deleteriously by, for instance, inducing Al^{3+} toxicity, reducing macronutrient availability, and diminishing nutrient-cycling rates. Estimates point to 30% of global ice-free soil area being affected by soil acidification, including 10.6% of farmland (Rautengarten et al. 1995). Though global estimates are empirically tenuous (Caspari et al. 2015), many instances of human-induced soil acidification have been documented (Sumner and Noble 2003). For the middle reaches of the Drava floodplain, long-term monitoring raises confidence in data reliability over regional expanses (Baranyai et al. Kovács 1987; Várallyay et al. 2000).

Explanations for acidification focus on estimating relative inputs from each factor, yielding overall assessments of principal causes in different situations. Human activities, where they are deemed causally important, are largely examined no further than their sheer existence as such (e.g., Barak et al. 1997). The contribution of fertilizer application, harvest-based cation removal, and acid precipitation accentuated by fossil fuel combustion sources, for example, are either left unexplained or they are deemed the result of generic processes like industrialization, poor management, or demographic expansion (e.g., Rautengarten et al. 1995; Sumner and Noble 2003). There is little to no social contextualization or exploration into the historical changes in society leading to diverse human impacts with the same level of industrialization or demographic change. Richter and Markewitz (2001: 43–48) have been exceptional in pointing out the acidification effects of land-use change tied to processes of colonization and slave plantation farming in the Southeastern US, but they understand such processes as historical background and legacy rather than ongoing settler colonial projects, and they fail to extend any critical lens to the analysis of current land use or to the context of the field sites relative to wider social phenomena and interlinkages. In other words, they miss the multiple-scaled social relations of power determining what sort of human impact occurs, where, and to what degree.

Class, gendered, and racialized dynamics subtending human impacts may lead to accelerating, attenuating, or reversing acidification trends. As many have already demonstrated from a variety of perspectives, power relations, manifested as structural social inequalities, imply (1) compulsion and/or incentives for different forms and intensities of environmental impact and (2) uneven benefit or harm from environmental change—human-induced or otherwise (e.g., Blaikie 1985; Heynen et al. 2007; Pulido 2015). Feminist approaches have been at the forefront in addressing the linkages among combined forms of relations of domination (including patriarchal), farming, and soils. Some have argued for direct connections between gendered farming practices and soil degradation or compromised soil conservation outcomes (Carney 1991; Sachs 1996), while others point to a more context-contingent connection (Gladwin 2002; Leach and Fairhead 1995). In political ecology, there have also been findings debunking, by way of soil analyses, institutional soil degradation or fertility narratives (e.g., Benjamin et al. 2010; Scoones 2001). Illuminating with respect to social causes, these studies exemplify the obverse problem identified in scholarship focused on the biophysical by eschewing analyses of factors external to society that explain how and what sort of environmental change may occur with given human impact. It is curious, for instance, how decisive pedochemical processes like pH and CEC are virtually ignored or treated as static, even in studies on soil fertility. The contingent outcomes noted in some studies (e.g., Leach and Fairhead 1995) may be explained by wider ecological processes or to shifting soil characteristics rather than mainly social processes. To do so, however, requires a refocusing of research that CPG offers, as attempted in this chapter.

The Study Area

The area investigated (Fig. 19.1) is located in the northern part of the Drava River floodplain (45° 49′ 9.582″ N, 17° 54′ 20.106″ E; 45° 53′ 47.808″ N, 18° 08′ 04.584″ E). Most of the plain is underlain by a series

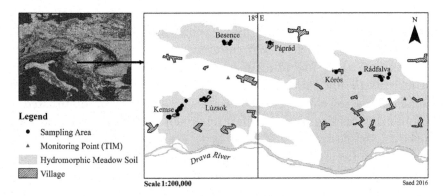

Fig. 19.1 Location of case study and sampling areas

of Early Holocene fluvial aggradation and degradation sequences. The eastern part of the plain reaches into Late Pleistocene paleodunes and loess terraces. A succession of cutoff meanders and oxbow lakes makes for variable soil texture and predominantly alluvial parent material. The area was more directly and frequently subjected to overbank deposition until nineteenth-century levee construction and stream canalization (Lovász 1977). Most rainfall that occurs in late spring and late summers are driest (mean annual precipitation 670–700 mm). Mean annual temperature is 11 °C, ranging between −3 °C and 27 °C. Over the past decades, autumns and winters have become milder, while total yearly precipitation has declined (Pongrácz et al. 2014; Trájer et al. 2013). During fieldwork years, 2008 precipitation was higher but less acidic (491 mm; average pH 6.01) than 2009 (474 mm; average pH 5.63). The 2009 growing season (April–October) was especially dry (226 mm) compared to the preceding year (329 mm). Data from the Soil Conservation Information Monitoring system (TIM) and its predecessor, the Agrochemical Information and Management System (AIIR), point to 54% of farmland characterized by Hydromorphic Meadow soils (*Öntés réti talajok*), which are well to poorly drained. Clay mineralogy is predominantly chlorite, illite, and smectite.

This largely agricultural area is among the poorest in Hungary and is composed of small settlements, most inhabited by ≤100 people. Land consolidation under cooperative farm management reached the area by the late 1960s. Within this wider historical context, the region initially witnessed the abandonment of some areas bordering what was then Yugoslavia, owing to tensions between regimes in the 1950s. Roma people (often called "Gypsies") were thereby able in part to secure housing and land access, such that some villages became majority Roma, yet socially marginalized and under-serviced by state institutions (Stewart 1998). The introduction of the New Economic Mechanism in 1968 ushered a push for small private household plot production, supported by cooperative farm infrastructure, including mechanization and guaranteed markets. During this period, household plot production became an additional income source (Swain 1985). Such ventures came to be dominated by economically better-positioned farming men, and Roma communities were largely by-passed through preexisting economic marginalization processes and discrimination by cooperative farm management. Women were increasingly excluded from such profit-oriented farming and found jobs in other economic sectors, while Roma men and sometimes women were often hired as lower-paid farm hands (Corrin 1994; Kende 2000). Most farming operations, and soil science and agronomy, were redirected toward increasingly mechanized, agrochemicals-intensive, export-oriented production. Soil monitoring reached its peak, alongside sampling programs and amendment campaigns, reaching six-hectare resolution (Engel-Di Mauro 2006a).

After 1989, privatization and cuts in state agricultural support led to the disappearance of most cooperatives and support mechanisms to smallholders, and cooperatives were either turned into private operations or disbanded. The soil monitoring system was simultaneously downsized and extension services now have to be purchased. In the study area, only eight stations remain out of hundreds covering cooperative farm parcels. These processes, part of a longer-term reabsorption into US-centered global capitalist relations (Böröcz 1992), were abetted by IMF and then EU pressures, especially in the run-up to EU accession talks in 1998 (Böröcz 2000; Melegh 2006). Parcels were subdivided into smaller private units redistributed through a voucher system (privileging pre-1948 title-holders and cooperative farm members and employees), formally restricted to citizens. Roma were particularly pauperized by these developments, since Roma were overwhelmingly denied land titles during the 1945 land reform, and Roma comprised most manual or low-level workers sacked through cooperative farm restructuring (Kende 2000). Often lacking in skills to run commercial ventures or unable to adapt to new workplace demands, elderly were often negatively affected as were men in manufacturing jobs, as many industries suddenly closed or were restructured when foreign companies took over. Women, who contributed mostly as manual laborers or office workers, had already been marginalized from commercial farming during the 1970s, with the promotion of farm mechanization. They were either further excluded with the restructuring of farming (through job loss or migration to other, often lower-paying office jobs) or, in the case of a minority, were able to obtain land titles through linkages via (mostly male) relatives or through wealth accumulated with employment in other economic sectors. Those in cooperative farm management tended to gain from the privatization process, as wealthier individuals who profited from privatization policies during or following the state-socialist period (Corrin 1994; Engel-Di Mauro 2006b). By 2005, 40% of arable land was owned by 1.3% of farm businesses (Varga 2010). Their social composition replicates previous tendencies for commercial farming to be a largely white male endeavor.

Methods

Site Characteristics and Data Gathering

The sampling frame consisted of farmed parcels within the extent of Hydromorphic Meadow soils. A soil map (1:100,000 scale) was used to determine soil area coverage (MÉM Országos Földügyi és Térképészeti Hivatal 1983). AIIR cooperative farm cartograms (1:10,000) from the 1980s and local agronomists were consulted to determine distances from past sampling areas. Digitized AIIR archives and data for the nearest TIM sampling locations were obtained from colleagues at the Crop Health and Soil Protection Station of Baranya County. Atmospheric deposition data were provided by the National Meteorological Institute.

Soil and organic fertilizer sampling, archival research, and semi-structured interviews occurred between 2008 and 2010. Thirty-three parcels, belonging to 20 households and three municipalities (Table 19.1), were selected on the basis of soil type, land use, and owner permission (Fig. 19.1). Each parcel number corresponds to a household or institution, and 25 parcels were under crop cultivation during the fieldwork period. Parcel numbers are subdivided using letter suffixes to distinguish different parcels under the same ownership. Parcels are located on flat plains, except for 8a and 15, which lie, respectively, on a shoulder/back-slope (3–5%, leading to an irrigation channel) and shoulder (4–6%, with lower-lying back- and foot-slope forest cover leading to a stream). Of these, 12 were privately owned under state socialism, and 18 are on former cooperative farmland, eight of which could be traced to AIIR records within 100 m of past sampling locations. Two sites had been sampled previously (Engel-Di Mauro 2003).

Three fields (two residential parks, one forest) functioned as comparative controls alongside TIM and AIIR data. Parcel areas range from 0.02 to 20 ha and comprise subsistence and/or commercial farms with varying soil treatments. Semi-structured interviews ($N = 17$, one per household), agrochemical fertilizer company data, and direct observation of farming practices served to gather information on agrochemical fertilizer content and use, crop type, and harvest removals. Interviews were completed with 6 women and 11 men (one whose land was not sampled), covering 16 farms and associated parcels (1–5, 7–10, 12, 14–19, and 21). The three remaining parcel-owning participants also answered interview questions but mainly on farming practices.

After preliminary field analyses to verify soil type, one composite single-tiered (30-cm depth) sample per ha or smaller was collected at each site

Table 19.1 Sampled parcel and land use characteristics (parcel numbers refer to single owners or municipalities; mixed use refers to both subsistence and commercial)

Place	Parcel	Area ha	Land use 2008	2009	Former status[a]	Purpose
Lúzsok	1	0.50	Watermelon	Fallow	P	Subsistence
	2	0.05	Potato, watermelon	Potato, watermelon	P	Mixed
Rádfalva	3a	1.50	Barley	Maize	C	Subsistence
	3b	2.00	Maize	Barley	C	Subsistence
Lúzsok	4a	3.00	Rapeseed	Wheat	C	Commercial
	4b	4.00	Rapeseed	Wheat	C	Commercial
Rádfalva	5a	12.00	Maize	Soy	C	Mixed
	5b	9.00	Maize	Barley	C	Mixed
Besence	6	1.02	Pasture	Ploughed	P	Subsistence
Kemse	7a	1.10	Maize	Maize	C	Subsistence
	7b	0.70	Maize	Maize	C	Subsistence
Besence	8a	0.37	Vegetables	Vegetables	P	Subsistence
	8b	0.85	Fallow	Ploughed	C	Subsistence
Rádfalva	9	0.29	Maize	Maize	P	Subsistence
	10	0.26	Maize	Maize	P	Subsistence
Kemse	11	0.03	Orchard	Orchard	P	Subsistence
	12a	8.00	Wheat	Maize	C	Commercial
	12b	10.00	Sunflower	Maize	C	Commercial
	12c	10.00	Sunflower	Maize	C	Commercial
	13	1.00	Fallow	Fallow	C	Commercial
Kórós	14	0.26	Maize	Wheat	P	Subsistence
Rádfalva	15	0.28	Wheat	Maize	P	Subsistence
Kemse	17	20.00	Sunflower	Reeds (biomass)	C	Commercial
	18a	5.33	Wheat	Rapeseed	C	Commercial
	18b	5.33	Maize	Soy	C	Commercial
	18c	5.50	Rapeseed	Wheat	C	Commercial
Kórós	19	0.13	Maize	Maize, vegetables	P	Subsistence
Besence	20	20.00	Wheat	Sunflower	C	Commercial
Páprád	21a	0.02	Lettuce	Lettuce	P	Subsistence
	21b	0.03	Orchard	Orchard	P	Subsistence
Besence	24	1.00	Forest	Forest	M	Public
	25	0.03	Park	Park	M	Public
Lúzsok	26	1.00	Park	Park	M	Public

[a]Under state-socialism: C former cooperative farm plot, P under private ownership, M municipal

(February–March) before fertilizer application (Baker et al. 1981; Tan 1996: 8–9). A hand-held Trimble Juno ST GPS receiver was used to ensure sampling at the same location. Surface bulk density samples were taken from each parcel with a core sampler, using a slide hammer. A corresponding sample was collected from C horizons by excavating and exposing soil profiles in the middle of each sampled field. Manure and compost samples were taken as applicable (N = 5). Samples were air dried, processed, and analyzed at the Crop Health and Soil Protection Station of Fejér County (details in Engel-Di Mauro 2018).

Data Processing and Analysis

Precipitation data were turned into H^+ mol kg^{-1} values (moles of hydrogen per kilogram) and grouped according to 11-month periods preceding sampling. NH_4^+ was counted as a source of acidification because of microbe-induced oxidation to NO_3^- in spring and summer, leading to net H^+ release (Blake 2005: 2). Because pH only partially captured such an outcome, net input was calculated by subtracting molar values of alkaline from acid compounds.

Variables affecting soil pH were grouped by sampling year and land use and temporal change was calculated, subtracting 2009 from 2010 data. Only 2010 pH data were based on both water and KCl extraction methods, so results from the latter are used to describe inter-annual change. Parent material influence was represented as base cation content, measured as the sum, in $cmol_c$ kg^{-1} soil (or meq 100 g^{-1}), of exchangeable bases (Ca, K, Mg, and Na). Clay content was represented as KA values (<30 = sand; >50 = clay).[1] CEC ($cmol_c$ kg^{-1} soil) served as proxy for organic substances, clay mineralogy, oxyhydroxides, and other reactive minerals (Richter and Markewitz 2001; Sumner and Noble 2003).

Fertilizer data were converted into H^+ $kmol$ ha^{-1} release or consumption. Over short periods, NH_3 and NO_3^-–N leads to OH^- production, while inputs of NH_4^+–N and SO_4^-–S results in the release of H^+ and $2H^+$, respectively (Bolan et al. 2003: 229; Fisher et al. 2003; Tarkalson et al. 2006: 371). Crop harvest base cation removal was estimated by known ash-alkalinity content (Fageria et al. 1997; Antal 1999). Figures were turned into mol kg^{-1} values of $2CaO$, $2MgO$, $2K_2O$, $2P_2O_5$, and $2N_2O_5$–N and summed. The latter two compounds' totals were subtracted from base cation totals for net base cation losses. This was done to prevent overestimations of the acidifying effects of harvest removals. Net base cation losses were multiplied by farmer-reported crop yield to determine total base cation removals.

Interview results on social status were used as transcribed or coded according to categorical or rank data according to response type. For example, gender was categorized (as self-reported) into nominal data of female (1) or male (2). Data such as yearly income levels were instead rank ordered into income brackets (ranks 1–4 with 4 > 5000 USD). Figures were otherwise unaltered when referring to ratio data, such as years of farming experience.

Data, analyzed using SPSS 13.0, display Gaussian distributions (Shapiro-Wilk test; D'Agostino and Stephens 1986) except all $CaCO_3$-related values, 2010 SOM, fertilizer inputs, harvest removals, and owned land area. To evaluate the relative importance of farming practices relative to other pedochemically altering processes, results from Paired Sample *t*-Tests were compared with correlations among all variables. These were also regressed for potentially predictive relationships. For multiple regression analyses, multicollinearity was addressed by regressing variables to each other, selecting against variance inflation factors' values above 3. Contingency tables were used for data on interviewees to detect relationships among social factors. The results helped identify connections between class, gender, and ethnicity, and farming practices. The following parameters were considered, with uncultivated plots and TIM data as controls: farming orientation, mechanization, fertilizer application, and harvest removals. One-Way ANOVA was applied to discern differences between groupings relative to social factors and associated cultivation effects. Nonparametric tests were conducted on non-Gaussian data to detect patterns linking categorical interview data to changing soil chemical properties relatable to soil use. Interviewees' scores were weighted according to parcels sampled (e.g., for an interviewee with three sampled parcels, scores were multiplied by three).

Results

Archival records and interviews revealed no liming for at least a decade, no major soil disturbances (profile truncation, mixing, or burial), and no observable human-introduced parent materials. Acid additions calculated from precipitation data were inconsequential relative to soil pH (from 1.957×10^{-5} to 2.283×10^{-5} mol kg^{-1}; compare, e.g., Helyar and Porter 1989). Below are described, in turn, (a) soil and parent material characteristics, (b) farming practices' impacts on soil pH (past and current), and (c) social status and farming practices.

Soil and Parent Material Characteristics

As shown in Table 19.2, texture varied from clay to sandy loam and tends toward moderate clay content (KA 45.85). Figures for pH varied from slightly

Table 19.2 Surface soil (S, 0–30 cm) and parent material (PM) properties at sampled sites ($N = 33$)

Parcel	Texture	SOM %	CEC	Exch. Ca^{2+}	Exch. acidity H$^+$ kmol ha^{-1}	KA S	PM	pH$_w$ S	PM	CaCO$_3$% S	PM
1	Clay loam	2.53	19.20	10.10	6.00	49	46	6.81	8.56	0.9	15.0
2	Clay loam	2.11	17.00	11.00	3.80	44	50	7.29	7.97	1.3	6.0
3a	Loam	1.69	12.40	5.27	5.95	38	43	6.32	6.96	0.9	0.9
3b	Sandy loam	2.13	15.90	7.59	6.51	37	43	6.61	6.96	0.9	0.9
4a	Sandy loam	1.55	9.85	3.83	4.99	33	29	6.54	6.12	1.3	0.0
4b	Clay loam	1.74	14.20	8.72	3.40	43	29	7.42	6.12	0.9	0.0
5a	Clay loam	1.89	20.60	13.00	4.50	45	37	7.48	6.76	1.3	0.0
5b	Clay loam	2.55	23.60	18.00	3.40	48	43	7.85	8.59	1.3	19.0
6	Clay loam	2.71	25.50	15.10	6.70	46	37	6.78	8.14	0.9	9.0
7a	Clay loam	2.33	18.60	12.90	2.90	45	38	8.11	6.94	3.2	14.0
7b	Clay	3.26	16.30	9.68	3.40	56	55	7.73	8.25	3.2	3.2
8a	Loam	2.26	16.40	8.21	6.64	38	33	6.33	7.10	1.9	1.3
8b	Clay loam	2.22	21.20	10.10	8.00	43	33	6.25	7.10	0.9	1.3
9	Clay	3.16	27.20	16.00	6.40	52	38	7.04	8.34	0.9	8.0
10	Clay	3.26	23.70	15.10	4.90	50	38	7.63	8.34	0.9	8.0
11	Loam	2.75	19.50	10.20	6.90	41	32	6.57	8.07	0.9	4.6
12a	Clay loam	1.63	13.90	10.90	1.70	43	42	8.23	8.32	2.3	3.2
12b	Clay loam	1.75	14.30	9.93	2.30	44	37	8.00	8.64	1.9	9.0
12c	Clay loam	2.27	15.10	11.20	2.00	46	24	8.07	7.57	3.2	0.9
13	Clay	1.85	15.90	12.30	1.80	52	47	8.17	8.13	4.6	8.0
14	Loam	2.02	21.40	12.10	6.00	41	33	6.92	7.48	0.9	1.3
15	Clay	2.53	16.60	10.10	4.40	52	29	7.15	7.90	0.9	0.0
17	Clay loam	2.97	18.70	15.50	1.60	45	33	8.06	7.87	3.6	0.0
18a	Clay	2.56	22.30	15.70	3.50	57	41	7.98	8.36	0.9	5.0
18b	Clay	2.26	12.30	6.33	4.51	50	45	7.46	8.00	1.3	8.0
18c	Clay	3.52	19.00	9.21	5.90	52	45	7.44	8.00	0.9	8.0
19	Clay loam	1.99	15.70	12.00	0.90	43	30	7.42	8.39	1.3	9.0

(continued)

Table 19.2 (continued)

Parcel	Texture	SOM	CEC	Exch. Ca^{2+}	Exch. acidity	KA		pH$_w$		CaCO$_3$%	
		%	cmol$_c$ kg^{-1}		H$^+$ kmol ha^{-1}	S	PM	S	PM	S	PM
20	Clay loam	2.34	21.50	12.60	6.20	43	30	7.10	7.24	0.9	0.0
21a	Sandy loam	1.70	15.00	8.75	3.90	37	31	7.51	7.42	0.9	0.9
21b	Sandy loam	2.08	11.90	5.56	4.77	35	33	6.66	6.90	0.5	0.0
24	Loam	2.57	21.00	9.23	9.00	41	41	5.89	6.94	0.5	0.5
25	Clay	3.01	23.90	12.30	8.50	54	42	6.25	8.41	0.9	12.0
26	Clay	4.56	33.90	18.10	9.30	70	64	6.79	8.06	0.9	0.9
Average		2.42	18.59	4.87	11.11	45.85	38.52	7.21	7.70	1.45	4.78
SD		0.65	5.02	2.22	3.51	7.42	8.38	0.66	0.71	1.00	5.14
SE		0.11	0.87	0.39	0.61	1.29	1.46	0.12	0.12	0.17	0.89

acid to alkaline (Table 19.2) and are nearer the TIM reported maximum (pH$_w$ 5.51–7.21). CaCO$_3$ and exchangeable Ca^{2+} were moderate relative to EA, and CaCO$_3$ averages fell within TIM data range (0.0–10.1). Values for pH correlated with CaCO$_3$, CEC, Exchangeable Ca^{2+}, and EA. Texture (KA) was similarly aligned with SOM, CEC, and exchangeable Ca^{2+}, which were in turn correlated with each other. Both pH and CaCO$_3$ varied inversely with EA. Parent material, found at no more than 100-cm depth, tends to be alkaline and high in CaCO$_3$ (Table 19.2). Surface soil pH, texture, CEC, and exchangeable Ca^{2+} varied significantly with parent material pH. Surface soil texture correlated with parent material texture, pH, and CaCO$_3$. Parent material texture and pH also correlated with exchangeable Ca^{2+} and CaCO$_3$, respectively.

Results align with TIM-reported soil characteristics but exhibit greater alkalinity due in part to parent material influence. Surface Ca^{2+} and CO$_3$ variability suggest a greater role for human impact. Acid-neutralizing capacity and SOM, though moderate to high, declined significantly over 2009–2010, along with CEC and Exchangeable Ca^{2+} (Table 19.3; Engel-Di Mauro 2018). SOM values also fell in uncultivated sites, except under forest, due to low rainfall (overcome by irrigation in cultivated areas). The rapid downward shifts in these variables are interrelated because buffering capacity is diminished by lower base cation content and cation exchange sites associated with SOM levels.

Table 19.3 Statistically significant changes in 2009–2010 surface soil properties (Paired Samples t-Test, two-tail; N = 33) and preceding fertilizer input and harvests at sampled sites

Parcel	SOM*** %	CEC* $cmol_c$ kg^{-1}	Exch. Ca^{2+}** $cmol_c$ kg^{-1}	Treatment[a] 2008	2009	2008 Input[c]	Removal[c]	2009 Input[c]	Removal[c]
1	−0.05	4	2.51	2	0	30.098	0.268	0.000	0.000
2	−0.55	−1.2	0.8	2	2	22.665	0.743	22.665	0.743
3a	−0.63	−3.9	−2.76	3b	3b	0.001	6.334	0.001	12.178
3b	0.44	2.3	1.02	3b	3b	0.001	12.178	0.001	6.334
4a	0.09	−0.95	−0.05	3b	3b	N.A.	6.454	N.A.	10.396
4b	−0.23	−2.1	0.36	3b	3b	N.A.	6.454	N.A.	10.396
5a	−0.04	1.4	0.9	2b	2b	0.001	13.538	0.001	6.334
5b	−0.13	−2.4	−2.2	2b	2b	0.001	12.178	0.001	6.334
6	−0.03	0.4	−0.2	0	0	0.000	0.000	0.000	0.000
7a	−0.41	−0.6	−1	3b	3b	0.408	6.334	0.408	6.334
7b	−0.12	−13.5	−13.32	3b	3b	0.408	6.334	0.408	6.334
8a	0.06	1.9	0.42	0	0	0.000	0.846	0.000	0.846
8b	−0.62	−1.6	−1.3	0	0	0.000	0.000	0.000	0.000
9	−0.24	−1.1	−1.3	2	3	0.001	6.334	20.801	6.334
10	0.37	−2.5	−1.9	2	3	0.001	6.334	20.801	6.334
11	−0.79	−0.7	−1.1	0	0	0.000	0.000	0.000	0.000
12a	−1.28	−2.1	−1.6	3b	3b	0.001	6.334	0.001	6.454
12b	−0.06	−1.4	−1.47	3b	3b	0.001	6.334	0.001	8.071
12c	0.10	−2.4	−2.9	3b	3b	0.001	6.334	0.001	8.071
13	−0.56	−2.1	−1.6	0	0	0.000	6.334	0.000	6.216
14	−0.36	0.6	−0.1	2b	2b	20.801	6.454	20.801	6.334
15	−1.01	−4.1	−1.6	3	3	0.001	6.334	0.000	6.454
17	−0.06	−3.1	−2.3	3b	0	0.000	0.001	0.000	8.071
18a	0.12	−2.6	−1.6	3b	3b	0.407	6.454	0.000	10.396
18b	−0.44	−11.2	−7.77	3b	3b	0.407	13.538	0.407	6.334
18c	0.15	−7.1	−8.19	3b	3b	0.001	10.396	0.407	6.454

(continued)

Table 19.3 (continued)

Parcel	SOM*** %	CEC* cmol$_c$ kg^{-1}	Exch. Ca^{2+}* cmol$_c$ kg^{-1}	Treatment[a] 2008	2009	Input[c] 2008	Removal[c] 2008	Input[c] 2009	Removal[c] 2009
19	−0.64	−5.2	−2.2	2	0	0.001	0.845	0.000	6.334
20	−0.02	−0.7	−3.2	3[b]	3[b]	0.001	8.071	0.0007	6.4543
21a	−0.38	−0.2	−0.11	1	1	29.395	0.846	29.395	0.846
21b	0.00	−2.3	−1.49	1	1	0.000	0.140	0.000	0.140
24	0.04	−0.5	3.31	0	0	0.000	0.000	0.000	0.000
25	−0.74	−2.6	−3.2	0	0	0.000	0.000	0.000	0.000
26	−0.47	−23	0.8	0	0	0.000	0.000	0.000	0.000
Average	−2.74	−1.65	−1.65			4.610	5.050	2.298	4.949
SD	4.98	3.12	3.12			9.871	3.690	7.410	4.328
SE	0.87	0.54	0.54			1.718	0.642	1.290	0.753

***Significant at the 0.001 level (two-tail).

*Significant at the 0.05 level (two-tail).

[a]0 = none; 1 = manure; 2 = mixed manure and agrochemical fertilizer; 3 = agrochemical fertilizer only; 4 = no fertilizer with harvest. Soils treated with fertilizer were also cropped and harvested

[b]Application of synthetic fertilizer > 100 kg

[c]Input refers to estimates of total acid additions from fertilizer applications; removal is an estimate of total loss of base cations through crop harvesting. Values are in mol kg^{-1}

Farming Practices' Impact on Soil pH

Smallholders grew a variety of vegetables on smaller plots (beans, potatoes, paprika peppers, celeriac, onions, garlic, herbs, etc.) and included maize or watermelon where parcels were larger than ca. 0.25 ha. Such plots were for subsistence only (six farmers) or for occasional sale (six farmers). Larger land owners (>5 ha) grew mainly cereal crops (wheat, barley, and maize), sunflower, and soy for commercial ends. Crop yields (data available upon request) showed no major shifts, despite a dearth of rainfall in the growing season. They were typically within FAOSTAT-reported national averages for farms using more than 100-kg agrochemical fertilizer (Table 19.1), but above average for maize.

Synthetic fertilizer is widely used (16 households), but only nine farmers can afford to apply more than 100 kg per hectare. Seven farmers mix it with different kinds of manure. Synthetic fertilizer included NH_4–NO_3, CAN (Ca–NH_4–NO_3), MAP ($NH_4H_2PO_4$), NPK mixtures (mostly 15:15:15), and potash. Ten farmers used CAN on 13 fields (1, 2, 3, 5, 9, 10, 14, 15, 17, 19), which can raise exchangeable Ca^{2+}. K^+-containing fertilizer, applied by 12 farmers applied on 18 fields (2, 3, 5, 9–10, 12, 14–15, 17–20) also adds alkalinity. Because of no fertilizer application in six farms, mean fertilizer-added acidity was halved from 2009 to 2010, but average harvest base cation losses declined marginally (Table 19.3), even when considering only cropped fields. Most households (14) keep pigs and/or fowl, irrespective of farm size or income levels, and they were largely for household consumption and fed from crops grown on the farm. Farmers 5, 9, and 10 (all women) use green compost and/or bird manure, while farmers 1, 2, 14, and 21 (two Roma women, one Roma man, one Hungarian man) apply one to three metric tons of cow, horse, and/or pig manure, which tends to release organic acids upon breakdown.

Overall, farming practices resemble those of cooperative farm management from the previous regime, except that access to agrochemicals is highly uneven and lime is largely beyond reach for most. Nevertheless, legacies from past practices may affect the current state of soil pH and hence fertilizer effectiveness. According to agronomic reports in AIIR archives, liming occurred from the early 1980s. However, data are not available until 1986–1990 and are in aggregated form. Figures in pH_w are also inconsistently reported. Such reports cannot be linked directly to parcels in this study, but they point to potentials for long-term human-induced changes in soil characteristics (Table 19.4). Acidification is relatively clear in eight of the parcels for which data could be

Table 19.4 Long-term change in sampled cultivated soils according to available AIIR information (post-1983 data aggregation do not permit direct linkage to 2010 parcel data)

Site	Parcel	Year	pH_{KCl}	SOM %	$CaCO_3$	Liming (100–300 kg ha^{-1})
Formerly under cooperative farm management						
Besence	8b	1979	5.66	1.84	0	No data
	8b	1982	6.83	1.7	1	None
	8b	2010	4.81	2.22	0.9	None
	20	1979	4.98	2.43	0	No data
	20	1982	6.50	2.93	0.8	None
	20	2010	5.81	2.34	0.9	None
Lúzsok	4a	1980	6.97	2.15	0	No data
	4a	1983	6.05	1.61	0	1986–1990, with manure (300–550 kg ha^{-1})
	4a	2010	5.17	1.55	1.3	None
	4b	1980	6.03	2.11	0	No data
	4b	1983	5.69	1.9	0	1986–1990, with manure (300–550 kg ha^{-1})
	4b	2010	6.31	1.74	0.9	None
Rádfalva	3a	1980	4.43	2.49	0	No data
	3a	1983	5.61	1.89	0	1985–1988
	3a	2010	4.60	1.69	0.9	None
	3b	1980	4.53	2.48	0	No data
	3b	1983	6.33	2.04	0	1985–1988
	3b	2010	5.14	2.13	0.9	None
	5a	1980	5.26	2.82	0	No data
	5a	1983	5.28	3.17	0	1985–1988
	5a	2010	6.29	1.89	1.3	None
	5b	1980	5.36	4.2	0	No data
	5b	1983	7.23	1.83	1.59	1985–1988
	5b	2010	6.86	2.55	1.3	None
Never under cooperative farm management						
Páprád	21a	1999	6.62	2.82	0	None
	21a	2010	6.49	1.70	0.9	None
	21b	1999	5.83	1.54	0	None
	21b	2010	5.48	2.08	0.5	None

found. These include instances of no lime application (Besence) and no cooperative farm involvement (Páprád). However, current average soil pH of former cooperative parcels (pH 7.58; SE = 0.144; N = 18) is higher than that of historically private plots (pH 7.15; SE = 0.182; N = 12) to a significant degree (One-Way ANOVA, α = 0.05, $F (1, 31)$ = 9.418, p = 0.004, significance at 0.001, two-tail). $CaCO_3$ levels are also higher by an average of 0.96% more in favor of former cooperative farmland (One-Way ANOVA, α = 0.05, $F (1, 31)$ = 9.545, p = 0.004, significance at 0.001, two-tail).

Social Status and Farming Practices

To explain the above-illustrated differential acid or basic additions and highly variegated farming practices, an exploration of farmers' social status is necessary. Interviewees on average had two or more decades of farming experience and mean household size was three to four people, with at most two other household members assisting in farming operations. Of 20 participants (17 of whom completed interviews and 3 who only provided information on farming practices), 12 farm for subsistence (Table 19.1). Six of these subsistence farmers have parcels smaller than 0.3 ha. They include two Hungarian widows, all three Roma women participants and a Roma man. Another two participants farmed both for subsistence and market. Six interviewees farmed solely commercially and owned at least 5 ha. Three of them sold their produce directly, while the other, larger landowners, had contracts with national commercial enterprises.

Land ownership was highly unequal (average 92.42 ha, s = 348.70, SE = 66.68) and seven interviewees received rent by letting part of their land. Women owned less land on average (40.83 ha, s = 98.56, SE = 40.24) compared to men (191.99 ha, s = 599.75, SE = 180.83). The largest landowner, a relatively wealthy Hungarian man, had 2000 ha but managed farming operations directly, compared to another Hungarian man who let land to a tenant farmer. Land ownership disparities were also large along ethnic lines. The three Roma farming men interviewed owned a total 8.8 ha, compared to a combined 38 ha owned by the three least propertied white male farmers. The situation was reversed by a slight margin among Roma farming women and the least propertied white women farmers (1.6325 compared to 0.5456 ha). Most land was acquired within the last 20 years, or earlier, if parcels were adjacent to interviewees' home. Roma participants generally did not possess land during state-socialism, and two men were able to buy formerly cooperative land (rather than inheriting it through the voucher system or claiming it as former cooperative farm member). One of the women farmers was able to acquire former cooperative farmland but only through her brother, a former cooperative member.

Generally, the six larger landowners had fully mechanized operations, while others rented machinery services (two) or owned some small motorized equipment (three). The rest (six) relied on their own and sometimes others' manual labor as well as free machine operation services (provided by the local government) for tilling and harvesting. The level of mechanization coincided with income and landholding size, with most Roma and women having the least mechanized operations. Most farmers (nine) did not hire anyone for agricultural tasks. This included all Roma and all but one of the women farmers. Loans and subsidies affected households at every income level, irrespective of gender and ethnicity, though to widely differing degrees.

Yearly income levels tended to be less than 250,000 HUF (1250 USD, in 2010 USD), with substantial disparities among households. Four of the six women farmers had the smallest parcels and the lowest income levels (less than 1250 USD per year). Two of those women were Roma, represented by another two men with slightly higher yearly income levels (less than 5000 USD). An Independent Samples Kruskal-Wallis test showed a significant difference in the medians of landownership area across income levels ($\alpha = 0.05$, $p = 0.027$). This suggests a close linkage between property, land-holding size, and income that, as indicated above, is marked by gendered and racialized stratification.

Discussion

Short-term decreases in major pedochemical variables are not reflected in pH but are inexplicable by intrinsic soil properties alone. When compared to longer-term data (see above), these may portend acidification problems, even if soils tend to be well buffered. The buffering seems to result from a combination of intrinsic soil properties and past practices under cooperative farms (see also Chambers and Garwood 1998). The effects of former cooperative farm practices are spatially uneven. Not all current parcels have been affected to the same degree. The outcomes of past cooperative farm management can nevertheless be deduced (Table 19.4). For example, where liming has not happened over more than a decade and of surface pH and $CaCO_3$ are significantly correlated, there is a likely influence of past cooperative farm inputs, especially if parent material influence can be excluded by a lack of relationship between surface and parent material $CaCO_3$. The findings also point to CEC, $CaCO_3$, and exchangeable Ca^{2+}, along with parent material $CaCO_3$ and exchangeable Ca^{2+} as predictors on pH (see also Prasad and Power 1997, 74). Hence, clay and other minerals may be playing a lesser role than SOM in affecting pH values, as would be expected over annual scales.

Regression operations among all pH-affecting variables led to their reduction to SOM, $CaCO_3$, and exchangeable Ca^{2+}, along with parent material $CaCO_3$ and exchangeable Ca^{2+}. A multiple regression model showed these five variables provide the most significant prediction of pH_{KCl} ($F(4, 28) = 10.095$, $p < 0.0005$, $R^2 = 0.651$). This finding, even if preliminary due to a relatively low sample number (Tabachnick and Fidell 2001), still indicates that soil pH involves the interplay between intrinsic soil properties and both past and present farming practices. This is because SOM and exchangeable Ca^{2+} are affected by farming (e.g., tillage, Ca-containing fertilizer additions, and liming). These factors can interact in mutually accentuating or dampening ways, explaining discrepancies between pH and related pedochemical variables. In this case, most fertilizer input has resulted in net alkali additions (i.e., none to negligible acidity added, Table 19.3). Thus, base cation removals are being more than offset in most fields, but in some parcels, net alkalinity is accentuating preexisting soil alkalinity and/or prior net alkaline farming inputs. This can explain the lack of correspondence between current pH levels and related pedochemical variables. However, this explanation does not address the variability in fertilizer quality and the uneven distribution of parcels with differing soil pH levels. These are due to social relations power.

The means of farming are distributed in extremely unequal ways because of skewed relations of power benefiting mainly Hungarian men. Manure was used most by mixed subsistence-commercial farmers and none by commercial farmers (Tables 19.1 and 19.4). It was also common in partially and non-mechanized farms and absent in fully mechanized operations or those renting farming equipment and services (Kruskal-Wallis Test, $\alpha = 0.05$, $p = 0.008$). Harvest removal rates were highest in large commercial farms (>50 ha), increasing with level of mechanization (Kruskal-Wallis Test, $\alpha = 0.05$, $p = 0.001$). Ethnicity was also found to be a significant factor in that Roma exerted six times less harvest-related cation removals than their Hungarian counterparts (Mann-Whitney U Test, $\alpha = 0.05$, $p = 0.001$).

Hungarian male farmer-owned, commercial operations with larger land-holdings and highest incomes feature the highest average pH. Mixed subsistence-commercial, smaller landholdings, and middle-income farms have the lowest pH (One-Way ANOVA, $\alpha = 0.05$, $F (2, 25) = 4.037$, $p = 0.030$). A similar pattern is evinced with exchangeable Ca^{2+} (One-Way ANOVA, $\alpha = 0.05$, $F (1, 31) = 4.364$, $p = 0.013$). This in part follows from pH and $CaCO_3$ averages being higher for formerly cooperative farm parcels, mainly owned by wealthier farmers. What is striking is that the poorest participants, producing solely for subsistence, had the next highest pH (One-Way ANOVA, $\alpha = 0.05$, $F (3, 26) = 3.119$, $p = 0.043$), especially women farmer operations (One-Way ANOVA, $\alpha = 0.05$, $F (1, 28) = 4.522$, $p = 0.042$). Poorer farming households also had the highest SOM levels (One-Way ANOVA, $\alpha = 0.05$, $F (3, 25) = 4.744$, $p = 0.009$). The inverse tendency emerged relative to EA, with the highest found with middle-income farmers (One-Way ANOVA, $\alpha = 0.05$, $F (1, 31) = 3.048$, $p = 0.046$).

These findings point to social causes for part of current soil pH distributions. To name but three examples, there is a lack of available means for some to counteract low pH in many parcels, some farmers' practice excessive harvest-induced base-cation removals, and the use of manures is highly uneven, affecting SOM and pH levels differentially. Most farmers in the case study area cannot afford liming or must weigh the matter against increasing indebtedness and production costs. Especially in the case of middle-income farmers, mostly Hungarian men, the combined pressure of increasing costs of production and decreasing profitability spurs contradictions in plant nutrient additions compared to removals. The tendency to use conventional methods while missing the economic means to sustain such practices seems to compel farmers in that specific gendered and racialized class position to engage in questionable farming practices. This is unlike larger commercial landowners with higher incomes and, for example, greater access to credit, farming equipment, fertilizers, agronomic extension services, and, potentially, lime. Subsistence farmers, on the other hand, tend to be land-poor, low-income, female, and Roma. Their demands on soils tend to be much lower, but unlike their relatively wealthier, often male Hungarian counterparts, they do not have access to former cooperative farm parcels and, as matters stand, to on average higher pH soils.

Conclusions

Soils involve multiple processes, including social ones when it comes to farming-affected soils, and soil pH is no exception. The evidence collected show in part an acidification and in part a countervailing trend linked to social position, where net alkali additions have occurred, as well as to strong soil buffering capacity traceable to processes related (in the case of past human inputs) as well as unrelated to social change. In the case study area, soil pH variability is then arguably related to three main and partly interrelated factors: (1) intrinsic soil properties, (2) wider environmental processes, and (3) social relations of power leading to different forms of human impact.

The first is a set of characteristics due to soil formation and development. In this case, several properties, like preexisting texture and $CaCO_3$ and exchangeable Ca^{2+} content in parent material, ensure that many local Hydromorphic Meadow soils are well buffered against acid inputs. The second factor is comprised of multiple sources of change not necessarily traceable to social processes, such as atmospheric deposition. In this case study, it turns out that such environmental processes are relatively inconsequential to local soil pH variability. It should also be acknowledged that it is increasingly difficult to extricate human from environmental sources, given, for example, long-range transport of contaminants or acidifying compounds. This brings into the fore a third major factor, which is what most research continues to miss, arresting the analysis of environmental change when human impact is demonstrated. Social processes are crucial to explaining current soil pH because, perhaps to state the obvious, different forms of human impact occur in some areas of the world and not others. Such processes shape, among other things, cropping decisions and the degree and type of fertilizer used, which result in differential impacts altering soil pH.

In this case, higher soil pH is related to social relations of power in two ways, through both greater wealth, mainly for Hungarian men, and abject poverty, mainly for Roma women. Wealthier farmers appropriate some benefits from state-socialist liming programs and can afford to maintain higher soil pH levels, while the poorest farmers accomplish similar pedochemical outcomes by means of lower agrochemical inputs and low-demanding cropping systems (agrodiverse, low-yield per crop). Middle-income, mostly male smaller-holding (5–50 ha) farmers tend to deal with lower soil pH and contribute more acidity overall.

In the background are past social relations that also have to be accounted for because they also lead to some pH-altering impacts. Both farming practices and soils have histories that should be studied together because they are increasingly intertwined, as in this case study.

Industrialized agriculture arrived relatively recently (the 1970s), during a period of export-led economic expansion and incipient land privatization, among other social processes that promoted a different, arguably more destructive use of soils and that largely continues to be practiced. These processes of social change are behind shifting human impacts and, in part, long-term effects on soils. This is evident, for example, with respect to liming under state-socialist cooperative farm management, whose pH-raising effect has been dwindling in some fields and may persist in others, and the absence of liming in the present.

The tendency for pH decline, however, may be more widespread through declining buffering capacity not evident in pH change. Some of the trend may also be associated with both short- and long-term acid additions resulting from SOM breakdown with sparse replacement and frequent base cation removal. Legacies from decades of fertilizer N applications cannot be ruled out on former cooperative plots, but intense manuring may be accelerating the process in some farms. In some fields, years of liming up until 1990 and intrinsic soil alkalinity may combine to sustain soil-buffering capacity relative to the impact of acid additions, for those farmers who were able to purchase such land.

These are the sort of subtle, gradual processes of ecosocial change that often escape academic and wider public scrutiny but which deserves greater attentiveness in both detecting early warning signals and explaining soil development and people-environment relations generally. Including social aspects in studying pedochemical change widens perspective and enables more thorough explanations on the basis of which appropriate solutions can be formulated and preventive actions can be taken. In this case study, addressing a decline in soil-buffering capacity necessitates at least a grasp of social relations (e.g., politics of land distribution and access; highly differentiated economic pressures) so as to formulate alternatives, such as land redistribution and reinstating national monitoring and support programs not beholden to a logic of profitability.

Acknowledgments The author wishes to thank Mazen Labban and Rebecca Lave. Without their efforts, encouragement, insights, and constructive critiques this manuscript would not have been conceived. The author is also indebted to two thoughtful, anonymous reviewers, whose careful critical reading helped markedly improve this manuscript. Many thanks also go to Zita Ferenczi (Hungarian Meteorological Service), Sándor Hajdú (Soil and Plant Protection Agency, Baranya County), Sándor Kucsera (Soil Testing Lab, Velence), Sándor Kurucz (Soil and Plant Protection Agency, Baranya County), Attila Melegh (Demographic Research Institute, Budapest), Gábor István Tóth, Tibor Tóth (Institute for Soil Sciences and Agricultural Research), and Kálmán Vörös for their generous and crucial assistance prior to and during the project. This manuscript is based on a project supported by the National Science Foundation (USA) under Grant No. 0615878.

Notes

1. KA (*Arany kötöttség*) values are derived from the Sándor Arany method, a 0–80 plasticity index. It is based on distilled water volume added to turn 100 g of soil into a near-saturation paste at low plasticity, using a mechanical mixer. The water volume added is divided by sample weight and multiplied by 100.

References

Antal, J. 1999. Fertilisation for crops. In *Nutrient management*, ed. Gy Füleky, 321–322. Budapest: Mezőgazda Kiadó. (in Hungartian).

Baker, A.S., S. Kuo, and Y.M. Chae. 1981. Comparisons of arithmetic average soil pH values with the pH values of composite samples. *Soil Science Society of America Journal* 45: 828–830.

Barak, P., B.O. Jobe, A.R. Krueger, L.A. Peterson, and D.A. Laird. 1997. Effects of long-term soil acidification due to nitrogen fertilizer inputs in wisconsin. *Plant and Soil* 197: 61–69.

Baranyai, F., A. Fekete, and I. Kovács. 1987. *The results of soil nutrient content analyses in hungary.* Budapest: Mezőgazdasági Kiadó. (in Hungartian).

Bell, M., and N. Roberts. 1991. The political ecology of dambo soil and water resources in Zimbabwe. *Transactions of the Institute of British Geographers* 16 (3): 301–318.

Benjaminsen, T.A., J.B. Aune, and D. Sidibé. 2010. A critical political ecology of cotton and soil fertility in mali. *Geoforum* 41 (4): 647–656.

Blaikie, P. 1985. *The political economy of soil erosion in developing countries.* Essex: Longman.

Blake, L. 2005. Acid rain and soil acidification. In *Encyclopedia of soils in the environment*, ed. D. Hillel, 1–11. Amsterdam: Elsevier.

Bolan, N.S., D.C. Adriano, and D. Curtin. 2003. Soil acidification and liming interactions with nutrient and heavy metal transformation and bioavailability. *Advances in Agronomy* 78: 215–272.

Böröcz, J. 1992. Dual dependency and property vacuum. Social change on the state socialist semiperiphery. *Theory and Society* 21: 77–104.

———. 2000. The fox and the raven: The European Union and Hungary renegotiate the margins of 'Europe'. *Comparative Studies in Society and History* 42 (4): 847–875.

Carney, J. 1991. Indigenous soil and water management in senegambian rice farming systems. *Agriculture and Human Values* 8: 37–58.

Caspari, T., G. van Lynden, and Z. Bai. 2015. *Land degradation neutrality: An evaluation of methods.* Dessau-Roßlau: Umweltbundesamt.

Chadwick, O.A., and J. Chorover. 2001. The chemistry of pedogenic thresholds. *Geoderma* 100: 321–353.

Chambers, B.J., and T.W.D. Garwood. 1998. Lime loss rates from arable and grassland soils. *The Journal of Agricultural Science* 131 (4): 455–464.

Conyers, M.K., D.N. Munns, K.R. Helyar, and G.J. Poile. 1991. The use of cation activity ratios to estimate the intensity of soil acidity. *Journal of Soil Science* 42: 599–606.

Corrin, C. 1994. *Magyar women. Hungarian women's lives, 1960s–1990s.* New York: St. Martin's Press.

D'Agostino, R.B., and M.A. Stephens. 1986. *Goodness-of-fit techniques.* New York, NY: Marcel Dekker.

Engel-Di Mauro, S. 2003. The gendered limits to local soil knowledge: Macronutrient content, soil reaction, and gendered soil management in SW Hungary. *Geoderma* 111 (3–4): 503–520.

———. 2006a. From organism to commodity: Gender, class, and the development of soil science in Hungary, 1900–1989. *Environment and Planning D: Society and Space* 24: 215–229.

―――. 2006b. Citizenship, systemic change, and the gender division of labour in rural Hungary. In *Women and Citizenship in Central and Eastern Europe*, ed. J. Lukić, J. Regulska, and D. Zaviršek, 61–80. Aldershot: Ashgate.

―――. 2018 (forthcoming). Short-term soil acidification detection through acid neutralising capacity (ANC) analysis along the Northern Dráva Floodplain, SW Hungary. *Agrókémia és Talajtan*.

Fageria, N.K., V.C. Baligar, and C.A. Jones. 1997. *Growth and mineral nutrition of field crops*. New York, NY: Marcel Dekker.

Fisher, J., A. Diggle, and B. Bowden. 2003. Quantifying the acid balance for broad-acre agricultural systems. In *Handbook of Soil Acidity*, ed. Z. Rengel, 117–133. New York: Marcel Dekker.

Gladwin, C. 2002. Gender and soil fertility in Africa: An introduction. *African Studies Quarterly* 6 (1–2). http://asq.africa.ufl.edu/files/Gladwin-Vol6-Issue-12.pdf

Helyar, K.R., and W.M. Porter. 1989. Soil acidification, its measurement and the processes involved. In *Soil acidity and plant growth*, ed. A.D. Robson, 61–101. Sydney: Academic Press.

Heynen, N., J. McCarthy, S. Prudham, and P. Robbins, eds. 2007. *Neoliberal environments: False promises and unnatural consequences*. New York: Routledge.

Kende, Á. 2000. The hungary of otherness: The Roma (Gypsies) of Hungary. *Journal of European Area Studies* 8 (2): 187–201.

Kiage, L.M. 2013. Perspectives on the assumed causes of land degradation in the rangelands of sub-Saharan Africa. *Progress in Physical Geography* 37 (5): 664–684.

Lave, R. 2015. Introduction to the special issue on Critical Physical Geography. *Progress in Physical Geography* 39 (5): 571–575.

Lave, R., M.W. Wilson, E. Barron, C. Biermann, M. Carey, M. Doyle, C. Duvall, et al. 2014. Critical Physical Geography. *The Canadian Geographer* 58: 1–10.

Leach, M., and J. Fairhead. 1995. Ruined settlements and new gardens: Gender and soil-ripening among kuranko farmers in the forest-savanna transition zone. *IDS Bulletin* 26 (1): 24–32.

Lovász, Gy. 1977. *Baranya Megye Természeti Földrajza* [The physical geography of Baranya county]. Pécs: Baranya Megyei Levéltár.

McClintock, N. 2015. A Critical Physical Geography of urban soil contamination. *Geoforum* 66: 69–85.

Melegh, A. 2006. *On the East-West slope. Globalization, nationalism, racism and discourses on central and Eastern Europe*. Budapest: Central European University Press.

MÉM Országos Földügyi és Térképészeti Hivatal. 1983. *Magyar Népköztársaság, Baranya M.-Somogy M., 04 Siklós, 1:100000*. Budapest: Kártográfiai Vállalat.

Phillips, J.D. 2001. Human impacts on the environment: Unpredictability and the primacy of place. *Physical Geography* 22 (4): 321–332.

Pongrácz, R., J. Bartholy, and A. Kis. 2014. Estimation of future precipitation conditions for hungary with special focus on dry periods. *Időjárás* 118 (4): 305–321.

Porter, W.M., C.D.A. McLay, and P.J. Dolling. 1995. Rates and sources of acidification in agricultural systems of Southern Australia. In *Plant-soil interactions at low pH: Principles and management*, ed. R.A. Date, N.J. Grundon, G.E. Rayment, and M.E. Probert, 75–83. Dordrecht: Kluwer Academic Publishers.

Prasad, R., and J.F. Power. 1997. *Soil fertility management for sustainable agriculture.* Boca Raton: CRC Press.

Pulido, L. 2015. Geographies of race and ethnicity I: White supremacy vs white privilege in environmental racism research. *Progress in Human Geography* 39 (6): 1–9.

Ramankutty, N., A.T. Evan, C. Monfreda, and J.A. Foley. 2008. Farming the planet: 1. Geographic distribution of global agricultural lands in the year 2000. *Global Biogeochemical Cycles* 22, GB1003. https://doi.org/10.1029/2007GB002952.

Rautengarten, A.M., J.L. Schnoor, S. Anderberg, K. Olendrzynski, and W.M. Stigliani. 1995. Soil sensitivity due to acid and heavy metal deposition in East Central Europe. *Water, Air, and Soil Pollution* 85: 737–742.

Richter, D., Jr. 2007. Humanity's transformation of earth's soil: Pedology's new frontier. *Soil Science* 172 (12): 957–967.

Richter, D., Jr., and D. Markewitz. 2001. *Understanding soil change. Soil sustainability over millennia, centuries, and decades.* Cambridge: Cambridge University Press.

Sachs, C.E. 1996. *Gendered fields. Rural women, agriculture, and environment.* Boulder: Westview Press.

Scoones, I. 2001. Transforming soils: The dynamics of soil-fertility management in Africa. In *Dynamics and diversity. Soil fertility and farming livelihoods in Africa*, ed. I. Scoones, 1–44. London: Earthscan.

Sparks, D. 2003. *Environmental soil chemistry.* 2nd ed. San Diego: Academic Press Publishers.

Stewart, M. 1998. *The time of the gypsies.* Boulder: Westview Press.

Sumner, M.E., and A.D. Noble. 2003. Soil acidification: The world story. In *Handbook of soil acidity*, ed. Z. Rengel, 1–28. New York: Marcel Dekker.

Swain, N. 1985. *Collective farms which work?* Cambridge: Cambridge University Press.

Tabachnick, B.G., and L.S. Fidell. 2001. *Using multivariate statistics.* 4th ed. Boston: Allyn and Bacon.

Tadaki, M., G. Brierley, M. Dickson, R. Le Heron, and J. Salmond. 2015. Cultivating critical practices in physical geography. *The Geographical Journal* 181 (2): 160–171.

Tan, K.H. 1996. *Soil sampling, preparation, and analysis.* New York: Marcel Dekker.

Tarkalson, D.D., J.O. Payero, G.W. Hergert, and K.G. Cassman. 2006. Acidification of soil in a dry land winter wheat-sorghum/corn-fallow rotation in the semiarid U.S. great plains. *Plant and Soil* 283: 367–379.

Trájer, A., J. Bobvos, K. Krisztalovics, and A. Páldy. 2013. Regional differences between ambient temperature and incidence of lyme disease in Hungary. *Időjárás* 117 (1): 175–186.

van Breemen, N., J. Mulder, and C.T. Driscoll. 1983. Acidification and alkaliniza-tion of soils. *Plant and Soil* 75: 283–308.

Várallyai, Gy, L. Pásztor, J. Szabó, and Zs Bakácsi. 2000. Soil vulnerability assess-ments in Hungary. In *Implementation of a soil degradation and vulnerability data-base for central and Eastern Europe (SOVEUR Project)*, ed. N.H. Batjes and E.M. Bridges, 43–50. Wageningen: FAO and ISRIC.

Varga, Zs. 2010. The post-socialist transformation of land ownership in Hungary. In *Contexts of property in Europe. The social embeddedness of property rights in land in historical perspective*, ed. R. Congost and R. Santos, 267–285. Turnhut: Brepols.

Weaver, A.R., D.E. Kissel, F. Chen, L.T. West, W. Adkins, D. Rickman, and J.C. Luvall. 2004. Mapping soil pH buffering capacity of selected fields in the coastal plain. *Soil Science Society of America Journal* 68: 662–668.

20

Questions of Imbalance: Agronomic Science and Sustainability Assessment in Dryland West Africa

Matthew D. Turner

Introduction

The critical physical geographic approach allows one to embrace the rich interplay between social and biophysical processes in environmental change, attracting practitioners and researchers from the physical, social science, and humanities traditions. Physical geographers are asked to be reflexive about the assumptions of society embedded within their analyses and to pay attention to the exercise of power in the knowledge politics that shape their research. Human geographers are asked to take the materiality of physical systems seriously with a deeper and detailed engagement with the truth claims made about physical systems than that provided by broad characterizations of epistemologies and ontologies. Knowledge politics that pervade environmental science operate through these truth claims—both through their unacknowledged assumptions and how some come to dominate others. These politics are not without material effect: our understandings of the physical world reshape that world.

Understandings of soil fertility dynamics in the agropastoral landscapes of dryland Africa could be advanced by a greater adoption of a critical physical geographic approach. In these dynamic environments, soil fertility and vegetative productivity display high spatiotemporal variability. This variability leads to equally dynamic farmer responses, as shaped by their access to

M. D. Turner (✉)
Department of Geography, University of Wisconsin, Madison, WI, USA

© The Author(s) 2018
R. Lave et al. (eds.), *The Palgrave Handbook of Critical Physical Geography*,
https://doi.org/10.1007/978-3-319-71461-5_20

resources (labor, knowledge, capital, land). These responses significantly affect soil fertility parameters. In these ways, the fertility and social landscapes are co-produced—the soil fertility landscape of the village territory mirrors the sociohistorical landscape of land rights and control. Understanding the heterogeneity of the soil fertility therefore requires an appreciation of the scale of assessment (farmer's field, farmer's land endowment, village territory, etc.), variation in land management practices, and how broader social dynamics affect these practices.

Unfortunately, the history of soil fertility assessment in an African dryland context deviates strongly from this vision. This work, dating back to the colonial era, is dominated by the work of agronomists whose research has sought to isolate the yield response to a single fertility parameter (e.g. N, P, K, pH) by controlling for all other growth factors. Such research, relying on controlled experimental trials, has largely been conducted at agricultural research stations. Most of such "on-station" research has been conducted using inorganic fertilizers with much less work conducted on manuring which is the dominant fertilization practice of subsistence farmers in the region (Powell et al. 1995). The focus by agronomists on chemical fertilizer treatments to assess soil fertility indeed reflects their interest to modernize smallholder agriculture. Chemical fertilizers also represent more useful soil supplements to distinguish yield response to variation in different soil fertility parameters (e.g. N, P, K, pH). Manure, on the other hand, is a substance that is variable in composition and less divisible with respect to soil fertility parameters. It is difficult to translate crop yield response to manure treatments to that which agronomists seek: crop yield response to a particular fertility parameter. In this way, the agronomist's experiment, focused on varying single fertility parameters, has become abstracted from farmers' manuring practices (concurrent inputs of plant nutrients through the application of heterogeneous material).

Efforts by agronomists to engage more seriously with the soil management realities faced by smallholders have involved a combination of "off-station" field research coupled with modeling work. It is this work that will be the initial focus of this chapter. As will be argued, some of the biases of "on-station" research are carried over into the "field"—with sampling frames shaped more by the physical landscape and coarse social measures (distance from village) than by an understanding of the diversity of farmers' practices. An interest in how farmers manage their fields is replaced by an implicit or explicit embrace of neo-Malthusian limits: variation in farmers' management practices means little, given the constraints that the whole "system" faces under population pressure.[1] In this way, the study of farmers' soil fertility management practices has been left to social scientists (and critical physical geographers?) with physical scientists focused on the ultimate constraints that are necessarily abstracted from the realities of smallholder practices. In the scientific culture of agronomy as practiced at national and international research centers, working with farmers is seen as less scientific than "on-station" work—more akin to extension than research.

In this chapter, I will focus on a particularly prominent form of analysis that seeks to evaluate the sustainability of smallholder agriculture through the calculation of nutrient balances: comparing the flows into and out of a defined area (field, district, country, or region). After first briefly describing the soils and soil management in the West African drylands, I will outline the basic features of this approach and its limitations. In so doing, I will describe how it developed as an "off-station" approach in the agronomic sciences to address questions of sustainability. As such, it has limitations including those briefly outlined above. As a result, the scientific assessments produced by this approach are not only incomplete but misleading. The chapter concludes by briefly presenting an alternative approach for understanding the heterogeneity of soil fertility and how it is produced by social relations within agropastoral communities. Taking such a critical physical geographical approach not only illuminates a socially differentiated view of sustainability but reveals new ways for addressing soil impoverishment as a problem.

Soils and Land Uses in Sahelian West Africa

In Sahelian West Africa (the area lying south of the Sahara Desert receiving 200–600 mm of annual rainfall on average), plant productivity is multiply constrained by interacting growth factors such as soil moisture and nutrient availability, particularly nitrogen and phosphorus (Penning de Vries and Djitèye 1982).[2] Soil fertility can vary significantly across agricultural landscapes. The amount of water at the soil surface (topographic position and slope), infiltration (soil texture), and water-holding capacity (soil texture and thickness) all affect water availability to plants. Soil texture is dominated by coarse-to-medium sands. Moreover, soils generally show limited horizon development and therefore their "structure" relates more to the presence/absence and depth of impermeable ferricrete crusts affecting effective soil thickness. Chemical fertility is shaped by the soil's parent material, its acidity, and its nutrient-holding capacity (or cation exchange capacity) which, due to low clay content, strongly reflects its organic matter content. On croplands and pastures, the chemical fertility of soils reflects the management of crop residues, fertilizers (inorganic and manure), and grazing.

Table 20.1 The mean, standard deviation (SD), minimum, and maximum of the soil analyses of 362 pooled samples (181 pooled samples for soil texture) collected for two 200-meter transects (pools of samples of 10 cm depth collected every 4 meters) in 181 fields located within two agropastoral territories of the Fakara region of western Niger

Parameter	Mean	SD	Min	Max
pH_H$_2$O (1:2.5)	5.46	0.38	3.54	6.84
pH_Kcl (1:2.5)	4.69	0.42	3.61	6.57
Total N (mg N/kg)	148.86	65.44	60.48	697.54
Organic Carbon (%)	0.18	0.07	0.09	0.78
Bray P1 (mg P/kg)	5.11	3.74	2.26	50.75
NH$_4$$^+$ (mg N/kg)	2.71	2.34	0.25	19.44
NO$_3$$^-$ (mg N/kg)	7.92	3.90	0.10	19.75
EB_H (cmol$^+$/kg)	0.05	0.03	0.01	0.22
EB_Al (cmol$^+$/kg)	0.03	0.05	0.00	0.38
EB_Na (cmol$^+$/kg)	0.05	0.03	0.01	0.46
EB_K (cmol$^+$/kg)	0.20	0.12	0.03	1.08
EB_Ca (cmol$^+$/kg)	0.36	0.20	0.02	2.03
EB_Mg (cmol$^+$/kg)	0.22	0.11	0.04	1.10
Total exchangeable bases (TEB)	0.83	0.37	0.28	4.17
Al saturation%	3.32	5.74	0.00	38.75
Sand % (2000-50 μm)	94.30	2.78	76.78	96.92
Silt % (50-2 μm)	2.60	1.31	0.94	9.68
Clay % (<2 μm)	3.09	1.57	1.72	13.91

See Turner and Hiernaux (2015) for field and analytical methods

Table 20.1 provides example data of texture and chemistry of Sahelian soils. The data are from a contiguous 412 km^2 area of the Fakara region in western Niger. The soils of 181 fields were sampled within two agropastoral territories.[3] Soil texture is dominated by medium and coarse particle sizes. Average macronutrient (N, P, K) concentrations and exchangeable bases are very low in these acidic soils. Still, the variation of chemistry values across the sampled fields is high. A consideration of the underlying factors that explain this variation is an important question raised by a Critical Physical Geography perspective that will be returned to near the end of this chapter.

Farmers recognize the multidimensional nature of the soil fertility and its variability across village territories (Osbahr 2001; Neimeijer and Mazzucato 2003). Still, except in areas where soils are thin, smallholders generally see chemical infertility as more limiting to crop production than the physical properties of soils (Warren et al. 2003). Without the application of fertilizers, the chemical fertility of a soil declines over time necessitating fallowing.[4] Due to the relatively high cost of inorganic fertilizers, farmers rely primarily on a mix of fallowing and manuring to maintain crop yields (Osbahr 2001; Neimeijer and Mazzucato 2003). Manuring can be seen as a strategy of redistributing nutrients from open pastures (including fallows) to cropped fields. Nutrients are ingested by livestock while grazing and are excreted in manure and urine (Turner 1998). Such transfers are leaky as a significant fraction of the nitrogen in urine is volatilized. While nitrogen losses occur, the passage of plant-bound nutrients through livestock results in nutrients being in more plant-available forms and, therefore, the nutrient cycle is accelerated.

Agronomic work on soil fertility changes has focused on changes in soil nutrient availability.[5] Anthropogenic declines in the chemical fertility of soils have been conceptualized as occurring through net losses of nutrients through human-managed exports (net removal of nutrient-containing organic material through cropping or grazing), soil erosion, or through acidification (changing nutrient availability) due to long-term net export of organic material. While our understandings of the importance of erosion on nutrient loss in the Sahel remain limited (Chappell et al. 1998), the export of nutrients through crop harvests and livestock grazing is generally seen as the major anthropogenic mechanism for chemical fertility loss. The very low inherent chemical fertility of Sahelian soils (Table 20.1) means that nutrient exports through cropping or grazing will have more dramatic impacts on productivity compared to what is observed on more fertile soils.

Measures and Models of Nutrient Management

Since the colonial era, less-than-systematic observations have been published about the effects of African land uses on soils in West Africa (e.g. Chevalier 1928; Collier and Dundas 1937; Harroy 1949).[6] Nye and Greenland's study (1960) figures prominently in the history of environmental assessment as a very important review of often unpublished work on soil fertility loss with cropping and the importance of fallowing (see also Allan 1965). Still, it wasn't until the 1960s and 1970s that chemical analyses of soils became more common (e.g. Charreau and Fauk 1970; Brams 1971; Jones 1973; Aina 1979; Pieri 1992). Since that time, the soil work performed by agronomists has been dominated by controlled treatment studies conducted at research stations. This is due to a number of reasons. First, as described earlier, it reflects the methodological commitments and traditions of agronomy that seek to statistically assess the yield response to particular fertility parameters (e.g. N, P, K, pH). In addition, on-station work is a response to the high spatial heterogeneity of soil chemical properties off station that necessitates the construction of more controlled conditions on station so as to be able to statistically observe any treatment effects.[7]

With growing concerns about the sustainability of agriculture, the 1990s brought new attention to soil fertility decline. At first glance, environmental assessment would require the longer-term monitoring of soil fertility in farmers' fields. This approach has in fact been extremely rare for a range of reasons including:

1. The availability of nutrients in soils change seasonally and total stocks may greatly exceed what is available to plants.
2. The time required for monitoring often exceeds the institutional and funding time frames.
3. The spatial heterogeneity of the soils, even with many samples taken within a single field, may complicate efforts to observe a temporal trend.
4. The monitoring of smallholder practices would require longer-term monitoring of farmers' fields not only with respect to the soil chemistry and yields but also to how they are managed. Given the high variation in soil fertility between fields (e.g. Table 20.1), an off-station monitoring program would necessarily require an understanding of the reasons for these differences which pull biophysical scientists into the uncomfortable terrain of social science.

Given these issues, environmental assessments in the region have avoided the monitoring of longer-term changes in nutrient stocks and instead focused on short-term flows through nutrient-balance assessments. While there was parallel work previously performed by others (e.g. Bertrand et al. 1972; Pieri 1985), expansion of this type of work is tied to the system modeling approach taken by a Dutch research group (Penning de Vries and Djitèye 1982). In 1990, two approaches within this intellectual lineage were described in publications (Stoorvogel and Smaling 1990; van Keulen and Breman 1990). The first by van Keulen and Breman (1990) assessed the sustainability of dryland cropping systems that rely on fallowing and manuring consistent with the much earlier observations of Nye and Greenland (1960: 139). Recognizing that the nutrients in the manure placed on farmers' fields come from outlying pastures, van Keulen and Breman (1990) developed estimates of the amount of rangeland required to support cropland through manuring. This calculation depended on estimates of the annual nutrient needs from manure required to offset the annual net losses of nutrients from croplands, the annual excretion of nutrients that could be captured on cropland, and the number of livestock that rangeland could support annually (carrying capacity). These and the other estimates that were required were highly uncertain with the authors relying on the upscaling of values measured in particular places and times to whole regions. Moreover, the estimates that they used were strongly affected by known variations in the ways in which farmers manage their fields and herders managed their livestock (Turner 1995; Mohamed-Saleem 1998).[8] Despite these uncertainties, van Keulen and Breman (1990) used their calculations to assess the sustainability of agriculture across broad bioclimatic zones (comparing cultivation fractions with sustainable cultivation fractions) pointing to a ubiquitous threat of land degradation and the need for inorganic fertilizers to sustain the West African region.

This early work focused on the sustainability of mixed farming systems (integration of crop and livestock production) without significant inorganic fertilizer inputs. In its attempt to scientifically assess the ultimate biophysical limits to agriculture in the Sudano-Sahelian zone, it necessarily abstracted from on-the-ground heterogeneities and ignored how its parameters are strongly shaped by choices made by dryland farmers and herders. In this way, it falsely and conveniently suggests that what smallholders do has little influence on agrarian futures—futures that are determined by demographic growth and neo-Malthusian limits. The predictions of this approach run counter to work in the region that provides evidence of the long-term persistence of the fertility of cropped fields through manuring (Krogh 1997; Harris 1998; Niemeijer and Mazzucato 2002; Mortimore and Harris 2005).

Despite the limitations of the rangeland-to-cropland ratio approach, it was relevant to a general category of dryland farming, that is, one dependent on the transfer of nutrients from rangelands to croplands in the form of manure and urine. Interestingly, this "specificity" proved to make it less influential compared to the other model introduced the same year. This model was simply a nutrient budget model without any connection to land requirements. The report authored by Stoorvogel and Smaling (1990) articulated the approach which was initially implemented in a series of analyses reported during the early 1990s (van der Pol 1992; Hafner et al. 1993; Smaling et al. 1993; Stoorvogel and Smaling 1993); was reviewed and promoted in a series of reports by the FAO (Roy et al. 2003; Food and Agriculture Organization 2004); and has since become one of the dominant approaches to assess the sustainability of African agriculture (Cobo et al. 2010).

In simple terms, such nutrient-balance assessments define an area (field, farm, village, district, nation, region, continent) and then subtract its nutrient exports from its nutrient imports (Fig. 20.1). The time period over which balances are assessed is most commonly one year or less (Cobo et al. 2010). More negative balances are seen as being less sustainable and more likely to lead to land degradation. As one would expect, except for some rare cases (plots experiencing high nutrient loading), nutrient balances calculated in this

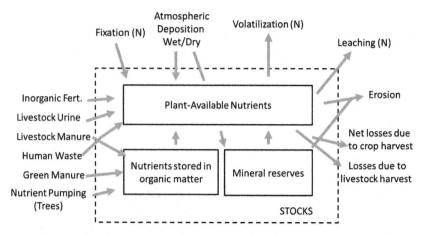

Fig. 20.1 Diagram showing inflows and outflows to and from nutrient stocks of the cropping system as typically presented in the nutrient-balance approach (adapted from Pieri 1985; van der Pol 1992)

way for Africa are generally negative (especially for nitrogen) and as with the rangeland-to-cropland ratio approach, these negative balances are used to express concern about expanding African populations, the sustainability of African agriculture, and the need to promote the use of inorganic fertilizers.[9] As argued by Scoones and Toulmin (1998), this follows a long tradition of "gap analysis" in the environmental sciences. Similar to Club of Rome "limits to growth" assessments, current resource use patterns (and their gaps) are projected out into the future.[10] In the case of the "soil nutrient mining" literature, this implicitly treats: (1) biophysical processes as static and (2) small farmers as either unable or uninterested in responding to soil infertility problems. Very few if any of these studies report on the levels of uncertainty tied to their estimates (Cobo et al. 2010). Given the large uncertainties of the estimates required to calculate these balances, it is likely that if rigorous error/uncertainty propagation were to be used, many of these studies would not be able to state whether their balances were positive or negative. More importantly, such uncertainty makes it very difficult to state whether one system is more sustainable or less prone to degradation than another.

Limitations of the Nutrient-Balance Approach

The methodological problems and uncertainties associated with nutrient-balance studies have been reviewed in a number of publications (Bationo et al. 1998; Roy et al. 2003; Schlecht and Hiernaux 2004; Cobo et al. 2010;

Mohamed-Saleem 1998). Therefore, I will not go into these in detail but simply provide a brief description of the major issues in the paragraphs below.

Problems of Measurement

Even a cursory review of Fig. 20.1 raises questions about how certain flows in and out of the bounded area can be measured. Even at an individual plot level, there are significant problems of measurement. First, measurements of losses associated with leaching, volatilization, and erosion or of gains from nutrient pumping, nitrogen fixation, and atmospheric deposition have proven daunting for soil scientists around the world. As a result, most balances are either "partial", calculating the balance solely between the gains from manure and fertilizer and the losses from grazing and crop harvests, or use model estimates (e.g. the Universal Soil Loss Equation) or measurements taken at different places and times. Even more tractable flows have measurement problems. Just as soil nutrient content/availability is highly variable at microscales so are the nutrient contents of organic materials (plant tissues and manure). Moreover, unlike inorganic fertilizer application, manure/urine inputs are episodic and quite variable within and across years. Urine deposition on fields, a major potential source of nitrogen from corralled and grazing livestock but one that is prone to nitrogen losses through volatilization, basically because of measurement difficulties, remains unmeasured in nutrient-balance studies (Schlecht and Hiernaux 2004). As a result of this, it is likely that human-managed nutrient inputs, other than inorganic fertilizer, are less likely to be adequately measured than outputs (leading to a bias toward negative balances).

System Analytical Assumptions and Issues of Scale

Questions of sustainability and land degradation are concerned with changes in the stocks of available nutrients over time. Nutrient-balance calculations are based on the assumption that geographical-defined stocks of (available) nutrients change in response to the net movement of nutrients across the stock's boundaries. Once defined, the stock, whether it be the available nutrients in the root zone of a field, several fields of a "farm", a village territory, district, or nation, is treated as a black box. Variation in the conditions that influence changes in the forms of nutrients from mineral, organic, and plant-available categories (Fig. 20.1) is assumed to be captured adequately during the study's monitoring period and that these conditions will remain the same into the future. For example, there are often residual positive effects of nutrient amendments in year 1 and in subsequent years (e.g. 2–3) even for stocks experiencing "negative" nutrient balances (in year 1). How could this be so? First, in approximately 50% of nutrient-balance studies (Cobo et al. 2010), stocks are not measured (just flows) and even if they are, residual fertility effects could be due to net changes in the form of nutrients <u>within</u> the nutrient stock.

A second major problem has to do with the choice of the bounded stock to calculate a nutrient balance. While recognizing that each scale is associated with different sets of methodological issues, nutrient-balance scholarship has tended to treat the choice of nutrient stock (or scale) as simply reflecting the analyst's intended audience (Cobo et al. 2010). But it is important to recognize that the choice of the nutrient stock determines what processes should be seen as internal (and black boxed) or external and producing flows into or out of the stock. As an example, let's take an agricultural system relying in part on manuring to maintain fertility. Determining nutrient balances for one or several (e.g. a "farm") fields was the focus of more than 90% of the studies surveyed by Cobo et al. (2010). This is understandable given that it is at these levels where empirical data on nutrient flows can be reasonably measured. Still, such measurements can lead to quite different estimates among fields given the wide variation, as shown in Table 20.1, in the fertility status of cropped fields (Gray 2005b; Ramisch 2005b; Cobo et al. 2010; Turner and Hiernaux 2015; Turner 2016).

So what is the appropriate scale of analysis for nutrient-balance work? In systems where manuring is practiced, one would want to capture the area within which nutrients are redistributed—crop fields receiving the manure as well as the areas where the livestock who are producing the manure are grazed. In mixed farming or agropastoral systems in West Africa, this is normally an area that approximates but is usually somewhat larger than the village territory (land controlled by village leadership for farming). A number of the commentators have raised problems with nutrient-balance studies failing to internalize certain flows as they scale up from plot-level estimates. Certain flows that are external to a plot should become internal to the stock (e.g. soil erosion, nutrient redistribution by livestock, local crop consumption) at wider scales. Schlecht and Hienaux (2004) have argued that failing to internalize nutrient losses by balance studies conducted on stocks that exceed the individual plot has led to published nutrient-balance estimates to become more negative as the scale of analysis expands (but see Cobo et al. 2010).

An unacknowledged problem with the widespread dependence on upscaling in nutrient budget studies is there seems to be little attention paid to the appropriate sampling frame to use. Whether at the village, district, nation, or regional scale, sampling should focus on the heterogeneity of the balance situations of land within the area within which most nutrient transfers occur (village territory). The sampling frame for most studies is not clearly articulated (Cobo et al. 2010). Village-scale studies are more likely to do so and they generally either ignore or use simplified models of the social geography of nutrient flows at village scales: either sampling plots based on different soil types within the village territory or across "infield"/"outfield" categories (distance from village).[11]

Ignorance of the Variation in Nutrient Management Practices

Nutrient-balance work has chosen to abstract from social factors that shape the nutrient landscapes across village territories let alone at broader scales. Empirical nutrient-balance work is performed on station (using simulated indigenous practices) or on farmers' fields. Most "off-station" work has been conducted on farmers' fields with little attempt to rigorously reconstruct the field's management history let alone place the field manager within the broader set of social relations that mediate nutrient transfers across the village landscape (exceptions including Gray 2005a; Mortimore and Harris 2005; Ramisch 2005a; Tittonell et al. 2010; Tittonell et al. 2013; Turner and Hiernaux 2015).[12] In fact, it has been my experience that often the management of farmers' fields is taken over by the researchers to ensure that nutrient inputs and outputs are sufficiently controlled and monitored. In this way, nutrient-balance field sites become what Kohler (2002) has described as "labscapes" (see also: Goldman et al. 2011; Fleming 2014).

This is surprising for a number of reasons. First, how farmers manage their crops, crop residues, and livestock have large effects on the underlying parameters affecting nutrient balances. This is not news to agronomists since they, through their on-station work, have demonstrated how sensitive these balances are to management variations (e.g. Powell and Mohamed-Saleem 1987; Bationo et al. 1993; Powell et al. 1996). Second, there are often large productivity differences between adjoining farmers' fields that are obvious when walking across cropland at the end of the rainy season. Third, it is not that the subject of cropland fertility and its management is something uninteresting or foreign to farmers (Lamers and Feil 1995; Krogh and Paarup-Laursen 1997; Osbahr 2001; Neimeijer and Mazzucato 2003; Fairhead and Scoones 2005). Farmers are quite prepared to discuss soil fertility variation and their management strategies.

Why then have agronomists avoided off-station work that engages with management variation among smallholders? While this is consistent with the long history of discounting smallholder management practices, there are other reasons tied to the agronomists' positionalities. First, scientific practice in agronomy has, as mentioned earlier, focused on controlling for variation for effective hypothesis testing. Therefore, there is a strong tendency to avoid situations of uncontrollable variations represented by heterogeneous village territories. In fact, work that engages with such heterogeneity would be seen by many agronomists as unscientific because of the lack of controls. Resistance to deviate from agronomic methodological tradition is conveniently explained by invocation of neo-Malthusian limits. Yes, there is heterogeneity but the whole system is limited without the importation of inorganic fertilizers (van Keulen and Breman 1990).

Interpretation of Different Balance Values

A fourth issue is how one should interpret balance values. As mentioned earlier, uncertainties are not rigorously analyzed nor reported. Therefore, it is difficult to compare results. Still, even if one were able to reduce uncertainties, how do we interpret different balance values? Although variable, the sandy soils of dryland West Africa have very low clay and organic matter contents. Therefore, the ability of these soils to "hold" nutrients is low (e.g. low cation exchange capacity). On-station experiments have found that over several years, it is difficult to increase soil organic matter with the incorporation of crop residues (e.g. Bationo et al. 1993). Given the poverty of the soils, plant-available nutrients are released in the root zone of soils by mineralization of organic material in moist soils. Both mineralization and plant uptake occur during the short rainy season. This raises the question of where the boundary over the "nutrient stock" is drawn in these studies. Nutrient "excesses" (inflows exceed outflows) generally would be in the form of crop residues, late growing-season manure deposition, and dust deposition lying <u>on</u> the soil surface. A small fraction of the organic material lying on the surface at the end of the cropping season will actually contribute to longer-term carbon stores in the soil and instead will be lost during the long dry season (termites, grazing, volatilization, erosion) or be mineralized with the onset of rains the next season.[13] Nutrients and carbon build up over much longer periods of time. In these highly nutrient-limited systems, less negative balances are most likely to occur for nutrients that are less limiting to plant growth and that are less mobile (less vulnerable to losses from volatilization or leaching). This is why nitrogen budgets tend to be more negative than those of other nutrients (Cobo et al. 2010). Where nitrogen significantly limits crop production, it is conceivable that heavily fertilized fields could show more negative balances than infertile fields of very low productivity. Nutrient loss (leakage) is inherent to the redistribution of nutrients to small productive areas of the landscape.

An Alternative Approach Offered by Critical Physical Geography

Despite its popularity, the nutrient-balance approach provides very few insights into the sustainability of smallholder agriculture beyond prior common knowledge that the chemical fertility of cropland soils will decline without sufficient fallowing or fertilizer inputs. More seriously, it provides little information useful for policy and extension which requires an understanding of how and who could change their livestock, soils, or cropping management practices to improve nutrient cycling.

A Critical Physical Geography approach could provide more insights by shifting attention toward the heterogeneity of nutrient availabilities across the social landscape of the agropastoral territory. Table 20.1 provides data from pooled soil samples collected in 181 fields of the 1900 fields mapped within two agropastoral territories of western Niger. Not only were these fields located within the broader geomorphology of these agropastoral territories (Turner and Hiernaux 2015) but also within the geographies of management through the collection of histories of manuring and of social difference with the lineage affiliation and endowments of labor, land and livestock wealth collected for the rural families tied to mapped fields as owners or managers. In short, this Critical Physical Geography approach embraces the complexity of soil fertility variation and seeks to understand how this complexity develops through the different practices of people who enjoy different access to productive resources.

Two papers present the results of this work (Turner and Hiernaux 2015; Turner 2016). In brief, this work reveals a highly heterogeneous landscape of soil fertility. While distributions of phosphorus availability are consistent with agronomists' infield/outfield model with higher phosphorus levels closer to village centers, the availabilities of other nutrients show more complex patterns reflecting recent management practices (Turner and Hiernaux 2015).[14] Other than phosphorus, the chemical fertility parameters are largely explained by manuring history (not cropping history) which varies, not by distance from village but from field to field. As a result, the relative importance of nitrogen and phosphorus as limiting factors varies significantly across a single agropastoral territory.

Variation in chemical fertility is explained in large part by manuring rates which are not related to land tenure security but instead to (livestock) wealth. Livestock remain, for farmers and herders alike, the major store of wealth. Those owning large numbers of livestock, no matter their ethnicity, manure the fields that they manage at higher rates. In many cases, manuring rates of the rich are higher than what would be recommended (falling rates of return to application). Thus, the social mechanism that underlies the soil fertility management is less one of incentives (e.g. tenure security) and more one of distribution of wealth within communities. Heterogeneities of chemical fertility are but another mechanism for social differentiation within agropastoral communities with the rich being able to gain significantly higher grain yields than the poor.

Conclusions

In studying soil fertility dynamics in dryland West Africa, the hypothesis-testing commitments of agronomists have led them to rely on controlled experiments on station or on borrowed farmers' fields off station. Their results reveal the yield consequences of changing levels of growth factors in isolation or interaction. Attempts to more directly consider smallholder practices off the station have led curiously to the rapid growth of nutrient-balance studies which rely on scaling up from tightly monitored farmers' fields to the farm, village territory, district, nation, and region. Such work has many problems as outlined in this chapter. Most importantly, such work ignores the wide socio-temporal variation in the flows being monitored. Yes, we would expect the incorporation of the crop residue in the soil to lead to less negative balances. Yes, we would expect that the balances of unmanured cropland to be more negative than manured cropland. The more important questions are: (1) How often does the farmer use these and other techniques over several years in order to maintain (or not) his yields? and (2) What are the constraints for him to use such techniques?

Such questions can't be answered simply through on-station work and nutrient-balance calculations. By not treating physical and social systems in isolation, a Critical Physical Geography approach shows much promise for addressing such questions by revealing how physical landscapes constrain human activities while also being transformed by them. Yes, the work done in western Niger was not controlled experimentation. It sought to learn about management effects through larger samples sizes, different statistical techniques (multiple regression, not analysis of variance), and by talking to farmers. Stocks of available nutrients were measured rather than flows. While the availability of nutrients in soils can change dramatically on a seasonal basis and are notoriously variable at the scale of meters, collecting pooled samples across a large number of fields addresses some of these concerns. Manuring and cropping histories relied on farmers' memories and therefore were prone to error, but the need for precision was not unduly high to evaluate the causal relationships of interest. Advances in soil analyses provide more opportunity to rapidly perform many soil analyses. The Critical Physical Geography project described above provides an example of how high sampling rates of material from physical landscapes coupled with fuller engagement with people using these physical landscapes has high potential.

The physical geography revealed by such work is not solely of limits but a heterogeneous one that is shaped by human difference with small areas of high nutrient loading controlled by the rich surrounded and supported by much larger areas of low nutrient availability. Understanding how these different parts of the village territory are related to each other socially and politically is necessary for understanding what we might actually mean by the phrase "nutrient mining" and a first step for addressing social imbalances in the control of village-scaled nutrient flows.

Notes

1. This characterization is based on decades of working with agronomists in the region. It is important to state clearly that it would be a gross simplification that the abstraction from farmers' practices is solely driven by an embrace of neo-Malthusianism. Prior training (controlled field experimentation) and an understandable interest to transform unproductive farming systems through modern inputs (why study something that will be replaced) also play important roles in the common practice of abstracting from the everyday constraints, decisions, and practices of smallholder farmers. Moreover, there is a long history of using biophysical signs of environmental decline or degradation without any consideration of how farmers and pastoralists engage materially with their environments to produce (or not) these symptoms (e.g. desertification).

2. These factors influence each other. For instance, moisture availability increases rates of mineralization particularly at the beginning of the rainy season (Powell et al. 1999).

3. Each agropastoral territory encloses not only the croplands of central villages and associated cultivation hamlets but also the pastures utilized by villagers' livestock.

4. Nutrient losses from cropping stem from the removal of the crop (grain or fruit) from the cropped area. The nutrients in the harvest are therefore lost to the local farming system either from actual physical export of crops, burial of nutrients in latrines and graves, or the loss of nutrients to the atmosphere (e.g. volatilization of nitrogen). Crop residue can be harvested and stocked elsewhere for animal feed, incorporated back into the soil, or grazed in place by livestock. These different management options have a significant effect on the flows of nutrients and the chemical forms in which they return to the soil (affecting their availability to plants).

5. Agronomists' focus on chemical rather than physical changes to soils reflects the perceived timescales of anthropogenic changes to soils (Duvall 2011). The lack of horizon development of these soils means that their vulnerability to structural changes is minimal. Their structure is strongly shaped by the development of impermeable ferricrete crusts which are seen as developing over long time periods. The texture of upland soils in the region is composed largely of medium-to-coarse sands, lowering their potential for compaction (e.g. trampling of livestock). Silt and clay contact increases in lower-lying depressions and active or fossil floodplains. These soils are more sensitive to compaction during short windows of time between inundation and full drying (Valentin 1985). There has been some work on micro-crust formation, which may reduce infiltration to some extent, on sandy soils with more silt/clay content (wide particle size distribution) associated with grain sorting with rain impact (Hoogmoed and Stroosnijder 1984). One could argue that human activities, by removing litter layer, increase the vulnerability to such micro-crusts. Still, the development of litter layers is rare (except under strong protection from termites, grazing, and fire), making most soils with the appropriate particle size distribution susceptible to such micro-crust forming during early rains. It could be argued that livestock trampling and cultivation, by breaking up these crusts, could improve rainfall infiltration.

6. For a general history of understandings of African soils, see Showers (2006).

7. An example is the influential work performed at ICRISAT's Sahelian Center at Sadoré, Niger, located south of the capital city of Niamey. Agronomists at the center conducted their experiments within the research station under the assumption that nitrogen was the major limiting factor affecting crop production in the region. To clarify the effect of nitrogen fertilization on crop yields, control and treatment plots were fertilized with all other nutrients that could influence crop response. With the realization that phosphorus is another macronutrient limiting crop productivity, agronomists and soil scientists had to look outside the station to perform their work while planting cover crops on test plots within the station to remove phosphorus from their soils.

8. Turner (1995) discusses how most of parameters in equations used to generate estimates of sustainable rangeland-to-cropland ratios show a wide range of variation with much of the variation due to known differences in how differ-

ent people manage their fields and livestock. Scoones and Toulmin (1998) likewise describe the many ways in which crop management can affect nutrient balances. Thus, to understand the sustainability of mixed farming systems, one needs to understand a broader set of institutional and political economic factors that affect resource management.

9. See Cobo et al. (2010) for an excellent review of the findings of nutrient-balance estimates across Africa.

10. These approaches are tied to the coupling of neo-Malthusian approaches for environmental assessment with the tradition of human ecological systems analysis of tracing nutrients through economies (e.g. Odum 1971; Moran 1990; Grote et al. 2005).

11. As described by Prudencio (1993), infields are those that tend to be more heavily manured and cultivated continuously within a short radius of the village. Outfields are areas that are not continuously cultivated and receive much more limited nutrient amendments.

12. Most of these researchers have mixed training in both the physical and social sciences with geographers constituting a large fraction of these.

13. This discussion also points to the sensitivity of such studies to the monitoring time frame (growing season, year, multiple years).

14. This finding is consistent with the conservative nature of phosphorus in the system with gradients of phosphorus availability positively associated with the long-term history of manure application which is negatively correlated with distance from village.

References

Aina, P.O. 1979. Soil changes resulting from long-term management practices in Western Nigeria. *Soil Science Society of America Journal* 43: 173–177.

Allan, W. 1965. *The African husbandman*. New York: Barnes & Noble.

Bationo, A., B.C. Christianson, and M.C. Klaij. 1993. The effect of crop residue and fertilization use on pearl millet yields in niger. *Fertilizer Research* 34: 251–258.

Bationo, A., F. Lompo, and S. Koala. 1998. Research on nutrient flows and balances in west Africa: State-of-the-art. *Agriculture Ecosystems and Environment* 71: 19–35.

Bertrand, R., J. Nabos, and R. Vicaire. 1972. Exportations minérales par le mil et l'arachide: Conséquences sur la définition d'une fumure d'entretien d'un sol ferrugineux tropical développé sur matériaux éolien à Tarna (Niger). *Agonomie Tropicale* 27: 1287–1302.

Brams, E.A. 1971. Continuous cultivation of West African soils: Organic matter dimunation and effects of applied lime and phosphorus. *Plant and Soil* 35: 401–474.

Chappell, A., A. Warren, N. Taylor, and M. Charlton. 1998. Soil flux (loss and gain) in southwestern Niger and its agricultural impact. *Land Degradation & Development* 9: 295–310.

Charreau, C., and R. Fauk. 1970. Mise au point sur l'utilisation agricole des sols de la region de Sefa (Casamance). *Agronomie Tropicale* 25: 151–191.

Chevalier, A. 1928. Sur la dégradation des sols tropicaux causée par les feux de brousse et sur les formation végétales régressives qui en sont la conséquence. *Comptes rendus de l'Academie des Sciences* 188: 84–86.

Cobo, J.G., G. Dercon, and G. Cadisch. 2010. Nutrient balances in African land use systems across different spatial scales: A review of approaches, challenges and progress. *Agriculture Ecosystems & Environment* 136: 1–15.

Collier, F.S., and J. Dundas. 1937. The arid regions of northern Nigeria and the French Niger colony. *The Empire Forestry Journal* 16: 184–194.

Duvall, C. 2011. Ferricrete, forests and temporal scale in the production of colonial science in Africa. In *Knowing nature: Conversations at the intersection of political ecology and science studies*, ed. M.J. Goldman, P. Nadasdy, and M.D. Turner, 113–127. Chicago: University of Chicago Press.

Fairhead, J., and I. Scoones. 2005. Local knowledge and the social shaping of soil investments: Critical perspectives on the assessment of soil degradation in Africa. *Land Use Policy* 22: 23–41.

Fleming, J. 2014. Political ecology and the geography of science: Lesosady, lysenkoism, and soviet science in kyrgyzstan's walnut–fruit forest. *Annals of the Association of American Geographers* 104: 1183–1198.

Food and Agriculture Organization. 2004. *Scaling soil nutrient balances: Enabling mesolevel applications for African realities*. Rome: Food and Agriculture Organization.

Goldman, M.J., P. Nadasdy, and M.D. Turner, eds. 2011. *Knowing nature: Conversations at the intersection of political ecology and science studies*. Chicago: University of Chicago Press.

Gray, L.C. 2005a. What kind of intensification? Agricultural practice, soil fertility, and socioeconomic differentiation in rural Burkina Faso. *The Geographical Journal* 171: 70–82.

———. 2005b. What kind of intensification? Agricultural practice, soil fertility and socioeconomic differentiation in rural Burkina Faso. *The Geographical Journal* 171: 70–82.

Grote, U., E. Craswell, and P. Viek. 2005. Nutrient flows in international trade: Ecology and policy issues. *Environmental Science and Policy* 8: 439–451.

Hafner, H., J. Bley, A. Batiano, P. Martin, and H. Marschner. 1993. Long-term nitrogen balance for pearl millet (*Pennisetum glaucum* L.) in an acid sandy soil of Niger. *Zeitschrift für Pflanzenernährung und Bodenkunde* 156: 169–176.

Harris, F.M.A. 1998. Farm-level assessment of the nutrient balance in northern Nigeria. *Agriculture, Ecosystems and Environment* 71: 201–214.

Harroy, J.-P. 1949. *Afrique terre qui meurt: La dégradation des sols africains sous l'influence de la colonisation.* Brussels: Marcel Hayez.

Hoogmoed, W.B., and L. Stroosnijder. 1984. Crust formation on sandy soils in the sahel. 1. Rainfall and infiltration. *Soil and Tillage Research* 4: 5–23.

Jones, M.J. 1973. The organic matter content of the savanna soils of West Africa. *Journal of Soil Science* 24: 42–53.

Kohler, R.E. 2002. *Landscapes and labscapes: Exploring the lab-field border in biology.* Chicago: University of Chicago Press.

Krogh, L. 1997. Field and village nutrient balances in millet cultivation in northern Burkina Faso: A village case study. *Journal of Arid Environments* 35: 147–159.

Krogh, L., and B. Paarup-Laursen. 1997. Indigenous soil knowledge among the Fulani of northern Burkina Faso: Linking soil science and anthropology in analysis of natural resource management. *GeoJournal* 43: 189–197.

Lamers, J.P.A., and P.R. Feil. 1995. Farmers' knowledge and management of spatial soil and crop growth variability in Niger, West Africa. *Netherlands Journal of Agricultural Science* 43: 375–389.

Mohamed-Saleem, M.A. 1998. Nutrient balance patterns in African livestock systems. *Agriculture Ecosystems & Environment* 71: 241–254.

Moran, E.F., ed. 1990. *The ecosystem approach in anthropology.* Ann Arbor: University of Michigan Press.

Mortimore, M., and F. Harris. 2005. Do small farmers' achievements contradict the nutrient depletion scenarios for Africa? *Land Use Policy* 22: 43–56.

Niemeijer, D., and V. Mazzucato. 2002. Soil degradation in West african sahel: How serious is it? *Environment* 44: 20–33.

Neimeijer, D., and V. Mazzucato. 2003. Moving beyond indigenous soil taxonomies: Local theories of soils for sustainable development. *Geoderma* 111: 403–424.

Nye, P.H., and D.J. Greenland. 1960. *The soil under shifting cultivation.* Harpenden: Commonwealth Bureau of Soils.

Odum, H.T. 1971. *Environment.* New York: Power and Society.

Osbahr, H. 2001. Livelihood strategies and soil fertility at Fandou Béri, Southwestern Niger. Ph.D. Dissertation, Department of Geography, University College London, London.

Penning de Vries, F.W.T., and M.A. Djitèye, eds. 1982. *La productivité des pâturages sahéliens.* Wageningen, The Netherlands: Centre for Agricultural Publishing and Documentation.

Pieri, C. 1985. Bilans mineraux des sytemes de cultures pluviales en zones arides et semi-arides. *L'Agronomie Tropicale* 40: 1–20.

———. 1992. *Fertility of soils. A future for the farming in the West African Savanna.* New York: Springer.

Powell, J.M., S. Fernandez-Rivera, P. Hiernaux, and M.D. Turner. 1996. Nutrient cycling in integrated rangeland/cropland systems of the Sahel. *Agricultural Systems* 52: 143–170.

Powell, J.M., S. Fernandez-Rivera, T.O. Williams, and C. Renard, eds. 1995. *Livestock and sustainable nutrient cycling in mixed farming systems of sub-Saharan Africa*. Addis Ababa, Ethiopia: International Livestock Centre for Africa.

Powell, J.M., F.N. Ikpe, and Z.C. Somda. 1999. Crop yield and the fate of nitrogen and phosphorus following application of plant material and feces to soil. *Nutrient Cycling in Agroecosystems* 54: 215–226.

Powell, J.M., and M.A. Mohamed-Saleem. 1987. Nitrogen and phosphorus transfers in a crop-livestock system in West Africa. *Agricultural Systems* 25: 261–277.

Prudencio, C.Y. 1993. Ring management of soils and crops in the West African semi-arid tropics: The case of mossi farming in Burkina Faso. *Agriculture, Ecosystems and Environment* 47: 237–264.

Ramisch, J.J. 2005a. Inequality, agro-pastoral exchanges, and soil fertility gradient in Southern Mali. *Agriculture Ecosystems & Environment* 105: 353–372.

———. 2005b. Inequality, agro-pastoral exchanges, and soil fertility gradients in Southern Mali. *Agriculture, Ecosystems, and Environment* 105: 353–372.

Roy, R.N., R.V. Misra, J.P. Lesschen, and E.M. Smaling. 2003. *Assessment of soil nutrient balance: Approaches and methodologies*. Rome: Food and Agriculture Organization.

Schlecht, E., and P. Hiernaux. 2004. Beyond adding up inputs and outputs; process assessment and upscaling in modeling nutrient flows. *Nutrient Cycling in Agroecosystems* 70: 303–319.

Scoones, I., and C. Toulmin. 1998. Soil nutrient budgets and balances: What use of policy? *Agriculture, Ecosystems and Environment* 71: 255–267.

Showers, K.B. 2006. A history of African soil: Perceptions, use and abuse. In *Soils and societies: Perspectives from environmental history*, ed. J.R. McNeill and V. Winiwarter, 118–176. Isle of Harris: The White Horse Press.

Smaling, E.M.A., J.J. Stoorvogel, and P.N. Windmeijer. 1993. Calculating soil nutrient balances in Africa at different scales: II. District scale. *Fertilizer Research* 35: 237–250.

Stoorvogel, J.J., and E.M.A. Smaling. 1990. *Assessment of soil nutrient depletion in sub-Saharan Africa, 1983–2000*. Wageningen, The Netherlands: Winand Staring Centre for Integrated Soil and Water Research.

———. 1993. Calculating soil nutrient balances in Africa at different scales: I. Supra-national scale. *Fertilizer Research* 35: 227–235.

Tittonell, P., A. Muriuki, C.J. Klapwijk, K.D. Shepherd, R. Coe, and B. Vanlauwe. 2013. Soil heterogeneity and soil fertility gradients in smallholder farms of the East African Highlands. *Soil Science Society of America Journal* 77: 525–538.

Tittonell, P., A. Muriuki, K.D. Shepherd, D. Mugendi, K.C. Kaizzi, J. Okeyo, L. Verchot, R. Coe, and B. Vanlauwe. 2010. The diversity of rural livelihoods and their influence on soil fertility in agricultural systems of East Africa—A typology of smallholder farms. *Agricultural Systems* 103: 83–97.

Turner, M.D. 1995. The sustainability of rangeland to cropland nutrient transfer in semi-arid West Africa: Ecological and social dimensions neglected in the debate.

In *Livestock and sustainable nutrient cycling in mixed farming systems of sub-Saharan Africa. Proceedings of an international conference, 22–26 November 1993*, ed. J.M. Powell, S. Fernandez-Rivera, T.O. Williams, and C. Renard, 435–452. Addis Ababa, Ethiopia: International Livestock Centre for Africa.

———. 1998. Long-term effects of daily grazing orbits on nutrient availability in sahelian West Africa: 1. Gradients in the chemical composition of rangeland soils and vegetation. *Journal of Biogeography* 25: 669–682.

———. 2016. Rethinking land endowment and inequality in rural africa: The importance of soil fertility. *World Development*. 87: 258–273.

Turner, M.D., and P. Hiernaux. 2015. The effects of management history and landscape position on inter-field variation in soil fertility and millet yields in Southwestern Niger. *Agriculture, Ecosystems & Environment* 211: 73–83.

Valentin, C. 1985. Effects of grazing and trampling on soil deterioration around recently drilled water holes in the Sahelian Zone. In *Soil Erosion and Water Erosion*, eds. S.A. El-Swaify, W.C. Moldenhauer, and A. Lo, 51–65. Ankeny, IA: Soil Conservation Society of America.

van der Pol, F. 1992. *Soil mining. An unseen contributor to farm income in Southern Mali*. Amsterdam: The Netherlands, Royal Tropical Institute.

van Keulen, H., and H. Breman. 1990. Agricultural development in the West African Sahelian Region: A cure against land hunger? *Agriculture, Ecosystems and Environment* 32: 177–197.

Warren, A., H. Osbahr, S. Batterbury, and A. Chappell. 2003. Indigenous views of soil erosion at Fandou Beri, southwestern Niger. *Geoderma* 111: 439–456.

21

Commodifying Streams: A Critical Physical Geography Approach to Stream Mitigation Banking in the USA

Rebecca Lave, Martin Doyle, Morgan Robertson,
and Jai Singh

Introduction

Market-based (or neoliberal) environmental management is based on the premise that traditional forms of environmental protection, such as wildlife preserves and command-and-control regulation, have failed to protect the environment from economic development. The way to save the environment from capitalism, advocates argue, is not to separate the two more strongly but to combine them more effectively, putting a price tag on nature and so forcing its inclusion in capitalist accounting and business decisions. This approach, pithily summarized as "selling nature in order to save it" (McAfee 1999), has been strongly promoted in international environmental policy circles since the late 1990s (Costanza et al. 1997; Daily 1997). There are now markets intended to promote environmental goods ranging from water quality to endangered species habitat to carbon sequestration, on every continent except Antarctica.

R. Lave (✉)
Department of Geography, Indiana University, Bloomington, IN, USA

M. Doyle
Nicholas School, Duke University, Durham, NC, USA

M. Robertson
Department of Geography, University of Wisconsin, Madison, WI, USA

J. Singh
Cbec Eco Engineering, West Sacramento, CA, USA

© The Author(s) 2018
R. Lave et al. (eds.), *The Palgrave Handbook of Critical Physical Geography*,
https://doi.org/10.1007/978-3-319-71461-5_21

It is not clear, however, what impacts these market-based approaches are actually having. We can say with certainty that they have not achieved either the financial or ideological reach for which their advocates had hoped. In monetary terms, these markets are quite small (Dempsey and Suarez 2016), particularly when compared with estimates of the overall value of the services nature provides us (Costanza et al. 1997, 2014). While there has been no protest against market-based approaches in the USA, the home of the earliest and best-established markets in ecosystem services, there has been sustained and successful resistance to the whole idea of monetizing the environment in some high-profile international settings, including EU policymaking and the Conference of Parties negotiations on the Convention on Biological Diversity (Lave and Robertson 2017; Dempsey 2016).

But what effects, if any, are market-based, neoliberal environmental management approaches having on the landscape? Are they an improvement over the preservationist approaches they aim to supersede? Until now, we have not been able to begin answering that question because no one has investigated whether selling nature in order to save it succeeds in practice. Most critical nature/society researchers do not have the physical science training to investigate the environmental impacts of the markets they study; for their part, most environmental scientists' focus on "pristine" field sites, and are not interested in investigating the outcomes of such a determinedly economistic manipulation of natural systems (Urban, this volume). This is unfortunate, because advocates of market-based environmentalism are persistent, and it is possible that these approaches may eventually dominate international environmental policy. We need to understand neoliberal environmental management not simply as a policy paradigm but as an eco-social system and to study the co-constituted landscapes it produces not from a purely social or physical science approach but via the integrated approaches and methods advocated by Critical Physical Geography (CPG).

This chapter takes a first step toward analyzing the success (or lack thereof) of market-based environmental management. The specific form we address here is Stream Mitigation Banking (SMB) in the USA, an offsetting market in which damage to streams is mitigated (or offset) via restoration of comparable streams elsewhere. We collected data on SMB using a wide range of qualitative and quantitative natural and social science methods including semi-structured interviews, policy document analysis, participant observation, Q method, data mining, and physical surveys. This combination of methods allowed us to investigate the interlinked political economies of environmental knowledge production (via the metrics used to define the commodity for sale in SMB markets) and market-based environmental management (via the implementation of those metrics and their physical impacts on the landscape). Taking a CPG approach allowed us to trace the social and physical forces that together co-constitute the fluvial landscape.

In the sections that follow, we first explain the regulatory framework and practice of stream mitigation banking. We then lay out our iterative physical and social science analysis of SMB as an eco-social system. We conclude by arguing that, at least in the case of SMB, market-based environmental management is magnifying the trend of environmental disruption it was intended to counteract.

Stream Mitigation Banking

Environmental protection for fluvial systems in the USA stems primarily from the Clean Water Act (initially titled the Federal Water Pollution Control Act), which was passed in 1972. Of specific relevance to this chapter, Section 404 of the Clean Water Act (CWA) regulates the discharge of dredged or other materials into waters of the USA with the intent of minimizing damage to aquatic ecosystems. US legislators assumed that the Environmental Protection Agency (EPA) and the US Army Corps of Engineers (Corps), to whom they gave joint responsibility to administer the CWA, would modify or deny permits to *prevent* harm to aquatic systems in the first place rather than allowing harm to take place and then asking permit applicants to make amends for it. In practice, however, neither agency has proved willing to deny permits except in the most extreme cases: for example, the Corps denied only 0.25% of all applications in 2004–2005 (Hough and Robertson 2009: 27), while the EPA has mostly refused to exercise its veto power, overturning 11 Section 404 permits in the first 36 years of the program's existence (ibid.: 22). But the CWA mandates the protection of aquatic resources. If regulators were unwilling to shoulder the political costs of preventing harm, there had to be a mechanism through which that harm could be counter-balanced after the fact.

Eventually, the EPA and the Corps developed a permitting process called the *mitigation sequence* (see Hough and Robertson 2009 for a very helpful historical overview). Projects seeking permits under Section 404 of the CWA are expected first to *avoid* impacts as far as possible. Once the options for avoidance are exhausted, applicants should *minimize* any impacts deemed unavoidable, and only after options for both avoidance and minimization are exhausted is *compensation* for unavoidable impacts via ecological restoration acceptable. The EPA's and the Corps' continued unwillingness to take on the political costs of denying permits, however, eventually converted what was intended to be a last resort—compensation—into the first and final step in the majority of cases, as it is often more palatable for regulatory agencies to ask permit applicants to pay compensation than to ask them for a thoughtful (and thus expensive) redesign of proposed projects. The end result is that, "Compensatory mitigation is so central to discussions of mitigation that 'compensation' is often mistakenly held to be synonymous with 'mitigation', even among the most experienced observers of the [Section 404] program" (Hough and Robertson 2009: 23). The practical and political considerations of implementation have transformed the Clean Water Act over time from a focus on prevention to a focus on compensating for harms committed. This shift in policy implementation is very important for the fluvial landscape of the USA because a wide range of projects trigger Section 404, from highways that arrow across the landscape on their routes between cities, intersecting the paths of multiple streams along the way, to new shopping malls and apartment complexes that affect a single reach of stream but cover broad swathes of adjacent upland habitat in asphalt and concrete in addition to affecting the channel itself.

There are several different ways that project proponents can provide compensatory mitigation, but as of 2008, the preferred way to do so is through a third-party provider, who produces compensation credits for sale through a stream mitigation bank (Corps and EPA 2008). By 2012, more than 75% of the stream mitigation banks in the USA were for-profit businesses (Doyle and Shields 2012), developed by mitigation banking companies. These companies are typically staffed by people with backgrounds in environmental science, GIS, and land acquisition, and while many are relatively small, with fewer than a dozen employees and a relatively local geographic reach, other mitigation banking companies have regional or even national scope.

To create a stream mitigation bank, a mitigation banking company acquires rights to land with a degraded stream on it in an area where there is sufficient development pressure that there is likely to be demand for mitigation credits. In some cases, mitigation bankers buy land outright; more often they purchase easements, as they are only interested in the stream and the land immediately adjacent to it, and property owners are rarely interested in splitting their land by selling a slice out of it. Once land rights of some form are established, the mitigation banking company hires restoration consultants to develop both a design to restore the stream and a proposal for monitoring particular aspects of that design over a specified number of years (typically three to five) after project completion.

When the proposed restoration design is ready, the mitigation banking company submits it to the state Interagency Review Team (IRT), which is led by the local Corps District Regulatory office, with representatives from the EPA, the US Fish & Wildlife Service, and relevant state agencies. The IRT reviews the bank proposal and eventually the bank as implemented, authorizing the release of stream mitigation "credits" gradually over time if the bank meets its specified success criteria. Once certified by the IRT, mitigation credits can be purchased at a negotiated price by anyone needing a permit under Section 404 for a project that would cause ecological damage somewhere within the "service area" of the bank: the drainage basins within which restoration at the bank site is viewed as compensating for ecological harm.

Stream mitigation banking (and its elder sister, wetland mitigation banking) are among the longest and most firmly established examples of market-based environmental management in the world, and thus are deployed as proof of concept in many current proposals to establish compensatory mitigation markets in other jurisdictions (e.g. ECOLOGIC 2006; European Commission 2007; eftec et al. 2010). But what is often glossed over by those who use SMB as a poster child is that getting a market in stream credits up and running is not a simple process. It is not obvious how to create equivalence between the stream reach to be damaged by a permit applicant and the stream reach previously restored by a mitigation banker; how can a particular, geomorphologically, chemically, and ecologically interconnected reach of stream be converted into an isolated, standardized, tradable commodity? The customary ways we define commodities—by weight, volume, or number—are irrelevant because they are discrete. Instead, regulators and bankers needed new metrics capable of paring away the complexity and specificity of both the damaged and restored reaches, abstracting them into easily measurable, comparable units (Cronon 1991).

In entrepreneurial wetlands mitigation banking, which began in the mid-1980s, the development of these metrics has been the work of decades of scientific labor, primarily on the part of university- and agency-based scientists. In sharp contrast, when stream mitigation banking began in 2000, there was already a widely used metric ready on hand: the stream classification system developed by consultant David Rosgen (Lave et al. 2010; Lave 2012).

The Rosgen classification system focuses on channel form, using the transport of water and sediment as a proxy for the desired improvements in water quality and aquatic ecology that also drive stream mitigation. Streams are allocated to different alphanumeric categories based on fairly straightforward measurements of multiple aspects of the physical form of the channel (Rosgen 1996). For the analysis that follows, it is important to note that the Rosgen classification system takes great care to render many measurements that feed into it dimensionless so that within any given category, there can be a fairly big range of channel sizes. Rosgen's system can thus be described as scale-neutral, since the physical size of the channel does not determine its categorization.

Despite the classification system's exclusively physical focus, it was widely employed by regulators and mitigation bankers as the core metric enabling the definition of the commodity for sale in stream mitigation banking: the *stream credit*. This was despite the fact that Rosgen himself raised concerns that his classification system was not intended to define stream credits (Lave 2012). In most of the USA, a stream credit (as defined in guidance documents for each state, prepared primarily by local Corps District Regulatory staff in consultation with other state and federal agencies) consists of linear feet of a specific Rosgen stream type. This produced two clear perverse incentives. First, because the Rosgen classification system is scale-neutral, there was a financial incentive to address damage to larger streams by restoring smaller streams of the same Rosgen classification because it was far cheaper to do so. Second, with credits defined by the linear foot and appropriate properties with degraded streams on them difficult to find and expensive to procure, the current system created a strong incentive to maximize the sinuosity of the restored stream in order to squeeze the most linear feet possible onto a given property. This seemed likely to lead to lots of stream mitigation projects destroying their restored channels in even moderate storms. Thus there was a strong potential for SMB to produce substantially worse outcomes for fluvial systems than were produced previously through non-market approaches to conservation under the CWA. Was that in fact what was happening?

Assessing the Outcomes of Stream Mitigation Banking

To answer this question, we compared the physical form of streams restored to create mitigation banks both with streams restored for other purposes and with unrestored streams. We concentrated our research in the US state of North Carolina, a hotspot for stream restoration more broadly and for mitigation banking in particular. Some of our geomorphic data on channel form were mined from required monitoring reports on SMB projects, and other data came from surveys of a random selection of unrestored streams. Stream reaches in the latter category were in no sense pristine; presumably all of those surveyed had been affected by human actions directly or indirectly, although we did discard blatantly manipulated and/or straightened streams. In the end, we compared 186 streams, creating one of the biggest data sets on the physical form of streams in the USA (Doyle et al. 2015).

We integrated the collection and analysis of these geomorphic data with the collection and analysis of social science data. We draw here on semi-structured interviews with 20 state and federal agency staff members, mitigation bankers, restoration designers, and stream scientists (some of whom we interviewed multiple times), and on analysis of a series of policy documents developed by the local Corps District to establish guidelines for stream mitigation banking.

Our overall goals were first, to investigate the political economy of SMB and how stream "credits," the commodity for sale in SMB, were defined in theory and in practice, and second, to determine whether the switch to market-based management of streams was in fact improving on previous command-and-control regulatory approaches by reducing anthropogenic impacts on fluvial systems. We began with the following hypotheses, based on what seemed to be the logical outcomes of the way stream credits were defined:

1. Streams restored as mitigation banks will be heavily altered to maximize economic return by:

   a. Disproportionately placing mitigation projects on headwater streams to reduce restoration costs, and
   b. Maximizing the number of credits produced from a particular piece of land by increasing sinuosity beyond thresholds typical of unrestored streams in NC, or "credit chasing" as our interview subjects called it.

2. This will result in many damaged mitigation projects on disproportionately small drainage basins.

As we describe in the remainder of this section, both of these hypotheses turned out to be incorrect.

Hypothesis 1a suggested that we would find a disproportionate percentage of mitigation banking projects in small drainage basins. To test this, we compared the drainage basin size of streams restored for mitigation banking purposes to that of streams restored for other purposes and also to the frequency of occurrence of drainage basins of different sizes across the North Carolina landscape. Our results (Fig. 21.1) showed no such thing. Instead, we found that although the percentage of headwater streams (0–1 km^2) restored for

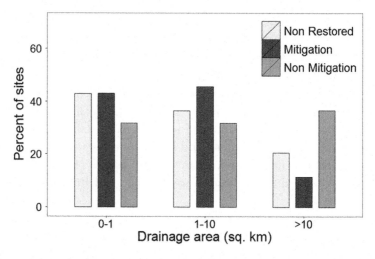

Fig. 21.1 Project site location as a function of drainage area. $N = 74$ for 0–1 km²; $n = 72$ for 1–10 km²; $N = 37$ for >10 km². The mean drainage area for mitigation was 4 km²; the mean for non-mitigation is 13.9 km²; and the mean for nonrestored is 7.4 km²; these differences in drainage area by stream type were significantly different (ANOVA; $p < 0.01$)

mitigation was indeed larger than the percentage of the smallest streams restored for other purposes, it was startlingly close to the percentage of headwater streams present in the North Carolina landscape. In the next largest category of drainage basins (1–10 km²), there were actually more streams restored for non-mitigation purposes. Clearly, mitigation banking did not cluster disproportionately in the smallest drainage basins.

There was a similar disconnect between our initial hypotheses and data in relation to sinuosity (Fig. 21.2). While there were statistically significant differences between the typical sinuosity of unrestored streams and restored streams, they were not particularly large. And given that the unrestored streams had very likely been affected by direct and/or indirect anthropogenic influences, it is likely that their sinuosity was artificially low rather than the sinuosity of restored streams being artificially high. Looking at the data for the Piedmont region, for example, there is nothing in the sinuosities of the restored streams to indicate credit chasing or to suggest in fluvial geomorphological terms that meanders were being "over" restored. While the differences in the Mountain region could be of more concern, the strong differences in slope between restoration sites and unrestored streams seems to explain those differences more plausibly than credit chasing (Fig. 21.3). What is perhaps most notable is the similar sinuosities of streams restored for any purpose, mitigation or otherwise.

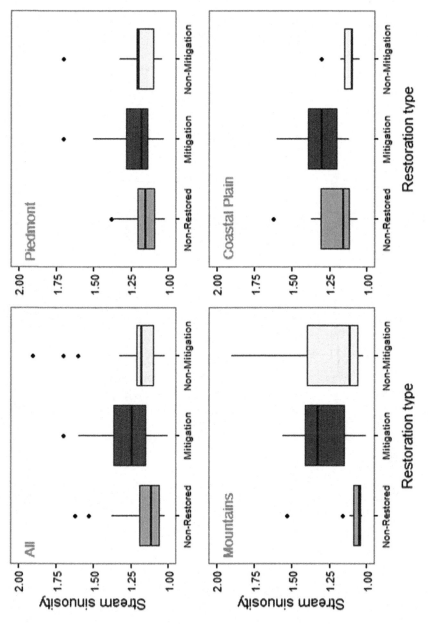

Fig. 21.2 Sinuosities of stream reaches by physiographic region and stream type

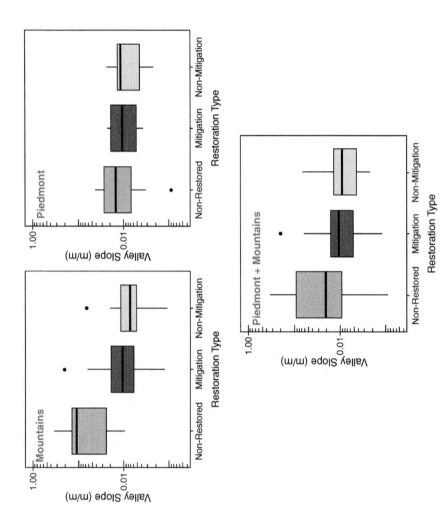

Fig. 21.3 Valley slopes of stream reaches by stream type in the mountain and physiographic provinces

What was going on here? Why were the very clear perverse economic incentives that seemed to be built into the definition of stream credits not visible on the landscape? With no way to answer that question from a physical science perspective, we turned to our social science data. In our interviews with restoration designers, they were, to a person, adamant that they used the same design approach (Rosgen's Natural Channel Design) for mitigation and non-mitigation projects. In the words of one designer, "No, there are no real differences in the design process .... [Y]ou take the same stuff into consideration as far as stream mechanics" (Author interview, December 11, 2012). Or as another explained in more detail:

> When it comes down to core design work, if you take two sites: Farm A and Farm B. They are both cattle farms, they are both 500 acres, they both have 12,000 feet of streams on them, and one is funded via mitigation and one is funded via grants. When it comes down to how are you going to look at the existing conditions, you know, assuming the goal is ultimate restoration as much as possible and assuming the budget is fairly reasonable on both, then our goal is going to be go to through the same design process that we would go through with any site ... and ultimately aim for the same end result, which [is] the highest level of lift that we can accomplish on it. (Author interview November 14, 2012)

There were several reasons for this refusal to pursue the obvious economic incentives built into stream credit definitions. First, the designers, like almost everyone else in the stream restoration community, care deeply about stream health and would see a design that obviously degraded a stream reach in order to maximize profit as professionally and environmentally unacceptable. Second, because the broader stream restoration community in North Carolina (including not just designers but also funders, regulators, etc.) strongly espouses Natural Channel Design, any proposed project that visibly departed from that approach's specified ranges of sinuosity would be obvious credit chasing, and thus subject to immediate regulatory crackdown, with long term implications for that restoration designer's professional reputation. And finally, as reflected in Hypothesis 2, above, it seems likely that credit chasing via increased sinuosity would result in very visibly failing restoration projects. In one designer's words:

> I've seen people try to maximize their length .... You know, if credit is tied to footage, then add as much footage as they can and put in a lot of sinuosity. However, I think you have to weight that with if you get too far outside of your design parameters is that you risk failure. To me that risk is not worth the reward

of the additional credit. Nature will let you know if you make that mistake! (*laughs*) You'd have to go back and make that repair [or lose the credits] .... The market kind of has a way to self-correct itself ... if you're being a little greedy on the front end, as far as trying to push the site beyond what credits it can really yield. (Author interview, December 7, 2012)

Somewhat at a loss with the refutation of our guiding hypotheses, we began asking our interview subjects what differences we might actually find between mitigation and non-mitigation restoration projects. Strikingly, both mitigation bankers and restoration designers pointed us to the physical implications of the real estate market and the difficulty and expense of procuring mitigation sites. For mitigation bankers, there were serious economies of scale in developing longer restoration projects, so that the upfront costs of obtaining rights to any property (not to mention design work and permitting) could be offset by a larger number of credits. For example, one mitigation banker pointed out that,

To do a mitigation project, you have got to do a mitigation plan, you have got to do all your monitoring, you have to got to do your permitting, you have got to get a conservation easement. All these things have a lot of fixed cost to them, and so you really need your project to be of a certain size ... so you can spread those fixed costs over more credits or more length. [This is important both economically and ecologically.] The whole point of mitigation banking is to accumulate money from a hundred different sources through credit sales, and instead of each one of those people doing little hundred foot ditches all over the place ..., you want to take all that and accumulate it into a large project where you are actually getting that sort of magnified improvement because you are doing seven or eight or ten thousand feet. If you are only doing like two thousand feet of restoration, ... it is sort of a marginal lift. (Author interview, November 16, 2012)

Another banker made a similar link between the project length and the ecological impacts of mitigation banking:

We typically won't do anything less than 2500 feet .... [H]ow much good are we doing these postage stamp sites? I mean half a mile?!? I prefer to work on 1st and 2nd order streams. I don't want to work on a 3rd, or 4th, or 5th order system because we're not really being able to make any contributions to improving water quality. Because when systems become that big they're basically conduits for stormwater .... [I]f you want to have impacts to water quality, you need to be working in the headwaters. (Author interview, July 9, 2012)

Thus multiple interview subjects told us to compare the project length of mitigation and non-mitigation restored streams.

Returning to our physical data, we indeed found that the overall length of mitigation projects was substantially higher than that of projects restored for other purposes (Fig. 21.4). However, what was more striking was how that greater length was achieved. Non-mitigation restoration projects almost always concentrated exclusively on the main stem of the stream, ignoring any tributaries that might intersect the project reach. In stark contrast, mitigation projects tended to restore not just the main stem, but any available tributary, including intermittent ones, in order to maximize the number of credits from difficult to procure restoration sites (Figs. 21.5 and 21.6).

When we showed these results to our interview subjects in a subsequent round of fieldwork, they confirmed this as an obvious and accepted facet of mitigation projects; the surprise for them was that non-mitigation projects did not address tributaries! As several people noted, if you want to control what happens in the reach of stream you are restoring, geomorphically and ecologically, you are better off controlling everything that flows into it. Looking at Fig. 21.6, for example, one mitigation banker said:

> I want to get everything [mainstem and tributaries]. I want at the end of the day, when I'm done with that project, I want to say every linear foot of stream or close to it, is under protection, or is being fixed. Because if you just do that one [mainstem] site and you leave this trib and this trib and this trib [*pointing to tributaries shown in* Fig. 21.6], you've got stuff coming right, you don't have control. I mean that's my goal, on every project we can. Now a lot of times you're too far down in the watershed and there's no way. But for mitigation projects, that's a big reason why we somewhat start up towards the headwaters, because you really want to have control of what's coming onto your main site. Partially because it feels like a better project [ecologically], it feels more holistic. (Author interview, October 5, 2015)

While revisiting our geomorphic data to examine project length, we noticed another, perhaps even more striking difference: the radii of curvature of stream reaches restored for any reason were far more homogenous than those of unrestored streams. The *radius of curvature* describes the relative tightness of a meander bed in a river or stream. A very sharp bend has a large radius of curvature, while a very gradual bend has a small radius of curvature. What we found was that unrestored streams had a very broad range of variation in radii of curvature, typically stretching over multiple orders of magnitude, while streams restored for any purpose typically had much less variation, with radii

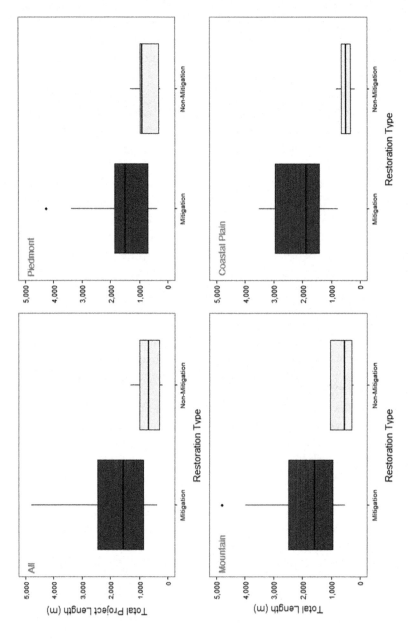

Fig. 21.4 Distribution of restoration projects site total project lengths for mitigation and non-mitigation projects. For all projects, median for mitigation = 1543 m; median for non-mitigation = 682 m ($p < 0.01$, Wilcoxon test assuming non-normal distribution)

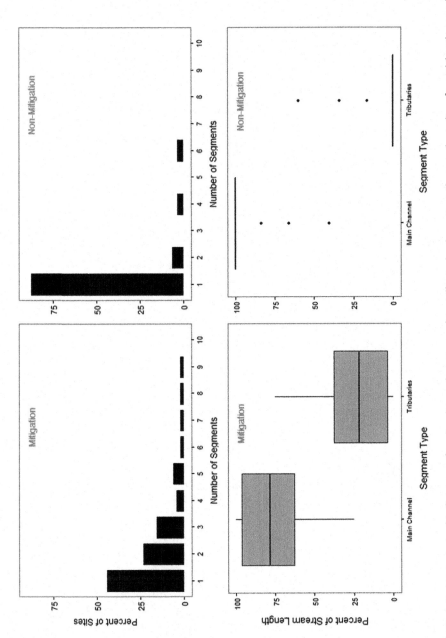

Fig. 21.5 Relative amount of influence of main channel compared to tributaries in restoration project sites for mitigation and non-mitigation. $N = 53$ project sites for mitigation, $N = 32$ project sites for non-mitigation. For lower graphs, relative amount of total restored stream length that was contributed from restored main channel compared to tributaries

Fig. 21.6 Example of mitigation project site showing the combination of restored main channel and six tributaries restored as part of the same project site. Identifying information removed from image

of curvature that stretched across only one or two orders of magnitude. Put differently, streams restored for any reason had far more homogenous forms than unrestored streams, dominated by large gradual curves; restored streams had more in common with sine waves than with the far more unpredictable and highly variable forms of unrestored streams (Fig. 21.7).

This surprising homogenization of channel form was another thing no one had suggested we look at in our initial round of interviews. Thus in our second round of interviews, we showed interview subjects versions of Figs. 21.6 and 21.7 and asked them how they would explain both the striking differences between restored and unrestored streams and the striking similarities between streams restored for mitigation and non-mitigation purposes. Unlike with the similarities in sinuosity in restored and unrestored streams, the answer does not lie in Natural Channel Design, which directs designers to develop their own ranges of radii of curvature for particular projects empirically by examining reference reaches in the area rather than specifying a fixed range to accompany each stream classification category. Instead, the answer

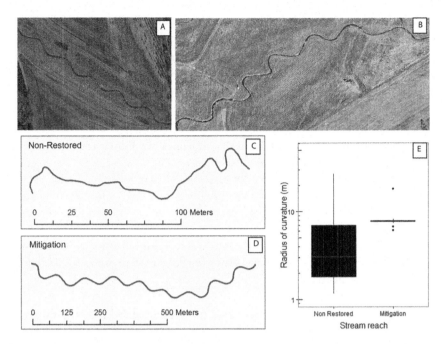

Fig. 21.7 (a and b) Examples of restored stream morphology (images from Google Earth; location and site name withheld). (c) Centerline of a nonrestored stream. (d) Centerline of a restored, mitigation stream. (e) Intrasite variability in radius of curvature for the nonrestored stream shown in (c) and the mitigation restored stream shown in (d)

lies to some extent in the limitations of AutoCAD (which makes it difficult to draw irregular curves) and in engineers' ingrained preference for symmetry. But in SMB, an even stronger impetus comes from the political economy of mitigation banking as a for-profit practice. In follow-up interviews, mitigation bankers consistently emphasized that mitigation projects in North Carolina are monitored for seven years. If a channel moves significantly during that time window, the Interagency Review Team certifying the bank will reduce the number of credits from the project, and fewer credits mean less profit. Sharp and irregular curves are more likely to move than smooth gradual curves (Odgaard 1987), so what we see on the landscape is gradual, sometimes disconcertingly symmetrical, curves with structural controls on the outsides of meander bends to more firmly hold them in place. This is notably inconsistent with the natural dynamics of rivers and streams. Meanders imply (somewhat tautologically) meandering, that is, movement. Yet the very movement that creates the channel form is considered via regulation as a sign of failure; in designing meanders to not move, the system has created artificially low, homogenous sinuosity streams—an artificial landscape bearing the signature of regulation rather than natural processes. In one regulator's words, "mitigation … in North Carolina is based on a lot of geomorphic measurements and otherwise to gain stability. Stability equals success …" (Author interview, July 12, 2010). Expanding on this blunt declaration, another regulator explained that:

> When we wrote these [stream mitigation guidelines] we didn't have good assessment tools, we didn't have good watershed assessment tools, we didn't have good stream assessment tools. And so a lot of decisions were made purely based on what the stream appeared to look like, stability-wise, … as a kind of proxy. And that can work in a lot of cases. So for example, you go out and look at a channel and it's deeply incised … so it's not connected with the floodplain … and it has lots of erosion … at least in the early days and still somewhat today is—people use that as a basis for saying, 'Wow, we can make that stream better.'" (Author interview, September 16, 2012)

From the designer's point of view, this means that stability is the target:

> So the bar in North Carolina pretty much up to current [times] … was that the channel … is going to be stable and for that monitoring period it's not going to move. There's going to be zero bank erosion, there's going to be no migration, there's going to be no incision. And so with those success criteria in mind I think there very well could be a tendency to design more conservatively …. It's how conservative you want to be is really the way I would phrase it, and where you've got success criteria and you or your client want—it really doesn't make a difference because it's your reputation either way—but when somebody is financially on the line for that system performing and somebody is going to have a scorecard in year five or seven when you did it or didn't do it, you know, that adds quite a bit of scrutiny to the process. (Author interview, July 11, 2012)

Conclusions

With these data, what can we say about the physical impacts of market-based environmental management? First, it is important to note than our initial hypotheses, based on the economic logic of stream credit definitions, were wrong. We found little increase in sinuosity, and that may actually have been preferable to what seems to have been a somewhat artificially low sinuosity on unrestored streams due to indirect and direct anthropogenic influences. We also found no over-siting of SMB projects on headwater streams proportionate to the overall presence of small drainage basins in the NC landscape.

What we did find were longer project lengths in streams restored for profit, the increased length resulting from the restoration of both main stem reaches and tributaries in mitigation but not in non-mitigation streams restoration projects. According to ecological theory, both of these things are more likely to produce a positive ecological impact (Bond and Lake 2003; Lake et al. 2007; Ardon et al. 2010 Beechie et al. 2010). This could suggest that SMB is actually better for the ecological health of fluvial systems than previous forms of non-market-based environmental management. There is a very substantive caveat here, however: to date, most types of stream restoration, including channel reconstruction, the type most commonly used in stream mitigation banking, have had limited success in delivering ecological improvement (Maron et al. 2012; Palmer and Filoso 2009; Bernhardt and Palmer 2011; Violin et al. 2011; Nilsson et al. 2015). Thus the ecological benefits of longer project lengths and tributary restoration are likely to be stronger in theory than in practice. If stream restoration is effective for generating ecological outcomes, then there are benefits to longer project lengths and tributary sites; but if restoration is ineffective, then these theoretically positive outcomes are unlikely to generate ecological lift.

Our most striking finding, however, was the homogenization of channel form in restored streams, regardless of why they were restored. At least in the case of stream mitigation banking, it seems that market-based environmental management is doing very little to restored streams that would not already be happening to them. Development that damages fluvial systems is continuing despite the explicit intention of the Clean Water Act to prevent harm, but this was already happening under the command-and-control regulatory paradigm that market-based environmental management aims to supplant. Further, while there are notable differences between stream restoration conducted for market and non-market purposes, it is not clear how much those differences matter given the overall uncertainty of restoration practice; the similarities, particularly the homogenization of channel sinuosity, seem more potentially consequential for fluvial health. Credit markets for streams thus appear to be intensifying trends that were already in place under an earlier implementation regime for the Clean Water Act rather than marking an inflection point that sends environmental conservation off in a sharply new direction.

Finally, we would also argue that this chapter demonstrates the utility of a CPG approach. There is no way to determine whether market-based environmental management is superior to command and control approaches to environmental policy without studying the former's physical and social impacts together, with an eye to political economic relations and the politics of knowledge production.

References

Ardon, M., J.L. Morse, M.W. Doyle, and E.S. Bernhardt. 2010. The water quality consequences of restoring wetland hydrology to a large agricultural watershed in the southeastern coastal plain. *Ecosystems* 13: 1060–1078.

Beechie, T.J., D.A. Sear, J.D. Olden, G.R. Press, J.M. Buffington, H. Moir, P. Roni, and M.M. Pollock. 2010. Process-based principles for restoring river ecosystems. *Bioscience* 60: 209–222.

Bernhardt, Emily, and Margaret A. Palmer. 2011. River restoration: The fuzzy logic of repairing reaches to reverse catchment scale degradation. *Ecological Applications* 21 (6): 1926–1931.

Bond, N.R., and P.S. Lake. 2003. Local habitat restoration in streams: Constraints on the effectiveness of restoration for stream biota. *Ecological Management & Restoration* 4: 193–198.

Corps [US Army Corps of Engineers] and EPA [US Environmental Protection Agency]. 2008. Compensatory mitigation for losses of aquatic resources. *Federal Register* 73: 19593–19705.

Costanza, Robert, Ralph d'Arge, Rudolf de Groot, Stephen Farber, Monica Grasso, Bruce Hannon, Karin Limburg, et al. 1997. The value of the world's ecosystem services and natural capital. *Nature* 387: 253–260.

Costanza, Robert, Rudolf de Groot, Paul Sutton, Sander van der Ploeg, Sharolyn Anderson, Ida Kubiszewski, Stephen Farber, and R. Kerry Turner. 2014. Changes in the global value of ecosystem services. *Global Environmental Change* 26: 152–158.

Cronon, William. 1991. *Nature's metropolis: Chicago and the great west.* New York: W.W. Norton & Company.

Daily, Gretchen C., ed. 1997. *Nature's services: Societal dependence on natural ecosystems.* Washington, DC: Island Press.

Dempsey, Jessica. 2016. *Enterprising nature: Economics, markets, and finance in global biodiversity politics.* London: Wiley.

Doyle, Martin W., and F. Douglas Shields. 2012. Compensatory mitigation for streams under the clean water act: Reassessing science and redirecting policy. *Journal of the American Water Resources Association* 48 (3): 494–509.

Doyle, M., Jai Singh, R. Lave, and M. Robertson. 2015. The morphology of streams restored for market and non-market purposes: Insights from a mixed natural-social science approach. *Water Resources Research* 51 (7): 5603–5622.

ECOLOGIC. 2006. *The use of market incentives to preserve biodiversity.* Brussels: DG Environment.

eftec, et al. 2010. *The use of market-based instruments for biodiversity protection: The case for habitat banking.* eftec: London.

European Commission. 2007. *Green paper on market-based instruments for environment and related policy purposes.* Brussels: European Commission.

Hough, Palmer, and Morgan M. Robertson. 2009. Mitigation under section 404 of the clean water act: Where it comes from, what it means. *Wetlands Ecology and Management* 17: 15–33.

Lake, P.S., N. Bond, and P. Reich. 2007. Linking ecological theory with stream restoration. *Freshwater Biology* 52: 597–615.

Lave, Rebecca. 2012. *Fields and streams: Stream restoration, neoliberalism, and the future of environmental science*. Athens: University of Georgia Press.

Lave, Rebecca, Martin W. Doyle, and Morgan M. Robertson. 2010. Privatizing stream restoration in the U.S. *Social Studies of Science* 40 (5): 677–703.

Lave, Rebecca, and Morgan Robertson. 2017. Biodiversity offsetting. In *Handbook of political economy of science*, ed. David Tyfield, Rebecca Lave, Samuel Randalls, and Charles Thorpe. London: Routledge.

Maron, Martine, Richard J. Hobbs, Atte Moilanen, Jeffrey W. Matthews, Kimberly Christie, Toby Gardner, David A. Keith, David B. Lindenmayer, and Clive A. McAlpine. 2012. Faustian bargains? Restoration realities in the context of biodiversity offset policies. *Biological Conservation* 155: 141–148.

McAfee, Kathleen. 1999. Selling nature to save it? Biodiversity and green developmentalism. *Environment and Planning D: Society and Space* 17 (2): 133–154.

Nilsson, Christer, Lina E. Polvi, Johanna Gardestrom, Eliza Maher Hasselquist, Lovisa Lind, and Judith M. Sarneel. 2015. Riparian and in-stream restoration of boreal streams and rivers: Success or failure? *Ecohydrology* 8 (5): 753–764.

Odgaard, A.J. 1987. Streambank erosion along two rivers in Iowa. *Water Resources Research* 23: 1225–1236.

Palmer, M.A., and S. Filoso. 2009. The restoration of ecosystems for environmental markets. *Science* 325 (5940): 575–576.

Rosgen, David L. 1996. *Applied river morphology*. 2nd ed. Pagosa Springs, CO: Wildland Hydrology.

Violin, Christy R., Peter Cada, Elizabeth Sudduth, Brooke A. Hassett, David L. Penrose, and Emily Bernhardt. 2011. Effects of urbanization and urban stream restoration on the physical and biological structure of stream ecosystems. *Ecological Applications* 21 (6): 1932–1949.

22

The Science and Politics of Water Quality

Javier Arce-Nazario

Introduction

Environmental justice examines how the distribution of environmental pro-
tections and risks relate to indicators of social power including race, ethnicity,
and economic status, and has become a popular conceptual framework
applied in many different disciplines from health to radical geography, political science, social science, the environmental sciences, history, and law.[1] The
environmental justice movement was launched to remedy the exclusion of
frequently disenfranchised groups in environmental decision-making: the
first discussions of environmental justice dealt with social movements and
inequities in exposure to industrial contamination. These early approaches,
termed by Walker (2009) as "first generation environmental justice research,"
mainly examined the frequency and proximity of environmental hazards to
disenfranchised groups. Subsequently, studies have shown that there are myriad environmental variables, methodologies, and theoretical frameworks that
can be used to observe an unequal distribution of environmental burdens or
benefits (Holifield et al. 2011).

It is easy to see the growing reach of environmental justice approaches as a
step towards reducing or removing inequalities. Certainly, it has provided a
more comprehensive toolkit of methods and theoretical frameworks to identify

J. Arce-Nazario (✉)
Department of Geography, University of North Carolina-Chapel Hill,
Chapel Hill, NC, USA

© The Author(s) 2018
R. Lave et al. (eds.), *The Palgrave Handbook of Critical Physical Geography*,
https://doi.org/10.1007/978-3-319-71461-5_22

465

problems of environmental justice, as well as useful models of how to organize collaboration among scientists, communities, and policymakers in order to resolve environmental justice problems (Martinez-Alier et al. 2016). The creation of an Office of Environmental Equity in 1992 within the United States Environmental Protection Agency (EPA) is representative of the degree to which the concept has transformed policy and institutions (EPA 1995). In turn, these institutions have shaped the concept of environmental justice and its use (Geltman et al. 2016). I argue in this chapter that the resulting policies can unintentionally reinforce inequity and injustice.

Specifically, in this chapter I explore how environmental justice has been conceptualized when applied to issues of drinking water quality, and the role of various institutions and agents in applying the concept. Non-compliance with environmental regulations plays a large role in the definition of environmental risk in relation to drinking water, and I describe some consequences of those choices. I argue that a Critical Physical Geography approach—specifically, including "critical attention to relations of social power" (Lave et al. 2014) as well as careful geographic methods that consider the inherent limitations of a selected scale in the evaluation of biological, chemical, and consumer-defined attributes of water—could help to define water quality without compromising the environmental justice goal of equal involvement in decision-making. I frame the discussion with a detailed history of environmental justice issues for household drinking water in small rural community-managed water systems in Puerto Rico.

Like most of the research in this volume, this study was shaped by my training and collaborations across disciplines, with training in qualitative social science and quantitative physical geographical and biological research methods reinforced by collaborations with microbiologists and community leaders in Puerto Rico. The data includes household surveys, informal interviews, oral histories, water sampling, and watershed analysis that my research group conducted in Puerto Rico between 2010 and 2015.

I have chosen to focus on the particular case of drinking water in Puerto Rico for several reasons. Importantly, issues of drinking water quality have not been frequently addressed from an environmental justice standpoint (VanDerslice 2011). This is especially surprising in light of the fact that water pollution was the most common concern among environmental justice groups in 1992 (Bryant 1995). The tragic case of lead contamination in Flint, MI (Hanna-Attisha et al. 2016) serves as a striking example of the fact that environmental risks in drinking water are not always equitably distributed.

It is also important to consider drinking water because it can challenge more standard methodologies in environmental justice research. In particular, the distribution, treatment, and consumption of drinking water involve important biophysical and social complexities that invite a Critical Physical Geography approach. Community-managed water systems in Puerto Rico provide an especially rich case study for this discussion, since household drinking water in Puerto Rico is governed by United States' regulations. Examining the superficially obvious case for environmental injustices in water systems where those regulations are not being followed will reveal a misreading of Puerto Rican ecology and demographics and indicate that the methodologies for identifying both environmental hazards and disenfranchised populations need updating in environmental justice research.

Moonshine Water: Rural Community Water Systems in Puerto Rico

Water systems in Puerto Rico can be classified by whether they are managed by the Puerto Rico Aqueduct and Sewer Authority (PRASA), or not, in which case they are collectively known as non-PRASA. Non-PRASA systems serve a minority of consumers in Puerto Rico: it is estimated that only 3% of the population in Puerto Rico receive water from non-PRASA systems (EPA 2001). Most of the non-PRASA systems are located in mountainous regions (Fig. 22.1) and are collectively owned and managed, although some are privately owned and managed, serving industrial consumers, hotels, or individual private homes. Rural non-PRASA systems are typically small. Among officially registered rural non-PRASA systems, the median number of families served is 340. But this number is almost certainly too high because the official registry of non-PRASA systems of the Puerto Rican Department of Health does not require systems with fewer than 15 connections or serving fewer than 25 people to be registered. A small number of larger non-PRASA systems with more than 15 connections are also covertly maintained without government oversight.

Non-PRASA systems exist in Puerto Rico for a variety of reasons. Many of these water systems were originally organized because the communities they serve were too far away to be conveniently connected to PRASA infrastructure. As the PRASA infrastructure has expanded, communities previously receiving non-PRASA water service have been connected to the PRASA system. Surprisingly, however, the PRASA system has not been uniformly embraced: some non-PRASA communities have decided not to connect to the PRASA system even when this would be relatively simple. Other communities receive water from both PRASA and non-PRASA systems, while in a few cases communities have made the transition from PRASA to non-PRASA systems to improve water service.

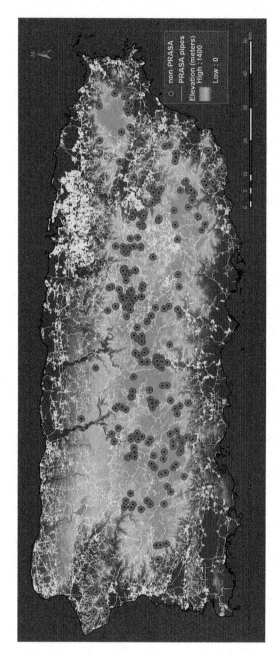

Fig. 22.1 Non-PRASA systems (red dots) and the PRASA major pipe infrastructure (yellow lines). Note that non-PRASA water systems are mostly located in the mountainous region (green areas in the map), and that many non-PRASA systems are contiguous to PRASA infrastructure. The precise location of non-PRASA points has been slightly altered to preserve the confidentiality of communities

Fig. 22.2 Examples of surface (A, B) and underground (C) rural water systems. In (A), a typical stream water-collection point, the community has constructed a small ditch and connected a PVC pipe. (B) is an example of a spring water collection system. The community constructs a box around a spring and the water is transferred to a collection tank. (C) is an example of a groundwater pump. Most non-PRASA systems transfer the water from the collection point to a tank where the water is chlorinated and distributed. However, some households within the community and some entire communities consume the water directly from the source

Small rural water systems may receive their source water from a stream, a spring, or from pumped underground water (Fig 22.2). The water source is a good predictor of many water quality characteristics, and partly determines the community management strategy and the applicable EPA regulations, derived from the Safe Drinking Water Act (SDWA).[2] For example, nearly all surface waters in Puerto Rico are required to be filtered, while groundwater is not. However, all water systems (large or small, surface or ground) are required to test for coliform presence. Among the tests mandated by EPA regulations, coliform tests are one of the most frequently performed: for example, a system serving fewer than 1000 consumers is required to perform coliform testing once per month, while it is required to perform lead testing only once per semester (Code of Federal Regulations, Coliform Sampling, title 40, sec. 141.21; Code of Federal Regulations, Monitoring Requirements for Lead and Copper in Tap Water, title 40, sec. 141.86). Most officially registered non-

PRASA systems serve fewer than 1000 people, so they are required to collect monthly data on the presence or absence of coliforms and also report frequent measurements of other water quality variables such as turbidity, free chlorine, and total chlorine. These data are used to determine which communities are in compliance with EPA regulations. Many non-PRASA communities persistently fail to comply because of missing tests, or because the tests detect the presence of *E. coli* or total coliforms. The level of coliforms in the source water determines whether or not systems must filter water, and since very few surface water non-PRASA systems have filtering capabilities, most are non-compliant. Other water quality measurements, which the regulations require to be performed more infrequently, are not measured in many non-PRASA systems. By contrast, most PRASA water systems complete all EPA required tests, and are in compliance with the coliform rule (EPA 2016a).

Environmental Justice Methods for Water Systems

Non-PRASA water systems have been identified as an environmental justice problem in both EPA and academic research. Researchers studying environmental justice across the United States and its territories often use compliance with the coliform rule or other contaminant rules as an indicator of risk (Balazs et al. 2011; 2012; Cory and Rahman 2009; Guerrero-Preston et al. 2008). Using regulatory compliance to define risk creates binary categories of safe and unsafe water, allowing researchers to make assumptions about whether or not an environmental risk is present without resorting to field-based techniques. The population exposed to risk is then defined in a straightforward way as the consumers of water from a non-compliant water system. Using regulatory non-compliance as a proxy for environmental risk, researchers have successfully identified systems that have exposed an overburdened population to contaminated water (Balazs et al. 2011; Hanna-Attisha et al. 2016). Besides facilitating data collection (compliance is a matter of public record), this definition may be especially useful because identifying non-compliance with the SDWA facilitates legal action on the matter (Ridley 2016).

Guerrero-Preston et al. (2008), in the first published academic work to address non-PRASA communities as an environmental justice problem, chose to use non-compliance as the measure of environmental hazard. The study investigated the possible characteristics of non-PRASA systems that might predict whether the system would be in compliance with regulations or not.

Using generalized linear models, the authors observed that small systems, surface water systems, and systems that had not installed treatment equipment had higher probabilities of non-compliance. The published model did not include demographic variables (such as income level or race) or spatial variables (rural or urban) but perhaps based on the demographic differences between the study site and the mainland United States, they determined that non-compliant non-PRASA systems should be classified as environmental justice communities.

In fact, demographic variables would not strengthen an argument for flagging non-PRASA systems as environmental justice concerns. The median household income of rural non-PRASA systems, estimated from census blocks including non-PRASA water systems is $15,085 (U.S. Census Bureau 2014). However, non-PRASA communities are not exceptionally poor relative to neighborhoods served by PRASA systems. The median household income for neighboring census blocks entirely served by PRASA water systems is $15,946; the median for all of Puerto Rico is $19,686. While the census block is not sufficiently granular to completely separate PRASA and non-PRASA consumers, these trends are corroborated by our household survey data, which shows that the median income of non-PRASA communities is statistically similar to that of neighboring PRASA communities.

Usually environmental justice studies of water quality analyze populations by water system first, instead of using other variables more conveniently linked to demographics (such as neighborhood) (Balazs et al. 2012; Cory and Rahman 2009). This is natural when risks are measured at the water system level, using a metric such as compliance or contaminant concentration at the treatment plant. However, it complicates the understanding of the characteristics of affected populations. The spatial distribution of water systems can be convoluted and so the origins of treated water are not always obvious, and water from different sources and water management systems can be consumed in nearby households. For example, many aging systems in the United States do not have up-to-date maps of the pipe system, making it difficult for a household to find out the specific source of its water (AWWA 2012; Uslu et al. 2016). In Puerto Rico, neighboring households can receive water from either the central authority or from community-managed systems, and the watersheds serving neighboring households in the metropolitan area can even be on opposite sides of the island. For those water systems managed by the central authority, local utility workers' knowledge of the pipe system at fine scales is gained through oral tradition or trial and error rather than a map (Rivera-Arguinzoni 2015). Perhaps because of these complications, in Puerto Rico the water system classification itself (PRASA or non-PRASA) is used to describe the populations affected by risk disparities.

The EPA has treated all non-PRASA systems as an environmental justice issue in its environmental justice reports and plans since the 1990s (EPA 1995). Like Guerrero-Preston et al. (2008), the EPA identifies non-PRASA communities as victims of environmental injustice based solely on the fact that they have high rates of non-compliance with the SDWA. The proposed solutions to the environmental justice issue in these reports are to help the communities to comply with regulations by changing their water management techniques (e.g., helping them to implement filtration or chlorination) or their water source. For surface water systems, the suggestion is often to change to groundwater or simply to connect the community to a PRASA water system. The local EPA, the Department of Health, local engineering groups, and academic groups have collaborated to provide the technical expertise to promote compliance for non-PRASA communities (EPA 2001, 2016b).

Reassessing the Case for Environmental Injustice

The preceding discussion leaves one very important question unanswered: is the fact that many non-PRASA systems are not compliant with the SDWA actually an environmental justice problem? The stated goals of environmental justice initiatives, including that of the EPA, are to make access to environmental goods and evils independent of factors such as income level, ethnicity, and national origin. Environmental justice studies of drinking water in Puerto Rico consistently describe non-PRASA communities as rural or poor, but do not use a comparative approach to explicitly establish the relationship between risk and income, and do not provide any other demographic characteristics of these communities. In studies conducted outside of Puerto Rico, census demographics are typically used to determine if the environmental risk represents an environmental justice issue. For example, exposure of populations to arsenic through drinking water has been analyzed in Arizona (Cory and Rahman 2009) and California (Balazs et al. 2012). In both studies, the researchers first analyzed whether a water system had arsenic above or below the established EPA standard of 10 µg/L. Variables related to social and racial composition from census data were then introduced to determine if the non-compliance outcome was correlated with any of these variables. The California study determined that consumers of water from non-compliant systems were positively correlated with not owning a house and not being White, while in the Arizona case no statistically significant correlations were found. The California case was thus interpreted as an environmental justice issue, and the Arizona case was not.

PRASA and non-PRASA communities in rural, mountainous areas have very similar income profiles, but to even explore an argument that non-PRASA communities suffer environmental injustice, it should be established that the non-PRASA community members receive riskier water than PRASA consumers. This is certainly implied by the difference in PRASA and non-PRASA system compliance rates. However, it is interesting to note that some PRASA systems have also had persistent problems with SDWA compliance (Hunter and Arbona 1995; EPA 2016a) and recorded cases of illness from drinking PRASA water, and yet PRASA communities and the subgroups within them have not been evaluated for environmental injustice impacts related to water quality.

More strikingly, there is no current field evidence that non-PRASA water consumers face a greater health risk from their water than PRASA system users, since non-pathogenic *E. coli* is not itself a risk, but rather an indicator of risk.[3] Defining "a healthy environment" in the case of drinking water as "compliant with federal regulations" is not surprising in research by a federal agency, but this definition deserves some examination. The criteria to determine water compliance is based on expert scientific advisory boards and other politically appointed entities who set the standard of compliance by determining the list of water quality variables to be monitored, acceptable or appropriate ranges for each monitored variable, and how often each variable needs to be measured in the water system. The expert group that makes decisions about the accepted values of water quality must therefore generalize how water systems work over a diverse landscape of both biophysical and social variations. Applying a general rule to a varied eco-social landscape means these determinations can systematically indicate risk where none is present while missing existing hazards elsewhere. Current standards for water quality are universal for the United States and its territories and thus should not be expected to fully characterize safe and unsafe water anywhere.

In the case of non-PRASA systems in Puerto Rico, the SWDA compliance standard is particularly likely to over-predict some risks in drinking water. Our study of 602 adults in 15 non-PRASA and 15 adjacent PRASA communities demonstrated that cases of self-reported gastrointestinal illness were not statistically higher in non-PRASA communities (reported by 37% of adults) than in PRASA communities (reported by 43% of adults). In fact, while the Puerto Rico Department of Health currently has no record of epidemics related to water consumption from non-PRASA waters, there are records of such epidemics in PRASA communities (Casado-Cruz 2007; Departamento de Salud de Puerto Rico 1999). Thus while the presence of coliforms in surface water systems is usually referenced as the health risk in non-compliant non-PRASA systems (EPA 2016b), this coliform presence is not obviously making consumers sick. The reason for this apparent discrepancy is climate and geography: environmental microbiological research since the 1980s has shown that unlike temperate waters in the mainland United States, Puerto Rico's pristine surface waters have naturally present non-pathogenic *Escherichia coli,* uncorrelated with fecal contamination (Rivera et al. 1988). A water sample of almost any stream in Puerto Rico will show the presence of coliform bacteria from non-fecal sources (Santiago-Rodriguez et al. 2016).

Emphasizing compliance with federal regulations for coliform management can also lead to underestimating risk. One ground water-supplied non-PRASA community was found to be highly contaminated with tetrachloroethylene (PCE), an industrial compound used as a solvent for organic materials (EPA 2012). While this system was partially compliant with regulations for coliform bacteria since it had an underground source and a chlorinator, the community was at greater risk of exposure to other kinds of contamination as a result of the land cover and land uses of the watershed and the activities within it. Better tools than public databases would typically be needed to capture these kinds of environmental justice issues. A more explicitly geographical approach could be more sensitive to these risks, since studies in Puerto Rico show that watershed composition predicts levels of nitrates, coliform bacteria, phosphorus, and turbidity (Uriarte et al. 2011). To date, no systematic studies in Puerto Rico have been performed to assess environmental justice in the distribution of water contamination risks correlated with watershed land-use.

Equity or Autonomy?

The effort to bring environmental justice to non-PRASA consumers in Puerto Rico has had mixed effects. Under the EPA 1996 non-PRASA action plan, the Department of Health encourages the conversion of non-PRASA surface water systems to groundwater systems in order to eliminate the requirement of filtration. This conversion improves compliance rates, and thus improves water quality according to the simplistic definition used in most studies. However, supplying underground water is typically most costly and requires more infrastructure. The communities have to dig a well, install a pump, and pay the continuing costs of its electricity and maintenance. The other common solution, to connect to PRASA-managed infrastructure, also leads to higher consumer costs, including a subscription fee and consumption-based usage fees.

Expense is not the only concern for non-PRASA communities. The elimination of surface non-PRASA by conversion to PRASA or groundwater systems also erodes the resilience and autonomy of community managed water systems. By connecting to large PRASA systems, consumers have less control over management of water service interruptions: non-PRASA community members note the fact that after hurricanes, the community can quickly respond to repair any damaged infrastructure, and loss of electricity does not imply loss of water for surface-supplied water systems and underground water systems with access to a generator.

During hurricanes, we would have water but no electricity. Other people from nearby communities would come to collect water. We would especially supply water to people that we understood were in great need of water, such as elderly people. Even though they (PRASA) had "aqueducts," they did not have the type of system that we had. (H. Martínez 2015, Interview with a non-PRASA community leader)

Availability during extreme climate events, water pressure, and even the taste of the water are among the reasons non-PRASA community members have given for preferring their surface water systems to the alternatives offered.

There are less immediate consequences as well, since the attitude of water consumers towards their water and their communities can also be shaped by these policies. Because the argument for environmental risk is heavily based on the presence of coliform bacteria, local officials and academics studying non-PRASA water who find positive results for coliform presence inaccurately inform community members that their source water or tap water has fecal contamination, thus creating an association between non-PRASA water and a health risk where none necessarily exists. The stigma thus associated with community-managed surface water systems due to these policies is in strong contrast to the pride of stewardship that non-PRASA system managers and consumers express in interviews:

We know how to deal with and administer our own water system. Why should we accept something that does not work, or experiment with something else, when we have something that has worked for 50 years? We have a working system in this community, something that they (the Aqueducts and Sewer Authority) have not been able to do. (H. Martínez 2015, Interview with a non-PRASA community leader)

The communal aspect of the resource management strengthens community bonds—in some cases, the source of the non-PRASA water is a community-gathering place.

> *The day that the community goes to clean the reservoir is something beautiful. You eat breakfast together and make jokes.* (Anonymous 2015, Interview with a non-PRASA community member)

Understanding the precise source of their drinking water, and the effect that watershed land-use decisions have on their daily lives, has even made PRASA community members more active in conserving natural areas near their watersheds. Non-PRASA communities have successfully fought the construction of waste sites supported by municipal authorities (Santiago 2010), and deterred the plans of individuals clearing forests for farms or houses.

Bottom Up: Revisiting Methods in Environmental Equity Research

While compliance methods provide an apparently objective view of the distribution of risk, they draw definitions of environmental risk from a narrow perspective. A subtler drawback of these approaches is also apparent in water research, where compliance methods often impose the scale of the water system both for identifying variations in water quality and for the demographic analysis of affected populations, leaving important questions unanswered: when a water system is non-compliant, are all demographic groups served by the water system equally affected? When it is compliant, is the water quality uniform throughout the system? As I elaborate below, asking these questions and refining our approaches is important for more comprehensive and more accurate identification of disparities in water quality and access.

Where Compliance Falls Short

Notable cases of disparities in water quality have occurred *within* water systems, which may be harder to detect using compliance-based methods. Contamination as a result of an interaction between water treatment decisions and the nature of the water distribution system have led to an unequal distribution of lead contamination risk (Hanna-Attisha et al. 2016; Edwards et al. 2009; Miranda et al. 2007). System-wide data collection regulations do include the results of household testing in order to capture hazards introduced in the distribution system, but in using these data the research design is constrained to the choices made by the agency or institution overseeing the water system. This limits the choice of an appropriate scale of analysis and data collection methods. In Flint, collaboration between the community and scientists was required to provide data more sensitive to lead contamination than the water system's schedule of tests (Itkowitz 2016).

The specific targeting of non-PRASA systems for existing environmental justice research in Puerto Rico also highlights the fact that non-compliance is a definition of hazard that cannot effectively challenge the institutions monitoring those hazards. When risk is defined as failure to meet an institutionally defined standard, the institutions become more important but remain unexamined: the conversion of non-PRASA systems to PRASA systems or systems with reduced water treatment requirements does not guarantee safer water, but nevertheless counts as progress towards environmental justice. Creating and resolving compliance-based non-PRASA environmental justice cases in Puerto Rico may even serve as a distraction from other injustices that are less visible using the compliance methodology, such as possible inequities in water availability during extreme climate events within compliant, government-managed water systems.

Finally, regulatory compliance can exclude many human and ecological dimensions of environmental service that have been important in environmental justice movements. For example, the idea that environmental justice entails the "political, economic, cultural and environmental self-determination of all people" (FNPCELS 1991) is not evident in the approach to environmental justice for non-PRASA systems. In general, the compliance approach cannot capture how people interact with their environment. In the specific case of tap water, this research methodology excludes other social valuations of water quality, the inherent value of a water system to a community, and people's relationship to watersheds and water sources. By not balancing community interests with the scientific evaluation, environmental justice research on water systems risks promoting policies that exclude community agency, and unintentionally replicating the unjust policies of contamination it aims to dismantle.

Using Critical Physical Geography for Environmental Equity

What would a practical, critical geography approach to water quality injustice look like? Recognizing that social and biophysical effects combine at the water source, through the constructed distribution systems, and in the health and satisfaction of consumers, a Critical Physical Geography approach would suggest a more geographic, multi-scale method in defining good drinking water. Such a framework would also question the institutional definitions in play. Since involving the water consumers is a practical way to balance the role of water regulations and water management agencies in the definition of equity and risk, a first step towards a critical multi-scale approach would be to design research that incorporates measurements at the household scale. Generating this fine scale data could be costly but would yield improved sensitivity to disparities and risks introduced or mitigated after the water treatment plant and would be less constrained in how demographic factors were analyzed. Incorporating social science techniques and household sampling in the field research would also provide an opportunity to involve water consumers in generating the data, leading to more stakeholder involvement in the results, a potentially different understanding of water quality, and better relationships between academia and the affected community (Kolowich 2016).

A multi-scale geographical approach also suggests undertaking an analysis at the watershed scale. Physical geographers have tackled the problem of intersecting physical, ecological, and cultural drivers of processes in a watershed (Rhoads et al. 1999). Like household interviews or in-home sampling, watershed-level research would have its complications: source water can be contaminated by agricultural, industrial or urban activities in a watershed, and contamination may vary with time as a result of changes in weather such as precipitation or temperature, or due to other anthropogenic factors which can be seasonal or otherwise time-dependent (Santiago-Rodriguez et al. 2016). The ultimate effect on consumers will depend on the current regulations and treatment policies for the contaminant (Richardson and Kimura 2015; Edwards et al. 2009). Incorporating methodologies that overcome these complications to build up a coherent description of water risks from watershed processes could eventually encourage a more holistic approach to solving water quality issues: while the current disciplinary approach of environmental engineers and epidemiologists is to manage the constructed system and not the watershed, a greater awareness of correlations between land-use and water quality by academics, policymakers, and household consumers could promote conservation.

Conclusion

Critical geographers working with issues of environmental justice have called out the limitations of focusing on the simplified view of environmental justice as a question of the distance between a hazard and a disenfranchised group

(Walker 2009). Critical Physical Geography can go further, clarifying that the physical science techniques applied to identify and quantify environmental hazards are shaped by a frequently overlooked power dynamic, and the interplay of political, biophysical, and spatial influences cannot be ignored when assessing and mitigating the hazard or even when characterizing the impacted group. In the case of water, the simple lens of chemistry or compliance correlated to census data provides a dangerously restrictive view of justice issues, which can be remedied by a synthetic revision of both the positivist metrics applied to identify hazard and the social construction of the populations at risk. The case study analyzed here illustrates how the use of compliance with regulations for coliform bacteria as a measure of water safety and the unquestioned assumption that consumers of community-managed water are a distinct demographic group from their neighbors both weaken the case for environmental injustice in non-PRASA systems. This weakness is somewhat masked by the fact that the conclusion does not challenge any prevailing assumptions: viewed by the standards of the mainland United States , nearly all of Puerto Rico's predominantly Hispanic, Black, and poor population is a non-dominant minority, and nearly all of its surface waters, where coliform bacteria can thrive and reproduce, are contaminated. However, this case study also points to ways to improve the analysis of environmental justice in water quality: involving both the finer scale of the household and the larger scale of the watershed could provide more salient data, while also integrating human and ecosystem elements of robust solutions for environmental injustice. The steps sketched here clearly require refinement by practitioners before they are implemented coherently. Thus, Critical Physical Geography can play a key role in the future of environmental justice by helping to fully specify methodologies that serve the original goals of the environmental justice movement.[4]

Notes

1. A brief review of books, articles, and special issues on environmental justice from 1990 to 2008 can be found in Holifield, Porter and Walker (2011), and between 2009 and 2016 additional works have appeared. Examples of the latter include the International Journal of Environmental Research and Public Health (Griffiths 2011; Chakraborty et al. 2016) and The Geographical Journal (Martin 2013). Since 2008, the Journal of Environmental Justice has published work on global issues of environmental justice.

2. The SDWA is the federal law that establishes the standards for managing public water system for different sizes and water sources, except private wells. The Environmental Protection Agency (EPA) defines a set of water quality standards, and together with state agencies enforces these standards.

3. In the mainland United States, *E. coli* has been useful in identifying fecally contaminated water because pathogenic and non-pathogenic *E. coli* and many other pathogenic microorganisms have similar sources and thrive in similar environments. In contrast, especially in the tropics, non-pathogenic *E. coli* can be endemic, rather than from a fecal source. Its presence is common in pristine water sources in Puerto Rico.

4. This work was supported in part by the National Science Foundation [grant #1151458] and the the National Institutes of Health [grant #P20 MD0006144]. Many thanks to the community members of non-PRASA systems for allowing us to visit their homes and water systems. Any opinions, findings, and conclusions or recommendations expressed in this chapter are those of the author and do not necessarily reflect the views of any of the granting agencies.

References

American Water Works Association. 2012. *Buried no longer: Confronting America's water infrastructure challenge.* Denver, CO: AWWA.

Balazs, Carolina, Rachel Morello-Frosch, Alan Hubbard, and Isha Ray. 2011. Social disparities in nitrate-contaminated drinking water in California's San Joaquin Valley. *Environmental Health Perspectives* 119 (9): 1272.

Balazs, Carolina L., Rachel Morello-Frosch, Alan E. Hubbard, and Isha Ray. 2012. Environmental justice implications of arsenic contamination in California's San Joaquin Valley: A cross-sectional, cluster-design examining exposure and compliance in community drinking water systems. *Environmental Health* 11: 84.

Bryant, Bunyan I. 1995. *Environmental justice: Issues, policies, and solutions.* Washington, DC: Island Press.

Casado-Cruz, Jorge. 2007. *Aplicación de las enmiendas de La Ley de Agua Limpia Segura a Sistemas de la Autoridad Acueductos y Alcantarillados en Puerto Rico.* Caguas, PR: Universidad del Turabo.

Chakraborty, Jayajit, Sara E. Grineski, and Timothy W. Collins, eds. 2016. Environmental justice research: Contemporary issues and emerging topics. Special issue. *International Journal of Environmental Research and Public Health* 13 (17).

Coliform Sampling. 2015. *Code of federal regulations, title 40, section 141.21 (2015).* https://www.gpo.gov/fdsys/pkg/CFR-2015-title40-vol23/pdf/CFR-2015-title40-vol23-sec141-21.pdf

Cory, Dennis C., and Tauhidur Rahman. 2009. Environmental justice and enforcement of the Safe Drinking Water Act: The Arizona Arsenic experience. *Ecological Economics* 68 (6): 1825–1837. https://doi.org/10.1016/j.ecolecon.2008.12.010.

Departamento de Salud de Puerto Rico. 1999. *Reporte de brotes de enfermedades aso-ciadas al agua de Puerto Rico 1976–1998*. San Juan, Puerto Rico: Programa Agua Potable, Departamento de Salud de Puerto Rico.

Edwards, Marc, Simoni Triantafyllidou, and Dana Best. 2009. Elevated blood lead in young children due to lead-contaminated drinking water: Washington, DC, 2001–2004. *Environmental Science & Technology* 43 (5): 1618–1623.

EPA. 1995. *Environmental justice 1994 annual report: Focusing on environmental protection for all*. EPA-200-R-95-003, US Environmental Protection Agency.

———. 2001. *State strategies to assist public water systems in acquiring and maintaining technical, managerial, and financial capacity comprehensive summary of state responses to section 1420(c) of the safe drinking water act*. EPA 816-R-01-019.

———. 2012. *EPA adds Corozal well site in Corozal, Puerto Rico to the superfund list*. 12-036. US Environmental Protection Agency. https://yosemite.epa.gov/opa/admpress.nsf/d0cf6618525a9efb85257359003fb69d/7346cebd7aed2918852579c0005f8dcb!OpenDocument

———. 2016b. *Drinking water: EPA needs to take additional steps to ensure small community water systems designated as serious violators achieve compliance*. 16-P-0108. US Environmental Protection Agency. https://www.epa.gov/sites/production/files/2016-03/documents/20160322-16-p-0108.pdf

———. 2016a. *Enforcement and compliance history online*. June 1. https://echo.epa.gov/?redirect=echo

FNPCELS. 1991. *Principles of environmental justice*. http://www.ejnet.org/ej/principles.html

Geltman, Elizabeth Glass, Gunwant Gill, and Miriam Jovanovic. 2016. Beyond baby steps: An empirical study of the impact of environmental justice executive order 12898. *Family & Community Health* 39 (3): 143–150. https://doi.org/10.1097/FCH.0000000000000113.

Griffiths, Charles W., ed. 2011. Advances in environmental justice. Special issue. *International Journal of Environmental Research and Public Health* 8 (6).

Guerrero-Preston, Rafael, José Norat, Mario Rodríguez, Lydia Santiago, and Erick Suárez. 2008. Determinants of compliance with drinking water standards in rural Puerto Rico between 1996 and 2000: A multilevel approach. *Puerto Rico Health Sciences Journal* 27 (3). http://prhsj.rcm.upr.edu/index.php/prhsj/article/view/74

Hanna-Attisha, Mona, Jenny LaChance, Richard Casey Sadler, and Allison Champney Schnepp. 2016. Elevated blood lead levels in children associated with the flint drinking water crisis: A spatial analysis of risk and public health response. *American Journal of Public Health* 106 (2): 283–290.

Holifield, Ryan, Michael Porter, and Gordon Walker. 2011. *Spaces of environmental justice*. Vol. 25. New York, NY: John Wiley & Sons.

Hunter, John M., and Sonia I. Arbona. 1995. Paradise lost: An introduction to the geography of water pollution in Puerto Rico. *Social Science & Medicine* 40 (10): 1331–1355.

Itkowitz, Colby. 2016. The heroic professor who helped uncover the Flint lead water crisis has been asked to fix it. *Washington Post*, January 27. https://www.washingtonpost.com/news/inspired-life/wp/2016/01/26/meet-the-heroic-professor-who-helped-uncover-the-flint-lead-water-crisis/WashingtonPost

Kolowich, Steve. 2016. The water next time: Professor who helped expose crisis in Flint says public science is broken. *The Chronicle of Higher Education*, February 2. http://chronicle.com/article/The-Water-Next-Time-Professor/235136

Lave, Rebecca, Matthew W. Wilson, Elizabeth S. Barron, Christine Biermann, Mark A. Carey, Chris S. Duvall, Leigh Johnson, et al. 2014. Intervention: Critical Physical Geography. *The Canadian Geographer/Le Géographe Canadien* 58 (1): 1–10.

Martin, Adrian. ed. 2013. Global environmental in/justice, in practice special issue. *The Geographical Journal* 179 (2).

Martinez-Alier, Joan, Leah Temper, Daniela Del Bene, and Arnim Scheidel. 2016. Is there a global environmental justice movement? *The Journal of Peasant Studies* 43 (3): 731–755. https://doi.org/10.1080/03066150.2016.1141198

Miranda, Marie Lynn, Dohyeong Kim, Andrew P. Hull, Christopher J. Paul, and M. Alicia Overstreet Galeano. 2007. Changes in blood lead levels associated with use of chloramines in water treatment systems. *Environmental Health Perspectives* 115 (2): 221–225.

Monitoring Requirements for Lead and Copper in Tap Water. 2015. *Code of federal regulations, title 40, section 141.86 (2011)*. https://www.gpo.gov/fdsys/pkg/CFR-2011-title40-vol23/pdf/CFR-2011-title40-vol23-sec141-86.pdf

Rhoads, Bruce L., David Wilson, Michael Urban, and Edwin E. Herricks. 1999. Interaction between scientists and nonscientists in community-based watershed management: Emergence of the concept of stream naturalization. *Environmental Management* 24 (3): 297–308.

Richardson, Susan D., and Susana Y. Kimura. 2015. Water analysis: Emerging contaminants and current issues. *Analytical Chemistry* 88 (1): 546–582.

Ridley, Gary. 2016. Gov. Rick Snyder target of RICO lawsuit over Flint water crisis. *MLive.com*, April 6. http://www.mlive.com/news/flint/index.ssf/2016/04/gov_rick_snyder_target_of_rico.html

Rivera, Susan C., Terry C. Hazen, and Gary A. Toranzos. 1988. Isolation of fecal coliforms from pristine sites in a tropical rain forest. *Applied and Environmental Microbiology* 54 (2): 513–517.

Rivera-Arguinzoni, Aurora. 2015. AAA pierde lo que invierte. *El Nuevo Dia*, August 5. http://www.elnuevodia.com/noticias/locales/nota/aaapierdeloqueinvierte-2082494/

Santiago, Helen. 2010. Acueducto Rural Guacio: Algo más que un acueducto. *Maguey*, September.

Santiago-Rodriguez, Tasha M., Gary A. Toranzos, and Javier A. Arce-Nazario. 2016. Assessing the microbial quality of a tropical watershed with an urbanization gradient using traditional and alternate fecal indicators. *Journal of Water and Health* 14 (5): 796–807.

Uriarte, María, Charles B. Yackulic, Yili Lim, and Javier A. Arce-Nazario. 2011. Influence of land use on water quality in a tropical landscape: A multi-scale analysis. *Landscape Ecology* 26 (8): 1151–1164.

US Census Bureau. 2014. 2014 American Community Survey.

Uslu, Berk, Yeun J. Jung, and Sunil K. Sinha. 2016. Underground utility locating technologies for condition assessment and renewal engineering of water pipeline infrastructure systems. *Journal of Pipeline Systems Engineering and Practice*, 7 (4) 04016011.

VanDerslice, James. 2011. Drinking water infrastructure and environmental disparities: Evidence and methodological considerations. *American Journal of Public Health* 101 (Suppl 1): S109–S114. https://doi.org/10.2105/AJPH.2011.300189.

Walker, Gordon. 2009. Beyond distribution and proximity: Exploring the multiple spatialities of environmental justice. *Antipode* 41 (4): 614–636. https://doi.org/10.1111/j.1467-8330.2009.00691.x.

23

Transforming Toronto's Rivers: A Socio-Geomorphic Perspective

Peter Ashmore

Introduction

The channel of Highland Creek in Toronto has undergone dramatic morphological change. The narrow, sinuous stream of the 1950s has been transformed into a channel that in some places is 20–30 m wide, with resistant banks of armour stone, constructed pools and riffles, and highly regular constructed meanders (Fig. 23.1). How do we understand the morphological transformation of this particular river and of urban rivers more generally, and in what larger contexts? Contemporary fluvial geomorphology is presented with a challenge when confronted by the physically functioning, but largely human-designed, morphology of urban rivers. Should we say that they are not "natural" and therefore not amenable to analysis using established universal geo-scientific principles and relations and consequently not admissible as objects of study? Should we understand them as a separate nature impacted and degraded by human action? Or should we see them as part of a socio-fluvial landscape and seek to understand how they function and how they came to have the form that they do as socio-natural systems (Ashmore 2015; Urban, this volume)? Adopting the latter position leads us into intriguing issues around the forces, ideas, and knowledge systems that shape urban rivers

P. Ashmore (✉)

Department of Geography, University of Western Ontario, London, ON, Canada

© The Author(s) 2018

R. Lave et al. (eds.), *The Palgrave Handbook of Critical Physical Geography*,
https://doi.org/10.1007/978-3-319-71461-5_23

with the potential to expand the current epistemological norms of geomorphology and bring these designed systems into the scope of fluvial geomorphic enquiry. Previous work on river restoration (e.g. Eden et al. 2000) has recognized the socio-natural hybrid nature of restored rivers but the focus is on environmental decision-making rather than on understanding changing river morphology in the context of physical geography.

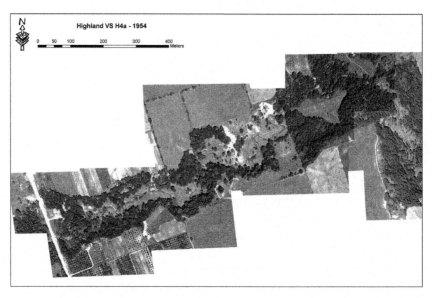

Fig. 23.1 (a–c) Aerial images of a reach of Highland Creek in 1954, 1999, 2015. Note the transformation of morphology from narrow highly sinuous channel (1954), through wider and less sinuous channel (1999), to wide, engineered meanders (2015). Source: City of Toronto ESM Web and Geospatial Competency Centre (https://web.toronto.ca/city-government/data-reports-maps/) and Triathlon Inc. Examples of urbanization effects: (d) Failed drainage infrastructure from erosion by large floods. (e) Deposition of mid-channel bar built from failed concrete channel lining. (f) Morphological transformation of Highland Creek by geomorphological engineering. Ground photos in the same reach as that depicted in aerial images in (a)–(c). 2006 image shows river affected by urban storm discharges but prior to major re-engineering. 2012 shows re-engineered channel and valley. Note change in channel dimensions, materials, and constructed "floodplain" on the left side of the 2012 image. (Photos (d)–(f): P. Ashmore)

Fig. 23.1 (Continued)

Fig. 23.1   (Continued)

Fig. 23.1   (Continued)

Urban rivers can, like all rivers, be analysed as physical systems with morphology related to watershed topography, geology, and hydro-climate. Many urban rivers are designed and engineered, and their watersheds have had extensive urban development. Consequently, the current state of many urban rivers can only be fully explained by understanding the combined physical system and the related sociopolitical events, processes, institutions, and actors, along with the nature of the scientific knowledge that has been used to manage and transform them (Ashmore 2015). The local contingencies and sequence of particular natural and sociopolitical events takes these rivers along particular paths of transformation influenced by larger processes and broader imperatives, such as global knowledge systems of fluvial geomorphology (Eden et al. 2000).

A CPG framework offers a way of thinking through these kinds of transformations with attention on the material biophysical aspects of rivers as socionatures. In this respect, CPG is a departure from previous analyses of river restoration (Eden et al. 2000) and socio-natures as primarily objects of political ecological analysis. Here I use of the example of the watersheds of (sub)urban Toronto from the 1950s to present and the specific contingencies of this case to exemplify the approach. A combination of natural events and conditions, institutional attitudes and projects, development practices, political influence, community action, and scientific intervention have all interacted to shape the trajectories of change and the current morphology of the rivers. Based on this case, I conclude with some ideas about the connections to, and implications for, developing socio-geomorphic and CPG analyses of urban and other rivers, and broader implications for understanding landscape design.

Toronto Watersheds: Geomorphology, Hydrology, and Change

While fluvial geomorphology has established universal generalizations about river forms and processes, the recognition of the role of contingency of particular locations and histories of particular fluvial landscapes and hydro-geomorphic processes also has a long-standing place in geomorphology (Simpson 1963; Schumm 1991; Lane and Richards 1997; Phillips 2007). Local contingencies affect the fluvial landscape and the interaction with human development of the landscape. The location of the City of Toronto on the north shore of Lake Ontario is one such contingency and the glacial landforms of the region are another, exerting a strong influence on current fluvial morphology (Phillips and Desloges 2014, 2015) (Fig. 23.2). The proximity of

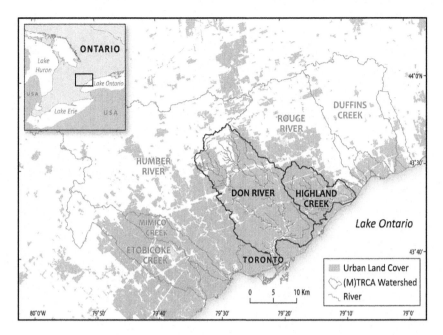

Fig. 23.2 Map of watersheds, major rivers, urban development and TRCA jurisdictions in Toronto region

the drainage divide of the Oak Ridges Moraine north of the City and the lakeshore to the south constrains the extent of the watersheds in the region. The watersheds are relatively small and consequently extensive urbanization expanding northward from the lakeshore into the headwaters of the rivers beginning in the late 1950s led to urban development covering almost the entire area of most watersheds (Fig. 23.2).

The physical characteristics of the rivers of the Toronto region are unusual. The headwaters are low-gradient, low-energy channels that steepen downstream and become progressively entrenched and confined in narrow valleys that are eroded into thick glacial deposits and are separated by flat inter-fluvial tablelands. In places, the river gradients approach those of rivers more typical of gravelly rivers in mountain regions (Vocal-Ferencevic and Ashmore 2012). Many of the channels have "semi-alluvial" characteristics (Ashmore and Church 2001; Phillips and Desloges 2014). They are eroded into cohesive, but highly erodible, glacial sediments but also have alluvial deposits in the channels and floodplains (Fig. 23.3). The erodibility of the cohesive glacial sediments makes the rivers as adjustable morphologically as fully alluvial rivers, unlike partially alluvial rivers in bedrock (Meshkova et al. 2012). An important consequence is that conventional understanding of alluvial river morphology and

Fig. 23.3 Example of semi-alluvial channel along Highland Creek. Bed and lower banks are exposed glacial clay, in-channel bar and upper bank are alluvial gravel and sand

mechanics, and the nature of morphological adjustment and mitigation, has to be modified in analysing these river systems and therefore a combination of both global and provisional local knowledge of these types of rivers has been important in the management and transformation of the rivers.

Development of the rivers of Toronto since the early 1950s is tied to rapid urban expansion across the watersheds. As an indirect consequence of the particular path and form of development, and of choices related to urban drainage, river discharge was, in a sense, inadvertently "redesigned" through the consequences of urban stormwater runoff processes (Trudeau and Richardson 2015). Recent application of stormwater retention and low impact development to mitigate these effects has led to further redesigning of the hydrology of the systems, but the effects have been minor. The response of fluvial hydrology to urbanization, while being a physical response to changes in surface hydrology, is also institutionally determined by planning and development regulations and norms, and varies historically and geographically because of those influences. The changes in river discharges were dramatic in some cases. Seasonality of flows was radically altered, peak flows increased by as much as five fold, and much of this occurred over a decade or

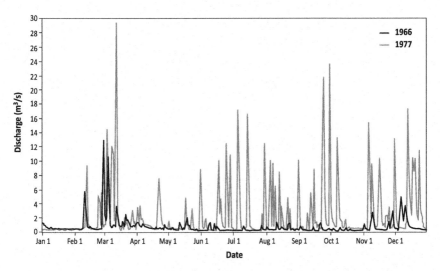

Fig. 23.4 Changes in river discharge and annual hydrograph following rapid urbanization, Highland Creek: comparison of 1966 and 1977

less (Vocal-Ferencevic and Ashmore 2012). The example of differences in annual hydrograph for Highland Creek in 1966 compared to 1977 (Fig. 23.4) is striking but is not unique among Toronto watersheds. Large floods that followed this hydrological change (e.g. in the mid-1970s and 1980s), and that were perhaps not anticipated in the initial phase of channel engineering in the 1950s and 1960s, were major drivers of the management response and priorities, and of the river transformations that followed. Interventions and changes that had begun following a singular catastrophic flood event associated with Hurricane Hazel in 1954 (see Fig. 23.4) were thus extended over a long period.

Changes to flood hydrology are a fundamental driver of fluvial form and change, and the morphology of Toronto rivers changed dramatically as a result of urbanization of the watersheds. Fluvial geomorphology has reliable conceptual, quantitative, and causal models of river morphology that can be used to predict the ways in which rivers should react to this change in hydrology (Ashmore and Church 2001). Quantitative analysis of river cross-section dimensions is one example (Eaton 2013). River widening, channel pattern changes, and incision were widespread following urbanization of several Toronto watershed and in river channels that were left to respond "naturally" to these effects; the extent of river channel widening is reliably predicted from existing hydraulic geometry equations for this type of river (McDonald 2011). At the same time, many of the rivers underwent considerable engineering intervention (channelization, straightening, bank protection) that constrained the natural response to the increased river flows. Where channel widening was constrained by bank protection and channelization, the models over-predict river widening in response to discharge (McDonald 2011). In effect, human intervention (itself a response to earlier flood events) constrained some of the river response to the peak discharges from urban runoff. At the same time, the constraints on river width are one cause of river channel incision occurring along many of the rivers and threatening urban infrastructure crossing and running along the valleys. Incision is an ongoing risk to which current river

Hydro-geomorphic Changes	1940	Policy Changes
		1946 Ontario Conservation Authorities Act: Four CAs covering 8 watersheds
1954 Hurricane Hazel, October – major flood damage along several rivers. Institutional and engineering response triggers phase of river transformation.		**1957** Merger of four Conservation Authorities into Metropolitan Toronto and Region Conservation Authority covering all major watersheds
		1958 MTRCA Plan for Flood Control and Water Conservation: flood control and channel 'improvement'
		1960s Beginning of extensive private urban development of watersheds.
1960s and **1970s** Extensive river improvement by engineering for flood and erosion protection. Urban development leads to increased discharges and peak flows in rivers. Incision and widening of un-engineered channels in response to increased flows. Increasing costs of 'traditional' channel engineering.		
		1980s MTRCA adopts new watershed planning policies based partly on ecosystem planning. Influence of ecosystem planning recommendations from 1988 Final Report of Royal Commission on the Toronto Waterfront including vision for revitalising Don River.
1990s River re-naturalization tied to Ontario Natural Channels Initiative and wider influence of river restoration ideas.		
		1994 City of Toronto Task Force and MTRCA Valley and Stream Corridor Management Program shift from flooding/erosion to green space, environment and community. Delineates and protects valley corridors.
2003 Wet Weather Flow Plan: > 100 km of planned re-naturalization of channels. Adoption of Geomorphic Master Plans. Geomorphological and engineering expert assessment and design procedures & ecological indicators. Additional impetus following effects of extreme 2005 storm event.		**2003** 25 year City of Toronto Wet Weather Flow Master Plan including geomorphic master plan concept.
	2010	

Fig. 23.5 Timeline of major hydro-geomorphic and policy events affecting Toronto watersheds and river transformation

design and restoration continues to respond and has provided impetus and justification for the institutional responses over several decades which has imposed particular fluvial forms in the landscape. An approximate timeline of events related to the transformation of Toronto's rivers is given in Fig. 23.5 to provide a reference for the following detailed discussion.

Urban Development, Institutional Policy and Power, Community Visions, and Fluvial Change

While fluvial morphological changes can be seen as a physical response to urbanization following a conventional hydro-geomorphic analysis, a variety of conditions, circumstances, and processes affected this response and the various interventions associated with it, which conditioned the river futures in particular ways.

The trajectory of river transformation is related to the way in which urban development was implemented and funded at this time (Harris 2004). In particular, the 1950s saw a shift from publicly controlled to privately financed development. The shift to privately controlled development necessitated (for financial feasibility) simultaneous construction in large subdivisions. Private developers were given responsibility for financing and servicing, and therefore for the scale, density, and layouts, including drainage plans. The financing needs drove the scale and rapidity of development (Harris 2004) and the commodification and transformation of the landscape. In Toronto, development occurred on the table lands because by the late 1950s valley lands had been set aside (see below). The valleys and rivers then became convenient conduits for uncontrolled stormwater runoff from the urban development. In this way, the overall institutional and economic conditions created a particular scale and rate of development arguably with important effects on fluvial hydrology and hazards that continue to drive much of the activity around river management and restoration.

There was nothing pre-determined about the form of post-war urban development in Toronto (Harris 2004). If housing policies, or state control, or powers of private developers, or changes in transportation, had been different, the form and density of development might have been different (Harris 2004) and consequently the hydrological and fluvial changes that followed from development might also have been different. If the nature of urban development is not pre-determined, then the extent and characteristics of fluvial change following urbanization is not pre-determined either. The urbanization processes and urban form are intrinsic to the socio-natural processes transforming the rivers, rather than external conditions imposed on the rivers. Furthermore, these kinds of socio-natural interactions are continual rather than discrete events. This is a perspective seldom taken in physical hydrology and geomorphology where the normal analysis is entirely physical, the form of development and the surface characteristics affecting hydrology are taken as given, and the hydrological impacts are clearly seen as being separate from issues of planning, politics, environmental assessment, and development. Yet the nature of development ultimately affects the hydrological and geomorphic effects and pursuit of this idea further illuminates the characteristics of the system.

Urban development also had more-direct effects on some rivers and valleys. Bonnell (2014) points this out with respect to 1960s visions of garden expressways, implemented in Toronto in the case of the Don Valley Parkway. The valley as park-like transportation corridor in which the road blends with the existing topography was a compelling vision promoted by planning and development interests. But the constraints of the valley also resulted in considerable channelization of the Don River. While implementing a vision for the road which saw it blending into the natural topography (even though considerable cut and fill was required in construction (Bonnell 2014)), this vision foreclosed other possible futures of the Don River.

The ravines and valleys of these socio-natural river systems have become part of the identity of the City. They are seen as part of the green infrastructure, natural areas, and recreational parkland in the City driven partly by a community vision and culture of "wilding" of some of the valleys (Desfor and Bonnell 2011). The rivers flowing down these valleys have a particular community status as well as distinctive geomorphic characteristics (see below). This ravine and tableland topography has also driven some of the development of infrastructure within the city, presented barriers to, and opportunities for, transportation and development and this has had important consequences for the changes in river channel characteristics. In this way, the rivers are an integral part of the city and the relationship with the city is multifaceted and has changed over time. This is also clear in watershed management approaches and framing.

In addition to the urban development processes, the post-1950s evolution of the rivers might be attributed to two initial and coincident path-setting events, which placed the futures of the rivers on particular trajectories and allowed particular institutions and actors to influence changes in river morphology and define the role and character of the rivers and valleys. The first of these events is primarily institutional. The Ontario Conservation Authorities Act (1946) established watershed management powers and responsibilities for Conservation Authorities (CAs) in the Province of Ontario. These CAs were established based on major watershed boundaries with management powers for particular watersheds. In the case of the Toronto region, four separate CAs were initially created covering eight watersheds. The second major path-setting event was Hurricane Hazel in October 1954. Significant property damage and loss along some of the rivers initiated an intensive institutional response and vision for the rivers. The storm modified many of the rivers by extensive bank erosion, but these immediate effects were short term relative to the longer-term institutional effects. The Hurricane Hazel flood did considerable short- and long-term work on the management, politics, and the communities of the watersheds with ongoing implications for river morphology and dynamics.

A major institutional response following Hurricane Hazel was the merging, in 1957, of the four separate Conservation Authorities into a single body— the Metropolitan Toronto and Region Conservation Authority (MTRCA) (renamed Toronto and Region Conservation Authority (TRCA) in 1994). The watershed boundaries for this merged CA encompass all of the City of Toronto and this coincidence, combined with Ontario provincial funding and oversight, gave MTRCA considerable powers, including crucial powers of land acquisition along the valleys in the City (McLean 2004). The geography and limited size of the watersheds enabled important institutional effects with significant consequences for river morphology. The combined MTRCA with its professional staff, associated powers, and funding moved quickly after its formation with a coordinated response to the mitigation of future flood effects. Political pressure to avoid a repeat of the Hurricane Hazel experience drove the early response in flood control and flood defence and the overall vision for the watersheds with its emphasis on engineered flood protection articulated in the 1958 MTRCA Plan for Flood Control and Water Conservation (McLean 2004).

One part of this strategy was to acquire land along the main ravines of the regional river system effectively removing existing development, and precluding any future development, along the valleys. This tied neatly with the development of the park system advocated by the City of Toronto, and the combined powers and cooperation of MTRCA and the City began the development of what is now an extensive park system along many of the valleys in the region. This then set a larger context in which the fluvial geomorphology of the region evolved. In some respects the setting aside of valley lands gave rivers some "freedom space" (Biron et al. 2014) which remains a factor in current design and engineering of river channels. At the same time, it gave the City and MTRCA primary influence over the characteristics of these recreational corridors and the rivers within them. Creating parks around the rivers was also influential in community and MTRCA attitudes to the rivers and subsequent visions of the nature of these valleys (Desfor and Kiel 2000) which had consequences for morphological changes. For example, the greening and wilding vision for the valleys, and the idea of grassroots community reclaiming of the valleys, provided the stage for re-naturalizing channels and undoing channelization of the 1960s and 1970s consistent with the greenways concepts promoted in the 1990s. One might even trace this back to planning visions of the City in the early twentieth century, partly stimulated by the City Beautiful movement which envisaged parklands along the river valleys of Toronto (Bassnet 2007).

The river valley policies also have led to some constraints that have affected subsequent river development. One example is removal of local water treatment plants from the valleys (Bonnell 2014) and construction of a sanitary sewer system along the main valleys that leads to water treatment plants located at the downstream end of the watersheds along the Lake Ontario shoreline. In several valleys, these sewers have become major constraints as river incision from increased flood discharges has threatened their integrity. This has become a central influence on the recent engineering and geomorphic design of the rivers. Recreational infrastructure (trails, bridges) in the valleys is also often at risk in large flood events, which has provided further justification for some of the engineering interventions and river designs.

Implementation of the 1958 Plan was focussed on channel improvement, erosion control, and flood control reservoirs. Some of these channelization works are still in place along the rivers and include defining elements such as hardened rock weirs and trapezoidal concrete-lined flood control channels. These represent the engineering-dominant approach and technical backgrounds of MTRCA staff as well as the prevalent control mentality of the time (see Karvonen (2010) for discussion of these "Promethean" approaches in the case of Seattle, and Waley (2000) for a similar discussion in relation to Japanese rivers and the associated national politics and economy). This was aided by federal financial support (McLean 2004) and the conditions for this support also had direct consequences for the nature of the work and the effects on channel morphology and valley corridors, including the need for stream improvement, erosion control, and buffer zones along high water lines. These early works also constrained later river channel adjustment to the effects of extensive urbanization, incurring new costs and also leading to rethinking these strategies and approaches to channel design after they resulted in costly and repeated repair work. This period also saw the beginnings of the long-standing community support of, and engagement with, MTRCA activities through promotion of the recreational facilities and heritage conservation. This is an important element of the collective vision of what the valleys and rivers should and can be.

The effects of Hurricane Hazel, the combined powers of the CA and City of Toronto and the coincidence of the watershed boundaries with their joint jurisdiction, the influence of levels of government, and the focus on flood and erosion control set the rivers along a path of engineering, bank protection, and channelization which continues today although with altered objectives and approaches. A change in approach can be seen beginning in the early 1980s, stimulated partly by changes in funding for the MTRCA, new watershed planning policy, and some disaffection with the earlier channel improvement approaches (McLean 2004). The final report of the Royal Commission on the Future of the Toronto Waterfront (1988) was particularly influential in articulating the need for ecosystem planning, understanding linked processes across watersheds and the "bio-region", and proposing valleys as part of an urban greenway system. It also emphasized the importance of affinity to local landscapes and landscape history to avoid "landscapes of anywhere". Taking cues from work such as McHarg's *Design with Nature* (McHarg 1969), the Royal Commission proposed that topography be allowed to define urban form and affirmed the importance of place making and the community experience of place. The re-engineering of channels and the regeneration of the Don River was an explicit part of the Commission's vision.

Some of the changes in approach were also initiated by citizen action groups, most notably "Bring Back the Don", a group dedicated to environmental improvements in the Don River valley, a river with an important place in the minds and lives of many Torontonians (Desfor and Kiel 2000; Bonnell 2014). This led to an official City of Toronto Task Force report in the early 1990s that called for changes to policies that had turned "… streams into sterile ditches in the name of flood control …" (quoted in McLean 2004: 218) and subsequent reports argued for a revitalized urban river flowing with life-sustaining water through regenerated habitat that would help to develop people's living connection to the (Don River) watershed (McLean 2004: 224). While partly fanciful, this vision remains influential. Although it is difficult to trace the direct consequences of these ideas and initiatives, there is no doubt that it was part of an overall shift in the ways in which the rivers were managed, and the way in which MTRCA (renamed TRCA in 1997) articulated its vision for the watersheds.

The 1994 "Valley and Stream Corridor Management Program" (MTRCA 1994) clearly moved away partly from the safety/protection mandate towards visions of continuous green space with refuge for wild life, vegetation, and people, and a renewed community partnership for regeneration of stream corridors. This is even more apparent in the 2014 "Living City" policy (TRCA 2014) that promotes the healthy sustainable coexistence of natural ecosystems with human communities. This increased emphasis on ecology is the most conspicuous part of the shift from the 1960s and 1970s to the early twenty-first century. A consequence of this from the perspective of river morphology has been increasing implementation of the redesign and naturalization of river channels. Although this work is still largely justified and funded through the flood control and erosion protection mandate, the channel naturalization and ecological benefits are a core element of the current thinking. These shifts in views and management plans for the river systems have had clear consequences for the morphology of the river channels. The powers (including funding) given to, and acquired by, the TRCA have also enabled it to articulate and proceed with these plans. Without the influence of TRCA and its powers, the vision for and futures of the rivers may have been very different even with the same set of geomorphic conditions, development processes, and hydrological events.

Natural Channel Initiatives and the Influence of Fluvial Geomorphology

Changes in the approaches to managing river erosion and ecology that occurred from the 1990s were also influenced by circumstances and ideas beyond the local management institutions and community. Explicit mention and inclusion of principles of fluvial geomorphology in watershed planning and management in the region appears initially in the early 1990s and in subsequent documentation becomes a more prominent element of management objectives and approaches. The reasons for this are not clear although the visibility of geomorphically based stream restoration more broadly (globally) is one likely influence. The early 1990s saw the beginnings of the Ontario Natural Channels Initiative that has consistently promoted ideas of river restoration, natural design, and eco-geomorphic science through regular conferences and publication of stream corridor management principles and guidelines. The ideas of natural channel design were promoted by the Provincial Ministry of Natural Resources through documents on river restoration and design (Ontario Ministry of Natural Resources 1994). Much of this was stimulated by promotion of similar ideas through, for example, the Rosgen natural channel design courses and literature (although these have been less influential than in the United States) and the rise of stream restoration as an activity, goal, and enterprise internationally (Lave 2009, 2012; Pasternak 2013). In this way, international influences in developing stream restoration, and fluvial science in general, played out in the local decision-making by TRCA and the City of Toronto.

The City of Toronto, partnered with TRCA, has been the leader in recent initiatives towards naturalization and channel design in the region. The most important policy event in this regard is the development of the Wet Weather Flow Master Plan adopted by City of Toronto Council in 2003 establishing a 25-year plan for watercourse management (City of Toronto 2003). The Plan has many facets but explicitly states that the City will take a proactive approach to the design and construction of new or replacement erosion control works consistent with best practices for natural channel design, sustainable design practices, and aquatic habitat protection. It laid out a series of stream restoration (re-engineering) projects in watersheds across the City taking in as much as 100 km of river length. Stream restoration projects are developed through a careful analysis of the hydrology and geology of the watershed, taking into account the impacts of past and future urban development and these are detailed in required "Geomorphic Master Plans" through formal environmental assessments. These master plans have elements of fluvial audits proposed more generally in applied fluvial geomorphology (Downs and Gregory 2014).

Geomorphic assessments and studies are now a routine part of long-term planning and response to local flood effects. Increased funding has been committed following several major flood events which caused erosion and infrastructure damage in the past decade, notably in 2005 and 2013, after which emergency funding facilitated extensive re-engineering along some of the channel systems, notably that of Highland Creek (Fig. 23.2). This is an echo of the influence of extreme events on the socio-natural system that began with Hurricane Hazel in the 1950s. However, the emphasis, and sources of funding, remains protection of infrastructure from erosion of watercourses resulting from earlier urban development. Downcutting of channel beds and widening of channels have been the dominant erosion impact on infrastructure (especially sanitary sewers running beneath and alongside the main channels), and policies require geomorphic studies which apply the principles of natural channel design at reach scale, although exactly what this means is not clearly articulated in the policy guidelines (see below). Works are to be sustainable and take into account the changes to the watershed and flow regimes caused by urbanization and climate change. Thus, there is clear insistence on a role for fluvial geomorphology as part of watershed scale planning, partly substituting for earlier piece-meal engineering works which have become increasingly expensive to maintain. Limits to public (mainly municipal) expenditures have thus become an important influence on river morphology. Costs for these designs run into millions of dollars per kilometre of channel and therefore the resilience of the design and construction, and the minimization of maintenance, is often paramount. Consequently, a particular type of channel morphology emerges because of these requirements, constraints, and geomorphological influence.

Explicit guidance about natural channel design and role of fluvial geomorphology is difficult to find in these plans and guidelines but particular styles of re-engineered channels have appeared. These seem to relate to the local circumstances and the expertise and experience of the public and private sector proponents and designers of these channels, as well as the physical conditions and constraints. The Conservation Authority set out guidelines (MTRCA 1994) for retaining, and formally delineating, valleys and watercourses as natural landforms. Guidelines for regeneration projects included requirements to use "proper morphology, geometry and other characteristics" including maintaining and enhancing channel length (and meander forms), using a range of bed material particle sizes and providing for formation of pools and riffles at appropriate intervals. River beds are often raised and regraded to mitigate the incision threat to sewers along the valleys. Beyond these guidelines, the actual design of restored channels has been left mainly to private consultants to devise and therefore has a strong commercial influence, although multi-agency approvals are required.

The exact rationale for designs is difficult to uncover. Lave's (2014) suggestion that private knowledge is valorized in this context and observations of Pasternak (2013) and Doyle et al. (2015) that commercialization may limit possible design solutions are both relevant here in understanding the drivers and consequences for river design and morphology. In most cases, assessment and design is in the hands of experts with graduate degrees in fluvial geomorphology and related fields of engineering and with local knowledge of the regional river systems. Although Rosgen-based analyses and designs are mentioned in some design reports, designs are usually based on broader fluvial geomorphic principles. These may be "textbook" principles of fluvial geomorphology and river mechanics along with techniques taken from river restoration manuals and guides. Design may be based on calculations of, for example, bed material entrainment conditions, stable channel widths, pool-riffle geometry and bank erosion, using established physical equations combined with knowledge of the local watersheds and rivers, including the semi-alluvial characteristics referred to in the introduction. Overall, a collective expertise has developed regionally among the professional staff at TRCA and City of Toronto and the local companies that are most frequently contracted for the work, along with other experts. This includes sharing of approaches, assessment of efficacy, and critiques of channel design principles and implementation (e.g. Ness and Joy 2002). In this sense, river morphology develops from a distinct epistemic community of technical expertise in various organizations supported to some extent by conservationists who have a shared (explicit and tacit) technical and landscape vision. Sandberg et al. (2013) have developed a more detailed discussion of this idea of epistemic communities related to conservation and development of the Oak Ridges Moraine that forms the northern boundary of several Toronto watersheds.

The geomorphic assessments from which design decisions are made use thorough reach-by-reach assessments of channel conditions, design alternatives and constraints, and watershed-based historical audits (e.g. Parish Geomorphic 2006; Aquafor Beech Limited 2008). Details of particular methods are sometimes difficult to know because of protection of commercial interests between competing consultants but technical design briefs (e.g. Parish Geomorphic 2008) provide some insight. Geomorphic details are compiled for channel dimensions and gradient, substrate characteristics, discharge, critical shear stress, and stream power along with whatever historical information can be found from aerial photographs and earlier surveys. In the case of some analyses, technical assessments evaluate current morphology relative to the expected morphology and so assess the extent to which channels have adjusted to prevailing urban runoff conditions (Aquafor Beech Ltd. 2008) based on hydraulic geometry equations applied to particular locations. Many assessments refer to a typical set of problems and issues: poorly developed pools and riffles, exposed glacial clay, disconnected floodplain and channel incision, and lack of expected planform, which are seen as deviations from how the river should be or from the characteristics of a stable channel. Decisions can be based on scoring systems for state of the river such as Rapid Geomorphic Assessment procedures developed for southern Ontario for calculating an index of channel stability. These are based on checklist inventories of evidence for aggradation (e.g. point bar accretion, medial bars, pool siltation), degradation (e.g. channel incised into non-alluvial material, exposed infrastructure), channel widening (large organic debris, exposed roots, basal scour on inner banks), and planform adjustment (single to multiple thread transition, cut-off channels). In some cases, these may be indicators of the normal characteristics of a type of river rather than indicators of channel change and instability. The insight with which these are applied varies between practitioners. Other physically based procedures have been proposed (Vocal-Ferencevic and Ashmore 2012) but are not yet widely adopted. Unstable reaches are prioritized in this process. Restoration options sometimes include the outcome of expert design charrettes that evaluate alternatives such as general options to "do nothing", make "selective" interventions, or adopt reach-based mitigation and design (e.g. Aquafor Beech Ltd. 2009). All of these analyses are accompanied by detailed reach-by-reach drawings and listings of proposed channel work. Multi-attribute scoring refers to criteria related to natural environment (e.g. flooding and aquatic habitat), the sociocultural environment (including aesthetic value), technical issues (e.g. approval of regulatory agency), and cost.

By identifying problems and concerns from a geomorphic perspective, the rivers are reimagined and re-engineered based on geomorphic calculations and principles (e.g. critical shear stress for erosion, stable channel dimensions for given discharges, and channel pattern relationships such as meander wavelength and pool-riffle spacing) and expectations, local constraints, and managerial decisions. River discharge for design calculations can be based on gauged data but more typically is derived from calibrated rainfall-runoff models. In-channel flows are also commonly modelled using widely available software such as HEC-RAS. In many cases, reaches that had adjusted "naturally" to increased urban stormwater flows (McDonald 2011) have been designed to accommodate particular peak flows. For example, one section of Highland Creek that had widened from less than 10 m in the 1960s to about 15 m in the early 2000s was subsequently constructed with a width of almost 25 m to accommodate the design flows (Fig. 23.1). Bed material for these channels is designed based on simple critical shear stress estimates to assure absolute stability, and features such as constructed riffles are based on empirical relations for length, slope, and amplitude that are often compiled by individual consultants. This results in singular morpho-sedimentary units such as riffles in channels with a 1% slope which have designed particle diameters of 0.4–0.8 m (Parish Geomorphic 2008), which is likely to be almost an order of magnitude larger than a natural riffle in this type of channel. Bedload transport calculations are rare and the channel stabilization may prevent significant downstream routing of coarse bed material. Many channels are confined within relatively narrow valleys and rivers often contacted the valley sides under pre-urban conditions. The hazard presented by this circumstance means that channel alignment is often constructed to avoid contact of the channel with unstable valley walls although former channels may be retained as back-water features along the valley.

The result is a largely novel river morphology based on general principles of fluvial geomorphology but constrained by considerations such as infrastructure protection, the compromises between engineering and geomorphology, the tendency to seek channel stability, and the altered hydro-geomorphic context (e.g. increased storm runoff) within the confined river valleys which calls for, in effect, a geomorphic reimagining of the rivers. But this is mainly morphology that would not otherwise exist and in which stability is partly forced by the design and larger objectives, for example, use of non-erodible armour stone blocks as bank material but also the overall hydro-geomorphic context. The natural materials lining the channels are then partly distally derived as is, ultimately, some of the knowledge base applied to the designs. Valley floors are sometimes partially filled in a manner that superficially reproduces floodplain morphology (Fig. 23.1f). At the same time, some original geomorphic features of the valleys are modified, destroyed, or hidden, including exposure of some glacial deposits that revealed the geomorphic history of the region and the valleys. This part of the story of the valleys and associated elements of the landscape are lost.

The goal is generally to create river form and function that are in keeping with the channel type that is suitable for the (urbanized) setting. The zoning and separation of valleys from the surrounding landscape give the designers some freedom space for implementing these designs, such as those employing constructed meanders, which also have some hydraulic advantages. At the same time, concerns for the stability of steep valley sides and encroaching development also limit that freedom. These professional consultants, and the science that they reference and develop, have had a substantial effect on current river forms in the restored reaches. In general, the resulting channel morphology and materials are hybrids of existing natural forms, channel engineering, geomorphically referenced river forms, and eco-geomorphic design. Seldom do they look much like the pre-existing channel (Fig. 23.1) or the channels that may have developed in response to hydrological change from urban development without channel engineering and design. It is debatable whether even the basic guidelines, for example, to create pool-riffle morphology, reflect the natural morphology of these channels under the urban-modified discharge conditions and semi-alluvial character of the rivers. The channels might be seen as appropriate features of such rivers taken by river restoration practitioners from the general fluvial geomorphology literature to meet the desire for "natural" channels while satisfying financial, commercial, and social constraints (Doyle et al. 2015). Another geomorphologist or another set of designers, using the same principles and information, with different experiences, may have differing perceptions of the problems with different outcomes for the river morphology. Added to this is the fact that these rivers are only partially geomorphically conceived. They are hybrids satisfying multiple influences and socio-natural processes. Their novel, hybrid characteristics and purposes include flood control, erosion mitigation, species diversity, habitat provision, natural morphology, and community resource. They materialize particular socio-natural history, conditions, and processes. The rivers are not so much restored as restoryed (Higgs 2003).

Synthesis and Linkages

A series of processes, constraints, and events have given the rivers of Toronto their current form. In a socio-geomorphic analysis within the context of Critical Physical Geography, the rivers are a hybrid form (Eden et al. 2000) related to the local hydro-geomorphic conditions and reflecting: the power of local agencies, the influence of developers, planning and development policy, the valley form that allows well-defined zoning and separation from development, the influence and application of international fluvial science, the changing global culture and science of river restoration, community needs and influences, the commercial and scientific interests of river design practitioners, and a particular sequence of socio-natural events. In almost all of these influences, local conditions, circumstances, and individuals play a core role in the outcomes and river futures. The local hydro-geomorphic conditions, including the coincidence of watersheds with institutional jurisdictions and the characteristics of the landscape, are an important element of the circumstances leading to the outcome for fluvial landscapes in this case.

Hybrid geographies articulated, for example, by Whatmore (2002), seek to transgress and displace nature-society boundaries, signifying new kinds of places and entities with multiple origins, performing and mobilized for multiple purposes. One aspect of this is that socio-natural entities have no fixed boundaries and are the products, in part, of the circulation of knowledges, money, and materials at multiple scales. Following these socio-ecological flows is an essential element of understanding how these entities are produced (Eden and Holloway 2013). Restored rivers can then be seen as hybrids in the sense that they are the product of the combination of scientific knowledge and sociopolitical processes and the interaction of the relevant actors (Eden et al. 2000; McDonald et al. 2004). The focus in analysis is the ways in which these actors are transformed, and the river morphology becomes somewhat incidental. Critical Physical Geography, while admitting these socio-natural circulations, puts the material features of the river at the centre of the analysis and the ways in which the combination of sociopolitical and biophysical processes and histories make the river. While it is possible to view rivers of this type as simply an "impacted" landscape from a physical perspective, this limits understanding of the processes leading to the current morphology, and in the context of deliberate design it misses crucial elements of the transformation processes. The alternative view sees these events as eco-social transformations and the rivers as partly the material manifestation of ways in which rivers are perceived, managed, imagined, and commodified. The intentionality of intervention and design is a crucial element of the river transformations and deeper examination of this process in particular contexts may be fruitful. Concepts of design are increasingly prominent in major urban river restoration projects, such as the Los Angeles River, in which architectural companies are taking on hydraulic and hydrologic design based on high resolution 3D data acquisition and new discourses and influences of design (Revitalization News 2016).

The phase of river reconstruction and restoration clearly shows that, as has often been recognized, river restoration is a strongly social-cultural process (e.g. Eden et al. 2000; Kondolf and Yang 2008). In the ecological restoration literature, there has long been concern around the types of interventions in ecological processes and these parallel those articulated around the issues of stream restoration. In particular, there is concern around the production of "expert landscapes" (Higgs 2003) developed from technological, top-down knowledge through the authority of science and backed up by the power of the associated institutions. This is essentially the outcome for rivers in Toronto and the concept of expert landscapes and novel landscapes are apt and useful in understanding the current river morphology. The rivers then reflect the institutional powers and the material effects of applying fluvial geomorphic science within the constraints of the local conditions, institutional power, and commercial control. Particular forms appear because river science has an important role in both policy and design, and differing scientific conceptions and approaches yield differences in morphology (Bouleau 2013; Tadaki et al. 2014; Lave 2016). The rivers reflect the prioritizing of particular knowledge systems and interests.

The way in which rivers materially manifest politics of river management has been pointed out previously (e.g. O'Neill 2006). The contingency of place arises in both the politics and the geomorphology. Waley (2000, 2005) and Waley and Arberg (2011) illustrate this very well in the case of Japan, proposing that there is a symbiosis of discourse and material practice reflected in the rivers. This extends the argument about the role of science and involvement of fluvial geomorphology in the processes of river transformation and breaks down the dichotomies of urban-rural, nature-culture, and real-discursive.

The outcome of the socio-geomorphic conceptions of river morphology in the case of Toronto is that the novel fluvial landscapes are the product of combined socio-natural processes playing out in the context of local landscape and the processes of urbanization, river management, and community, as they occurred in post-1950s Toronto. Placing this in a larger context of urban physical geography (Ashmore and Dodson 2017), cities are dense networks that are simultaneously global and local, human and physical, cultural and organic (Heyman et al. 2006). City and nature are inextricably woven through socio-ecological networks and feedbacks, and nature is inescapably political (Desfor and Laidley 2011). Events are local and distal, social and natural (McClintock 2015), playing out in particular places and times and with material effects in the physical landscape.

Acknowledgements My research on urban rivers in Toronto has been supported by grants from Natural Resources and Engineering Research Council, and by City of Toronto, Toronto and Region Conservation Authority, Parish Geomorphic and Aquafor Beech Ltd. Thanks to John McDonald for preparing the aerial photographs for Fig. 23.1 and Karen van Kerkoerle for Figs. 23.2, 23.4, and 23.5. John McDonald and Mariane Ferencevic's thesis work on Highland Creek is one of the things that first got me thinking about the ideas developed here and provided empirical support for some of the morphological and process transformations that I discuss. Belinda Dodson encouraged my thinking on this topic.

References

Aquafor Beech Limited. 2008. *Geomorphic systems master implementation project, highland creek, valley segment characterization report.* Report submitted to City of Toronto.

———. 2009. *Geomorphic systems master implementation project, highland creek watershed design charrette #2.* Report prepared for City of Toronto.

Ashmore, P. 2015. Towards a socio-geomorphology of rivers. *Geomorphology* 241: 149–156.

Ashmore, P., and Church, M. 2001. *The impact of climate change on rivers and processes in Canada.* Geological Survey of Canada, Bulletin 555.

Ashmore, P., and B. Dodson. 2017. Urbanizing physical geography. *The Canadian Geographer* 61: 102–106.

Bassnet, S. 2007. Visuality and the emergence of city planning in early twentieth-century Toronto and Montréal. *Journal for the Study of Architecture in Canada* 32: 21–38.

Biron, P., T. Buffin-Bélanger, M. Larocque, G. Choné, C.A. Cloutier, M.A. Ouellet, S. Demers, T. Olsen, C. Desjarlais, and J. Eyquem. 2014. Freedom space for rivers: A sustainable management approach to enhance river resilience. *Environmental Management* 54: 1056–1073.

Bonnell, J. 2014. *Reclaiming the don: An environmental history of Toronto's don river valley.* Toronto: University of Toronto Press.

Bouleau, G. 2013. The co-production of science and waterscapes: The case of the Seine and the Rhône Rivers, France. *Geoforum* 57: 248–257.

City of Toronto. 2003. *Wet weather flow master plan.* Toronto: City of Toronto.

Desfor, G., and J. Bonnell. 2011. Socio-ecological change in the nineteenth and twenty-first centuries: The lower don river. In *Toronto's waterfront*, ed. G. Desfor and J. Laidley, 305–325. Toronto: University of Toronto Press.

Desfor, G., and R. Keil. 2000. Every river tells a story: The don river (Toronto) and the los angeles river (Los Angeles) as articulating landscapes. *Journal of Environmental Policy and Planning* 2: 5–23.

Desfor, G., and J. Laidley. 2011. Introduction. In *Toronto's waterfront*, ed. G. Desfor and J. Laidley, 3–19. Toronto: University of Toronto Press.

Downs, P., and K.J. Gregory. 2014. *River channel management: Towards sustainable catchment hydrosystems.* New York: Routledge.

Doyle, M.W., J. Singh, R. Lave, and M.M. Robertson. 2015. The morphology of streams restored formarket and nonmarket purposes: Insights from a mixed natural-social science approach. *Water Resources Research* 51: 5603–5622.

Eaton, B.C. 2013. Hydraulic geometry: Empirical investigations and theoretical approaches. In *Treatise on geomorphology*, ed. J. Shroder (Editor in Chief) and E. Wohl, vol. 9., Fluvial Geomorphology, 313–329. San Diego, CA: Academic Press.

Eden, S., and L. Holloway. 2013. More-than-urban ecosystems, socioecological flows and the problems with boundaries: A response to Francis et al. (2012). *Transactions of the Institute of British Geographers* 38: 678–681.

Eden, S., S.M. Tunstall, and S.M. Tapsell. 2000. Translating nature: River restoration as nature-culture. *Environment and Planning D: Society and Space* 18: 257–273.

Harris, R. 2004. *Creeping conformity: How Canada became suburban, 1900–1960.* Toronto: University of Toronto Press.

Heynen, N., M. Kaika, and E. Swyngedouw. 2006. Urban political ecology: Politicizing the production of urban natures. In *In the nature of cities: Urban political ecology and the politics of urban metabolism*, ed. N. Heynen, M. Kaika, and E. Swyngedouw, 21–40. London: Routledge.

Higgs, E. 2003. *Nature by design.* Cambridge: MIT Press.

Karvonen, A. 2010. Metronatural™: Inventing and reworking urban nature in seattle. *Progress in Planning* 74: 153–202.

Kondolf, G.M., and C.-N. Yang. 2008. Planning river restoration projects: Social and cultural dimensions. In *River restoration: Managing the uncertainty in restoring physical habitat*, ed. S. Darby and D. Sear, 43–60. Chichester: Wiley.

Lane, S.N., and K.S. Richards. 1997. Linking river channel form and process: Time, space, and causality revisited. *Earth Surface Processes and Landforms* 22: 249–260.

Lave, R. 2009. The controversy over natural channel design: Substantive explanations and potential avenues for resolution. *Journal of the American Water Resources Association* 45: 1519–1532.

———. 2012. *Fields and streams: Stream restoration, neoliberalism, and the future of environmental science.* Athens: University of Georgia Press.

———. 2014. Freedom and constraint: Generative expectations in the US stream restoration field. *Geoforum* 52: 236–244.

———. 2016. Stream restoration and the surprisingly social dynamics of science. *WIREs Water* 3: 75–81.

McClintock, N. 2015. Critical Physical Geography of urban soil contamination. *Geoforum* 65: 69–85.

McDonald, J. 2011. Response of river channel morphology to urbanization: The case of highland creek, Toronto, Ontario, 1954–2005. M.Sc. Thesis, University of Western Ontario.

McDonald, A., S.N. Lane, N.E. Haycock, and E.A. Chalk. 2004. Rivers of dreams: On the gulf between theoretical and practical aspects of an upland river restoration. *Transactions of the Institute of British Geographers* 29: 257–281.

McHarg, I.L. 1969. *Design with nature*, 197. Garden City, NY: The Natural History Press, American Museum of Natural History.

McLean, B. 2004. *Paths to the living city: The story of the Toronto and region conservation authority.* Toronto: The Toronto and Region Conservation Authority.

Meshkova, L.V., P.A. Carling, and T. Buffin-Belanger. 2012. Nomenclature, complexity, semi-alluvial channels and sediment-flux-driven bedrock erosion. In *Gravel-bed rivers: Processes, tools, environments*, ed. M. Church, P.M. Biron, and A. Roy, vol. 424, 432. Chichester: John Wily and Sons, Ltd.

Metropolitan Toronto and Region Conservation Authority. 1994. *Valley and stream corridor management plan.* Toronto, ON: MTRCA.

Ness, T., and D.M. Joy. 2002. Performance of natural channel designs in Southwestern Ontario. *Canadian Water Resources Journal/Revue canadienne des ressources hydriques* 27: 293–315.

O'Neill, K.M. 2006. *Rivers by design: State power and the origins of U.S. flood control.* Durham: Duke University Press.

Ontario Ministry of Natural Resources. 1994. *Natural channel systems: An approach to management and design.* Toronto, ON: Ontario Ministry of Natural Resources.

Parish Geomorphic. 2006. *Highland creek valley segment 4A study; Characterization report.* City of Toronto Report 2005–50, December 2006.

———. 2008. *Highland creek valley segment 4/4A technical design brief.* Report to City of Toronto, October 2008.

Pasternak, G. B. 2013. Geomorphologist's guide to participating in river rehabilitation. In *Treatise on geomorphology, V.9, fluvial geomorphology*, eds. J.F. Shroder

(Editor in Chief), Wohl, E. (Volume Editor), 843–860. San Diego: Academic Press.

Phillips, J. 2007. The perfect landscape. *Geomorphology* 84: 159–169.

Phillips, R.T.J., and J.R. Desloges. 2014. Glacially conditioned specific stream powers in low-relief river catchments of the Southern Laurentian great lakes. *Geomorphology* 206: 271–287.

———. 2015. Glacial legacy effects on river landforms of the Southern Laurentian great lakes. *Journal of Great Lakes Research* 41: 951–964.

Revitalization News™. 2016. LA releases new planning tool for large, complex urban river restoration. *Revitalization News™: The Journal of Economic, Social and Environmental Renewal*, 30. http://revitalizationnews.com/article/new-design-planning-tool-complex-large-scale-urban-river-restoration/

Royal Commission on the Future of the Toronto Waterfront. 1988. Regeneration: Toronto's waterfront and the sustainable city. Final Report.

Sandberg, A., G.R. Wekerle, and L. Gilbert. 2013. *The oak ridges moraine battles: Development, sprawl, and nature conservation in the Toronto region*, 336. Toronto: University of Toronto Press.

Schumm, S.A. 1991. *To interpret the earth: Ten ways to be wrong*, 133. Cambridge: Cambridge University Press.

Simpson, G.G. 1963. Historical Science. In *The fabric of geology*, ed. C.C. Albritton, 24–48. Stanford, CA: Freeman, Cooper and Co.

Tadaki, M., G. Brierley, and C. Cullum. 2014. River classification: Theory, practice, politics. *WIREs Water* 1: 349–367.

Toronto and Region Conservation Authority. 2014. *The living city® Policies for planning and development in the watersheds of the Toronto and region conservation authority*. Toronto: TRCA. https://trca.ca/planning-permits/living-city-policies/

Trudeau, M.P., and M. Richardson. 2015. Change in event-scale hydrologic response in two urbanizing watersheds of the great lakes St Lawrence Basin 1969–2010. *Journal of Hydrology* 523: 650–662.

Vocal-Ferencevic, M., and P. Ashmore. 2012. Creating and evaluating DEM-based stream power maps as a stream assessment tool. *River Research and Applications* 28: 1394–1416.

Waley, P. 2000. Following the flow of Japan's river culture. *Japan Forum* 12: 199–217.

———. 2005. Ruining and restoring rivers: The state and civil society in Japan. *Public Affairs* 78: 195–215.

Waley, P., and E.U. Aberg. 2011. Finding space for flowing water in Japan's densely populated landscapes. *Environment and Planning A* 43: 2321–2336.

Whatmore, S. 2002. *Hybrid geographies: Natures cultures spaces*, 225. London: Sage Publications.

Part III

Conclusion: Reflecting on Critical Physical Geography

24

Proliferating a New Generation of Critical Physical Geographers: Graduate Education in UMass's RiverSmart Communities Project

Nicole Gillett, Eve Vogel, Noah Slovin, and Christine E. Hatch

Introduction: A Needed Model of CPG Graduate Education

To build our collective capacities to engage in CPG research and practice, Lave (2015a) calls for efforts to enhance diverse and progressive knowledge production. We need to proliferate[1] a new generation of critical physical geographers. To do this, we must train students who can think, conduct research, and share findings in ways that are critical, open-ended, and transdisciplinary. But what does this look like in practice?

In this chapter, we use our experience as two faculty and two Master's students working in an innovative and transdisciplinary project called RiverSmart to answer this question. We feel we achieved a remarkably integrated research project that transcends conventional disciplinary divides—and with it, a

N. Gillett (✉)
Tucson Audubon Society, Tucson, AZ, USA

E. Vogel
Department of Geosciences, University of Massachusetts, Amherst, MA, USA

N. Slovin
Milone & MacBroom, Cheshire, CT, USA

C. E. Hatch
Department of Geosciences, University of Massachusetts, Amherst, MA, USA

© The Author(s) 2018   **515**
R. Lave et al. (eds.), *The Palgrave Handbook of Critical Physical Geography*,
https://doi.org/10.1007/978-3-319-71461-5_24

graduate education that was rich in the principles of Critical Physical Geography.

This chapter proceeds in three parts plus a conclusion. In Part 1, the faculty members introduce RiverSmart and highlight five factors that they used to align the project and its graduate education with CPG. In Part 2, the graduate students describe how these factors played out in practice for them. In Part 3, we collectively reflect on the successes and challenges of RiverSmart as CPG graduate education and offer some lessons. We conclude by inviting others to borrow from our initiative in ways that fit their own aims and contexts.

Part 1, Faculty: RiverSmart Communities as a CPG Research and Graduate Education Project

In this part the two faculty, Eve Vogel and Christine Hatch, introduce the RiverSmart research project and its graduate education component. We identify and explain five factors that helped make the project work as a CPG project and a CPG graduate student experience. This faculty section sets out our conceptions and approach, leaving many of the details of what this looked like to Part 2, to be told from the graduate student perspective.

Background: RiverSmart Communities

RiverSmart (https://extension.umass.edu/riversmart/) is a program based in the UMass Amherst Department of Geosciences that combines social and river sciences, institutional and policy research, and community outreach to investigate river floods in New England.

This chapter primarily features the first project of the RiverSmart program, RiverSmart Communities, which was funded by a grant from the UMass Center for Agriculture Food and the Environment (UMass CAFE) from October 2012 through September 2016. The outline of the RiverSmart Communities project was:

1. Our science team, led by hydrogeologist Christine Hatch, used scientific investigations and fluvial-geomorphological understandings to develop a set of best management practices for reducing flood damage in New England that aligned with natural river dynamics.
2. Our institutional and policy team, led by human geographer Eve Vogel, highlighted the challenges and constraints caused by New England's distinct array of jurisdictional and institutional fragmentation, and investigated

and analyzed case studies of institutions that had successfully overcome these challenges and constraints.

3. While both teams had a measure of independence on the two parts of the project, group meetings always took place as one group. Project questions, goals, and methods were discussed together. Field trips and interviews were open to all team members.

4. Our extension work took our findings and draft products and disseminated them among towns, government officials, landowners, businesses, environmental organizations, road crews, and others.

5. We used community-based assessment and evaluation to ensure that our efforts were useful, comprehensible, and effective.

Our main products have been (some still in progress):

- A database of fluvial geomorphic assessment techniques;
- A series of factsheets on flooding, fluvial geomorphology, and mitigation for communities, landowners, and other publics;
- Profiles of several institutions that are successfully helping communities become more flood resilient;
- Presentations to and interactive activities with communities, at scientific meetings, for the public, and at other meetings; and
- A policy recommendation report.

To help with the work, we brought on both undergraduate and graduate research assistants. A human geography graduate student, Nicole Gillett, was supervised by Vogel, and a hydrogeology/geomorphology graduate student, Noah Slovin, was supervised by Hatch.

Five Factors That Helped Us Set Up RiverSmart Communities as a CPG Research and Graduate Education Project

We detail here five factors that made our project and graduate education fundamentally aligned with the principles of CPG (Lave et al. 2014; Lave 2015b): (1) an interdisciplinary setting in which we cross-pollinated Geosciences, Geography, Extension, and Water Research; (2) an ontology or conception of the biophysical and social/policy worlds as inherently messy, multilayered, and dynamic; (3) an epistemology that viewed causal factors and knowledge as diverse and multiple; (4) an applied problem, a commitment to outreach, and iterative feedback-based evaluation and revision; (5) an inclusive approach to pedagogy and collaboration.

An Interdisciplinary Setting: Cross-Pollinating Geosciences, Geography, Extension, and Water Research

CPG requires a "sustained integration of physical and critical human geography" (Lave et al. 2014). Our interdisciplinary institutional setting was a crucial aspect that helped us shape our project and graduate student mentoring as a CPG effort (cf. Baerwald 2010 re. institutional arrangements for interdisciplinarity). First, we (faculty members Vogel and Hatch) are in the same department, a Department of Geosciences, housing both Geology and Geography. The Geology program includes strengths in water and climate systems, while the Geography program has an environmental focus, creating a wide area of potential overlap.

Second, both of us have positions designed to promote interdisciplinary work. Vogel was hired in 2008 as a human geographer with a science background in a position that was intended to build linkages with the department's geoscientists. Hatch was hired in 2011 in an innovative position co-funded between Extension and Geosciences, to focus on connecting research and outreach regarding Massachusetts water resources and regional climate change. Importantly, our personnel incentives and rewards were also based on building inter- and transdisciplinary bridges.

Third, we are at the state's land-grant institution, which houses its state Extension office, state Geologist office and Water Resources Research Center. These have longstanding experience translating results of research to disseminate outreach products in a meaningful way into broader communities.

None of these *required* us to build an integrative project or graduate experience, and indeed, RiverSmart was the first of its kind in our department. Still, these institutional contexts provided crucial resources, encouragement, and support.

Ontology: A Conception of the Biophysical and Social/Policy Worlds as Inherently Messy, Multilayered, and Dynamic

The second factor that helped align our project and graduate mentoring with CPG was our conception of the biophysical and social worlds—what might be called our ontology. Like other CPG authors, we see human-environmental systems and landscapes as diverse and contingent socio-natural hybrids, variable and coevolving, messy in ways that belie straightforward measurement or theorization (Ashmore 2015; Blue and Brierley 2016; Brierley et al. 2013).

In RiverSmart, our starting point for this was our interest in understanding and supporting the dynamic change of river systems. This approach to rivers, like much of physical geography today, leaves behind the hope of identifying one dominant process, form, or static equilibrium, and instead sees change as constant, with new or uncommon events often disrupting existing forms and processes, but also creating room and connections for new ones (Aspinall 2010; Blue and Brierley 2016; Murray et al. 2014). We recognized this complexity as a part of the rich diversity and functioning of a range of ecological and physical systems on which we and other species depend (see e.g. Bisson et al. 1997; Wohl et al. 2015; selections from Boon and Raven 2012; Rhoads and Fonstad 2016).

We took a similar approach to governance and policy. Rather than looking for one dominant policy, or systematically critiquing fragmented governance and recommending a new streamlined order, we sought to understand a range of governance structures, policies, and programs, recognizing that they interacted in complex ways, as a multilayered, dynamic, messy system that nonetheless was productive and useful to a range of people in a variety of ways. Critical Physical Geography aims to bring needed attention to the issue of power relations; our approach was borrowed from Gibson-Graham and Berk et al. to bring attention to multiple processes and paths of cause and effect, and a variety of sites of interest and agency, in which a proliferation or "mangle" of possibilities is recognized and embraced (Gibson-Graham 2006; Gibson-Graham et al. 2013; Berk et al. 2013; cf. Andersson and Ostrom 2008; Ostrom and Cox 2010).

Linking these complex understandings of river science and governance and policy together, we recognized the complex and contingent ways rivers embodied a coevolution of biophysical and policy/management systems over time and space (Ashmore 2015; Harden 2012; Rhoads and Fonstad 2016).

This ontology shaped the training we offered our students. We sought to guide our students to be open to trajectories of change and to a range of influential and interacting factors and initiatives. We urged them to respect and appreciate both history and novelty, whether in rivers, landforms, or policy, and to consider their many interactions (cf. Trafford 2012).

Epistemology: A View of Causal Factors and Knowledge as Diverse and Multiple

To understand messy hybrid socio-natures, CPG embraces critical, reflective practices in which a researcher's own frame of understanding is seen as one among many, always partial and always constructed, and researchers deliberately seek to understand others' perspectives (Ashmore 2015; Tadaki et al. 2014; 2015). In researching river science and policy, and in making recommendations, we accordingly sought to recognize, hear, and under-stand a wide range of perspectives as well as the results of different actors' actions. For example, we aimed to understand the forces both experienced and exerted by a rock in a streambank, by moving water, by a municipal road foreman re-enforcing an embankment, by a state agency staffer pro-moting conservation, and by a federal policy-maker cutting expenditures (see also Castree 2015; Lave 2015a; Popke 2016; Tadaki and Fuller 2014).

We guided our students, too, to seek and assume a range of voices. As part of the initial creation of RiverSmart, we brought together diverse advisory teams to help guide us. Later, we conducted interviews with a range of people, from leaders of tiny municipalities to environmental contractors to state agency heads and scientists to nonprofit staff to landowners. Our students sought information, approaches, and ideas from an even wider network of scholars, practitioners, and community members.

An Applied Problem, a Commitment to Outreach, and Feedback-Based Evaluation and Revision

The fourth factor that helped align our project and graduate mentoring with CPG was that we had an applied problem and we wanted our solutions to be practically useful to non-academics such as landowners, municipal leaders, scientists, consultants, federal and state agency staffers, and policy-makers. This helped with integration of science and policy, grounded an ontology of hybrid interactive socio-natural systems, and guided our learning from mul-tiple perspectives.

As academics, we often talk about how hard it is to integrate across disci-plines and the public, how novel it is to see landscapes as hybrid socio-natural systems, and how important it is to realize scientific research has real impacts on environments and communities. However, we did not find any of these all that difficult or novel in the context of an applied problem. In our experience, on-the-ground managers routinely cross science-policy-outreach divides. It is a fundamental understanding for applied researchers and on-the-ground managers that research has tangible impacts, both during the process of con-ducting research and also in using research results to make management and policy recommendations. And, for those who think at all historically about actual rivers, it seems manifestly clear that river processes, management and policy shape one another over time, in contingent ways dependent on the specifics of time and place, collectively and interactively shaping river landscapes.

Certainly, applied work can be narrow, instrumental, and formulaic. It can lead to vast oversimplification of conclusions, management lessons, and landscapes (e.g. Sayre 2015). However, an applied project buttressed by a CPG ontology and epistemology led us to ask interconnected questions, seek feedback, follow up with further questions, and learn iteratively (cf. Tadaki et al. 2014). We worked, and we guided our students, to take queries and investigations where they went, and to reflect continually on feedback to understand nuance and interconnections.

An Inclusive Approach to Pedagogy and Collaboration

Our approach to teaching and supervision was in many ways just as messy, open-ended, contingent, and coevolutionary as the river and policy systems we were studying. This pedagogical approach grew out of and also contributed to the other factors listed above that made RiverSmart a CPG project: its interdisciplinary focus; our ontology that sees rivers and policy as messy, contingent hybrid systems; our openness to multiple epistemological positions; and our applied, feedback-informed research, thinking and learning. Perhaps unsurprisingly, our approach resonates with a variety of creative approaches to supervision in CPG and beyond (Trafford 2012).

From the beginning, we emphasized that students were going to experience not an idealized research experience of hypothesis testing through linearly gathered data, but rather an iterative process of learning; revising the questions being asked, the research targets and sometimes the approach itself; conducting new investigations; and learning more. To be candid, the research process was sometimes not just messy but also disorganized. Partly this was because we faculty were new and not used to managing projects or assistants. But it was also because what we learned changed what we wanted to do, due to our team's openness to reflection and iterative learning.

Part 2, Graduate Students: Experiencing and Conducting RiverSmart Communities as a CPG Research and Graduate Education Project

In this section the two graduate students, Nicole Gillett and Noah Slovin, detail our experience of CPG research and education as we experienced and practiced it in our work with RiverSmart. To provide context, we briefly outline our own backgrounds and how we both came to be part of RiverSmart at the University of Massachusetts. The focus of this section is to describe how the two RiverSmart graduate students experienced and practiced the factors that helped make RiverSmart a CPG project and graduate training endeavor, as explained in the previous section.

Student Backgrounds

As new graduate students, neither of us (Gillett or Slovin) had strong backgrounds in either physical or human geography. As undergraduates, we both majored in a scientific field (Slovin in geology and Gillett in environmental science). However, we had experience integrating science and social applications due to our backgrounds in environmental education. Gillett had worked with human-environmental geographer Eric Perramond in her undergraduate thesis work on Costa Rican fishing communities. Slovin had a range of experience as an educator, including work with the Cornell Cooperative Extension's Energy Corps on home energy efficiency as an undergraduate, and outdoor education with the Teva Learning Center between college and graduate school.

An Interdisciplinary Setting: Cross-Pollinating Geosciences, Geography, Extension, and Water Research

RiverSmart's interdisciplinary setting was a key part of our graduate experience. First, our tenure was spent in the same department and same space. We often shared a graduate student office, and it became second nature to bounce ideas off one another, swap files, co-write papers, and so on. In our shared department and building, we consulted experts in hydrology, local geology, mapping, and political ecology as a regular part of our routine. The Massachusetts State Geologist, also housed in our department, proved to be an exceptionally useful resource thanks to his scientific knowledge, his experience with state history and regulatory structure, and his own concurrent work on river flood hazard assessments.

The project's collaborations with other offices around campus, including UMass Extension, the Massachusetts Geologic Survey, and the Water Resources Research Center, also enriched our experience. Working with staff from these units, we learned about public outreach, community experiences with river floods and management, and the inner workings of the complex grant process and funding realities of research.

We valued this interdisciplinarity in part because it was so rare. Despite the possibilities for interdisciplinary work in departments and universities like ours, and the encouragement of interdisciplinary research by many funders, this project was the only one of its kind that we knew about. When we presented our research in a department colloquium, professors and students expressed interest and surprise at how well we could integrate the physical and social science aspects of our research. At a multi-disciplinary water resources conference on campus, we heard enthusiastic feedback about the uniqueness of our interdisciplinary approach. Even at a CPG conference, we drew attention as rare examples of fully engaged and integrated CPG graduate student researchers.

Ontology: A Conception of the Biophysical and Social/Policy Worlds as Inherently Messy, Multilayered, and Dynamic

Seeing both biophysical and social systems as messy, multilayered, and dynamic was fundamental to our learning and thinking as graduate students. This understanding derived from both our training and our research experience.

Tropical Storm Irene in 2011 was described as a "wake-up-call" by many of the residents we interviewed. Many New Englanders had assumed their river channels were static features in the landscape and were horrified when this assumption was proven wrong by ripped-out streambanks and bridges, multiple feet of deposited sediment, and rivers that carved entirely new channels. Our training in fluvial geomorphology allowed us to see the storm's effects not as a unique catastrophe, but as part of a long-term pattern and process of dynamic river movement. We could see that decades of dams, riprap, and other intrusive forms of flood and flow control had narrowed channels and concentrated the flow and force of rivers. We understood that this altering of stream power patterns had allowed streams to burst through barriers during Irene, causing new and unexpected interactions and results, both destructive and creative.

In our field visits and interviews, we learned from municipal leaders and government agency staff that the social realities of river management were just as messy and complex. For example, many interviewees noted a key conundrum: towns in New England traditionally have the authority to regulate land use and the responsibility to assess and maintain local infrastructure, but many lack the financial and technical resources to do this work adequately. Federal and state government agencies in turn offer programs of technical and financial support, but assistance was sometimes extremely slow in arriving, limited in funding, or unavailable because of local or state circumstances.

Epistemology: A View of Causal Factors and Knowledge as Diverse and Multiple

It quickly became apparent to us that it would be impossible to find singular answers. One reach of river may require careful management while one just downstream could be allowed to flow freely, while the policies and programs that worked well to manage flood risk in one town may not be applicable in another. Simplified solutions would not adequately serve those we were trying to help, nor reflect the multiple factors, processes, priorities, and experiences in human and environmental systems.

We would not have reached this conclusion so easily if we had taken a traditional research approach. Our advisors guided us to, and modeled for us, an openness to learning that led us to seek out a diverse range of sources. We spoke not only with academics and agency scientists but also with volunteer community leaders, farmers, staff of small nonprofits, and so on. We read with equal attention academic papers and community-based after-action reports, and used information about river changes gathered from community members as well as our own field and GIS data.

We initially planned to develop sets of recommended "best practices" either for science-based management or for policy, but our open-ended research approach produced open-ended research results. We found that practitioners did not need further prescribed solutions; rather, they needed more data and options. We instead developed respect and understanding for a host of different approaches, and a commitment to highlighting positives of a range of options. Our publications and recommendations reflect this.

For example, Slovin researched different stream geomorphic assessment methods with the goal of creating a best-practice assessment recommendation. However, the assessment methods he uncovered varied enormously in their purpose, as well as in their content, procedures, and outcomes. This brought Hatch as well as Slovin up short. Slovin instead put together a database of assessments, highlighting how each method worked and what it measured, and noting the advantages and challenges of each.

Similarly, Gillett discovered a great diversity of policies, authorities, and actors dealing with river management across New England. She and Vogel became convinced that a single policy change recommendation was not the direction to take. Instead, we could offer a wide range of options highlighting a variety of activities that can lower flood risks.[2] Our variety of audiences were appreciative of this as it allowed them to use the information flexibly and according to their own needs rather than a prescribed use.

An Applied Problem, a Commitment to Outreach, and Feedback-Based Evaluation and Revision

Our graduate student experience was both more integrated and more critically reflective because of the applied nature of our project. Our end goal was to produce usable information and tools for the communities we were studying. In our research, we attended town meetings, met with local and state officials, spoke with residents in their homes and on their land, and visited the offices and field sites of nonprofit agencies. We asked and learned about not only the science and institutional case studies we were investigating, but also about what people's practical needs were and how our research and our publications might best help them. We iteratively developed our findings and materials based on this input.

As an example, initially we intended to create a suite of educational factsheets to inform the public about flood management. After receiving feedback from municipal leaders, emergency managers, academics, and educators, we found ourselves rethinking our educational goals, target audiences, and vision of successful outreach. We found that our factsheets needed to be "messy" and multileveled, to provide information in language accessible to different groups, and presented in a more appealing format. The factsheets evolved from factsheets designed for municipal leaders, into educational pamphlets designed for public consumption, into poster-sized infographics designed for emergency managers and a set of detailed instructions and resources for municipal leaders.

An Inclusive Experience of Pedagogy and Collaboration

In its own process, RiverSmart was not only an integrated project open to feedback and evaluation from external sources; it was also a research team, and it functioned as an open-ended iterative discussion in which all voices, including ours, played an integral and integrated part.

We graduate students had some specific tasks to accomplish by the end of the project, but we were encouraged to explore what interested us most and what we considered to be the most important. We were also brought into both "sides" of the project, scientific and institutional, and invited to contribute to any part of RiverSmart.

This research process required developing the flexibility and willingness to toss ideas, previous work, and product ideas out the window. Group meetings were long and sometimes tedious as we often followed a non-linear process of questioning, rebuttal, and then adjustment. We would come up with one concept, present it to a community meeting and promptly be told all the reasons it would not work. We then had to go back to the drawing board. We had to learn to adapt and when to hold our ground. No matter what, we always had the support of our advisors and the benefit of having access to both advisers. It helped in some ways that while our advisors certainly had much more extensive knowledge and experience than us, neither of them were experts in the specific fields being addressed in RiverSmart. Therefore, we never felt patronized, but rather as though we were all learning together.

It was not always easy to work in this ever-changing environment, in which research was as messy as the environmental and social systems we were studying. As a short-term student, it can be frustrating to see several months of work be shelved. However, this was the reality of the research. For us as graduate students, it was a central part of our learning to be part of this critical, reflective, integrated, and collaborative research.

Part 3: Successes, Challenges, and Lessons Learned: Proliferating CPG by Mentoring a New Generation

We believe that RiverSmart's approach to research and pedagogy successfully mentored and nurtured its two graduate students to become part of a new generation of critical physical geographers. In this section, we consider what that has meant for the now graduated RiverSmart students Gillett and Slovin, some challenges all of us faced, and lessons for others who might want to institute a similar pedagogical effort in their own institution.

RiverSmart's Graduated Students: What Are the Results of a Successful CPG Education?

For the two now graduated students, Gillett and Slovin, our experience in RiverSmart changed our entire mind-sets in the way we approach human-environmental problems and research. We both already had an ability to think across science, social issues, and education. However, from our conversations with other graduate students, we have come to think that graduate education is often about diving deeply into a singular specialty. By contrast, we became immersed in the interdisciplinary perspective to the point that our personal theses were no longer siloed within a single discipline. Each thesis's background and conclusion sections delved into the implications of the research in the context of the other's field. While this may not have been deemed necessary by some outside of our research team, we felt our research was incomplete without that essential connection.

In addition to changing the way we approach research, this experience changed the way we approach our own professional lives. Upon graduation from UMass with his Master's degree, Slovin took a job at Milone & MacBroom, Inc., a consulting company with which RiverSmart worked previously, where his experience with both fluvial geomorphology and policy is valued. Similarly, Gillett completed her Master's and took a job in the nonprofit field working for Tucson Audubon Society where her interdisciplinary background and experience working with people on environmental issues on the local level is applied every day.

What Worked: What Most Helped to Make This Learning and Self-Transformation Possible?

To consider how this practically came about we can turn back to the five factors explored in Parts 1 and 2. For Gillett and Slovin as students, the most important factor for the success of the graduate experience was an inclusive experience of pedagogy and collaboration (factor #5). First and foremost, our interdisciplinary partnership with each other was critical. We each had a collaborator with whom to share our training and practice, explore ideas, and see and understand broader connections. We always had someone to turn to without fear of scorn or misunderstanding. We had someone who could act as the first level of review and screening who was roughly at the same knowledge level and so understood where the other was coming from. Also, as each of us began to focus more on our own fields, fluvial geomorphology and human geography, we could mine each other's learning and knowledge for more detail.

The relationships between the graduate students and the lead investigators were also very important, as well as others in the RiverSmart team. As our advisors and teachers, Vogel and Hatch were supportive and motivating. The RiverSmart team was always ready to listen to the ideas of all the members and take those ideas seriously. It became second nature to run our ideas by the rest of the group to gain new insights. As graduate students, we were allowed an amount of freedom much greater than our peers within the department. While that could be frustrating at times, by the end of the project we felt a much stronger sense of ownership and pride for our work.

These team partnerships were that much more meaningful and effective, though, because they were linked to an interdisciplinary setting (1), an applied problem with a commitment to outreach, and feedback-based evaluation and revision (4). RiverSmart placed us in the middle of an interdisciplinary problem, with real people to report to. We felt our work truly had meaning and impact on those we were studying and this deepened our commitment both to the project and to our advisors' applied, interdisciplinary, open-ended approach to research and education.

Neither the project's ontology, that is, its scientific and social conceptualizations of the biophysical and social/policy worlds as inherently messy, multilayered, and dynamic (2), nor its open-ended and multiple epistemologies (3), were as fundamental to our experience as the other three aspects. Nonetheless, they became deeply important to how we approached our research, and gradually percolated within our understandings to influence how we see the world and think about our future. In that sense, they may be the most lasting aspects of our graduate education. Fundamental to us were: experiential learning through trial and error; our frequent presence in the field, not only at sites of physical river changes but also talking to many different people; our exploration of multiple ideas; and our delving into a wide range of fields and topics.

Challenges for RiverSmart in Achieving CPG Education

RiverSmart experienced a set of challenges, some unique to our own situation but others which are shared by similar research. This section is written partly by faculty and partly by graduate students and includes challenges for both.

First, for the faculty, though there were advantages to our interdisciplinary leanings, we were both on the margins of our broader departments and colleges, navigating uncharted ground with little guidance. We were forging new ground as junior faculty, managing the first large grants of our careers, and trying to mentor students along the way. Occasionally we each felt deep insecurity about whether we were doing what we needed for tenure or promotion, or whether, if we ever wanted to leave UMass, we could be competitive in any other institution. Partly because of this insecurity, and partly just because of our own interests, each of us also felt committed to other, more clearly disciplinary and "scholarly" projects outside of RiverSmart as well. For our students, a key upshot was that our time and attention was often fragmented (a condition further exacerbated by the fact that we both had young children).

Partly but not only because of this, both faculty and students felt an ever-present concern of time. Especially under the pressures and limitations of grants, we simply never seemed to have enough time to complete all our work. With the need to receive input from everybody in the group as well as from those in the communities in which we worked, our review times for any product were much greater than we anticipated. We found ourselves needing to shrink our scope and the number of case studies to meet deadlines. By the time Slovin and Gillett graduated and the original RiverSmart Communities project ended, we had completed only a fraction of our planned projects.

The issue of time is ever-present in research, of course, but there is no denying it was compounded by our CPG approach, particularly our commitments to ontological and epistemological openness and to iterative learning and collaboration. However, this open thinking and iterative process were perhaps the most important reasons this project was so valuable. We counsel that if others are to work under a conception of biophysical and social systems as messy and dynamic, an open and inclusive epistemology, and iterative learning and collaboration, an acceptance of longer timelines should likely be factored in.

For the graduate students, too, there was a certain amount of isolation from our peers and other members of our fields. We took some overlapping courses, but we never worked in the same labs or classrooms as others in the department. Our research was so unfamiliar it was difficult to hold academic discussions with our peers. When attending conferences, it was difficult to find a place to present our research. While we could turn this into a positive by attending conferences and sessions we would most likely have otherwise missed, we never found similar work we could compare closely to ours. It took a dose of self-assurance and perseverance to not be negatively affected by this.

Despite our openness, true interdisciplinarity was sometimes challenging. For example, as a physical scientist, Hatch had to be willing to hear and integrate new meanings for "known" concepts, understand foreign social science methods like participant observation, and accept Vogel's pushbacks against applications of science into policy that Vogel felt were insufficiently attuned to diverse perspectives and authorities. Then, Hatch had to internalize all of this as she supervised Slovin's work. Conversely, Vogel had to work closely with Gillett as they sought terms from Hatch and Slovin that the scientific community would accept as "correct" for the processes they aimed to describe, and carefully vet her explanations of scientific principles. These kinds of mutual learning required extensive workshopping, brainstorming, and revisions of each project product.

Lessons for CPG Education

For other programs or individuals wanting to build CPG education, there are a few key take-home lessons from RiverSmart about what can help make this possible:

How to Achieve Transdisciplinarity in Education

By transdisciplinarity, we mean deep integration of students' thinking and work so there is no longer a firm boundary between disciplines. Transdisciplinary research and education require a foundation of interest, openness, mutual respect, and ongoing dialogue. Institutional interdisciplinarity helps significantly, as does an applied project.

- Faculty leaders must be interested in and truly open to the expertise and insights of the other disciplines with which they collaborate; and all members of the research team must accept and respect the methods, theories, and ideas of the others. It is helpful, though not essential, for faculty, students, and other team members to have background in each other's fields.
- There need to be regular, ongoing meetings and discussion with all parties, and all must be willing to modify their ideas, methods, plans, and products.
- There needs to be institutional support for interdisciplinarity and transdisciplinarity. At the very least, faculty need to know that the committees who decide their tenure and promotion will look favorably on their transdisciplinary research and publications, and graduate students need to know the same about their graduate committees and graduate program directors. Our experience suggests it can help if the departments in which faculty and students are housed are interdisciplinary (like UMass Geosciences), or if the personnel positions they inhabit are interdisciplinary (like Hatch's Geosciences/Extension position). It is not necessary for students to earn a transdisciplinary degree; a degree in a broad discipline like Geosciences or Geography leaves plenty of room for CPG work.
- Faculty in single-degree departments will likely need to build one or more partnerships around campus. Two rich potential resources for faculty who

work in the traditional agricultural college of their state are their state's Extension program and Water Resources Research Center.

- Scholarly training too often shuns applied projects, but applied projects are rich opportunities for transdisciplinary work that is relevant to wider communities. If approached with an ontology that recognizes and embraces messy, contingent and productive systems, and an epistemology that is open to multiple knowledges and perspectives, applied projects can be rich sources of multiple and imbricated insights and conversation.

Building and Sustaining a Team

Our experience in RiverSmart suggests that the single most important ingredient to a successfully transdisciplinary research project and graduate education lies in positive relationships among the researchers.

- The team mind-set needs to be established from the beginning. Faculty must set a tone of trust and respect among *all* team members, interest and willingness to learn from everybody, enjoyment of the research, equal validation of all ideas, and of multiple kinds of learning and knowledge.
- A multi-day research tour at the start of the academic year serves as a team bonding experience and motivates new students and staff with excitement about the new project.
- A partnership between two or more graduate students helps them develop the interdisciplinary skills they need, and gives them someone they can learn, experiment, fail, and grow with. We note that Gillett and Slovin did not start at the same time (one year apart in graduate study, one semester apart in RiverSmart) so there is some flexibility in timing, but a close overlap is important. It is very useful if they can have close or ideally shared office space.
- As the project moves along, sometimes it becomes more difficult to keep the parts connected and in active dialogue. We achieved this through biweekly all team meetings, and several shared activities throughout the year. Occasional celebrations and informal gatherings were a boost as well.

Embracing Messy, Dynamic Reality and Multiple Types of Knowledge

By conceptualizing both biophysical and social worlds as messy and dynamic, and by understanding this messiness and dynamism as productive, scholars can open research, understanding, and graduate student learning. They can also foster a tone of positive acceptance and affect which can help motivate students and sustain their well-being while still allowing room for critical thinking.

- Physical geography, and geomorphology more broadly, today embraces messy, dynamic processes and landscapes. This can be a profoundly insightful and liberating base upon which to rest a research project on human-environment interactions. Besides highlighting a range of physical, biological, and social processes and interactions, this sets the tone for students to become comfortable with the unexpected, and supports their interest in diverse forces and interactions.
- The embrace of messy, dynamic processes can be especially liberating if it is extended to social and policy worlds. With this stance, students can develop an open, positive attitude toward a diversity of people and possible policy approaches. This conceptual stance also fit well with Hatch's Extension orientation and the project's applied focus.

Addressing Challenges

We can't say with full confidence how to minimize all the challenges we faced, but here we suggest a few ideas:

- Especially in a new project or transdisciplinary context, it is helpful to have confident, self-directed students who already have experience with interdisciplinary research and work. Both Slovin and Gillett had the confidence and ability to step into leadership roles. Students requiring more explicit direction may have become lost or unproductive.
- It is helpful in early meetings to lay out concrete timelines with small, achievable pieces that can generate early feedback. Iterative learning is essential in this kind of project, but there is less inefficiency and confusion for students (as well as faculty and staff) if it happens earlier and after smaller pieces.
- Faculty may need to play an active role helping students to integrate into scholarly and professional communities. The AAG proved one good outlet for both students to present their work. A reading group might help familiarize both students and faculty with the wider scholarship that informs their partners' thinking.

Conclusions

This chapter profiled UMass's RiverSmart, a research project that has had considerable success in training graduate students in CPG-allied thinking and practice. Political geographer Eve Vogel and hydrogeologist Christine Hatch came together to conduct research and outreach on river flood hazards in New England. They each brought on a graduate student, respectively, Nicole Gillett and Noah Slovin. The four of us worked together, along with others in the RiverSmart team, for over two years. By the time Slovin and Gillett earned their Master's degrees in 2015 and 2016, it had become deeply ingrained in their thinking, learning, and work to join critical human geography—particularly a recognition and pursuit of multiple types of knowledge and multilayered interactions—with a physical geography oriented to dynamic processes and complex systems. This transdisciplinary mind-set influenced their theses and continues to influence their thinking and work postgraduation.

What made RiverSmart work as a graduate education project? We emphasized in this chapter five factors: (1) an interdisciplinary setting; (2) our ontology, a conception of the biophysical and social/policy worlds as inherently messy, multilayered, and dynamic; (3) our epistemology, a view of causal factors and knowledge as diverse and multiple; (4) an applied problem, a commitment to outreach, and feedback-based evaluation and revision; and (5) an inclusive approach to pedagogy and collaboration.

Of these, most important in terms of a successful graduate student experience was the inclusive team approach (5). Our interdisciplinary setting (1) and the applied nature of the project (4) offered crucial support, context, and motivation. Our ontological and epistemological stances (2 and 3), embracing "messy" natures and multiple types of knowledge, were less fundamental to Gillett and Slovin in their student experience, but became long-lasting orientations to thinking and research for them as graduates.

The transdisciplinary skills and the critical, diverse thinking that students can gain through a CPG education open layers and levels of understanding and communication that are unavailable to many more narrowly focused graduate students. Also, diversified skills and wider experiences are becoming more appealing in the modern job market where graduates with these attributes stand out among their peers. We hope that the insights we offer here from RiverSmart can help others launch and expand other CPG graduate training efforts to help proliferate a new generation of CPG researchers.

Acknowledgments RiverSmart was made possible by grants from the UMass Center for Agriculture, Food and the Environment (McIntire-Stennis Project 231297); the

US Army Corps of Engineers Institute for Water Resources (grant 11488334 with administrative help from the USGS); and the USDA National Institute of Food and Agriculture (grant 11447848).

Notes

1. We use "proliferate" here drawing on inspiration from J.K. Gibson-Graham (2006). We intend to evoke the sense of a multiplication of small, individual efforts, with no requirement for either exact replication or expansion of one particular kind of effort. While Critical Physical Geography is critical, it is diverse and its boundaries and its definitions are unpoliced; we support that spirit in our vision of proliferation, in which each new iteration can be started or nurtured with a seed or encouragement or resources from others but grows in its own place and context in its own way.
2. We note that our recommendations (Vogel et al. 2016) aimed to guide policy, not management. Complex and open-ended recommendations might be frustrating for landowners, small town governments, and other on-the-ground managers who often find it easier to work with something more formulaic. To guide policy, however, there was no good way to be formulaic: there are so many different policy-making institutions—six states just in New England, each with 5–10 relevant agencies, and about 12 relevant federal agencies—that we could not prescribe one thing that would work for all of them. The few things that would make a difference for all of them (e.g. FEMA maps that would include fluvial hazards, or changing FEMA flood mitigation funding requirements and processes) were simply not politically viable in the near term, and hammering too hard might make our entire effort simply dismissed. Thus in our policy recommendations we came up with fundamental things like "develop and implement fluvial hazard assessment, mapping and user access systems" which different state or federal agencies, nonprofits, or legislators could develop in their own way. Any specifics and formulas needed by on-the-ground managers, in other words, would ultimately come from policy-makers, not us. We offered ideas about specific approaches by finding and describing examples of how one state, or community, or set of agencies, or a nonprofit, made something work.

References

Andersson, Krister P., and Elinor Ostrom. 2008. Analyzing decentralized resource regimes from a polycentric perspective. *Policy Sciences* 41 (1): 71–93.
Ashmore, Peter. 2015. Towards a sociogeomorphology of rivers. *Geomorphology* 251: 149–156.
Aspinall, Richard. 2010. A century of physical geography research in the *Annals*. *Annals of the Association of American Geographers* 100 (5): 1049–1059.

Baerwald, Thomas J. 2010. Prospects for geography as an interdisciplinary discipline. *Annals of the Association of American Geographers* 100 (3): 493–501.

Berk, Gerald, Dennis C. Galvan, and Victoria Hattam, eds. 2013. *Political creativity: Reconfiguring institutional order and change.* University of Pennsylvania Press.

Bisson, Peter A., Gordon H. Reeves, Robert E. Bilby, and Robert J. Naiman. 1997. Watershed management and Pacific Salmon: Desired future conditions. In *Pacific salmon & their ecosystems*, ed. Deanna J. Stouder, Peter A. Bisson, and Robert J. Naiman, 447–474. New York, NY: Springer.

Blue, Brendon, and Gary Brierley. 2016. 'But what do you measure?' Prospects for a constructive Critical Physical Geography. *Area* 48 (2): 190–197.

Boon, Philip, and Paul Raven, eds. 2012. *River conservation and management.* Hoboken, NJ: Wiley.

Brierley, Gary, Kirstie Fryirs, Carola Cullum, Marc Tadaki, He Qing Huang, and Brendon Blue. 2013. Reading the landscape: Integrating the theory and practice of geomorphology to develop place-based understandings of river systems. *Progress in Physical Geography* 37 (5): 601–621.

Castree, Noel. 2015. Geography and global change science: Relationships necessary, absent, and possible. *Geographical Research* 53 (1): 1–15.

Gibson-Graham, J.K. 2006. *A postcapitalist politics.* Minneapolis, MN: University of Minnesota Press.

Gibson-Graham, J.K., Jenny Cameron, and Stephen Healy. 2013. *Take back the economy: An ethical guide for transforming our communities.* Minneapolis, MN: University of Minnesota Press.

Harden, Carol P. 2012. Framing and reframing questions of human–environment interactions. *Annals of the Association of American Geographers* 102 (4): 737–747.

Lave, Rebecca. 2015a. The future of environmental expertise. *Annals of the Association of American Geographers* 105 (2): 244–252.

———. 2015b. Introduction to special issue on Critical Physical Geography. *Progress in Physical Geography* 39 (5): 571–575.

Lave, Rebecca, Matthew W. Wilson, Elizabeth S. Barron, Christine Biermann, Mark A. Carey, Chris S. Duvall, Leigh Johnson, et al. 2014. Intervention: Critical Physical Geography. *The Canadian Geographer/Le Géographe Canadien* 58 (1): 1–10.

Murray, A. Brad, Giovanni Coco, and Evan B. Goldstein. 2014. Cause and effect in geomorphic systems: Complex systems perspectives. *Geomorphology* 214: 1–9.

Ostrom, Elinor, and Michael Cox. 2010. Moving beyond panaceas: A multi-tiered diagnostic approach for social-ecological analysis. *Environmental Conservation* 37 (4): 451–463.

Popke, Jeff. 2016. Researching the hybrid geographies of climate change: Reflections from the field. *Area* 48 (1): 2–6.

Rhoads, Bruce L., and Mark A. Fonstad, eds. 2016. The natural and human structuring of rivers and other geomorphic systems: A special issue in honor of William L. Graf. *Geomorphology* 252: 1–184.

Sayre, Nathan F. 2015. The coyote-proof pasture experiment: How fences replaced predators and labor on US rangelands. *Progress in Physical Geography* 39 (5): 576–593.

Tadaki, Marc, Gary Brierley, Mark Dickson, Richard Le Heron, and Jennifer Salmond. 2015. Cultivating critical practices in physical geography. *The Geographical Journal* 181 (2): 160–171.

Tadaki, Marc, and Ian C. Fuller. 2014. Freshwater geographies: Prospects for an engaged institutional project? *New Zealand Geographer* 70 (1): 1–6.

Tadaki, Marc, Jennifer Salmond, and Richard Le Heron. 2014. Applied climatology: Doing the relational work of climate. *Progress in Physical Geography* 38 (4): 392–413.

Trafford, Julie. 2012. Research supervision practices in New Zealand postgraduate geography: Capacity-capability potentialities. PhD Dissertation, The University of Auckland, Auckland, New Zealand.

Vogel, Eve, et al. 2016. *Supporting New England communities to become river-smart: Policies and programs that can help New England towns thrive despite river floods.* Amherst, MA: UMass Amherst Center for Agriculture, Food and the Environment. https://extension.umass.edu/riversmart/policy-report

Wohl, Ellen, Stuart N. Lane, and Andrew C. Wilcox. 2015. The science and practice of river restoration. *Water Resources Research* 51 (8): 5974–5997.

25

Charting a Critical Physical Geography Path in Graduate School: Sites of Student Agency

Lisa C. Kelley, Katherine R. Clifford, Emily Reisman,
Devin Lea, Marissa Matsler, Alex Liebman,
and Melanie Malone

Overview

What can graduate students do to set themselves up for success in Critical Physical Geography (CPG)? The epistemological, methodological, logistical, and institutional challenges of interdisciplinary research have been discussed for collaborative research teams and institutions (Lattuca 2001; Lélé and

L. C. Kelley (✉)
Department of Geography, University of Hawaii-Mānoa, Honolulu, HI, USA

K. R. Clifford
Department of Geography, University of Colorado-Boulder, Boulder, CO, USA

E. Reisman
Department of Environmental Studies, University of California at Santa Cruz, Santa Cruz, CA, USA

D. Lea
Department of Geography, University of Oregon, Eugene, OR, USA

M. Matsler
Cary Institute of Ecosystem Studies, Millbrook, NY, USA

A. Liebman
Department of Horticultural Science, University of Minnesota, Saint Paul, MN, USA

M. Malone
Earth, Environment, and Society, Portland State University, Portland, OR, USA

© The Author(s) 2018
R. Lave et al. (eds.), *The Palgrave Handbook of Critical Physical Geography*,
https://doi.org/10.1007/978-3-319-71461-5_25

Norgaard 2005; Lowe and Phillipson 2009; Donaldson et al. 2010); for individuals (Öberg 2009, 2010; Ray 2006; Trompf 2011); and for specific research domains (Wodak and Chilton 2005; Bardhan and Ray 2006; Hiwasaki and Arico 2007; Beder 2011; Lele and Kurien 2011; Tacconi 2011). These barriers can be intimidating, particularly in the field of CPG and particularly for graduate students (Graybill et al. 2006; Borrego and Newswander 2010). Our focus in this chapter is on how to navigate them. CPG research may be more difficult, but it is doable. Building off Hedberg et al.'s (2017) discussion of institutional "seedlings," and drawing on our own experiences in graduate school, we describe ways we have found to engage in CPG research and training at the graduate level despite institutional barriers.

As Lave et al. (2014) develop, CPG combines "critical attention to relations of social power with deep knowledge of a particular field of biophysical science or technology in the service of social and environmental transformation" (2–3). We are drawn to this approach because it would problematize, for example, a purely "social" or "natural" analysis of toxic drinking water in Flint, Michigan. A CPG approach might instead highlight how corroded pipes and hydrological dynamics converged with legacies of divestment, political greed, and racial segregation to co-produce lead-contaminated water. Through its focus on these connections, we see CPG as an opportunity to produce both novel scholarship and a lens of analysis vital for grassroots organizing, social movements, and political change.

Pursuing CPG in graduate school, however, can involve complex considerations. Simultaneously gaining expertise in critical social science and biophysical science is labor and time intensive. Undertaking CPG research generally requires building multiple intellectual communities, lab groups, mentoring arrangements, conference groups, and research partnerships. Further, most universities and graduate research positions now emphasize metric-oriented scholarship (Slaughter and Rhoades 2000; Lane 2017), incentivizing article quantity over quality in some cases and compounding academic time management challenges. Ultimately all of us need to keep the lights on and vegetables in the fridge too. Like many graduate students working across disciplines, we worry about our positionality professionally. These concerns are particularly important to recognize for people of color, LGBTQ students, women, and others who have historically been marginalized or delegitimized within academia, in part through an institutional emphasis on accepted practice.

Further, whereas some fields are relatively easily married (e.g., ecology and economics), physical and Critical Human Geography are characterized by different research foci, research methodologies, approaches to integrating data and theory, and conceptualizations of knowledge and scientific practice.

Integrating across these two fields involves careful decisions about the way data will be collected, analyzed, and presented. This makes undertaking CPG research in graduate school particularly challenging. Ideally these considerations are ironed out long before the research begins. Graduate students, however, are often still in the process of developing relevant expertise at the time they design, defend, and seek funding for their research.

This chapter focuses on strategies that have helped us to pursue our interest in CPG research and training in graduate school *despite* barriers we have confronted. It emerges from two workshops we participated in under the mentorship of two advanced CPG scholars. This workshop brought together nine graduate students at different points in their graduate education who shared an interest in CPG. Over the course of the two workshops, we provided feedback on one another's research proposals and drafts, discussed the highs and lows of our graduate training and research, and strategized challenges we were facing in our CPG work. We have since stayed in touch, exchanging emails, sharing data and references, and organizing two conference sessions to feature graduate research in CPG. This chapter consolidates our discussions over the past two years, providing concrete recommendations for other early scholars interested in this type of intellectual work.

Key Milestones for Engaging CPG in Graduate Research and Training

What do you need to conduct CPG research in graduate school? We've found that the answer to that question is different at different stages of graduate work. In the sections below, we describe strategies for negotiating seven key milestones in a graduate education.[1]

Selecting a Program and Advisor

It is difficult to do CPG research in graduate school if you find yourself in a department or funding context that discourages it. Selecting a program and an advisor that will facilitate—or at least not fight—a CPG research agenda is critical. Without this support, graduate students may find themselves forced to temporarily set aside some CPG aspirations.

Because CPG research projects span both physical and human geography, it is useful to consider programs which provide access to multiple forms of scientific expertise and training. One member of our group interested in researching agricultural production, for example, chose a program with strong institutional support for interdisciplinarity and programmatic strength in agrarian political economy and agroecology. It is also helpful to consider how much flexibility you will have in project development, course selection, and your training in different programs and under different advising relationships. For each of us, flexibility has been pivotal in conceptualizing creative (but also cohesive and manageable) CPG research. Being able to sit in on undergraduate courses in political ecology and development studies while training in a quantitative ecology lab, for example, helped one of us to raise questions about the simultaneous ecological and socio-political shortcomings of climate change mitigation policy in her research site.

Open-minded mentors willing to entertain unconventional research paths and questions are essential regardless of institutional context. Before selecting a program or advisor, reach out to a range of prospective mentors. Use meetings to assess not only how their expertise and interests complement yours but how open-minded they are of mixed-methods, integrative research. Do you sense they will be able and willing to act as an advocate for you? Supportive professional relationships and mentorship from advanced scholars are particularly essential for people of color and other marginalized groups struggling to gain credibility in the ivory tower. Advocates are needed to fight for students when concerns about "objective knowledge," "accepted practice" and/or disciplinary norms and conventions are used to delegitimize their work and ideas.

CPG research also usually diverges from the traditional advisor-advisee apprenticeship model. In CPG, students must synthesize across several academic lineages, whereas traditional scholarship largely follows the single well-established research traditions of an advisor. To address this, many of us adopted a co-chair approach, with one advisor in both biophysical and critical social science. Many institutions allow students to formally include external committee members from other universities, research institutions, or government agencies. Informal "shadow committee" members provide an additional leg of support. "Shadow committees" refer here to more advanced scholars who fill gaps in formal advising arrangements and who provide advice and mentorship under an informal (and often institutionally unrewarded) model. Not everything needs to be ironed out from the first moment you begin graduate school. Mentoring arrangements can evolve over time as you identify and refine your research objectives.

In selecting programs and advisor, it is also important to consider how much ownership you will have over your research in graduate school. This can be hard to control, as it is closely linked to funding arrangements. Funding sources tied to highly structured research opportunities or course sequences generally limit the experimentation needed to envisage and undertake CPG research. Where possible, and where given the choice, we recommend selecting programs or advisors which offer greater support (financial and institutional) for independent dissertation research. Where this is not possible, we suggest pursuing early career fellowships that fund researchers rather than specific research projects. Such fellowships allow for breadth and exploration, often freeing recipients from other paid work obligations. While fellowship applications are difficult to navigate early in graduate school, we have found that other members of our cohorts or lab groups have often been willing to workshop application materials.

In many cases, however, your program or funding arrangements may be less than ideal. As we discuss throughout this chapter, it is still possible to plant the seeds for future CPG research or to compensate for gaps in your program by drawing on resources from elsewhere. Faculty who work across disciplines or unite traditionally separate areas of expertise, even if it is not CPG-specific, have offered many of us helpful advice on mixed-methods research design, data integration and synthesis, and on how to navigate specific institutional barriers. All of us have also benefited from developing connections to fellow students in both biophysical and critical social science circles. Peer groups facilitate CPG training through formal or informal research groups, reading and writing groups, and inter-institutional working groups or conference sessions. Peer groups also provide a more relaxed context in which to workshop early ideas, develop side projects and gain expertise with a particular method or in a particular subject.

Where possible, we suggest the following characteristics are key when thinking about programs or advisors that will best support CPG research:

- Programmatic flexibility with course guidelines and requirements;
- Institutional support and funding arrangements that encourage experimentation and exposure to new fields early in the research process;
- Opportunities to engage with disparate intellectual communities or research groups;
- Mentorship from biophysical scientists, critical social scientists, *and* experts in data integration.

Developing Research Questions and Projects

Developing research questions and projects can be daunting in any field. Research questions serve as the basis for funding applications, human subjects approval, and detailed research plans. Good questions are also the driving purpose for research. Developing research questions for CPG research, however, can be especially challenging for graduate students. CPG projects rarely follow an established research approach. Advanced scholars with long-term interest in a particular research tradition may be skeptical about why diverse theories, data, and approaches are needed. Further, CPG students are often in the midst of learning one or more disciplinary "vernaculars" at the time they are developing their questions (Johnston 1986). As Öberg et al. (2013) describe, this involves a process of "strangification," in which people learn that *a priori* understandings of other disciplines may have been overly simplistic or that particular questions or foci, not originally interesting, take on new dimension or research importance when viewed through other categories, concepts, and models.

In approaching the development of CPG research questions, we have found it helpful to think in terms of what questions *require* a CPG approach. In other words, what questions can only partially be answered through a purely biophysical approach or are incompletely explained using critical theory? Some topics may be more intuitive than others to approach with a CPG lens. For example, issues already debated across divergent human-environment research communities (such as climate change, food systems, water and energy governance) can be ideal for enabling research questions that demand both biophysical science and a critical analysis of power. Further, existing datasets, where possible to obtain, can mitigate some of the costs of independently undertaking mixed-methods research. Early exposure to unfamiliar domains is also an important means of generating research ideas. One member of our group identified undetected dust storms as the central puzzle of her research after attending a geology colloquium where a dust scientist presented a research problem: scientific instruments were missing an obvious pollutant. Her foundation in science studies led her from this presentation into research questions on the politics of dust science.

Topical specialists who have been able and willing to rigorously interrogate components of our research designs have also helped many of us to reduce the intellectual labor associated with CPG research. Their insights focused on our research questions and identified relevant methodologies early in the process. Support from scholars specifically invested in CPG research has also been invaluable, for example, where they have exposed possible short-comings and unforeseen challenges in the integration of different methodological approaches and data types. Support from CPG scholars is particularly important in getting research questions past your dissertation committee. Committee members may or may not agree as to which questions are interesting and important, or even about how questions should be phrased (Öberg, 2010; Lélé and Norgaard 2005). Including a CPG scholar on the dissertation committee can help alleviate these tensions. Their work as a scholar can remind other committee members of an established, if evolving, intellectual space for critical interdisciplinary research.

Many people do not know exactly how they will approach research before they begin graduate school. Education is an iterative process and many of us learned of CPG after choosing our programs. CPG research projects themselves also speak back, not only through initial observations that confound initial research conceptions but through collaborative relationships with people whose demands and needs change over time. Not all institutions will have sufficient expertise (or even support) for proposed research questions or projects in-house. After several years of training, some of us found ourselves lacking key support within our current departments and institutions. As is true when thinking about program selection, no school is an island. Extending networks of colleagues beyond a home institution is essential. Conferences, workshops, email correspondences, and research collaborations fill training gaps and open new opportunities. Through reciprocal relationships we have benefitted from feedback on research questions and design, creative insight, mentorship on how to navigate professional challenges, help finding an appropriate audience or framing for CPG work, and reassurance in moments of confusion.

Inter-institutional peer connections are also powerful. In our case, a pre-conference workshop organized by CPG faculty sparked sustained engagement with a supportive cohort. Engaging with students from diverse universities and perspectives,[2] with varying levels of critical theory and biophysical science training, allowed us to engage in creative discussions about each others' seed ideas and proposed research questions. Providing support and being supported by other graduate students at this interface (whether over Skype, or at workshops and conferences), has provided an intellectual home, creating space for workshopping creative and integrative research questions. We have found that the benefits of inter-institutional connections can also spill over into home institutions. For example, one member of our group found that CPG workshop discussions bolstered his efforts to establish a reflexive research community within his agronomy department. Discussing CPG literature with others in his cohort helped to clarify how critical social science can help to reframe agronomic problems while simultaneously initiating conversations about how corporate funding had influenced the department. While this group member had trouble finding faculty support for his own CPG research, this cohort provided him a space to develop CPG ideas.

As is true throughout graduate school, it is also worth remembering that while the dissertation is a research product, it is also a launching-off point. A dissertation may be an imperfect first iteration of future research aspirations. One member of our group intentionally chose an interdisciplinary program, and a mentor with training in both physical and social science. However, her committee ultimately found the scope of her work too broad for a dissertation. They were not comfortable with allowing her to collect ecological data as part of her PhD, so she removed it; a decision she still regrets. Her CPG aspirations however did not end with her dissertation. She accepted a postdoctoral position working under an ecologist to develop the biophysical side of her research and round out her CPG training.

To recap, when developing your project and research questions, we recommend:

- Examining whether your research question requires both a biophysical approach and critical theory;
- Engaging with disciplinary scholars and scholarship to identify relevant methodologies and expedite the process of "strangification";
- Planning for an iterative process that responds to early fieldwork and allows the research to evolve overtime;
- Finding a CPG mentor or advocate to provide support for projects that look different than other dissertations in your department; and
- Identifying (and collaborating with) other CPG graduate students who may be tackling similar challenges of integration or having difficulty finding mentors and peers.

Funding Research

A great project without funding remains a great idea. Even if a CPG project has committee and department support, it can falter if there is not funding to execute it, and the unique qualities of CPG research do not always fit the structure of standard funding sources. Many funding calls continue to separate social science and biophysical research. Additionally, CPG's critical attention to scientific practice can be perceived as a threat to some institutions. Though CPG methodological approaches often track between "radically different knowledge practices" (Whatmore 2013: 162), the integration of qualitative or ethnographic approaches with biophysical data may be less preferred than strictly quantitative methodologies.[3]

Despite this, CPG research benefits from an increasing enthusiasm for interdisciplinary research in the US and Canadian funding context (this may differ in the European context where PhD positions usually come with funding and do not require securing external grants). Integrative work is on the rise, with many institutions looking to build bridges across historically compartmentalized "natural" and "social" sciences. Funding calls now often support novel integrative research (or researchers). Much as is true for scholars within established disciplinary lineages, our capacity to generate funding through such opportunities has been linked to our capacity to demonstrate the specific *need* for the proposed research. Citing established CPG research can help to quickly situate work in a broader community of practice and provide proof of concept.

For some, disciplinary grants can also help elaborate a rigorous basis for sub-domains of a research project. One group member with a soil science background pieced together significant funding from state and federal agencies for environmental contaminant sampling and sediment source sampling. The majority of money she received was not from formal grants, but rather came through working collaborations where she agreed to share her findings with relevant agencies. While the environmental and geophysical findings of her research were driven by a variety of social and institutional processes, she did not include the social and critical components of her project in her agency-funded collaborations until a later phase of her research. At this later phase, she was able incorporate social methodology and analysis into her research, which helped to explain her findings.

Furthermore, CPG invites researchers to not privilege theory over empirical engagement. Engaging the materiality of a phenomenon may mean not only including material qualities in data collection but allowing these aspects to change how we think about and approach a topic. While it can be frustrating to divide carefully stitched together research questions, even disciplinary grants may thus produce cross-disciplinary insights. Disciplinary grant proposals can also benefit from cross-disciplinary training. For example, one of our group members with river modeling experience has been able to integrate technical concepts in his funding applications to interdisciplinary and social science grants by employing science studies literature.

We recommend the following three strategies for funding CPG dissertation research:

- Exploring interdisciplinary funding calls that ask for attention to both social and physical processes;
- Compartmentalizing your larger project into smaller, fundable parts that can fit within disciplinary funding calls; and
- Looking for collaborations beyond explicit funding calls where interests in data align.

Developing Core Competency

Ideally, graduate school affords time and opportunity to develop core competencies in both biophysical science and critical social science while simultaneously developing expertise in questions of data integration and research design. However, no graduate student can gain expertise in all theories or methods, particularly because graduate students often have other responsibilities outside their dissertation work. This means considering which theories and skill sets are most important to master to complete the research, which can be developed more slowly over time and which dimensions of the research might be effectively addressed through collaborative work. With these considerations in mind, we highlight strategies we have developed to identify intellectual gaps quickly and begin building mastery in new domains. These include strategies for identifying broad intellectual gaps in unfamiliar academic terrain (e.g., as previously trained biophysical scientists venturing into critical social theory, or vice versa) and how to fill these gaps early enough in the research process to develop coherent integrated research plans.

First, students need to find and address important intellectual gaps. This can happen by attending research colloquia in diverse departments and engaging with the mish-mash of unfamiliar terms and fresh ideas. For example, attending a presentation on more-than-human geographies[4] (and wondering what this term and associated disciplinary jargon meant) gave one of us headway into eventual CPG research. While graduate seminars provide excellent in-depth analyses, they often assume a theoretical foundation that has not yet been established, are highly specific in scope, and lack the overarching narrative desired when entering an unfamiliar field. Upper-division undergraduate coursework can be a better fit for laying out a range of key concepts before taking a deeper dive. For example, one member investigating the historical geography of almond production audited an upper-division undergraduate course in plant physiological ecology. Fine-grained attention to the interplay between soils, climate, and plant growth provided material details that deepened her engagement with theorizations of relationality and distributed agency coming from science studies. It also provoked questions about how agronomists prioritize different interacting variables in investigations of crop production, prompting her to conduct a comparative analysis of agronomic texts between two regions that might not otherwise have been considered or possible. Intellectual gaps (and good research questions) also often pop-up where theories and experience come into friction, provoking deeper analysis. Gaps in our training, for many of us, have been identified during our first phases of field research.

Once an important gap has been identified, there are several individual, peer-based, and institutional ways of building competence within existing time and resource constraints. For many of us, offering to work as a teaching assistant for a course slightly outside our comfort zones fulfilled a professional mandate and provided financial support to develop in new intellectual directions. Peer groups are key here too. Reading or writing groups addressing questions of data integration or specific research methodologies helped us grasp unfamiliar terrain in a supportive environment. Peer groups also lessen the individual burden of identifying related scholarship. Looking at conference session schedules and gathering seminar syllabi have also been valuable time-saving tools for surveying broad research areas, targeting our reading, and situating our scholarship.

Ideally, these approaches not only expedite the process of identifying and filling particular research gaps but also help to identify relevant and supportive mentors. Working alongside these mentors during qualifying exams, oral exams, or similar tests required for doctoral candidacy (in the US context) can then provide a means of gaining formal institutional feedback on any remaining gaps in disciplinary competency and preparedness for CPG research. Including CPG literature explicitly as one domain of a qualifying exam is the most direct and obvious way to show engagement with and preparedness to conduct CPG research. However, this may not be possible for all graduate students due to committee, department, and institutional priorities. When choosing an explicit CPG, reading list was not an option, some of us developed lists that let us explore dimensions of CPG such as "Integrating Science and Technology Studies with Political Ecology." This statement covered foundational literatures across these two traditions allowing for an engagement in the politics of environmental science that acknowledges how power relations structure systems we study, how we study them, and are co-produced through material relations with real consequences for people and landscapes. Another strategy is to scatter relevant scholarship throughout all of our reading lists, with CPG persisting as an unnamed theme. Whether explicitly named or mixed among other themes, working through CPG materials in the process of an oral or written exam will expose non-versed members of your committee to this approach. Ideally, this will allow them to see your work has a sufficient institutional home, even if different from their own.

All of these approaches involve trade-offs and have to be calibrated to specific research goals and the different reasons we engage in research as both scholars and activists. All, however, have allowed us to build emerging expertise in CPG research and related domains even if our chair or home departments have not been able to support all dimensions of our research.

In summary, we recommend establishing core competencies by:

- Finding strategies to quickly identify broad intellectual gaps in unfamiliar or neighboring fields;
- Filling gaps and developing understandings and skill sets early in the research process to allow them to evolve in the course of research; and
- Taking upper-level undergraduate courses, building peer groups and mentoring relationships, and possibly teaching new material to structure your training in theories and methods outside your core expertise.

Undertaking Research

Even with well-developed research questions, supportive advisors, ample funding, and relevant expertise, undertaking CPG research can be challenging when the rubber hits the road. Managing the time and capital intensity of mixed-methods and multidisciplinary research is no small undertaking, particularly because CPG research often involves spatial and temporal mismatches between biophysical and social scientific research timelines and methods. Thinking and writing about this type of work is one thing; *doing* it is another (this is another reason we see it as important to conceptualize dissertation work as a beginning rather than an end)!

Areas of graduate student agency in pursuing and completing CPG research nonetheless exist. A manageable dissertation research plan is key. For example, because biophysical analyses often require a time series analysis, some of us began biophysical data collection early in our PhD process. This can feel like putting the cart before the horse if it is not yet possible to fully articulate a plan for eventual data integration. However, collecting such data often provides a strategic opportunity to conduct (and inform) preliminary participant observation, interviews, or other qualitative field methodologies. One member of our group, for example, used fieldwork assessing the accuracy of remotely sensed land cover change as an opportunity to collect geo-located oral histories. This work provided initial foray into a socio-political assessment of landscape change and the politics of mapping land cover change. Why, for example, had forest cover loss peaked in the late 1990s? What histories would always remain invisible to the technique, despite growing data availability and resolution?

Working at this scholarly intersection, however, should not require the research of two dissertations. While CPG integrates across disciplinary boundaries, no scholar will have equal expertise in different fields, and this means that graduate students will likely privilege one element of research, in terms of time, expertise, and resources, over another. Graduate students who have greater expertise in one set of theories or methods may require help from a peer, research assistant, or committee member to better bring biophysical perspectives to bear on social questions or vice versa, depending on disciplinary training. The level of commitment to a given area depends upon the question being addressed, as well as the researcher's short- and long-term goals.

Students can also explore creative collaborations. Two members of our group from separate departments at the same school designed their dissertations around the same topic with the goal of producing a co-authored dissertation chapter. The pair worked on questions surrounding green infrastructure. One student worked on testing biophysical processes of green infrastructure at different sites while the other explored how expertise was formed in knowledge systems at these same sites. By intentionally designing tandem inquiries, they produced data that could more easily be compared and integrated. They scheduled regular check-in points throughout data collection/analysis and scheduled site visits to coincide. This coordination helped them share disciplinary perspectives in real time as research and results developed. Other graduate students engaged in interdisciplinary geography have called for this type of innovative solution (see Gillett et al.'s chapter within this volume). Hedberg et al. (2017) have also suggested dissertation models that range from single author articles on collaborative questions to a "middle-spectrum" where a dissertation housed single and co-authored dissertation chapters. While the power to make this decision lies beyond the individual graduate student, collaborative approaches are worth exploring in supportive contexts.

To recap, we make the following suggestions for proposals and undertaking dissertation research:

- Recognizing that your dissertation is only the entry point rather than the culmination of your development as a CPG scholar. It will be imperfect but can inform a set of themes you continue to work on throughout your career;
- Creating a manageable dissertation plan that is realistic and has fair expectations in terms of labor-intensity and outcomes;
- Mapping out the timing of different processes (biophysical and social) and developing a research plan that addresses these temporal dynamics to the extent possible; and
- Exploring possibilities for collaborative work, particularly where formal institutional support allows for co-authored dissertation work.

Data Integration

Much of the intellectual strength of CPG research comes from its capacity to integrate diverse types of data. This crucial process is also one of the most difficult dimensions of research however, with much research that aims to be integrative resulting in isolated biophysical and social analyses. To be successful CPG research must maintain an integrated approach not only in envisaging the questions but in undertaking the analysis and writing.

Whether individually or in collaboration, CPG projects also require balancing expectations among disparate research communities. How researchers synthesize information is highly specific to their goals. Important determinants of our own approaches to data integration have included (among other things): what forms or types of data are accessible; what debates are seen as most resonant or important with respect to the research project and goals; trade-offs between internal and external validity and the particular gap the research is addressing; our own positionality along a spectrum of possible epistemological stances (e.g., strongly constructivist approaches vs. critical realist approaches); intended future audiences or communities of practice; particular methodological or theoretical skill sets (both pre-established and in progress); and considerations about what is most important and relevant to the broader communities among whom the research is situated. Our own approaches to data integration have evolved over time as our depth of familiarity with the data and ourselves as scholars has grown in resolution. While it is difficult to rush this process, students can anticipate promising directions by considering, for example, their tendency to engage with certain research journals, the scholarship they find most resonant, and the conversations where they hope to contribute.

As early stage scholars we draw inspiration from the growing body of CPG examples (e.g., Lave and Lutz 2014; McClintock 2015; Sayre 2015; Arce-Nazario 2016; Blue and Brierley 2016) and from other bodies of interdisciplinary work, particularly political ecologies which deeply engage biophysical science and questions of resource materiality. Not every CPG wheel has to be reinvented. Approaches taken by other scholars, even in disparate research domains, offer insight into different modes of integrating across divergent data types and disciplinary norms. An extensive literature on "qual-quant" integration provides specific methodological guidance for how to conceptualize the qualitative dimensions of quantitative data (and vice versa) while simultaneously conceptualizing the merits and costs of mixed-methods research designs (e.g., Bardhan and Ray 2006; Ray 2006; Morgan 2007, 2013). An emergent literature also speaks to how approaches from science and technology studies can be integrated with political economic approaches (Castree 2002; Gareau 2005; Braun 2008).

Two key strategies for data integration are sequential organization (i.e., with one research method or approach informing or guiding the next in some regard) and triangulation (i.e., using multiple research methods or data sources to inform a particular analytical claim). Even when ideal sequences of data collection and analysis cannot be achieved for reasons of limited time, money, or data inaccessibility, we have found structuring research phases sequentially as best as possible can enable creative sampling strategies as well as insight into interconnections in the data. For example, one student's project began with her detection of herbicides in streams that were supposed to be protected from agrochemical runoff by conservation corridors. After identifying concentrations of herbicides that were of risk to human and ecological health, she examined how farm management practices had shifted under neoliberal policy regimes, increasing herbicide use among farmers in her study area.

Of course, not all data necessarily tells the same story. In these cases we have found it helpful to think of triangulation as a tool for identifying discordance or productive tensions that merit further inquiry. These epistemic frictions have, for many of us, become the most interesting moments in our research process, often generating the most important next questions. For example, one student compared how different scientific disciplines came to know a geologic phenomenon—dust. She found convergences and divergences in how disparate approaches rendered dust visible (or invisible). Opposing knowledge claims provided some of the clearest evidence of the interlinkages between the science and the politics of dust. Examining divergence can also be a research approach, as for one member of our group who asked how physical models of flood hazard diverge from peoples' experiences living in floodplains. Not all data integration produces crystal clear results. Nonetheless, as is true of the process more broadly, each stumbling block is part of the ongoing reflexive research praxis that persists well beyond the dissertation defense.

In conclusion, we recommend the following strategies for data integration:

- Taking an iterative approach that begins by identifying the scholarship and approaches most resonant to you and/or most meaningful to the communities with whom you work;
- Looking to CPG exemplars to see how they have integrated diverse types of data and see if you can use similar or aligned strategies;

- Exploring the literature on mixed-methods, quant-qual, or interdisciplinary research design to identify compelling and problematic approaches; and
- Considering using strategies of sequential analysis or triangulation to guide your analysis.

Writing, Publishing, and Getting Feedback

Research can only circulate, get read, change minds, and contribute to the larger CPG project if it is first written (and written in a way that tells a cohesive story and effectively showcases results). This is not necessarily an end point in the research process. Writing, publishing, and soliciting feedback on findings can take many forms and, for many of us, is part of a cyclical process within collaborative or participatory projects. We focus here, however, on one of the most central writing tasks for graduate students: the dissertation. Dissertation writing can be particularly challenging because the "innate centrifugal tendencies in academia" (Lele and Kurien 2011: 1) reassert themselves in the process of finding an audience, finding relevant publication fora, and even finding reviewers that can provide rigorous feedback on all dimensions of the work. These hurdles are closely related to developing a voice and building an intellectual community. This process can be particularly murky for those students not directly following their advisor or a specific academic lineage.

One option is to treat the dissertation as a book manuscript with space to fully explicate different data sources and explore creative synthesis. Here the largest challenge might be identifying an appropriate audience for feedback or deciding how to parse the presentation of data and methods. As with designing research questions, feedback from both biophysical scientists and critical social scientists is essential and best achieved by guiding readers to the areas most in need of their expertise (e.g., methods, theoretical framework, etc.). Whenever possible, bring these people into the same room. Positive feedback attests to the rigor of the scholarship in areas with which other committee members are less familiar. Fundamental disagreements (e.g., around the ways society-nature linkages are theorized) are also a learning process for disciplinary members of committees. Take advantage of the rare occasions an entire committee is in the room together and ask questions that encourage them to engage with each other (this process is also very important during research design).

Increasingly graduate students are producing article-based dissertations. In this model, CPG students can target the growing number of journals which accept cross-disciplinary work. When uncertain about whether a particular journal is open to publishing CPG research, write to the editor, ask colleagues who have experience with the journal, or look into where more advanced CPG scholars have been publishing integrative work. Focused articles on a subset of the research can also highlight the strength of a specific dataset in disciplinary terms. Building distinct voices for divergent disciplinary audiences and growing distinct intellectual communities (e.g., within both agronomy and critical agrarian studies) is a slow process that progresses in fits and starts. Doing so can allow researchers to channel different dimensions of intellectual and political projects at different moments, while still engaging in cohesive CPG work through the dissertation or broader body of research.

Identifying other graduate students working at this interface for reciprocal feedback and support is invaluable. With little pre-existing infrastructure, this means building it from the ground up. Organizing conference sessions, particularly if the call is shared widely, can bring together graduate students from different institutions. Try tacking on mini-conferences and workshops to existing conferences, including the AAGs, DOPE, 4S, and AGU conferences,[5] all of which have relevance for the CPG community. Our group specifically leveraged time prior to DOPE and AAG conferences to convene focused sessions on CPG scholarship. Of course feedback isn't just restricted to academic circles. Learning to link our research with diverse audiences has built our capacity to establish collaborations with non-academics bridging biophysical and socio-political fields. Ideally CPG can foster reciprocal relationships in which non-academic communities are a key partner throughout the research process.

To summarize, when in the writing phase and working toward publications, we suggest:

- Considering whether a book model dissertation or an article format will better support your specific goals for synthesis and eventual integration;
- Giving your written work—or portions of it—to both social *and* biophysical scientists;
- Developing disciplinary voices for different fields that will allow you to publish articles in a broad set of journals, not just interdisciplinary ones; and
- Finding peers who work across similar disciplinary and methodological boundaries, whether at your home institution, at conferences, or beyond.

Concluding Thoughts on the Power of the Cohort and the Long-Term View

While this chapter has emphasized spaces for graduate student agency, it is worth recognizing that many of the barriers to CPG are beyond graduate student control and could be ameliorated by institutional shifts (Pain 2014; Mountz et al. 2015; Meyerhoff et al. 2011). Lack of access to secure funding, rigid institutional and advising arrangements, and minimal contact with CPG students or experts have impeded full engagement with CPG for many of us at different points in the process of obtaining our degrees—as have norms about what constitutes accepted practice within the academy and what does not. However, while not all the institutional barriers to a CPG dissertation are readily surmountable, we have found that a long-term view and supportive collaborative relationships are most important in helping to ameliorate them.

We often remind ourselves that not all aspects of our expertise must be fully achieved in graduate school and that not all components of our ideal project must be completed independently. Graduate school is an opportunity to develop the core competencies that will facilitate a future career in CPG. Even gaining basic conversancy can be an important step in identifying future research directions or in piquing the interest and potential support of future research collaborators. Similarly, our own convergence and shared work as scholars has been a keystone of our graduate experience. Our cohort, initiated as a simple pre-conference workshop, continues to deepen our intellectual connections more than a year later. Gaining guidance from senior scholars and developing a peer group emboldens graduate students to engage CPG research agendas at their home institutions and provides invaluable professional development and networks.

We are thrilled to be part of the handbook and among the growing number of scholars embarking on the intellectual journey of CPG research. We hope these reflections may help others to proactively layout a successful CPG path both by navigating their own institutions and forging connections beyond. CPG scholarship is an ongoing intellectual odyssey, but it need not be a solo voyage, and we look forward to growing and strengthening this community.

Notes

1. The milestones we identify are not distinct moments in time, and each may also differ in specifics (and/or order of completion) across different institutions, degree programs, and individuals.

2. Our workshop had graduate students with different pre-existing skill sets and forms of expertise and drawing from diverse departments and programs, including Urban Studies and Planning, Geography, Environmental Studies, Horticultural Science, and interdisciplinary programs in Environment and Society.

3. For example, despite tremendous growth in agroecological science and food and agricultural social movements over the past several decades, research funded by the USDA continues to overwhelmingly fund traditional agronomic research organized around a productivist ideal (DeLonge et al. 2016).

4. More-than-human geography is an approach that decenters human agency by foregrounding the dynamic influence of nonhumans; it challenges the divide between social and natural, instead seeing the world as emergent and co-produced through webs of relation (Whatmore 2002; Braun 2008; Panelli 2010; Robbins and Marks 2010; Tsing 2014).

5. These are a few potential conferences that other CPG scholars are especially likely to attend and include annual meetings of the American Association of Geographers (AAGs), Dimensions of Political Ecology (DOPE), Society for the Social Studies of Science (4S), and the American Geophysical Union (AGU). CPG scholars participate in a broad array of meetings and conferences; these four have been useful to us in building research connections.

References

Arce-Nazario, J.A. 2016. Translating land-use science to a museum exhibit. *Journal of Land Use Science* 11 (4): 417–428.

Bardhan, P., and I. Ray. 2006. Methodological approaches to the question of the commons. *Economic Development and Cultural Change* 54 (3): 655–676.

Beder, S. 2011. Environmental economics and ecological economics: The contribution of interdisciplinarity to understanding, influence and effectiveness. *Environmental Conservation* 38 (2): 140–150.

Blue, B., and G. Brierley. 2016. 'But what do you measure?' Prospects for a constructive Critical Physical Geography. *Area* 48 (2): 190–197.

Borrego, M., and L.K. Newswander. 2010. Definitions of interdisciplinary research: Toward graduate-level interdisciplinary learning outcomes. *The Review of Higher Education* 34 (1): 61–84.

Braun, B. 2008. Environmental issues: Inventive life. *Progress in Human Geography* 32 (5): 667–679.

Castree, N. 2002. False antithesis? Marxist, nature and actor networks. *Antipode* 34 (1): 111–146.

DeLonge, M.S., A. Miles, and L. Carlisle. 2016. Investing in the transition to sustainable agriculture. *Environmental Science & Policy* 55 (1): 266–273.

Donaldson, A., N. Ward, and S. Bradley. 2010. Mess among disciplines: Interdisciplinarity in environmental research. *Environment and Planning A* 42 (7): 1521–1536.

Gareau, Brian J. 2005. We have never been human: Agential nature, ANT, and marxist political ecology. *Capitalism Nature Socialism* 16 (4): 127–140.

Graybill, J.K., S. Dooling, V. Shandas, J. Withey, A. Greve, and G.L. Simon. 2006. A rough guide to interdisciplinarity: Graduate student perspectives. *BioScience* 56 (9): 757–763.

Hedberg, R.C., A. Hesse, D. Baldwin, J. Bernhardt, D.P. Retchless, and J.E. Shinn. 2017. Preparing geographers for interdisciplinary research: Graduate training at the interface of the natural and social sciences. *The Professional Geographer* 69 (1): 107–116.

Hiwasaki, L., and S. Arico. 2007. Integrating the social sciences into ecohydrology: Facilitating an interdisciplinary approach to solve issues surrounding water, environment and people. *Ecohydrology & Hydrobiology* 7 (1): 3–9.

Johnston, R.J. 1986. Fixations and the quest for unity in geography. *Transactions of the Institute of British Geographers* 11: 449–453.

Lane, S.N. 2017. Slow science, the geographical expedition, and Critical Physical Geography. *The Canadian Geographer/Le Géographe Canadien* 61 (1): 84–101.

Lave, R., and B. Lutz. 2014. Hydraulic fracturing: A Critical Physical Geography review. *Geography compass* 8 (10): 739–754.

Lave, R., M.W. Wilson, E.S. Barron, C. Biermann, M.A. Carey, C.S. Duvall, L. Johnson, et al. 2014. Intervention: Critical Physical Geography. *The Canadian Geographer/Le Géographe Canadien* 58 (1): 1–10.

Lattuca, L.R. 2001. *Creating interdisciplinarity: Interdisciplinary research and teaching among college and university faculty.* Nashville: Vanderbilt University Press.

Lele, S., and A. Kurien. 2011. Interdisciplinary analysis of the environment: Insights from tropical forest research. *Environmental Conservation* 38 (2): 211–233.

Lélé, S., and R.B. Norgaard. 2005. Practicing interdisciplinarity. *BioScience* 55 (11): 967–975.

Lowe, P., and J. Phillipson. 2009. Barriers to research collaboration across disciplines: Scientific paradigms and institutional practices. *Environment and Planning A* 41 (5): 1171–1184.

McClintock, N. 2015. A Critical Physical Geography of urban soil contamination. *Geoforum* 65: 69–85.

Meyerhoff, E., E. Johnson, and B. Braun. 2011. Time and the university. *ACME: An International E-Journal for Critical Geographies* 10 (3): 483–507.

Mountz, A., A. Bonds, B. Mansfield, J. Loyd, J. Hyndman, M. Walton-Roberts, R. Basu, et al. 2015. For slow scholarship: A feminist politics of resistance through collective action in the neoliberal university. *ACME: An International Journal for Critical Geographies* 14 (4): 1235–1259.

Morgan, D.L. 2007. Paradigms lost and pragmatism regained methodological implications of combining qualitative and quantitative methods. *Journal of Mixed Methods Research* 1 (1): 48–76.

———. 2013. *Integrating qualitative and quantitative methods: A pragmatic approach.* Thousand Oaks, CA: Sage Publications.

Öberg, G. 2009. Facilitating interdisciplinary work: Using quality assessment to create common ground. *Higher Education* 57 (4): 405–415.

———. 2010. *interdisciplinary environmental studies: A primer.* New York: Wiley.

Öberg, G., Fortmann, L., and Gray, T. 2013. *Is interdisciplinary research a mashup?* IRES Working Paper Series No. 2013–02. Institute for Resources, Environment and Sustainability.

Pain, R. 2014. Impact: Striking a blow or working together? *ACME: An International E-Journal for Critical Geographies* 13 (1): 19–23.

Panelli, R. 2010. More-than-human social geographies: Posthuman and other possibilities. *Progress in Human Geography* 34 (1): 79–87.

Ray, I. 2006. Outcomes and processes in economics and anthropology. *Economic Development and Cultural Change* 54 (3): 677–694.

Robbins, P., and B. Marks. 2010. Assemblage geographies. In *The SAGE handbook of social geographies*, ed. S. Smith, R. Pain, S. Marston, and J.P. Jones. Los Angeles: Sage Publications.

Sayre, N.F. 2015. The coyote-proof pasture experiment: How fences replaced predators and labor on US rangelands. *Progress in Physical Geography* 39 (5): 576–593.

Slaughter, S., and G. Rhoades. 2000. The neo-liberal university. *New Labor Forum* 6: 73–79.

Tacconi, L. 2011. Developing environmental governance research: The Example of forest cover change studies. *Environmental Conservation* 38 (2): 234–246.

Trompf, G.W. 2011. The classification of the sciences and the quest for interdisciplinarity: A brief history of ideas from ancient philosophy to contemporary environmental science. *Environmental Conservation* 38 (2): 113–126.

Tsing, A. 2014. More than human sociality. In *Anthropology and nature*, ed. K. Hastrup. New York: Routledge.

Whatmore, Sarah. 2002. *Hybrid geographies: Natures, cultures, spaces.* London: Sage Publications.

Whatmore, S.J. 2013. Where natural and social science meet? Reflections on an experiment in geographical practice. In *Interdisciplinarity: Reconfigurations of the social and natural sciences*, ed. A. Barry and G. Born, 161–177. London: Routledge.

Wodak, R., and P. Chilton, eds. 2005. *A new agenda in (critical) discourse analysis: Theory, methodology and interdisciplinarity.* Vol. 13. Amsterdam: John Benjamins Publishing.

26

Critical Reflections on a Field in the Making

Christine Biermann, Stuart N. Lane, and Rebecca Lave

Together, the chapters of this Handbook begin to flesh out a Critical Physical Geography (CPG) that is topically, theoretically, and methodologically diverse, held together by a shared impulse to integrate across the human-physical divide in order to produce knowledge that transforms our communities, environments, and worlds. Toronto's river systems are explored as functions of watershed topography and development practices, hydroclimate, and institutional attitudes (Ashmore, this volume); mosquitoes in West Baltimore are embedded in histories of disinvestment and environmental injustice (Biehler et al., this volume); and soil chemical properties in Hungary's Drava River floodplain are considered in relation to legacies of state-socialist land ownership and socially differentiated farming practices (Engel-DiMauro, this volume). This type of careful integrative work is crucial if we are to develop thorough and rigorous accounts of the changing social and biophysical worlds we inhabit and research. But CPG is not unique in its quest for interdisciplinary and transformative scholarship. Within Geography, calls for research across the human-physical divide have recurred with such frequency that they seem to be a defining feature of the discipline. Within the academy

C. Biermann (✉)
Department of Geography, University of Washington, Seattle, WA, USA

S. N. Lane
Institute of Earth Surface Dynamics, Université de Lausanne,
Lausanne, Switzerland

R. Lave
Department of Geography, Indiana University, Bloomington, IN, USA

© The Author(s) 2018
R. Lave et al. (eds.), *The Palgrave Handbook of Critical Physical Geography*,
https://doi.org/10.1007/978-3-319-71461-5_26

559

more broadly, the declaration that we are living in the Anthropocene has spurred attention by social science and humanities scholars to earth system processes and by geoscientists to human processes and impacts. Indeed CPG's impulse to integrate is increasingly common.

At the same time, the academy is rife with calls for both natural scientists and social scientists to translate their work for public audiences, to perform socially and politically meaningful research, and even to embrace the roles of public intellectual, advocate, or activist (Crowley 2016; Kristof 2014). Despite heated debates around the proper relationship of science to politics, the 2017 March for Science saw 1.3 million scientists and supporters take to the streets in more than 600 cities around the world to champion the practical and political import of science. For environmental researchers, frustrations about climate change inaction and denial, in particular, are spurring many scientists to become more politically active and engaged. So here, too, is another core driver of CPG (the desire to perform transformative research) that is increasingly common.

Given these contexts, one might reasonably wonder what, if anything, CPG offers to existing efforts to produce integrative and transformative scholarship. If the value of integrating natural and social science perspectives and methods is already so widely recognized, is there a benefit to forming a new field in which to pursue this integration? By labeling integrative research as CPG, might we actually run the risk of marginalizing this work in the broader environmental science field? On the political front, if so many scientists are already stepping into the public and political spheres, are we taking down a straw man when we insist that science must be attentive to politics and power relations? Worse yet, might an emphasis on reflexivity, politics, and coupled biophysical-social transformation actually further erode public trust in science and render it a partisan issue, particularly at a time in which scientific knowledge on environmental issues, from the local to the global, is absolutely vital?

Weaving together insights from individual chapters, we conclude this Handbook by reflecting on the distinctiveness of CPG, the values and politics embedded in the work presented here, and the risks and benefits of endeavoring to produce transformative research. We do so by reflecting on a series of interlinked questions that have been raised in the writing of this Handbook, including those in the previous paragraph. We do not aim to provide exhaustive answers to these questions or solutions to the issues they raise. Rather we consider them in the spirit of self-criticism and reflexivity.

Some physical and social scientists are already working together to examine interconnections among human and natural systems. What, if anything, does CPG offer that is different?

Certainly, many natural and social scientists work together already, and interdisciplinarity has been institutionalized to some extent through funding mechanisms like the US National Science Foundation's Coupled Human and Natural Systems (CNHS) and Interdisciplinary Graduate Education, Research, and Teaching (IGERT) programs and the UK's interdisciplinary Rural Economy and Land Use research program. There are numerous examples of path-breaking integrative research predating CPG and extending well beyond Geography (e.g. Altieri 1989; Klepeis and Turner 2001; Lambin et al. 2001; Fischer-Kowalski and Haberl 2002; Lahsen 2005; Sundberg 2009; Dyer 2010) Yet, most integrative environmental research has involved a relatively narrow picture of the social, as scholars such as Ron Johnston have been pointing out since the early 1980s (Johnston 1986). Such research is often pursued and framed through a quantitative analytical approach in which social relations are reduced to variables such as population density, income, or land use (Castree 2015; Lane et al. this volume). Similarly, attempts among social scientists and humanities scholars to engage physical science risk oversimplification by treating the environment as either an inert platform upon which social processes unfold or as merely a set of representations about the world rather than a material reality. And while social scientists may acknowledge a co-constituted, biophysical-social world, social science research is rarely expanded to address the specific biophysical dynamics of this co-constitution. Indeed, it is far easier to claim biophysical-social co-constitution than to design and perform empirical research that reveals the bases of the claim being made. Thus while there unquestionably is existing integrated research, the nature of that integration commonly oversimplifies one or the other: natural or social.

But is integrated research actually what we need? Do integrative projects inevitably yield richer, more nuanced, and more thorough findings than projects which are more narrowly constituted? This claim was central to earlier framings of CPG:

> [CPG's] central precept is that we cannot rely on explanations grounded in physical or critical human geography alone because socio-biophysical landscapes are as much the product of unequal power relations, histories of

colonialism, and racial and gender disparities as they are of hydrology, ecology, and climate change. CPG is thus based in the careful integrative work necessary to render this co-production legible. (Lave et al. 2014)

In line with this emphasis on rigorous integrative work, the authors of the chapters of this Handbook sampled soils; trapped mosquitoes; cored trees; mapped wildlife; analyzed geomorphic data; administered a survey of tree-ring scientists; performed interviews with urban Baltimore residents, Hungarian farmers, Maasai herders, Puerto Rican water users, and US stream restorationists; and delved into archives on Colombian soil science, Mediterranean ecology, and range management in the American West (and more). Clearly, part of what defines CPG is a commitment to understanding environments as unique products of social and biophysical dynamics and not simply either one or the other (Urban, this volume).

Yet taken together, the chapters of this Handbook also pose a challenge to the blind faith in integration, interdisciplinarity, and transdisciplinarity that has become especially prevalent in academic discourses around the Anthropocene. The chapters of this Handbook ask researchers to re-examine our commitments to integration and associated narratives of co-constitution and to be cautious about *what* is being integrated, *how*, *by whom*, and *toward what ends*. The distinctiveness of CPG, then, perhaps lies less in its commitment to integrate across the human-physical divide and more in its self-conscious recognition that integrative research necessitates consideration of where concepts, frameworks, and methods come from and what types of insights are likely to be privileged by particular approaches. Bluntly, the integrated research that CPG envisages is not an intellectual advance if it simply brings a wider range of tools to answer the same old questions, or if it fails to question the concepts and theories that limit current understanding, not to mention the political commitments that undergird them.

Several chapters of this Handbook exemplify CPG's resistance to narrow and un-reflexive modes of integration. For example, we can identify a strand of work that unpacks the assumptions of established precepts and critiques the dissociation of scientific findings from their particular contexts. In their respective chapters, Diana Davis, Chris Duvall, and Nathan Sayre re-embed seemingly objective, universal scientific perspectives (Mediterranean plant ecology, savanna biogeography, and range science) within the biophysical and social relations out of which they have emerged. Davis' chapter, for example, reveals that the dominant vegetation classification scheme of Mediterranean ecology was based largely on colonial misunderstandings of North Africa as deforested and desertified. This misunderstanding has informed a series of unsuccessful agricultural and environmental projects in the Mediterranean region, which have in turn further undermined traditional livelihoods. Meanwhile, Duvall shows that biogeographic categories like "savanna" may actually impede knowledge of location-specific biophysical conditions while also perpetuating simplistic generalizations that serve particular political-economic interests. Finally, Sayre analyzes early experiments on grazing in western US rangelands, finding that hierarchies of race and class shaped range science as it developed around the turn of the twentieth century.

From these contributions we can extract one of CPG's crucial insights: integrative work needs to be keenly aware of what exactly is being integrated and from where its assumptions and categories derive. Failing to do so risks constructing deeply flawed accounts of the world, setting up environmental practices and policies for failure, and perpetuating social and environmental injustices and problems. In this way, CPG acts as a brake on "interdisciplinarity as usual" and encourages a deepened sense of curiosity and humility about the particular conditions under which disciplinary constructs and taken-for-granted concepts have been forged. But as Robertson et al. (this volume) note, the kind of reflexivity that CPG encourages goes against established scientific norms and may thus limit its attractiveness and accessibility as a framework, an issue we discuss later in this conclusion.

What is the value of reflexivity? Can engagement with the politics of knowledge production strengthen rather than undermine scientific inquiry?

Debates about the value (or lack thereof) of critical social theory for environmental scholarship have been ongoing since at least the mid-1990s (e.g. Soulé 1995; Proctor 1998), particularly around constructivist arguments that highlight the myriad ways in which science is political, value-laden, and contextual. In response, some scientists have characterized postmodern and poststructural deconstructions of environmental issues as anti-scientific rejections of objectivity, with "social theory" criticized as at best jargon-laden and at worst a dismissal of the unique value of science to society. In an uncomfortable twist, critiques of science have now entered the mainstream; politicians wield "alternative facts"; skepticism and denial, strongly reminiscent of social constructivist arguments, have become fairly common, even anticipated, reactions to scientific findings.

One reaction to this new intellectual and political climate is to double down on claims to objectivity in an attempt to re-seat science on its authoritative pedestal. There may indeed be some value to this approach; if policymakers and the public unquestioningly trusted scientific consensus on climate change, for example, perhaps political action on greenhouse gas emissions would be more tractable. But at the same time, there are very real risks to this approach, and we would suggest a different option, drawing upon the work of feminist and postcolonial philosopher of science Sandra Harding, among others. This option calls for science to internalize and expand rigorous self-critique, to swing the door open to diverse publics and ways of knowing, and to be more reflexive about the values embedded in research framings, questions, methods, and problem resolutions.

Might scientific authority and public trust in science be achieved through self-critique and openness rather than through claims to neutrality and objectivity (Blue and Brierley 2016)? We will return to this question later in the conclusion; for now, let's consider what might be achieved by turning critiques of science onto our own work rather than dismissing them out of hand. What if we consider how scientific assumptions and "facts" are shaped by particular social and political contexts as a means to formulate novel frameworks, methods, research questions, and hypotheses? Might bringing such critiques into the fold, so to speak, actually expand the intellectual and practical value of science (Lane 2017)? Might it not allow scientists to be more creative, if those critiques enable us to see what we study differently?

In this Handbook, chapters by Simon, Turner, Arce-Nazario, Goldman, and Lave et al. attempt this. While the chapters discussed in the previous section unpack dominant concepts and theories to clear the way for new science, these chapters begin to build, and in some cases test, new explanatory frameworks and methods. Reading these chapters together, they demonstrate that attention to social power relations, both within and outside of science, can prompt innovative modes of inquiry. Studying fire in the American West, Simon questions the assumptions implicit in the concept of the wildland-urban interface (WUI). Even as the WUI provides a framework within which to understand fire as a function of biophysical and social factors, it fails to "reveal the forces behind its own creation" (p. 161). According to Simon, popular understandings of wildfire at the WUI actually depoliticize and conceal the role of planning and development practices in producing costly and destructive fire events. What is especially significant is that Simon is not content with mere critique but instead proposes an alternative framing to remedy the identified limitations. Adopting the affluence-vulnerability interface (AVI) as a framework allows Simon (and others) to analyze development policies, economic incentives, and environmental changes together, thereby grappling with systemic causes of change, risk, and vulnerability rather than continuing to view fire as the result of inevitable and unquestionable exurban and suburban development.

While Simon's chapter builds on critique to enact an alternative framing, chapters by Turner and Arce-Nazario show how research methods are politically embedded, such that certain questions, themes, and scales are privileged while others are overlooked or ignored. Building on their analyses of the politics of knowledge production, these chapters begin to formulate new combinations of methods in order to overcome some of the issues they identify. Studying water quality and access in Puerto Rico, Arce-Nazario demonstrates that conventional environmental justice (EJ) approaches to water quality have relied heavily on governmental regulations and compliance standards, which "not only neglect to incorporate variables quantifying social, cultural, and ecological value, but can also be in conflict with these values" (p. X). To deal with these issues, Arce-Nazario's CPG approach entails field-based data collection at both the household and watershed scale and a broader reconceptualization of "water quality risk" as something other than mere non-compliance with environmental regulations, particularly because regulations created in one social and ecological context (the continental US) do not always travel well to others (Puerto Rico). Turner also attends to social relations, both within science and among scientists and West African farmers, to suggest and enact new methods for soil science in the region. By augmenting biophysical data on soils with conversations with farmers about farming practices and outcomes, researchers can overcome some of the limitations of nutrient budgeting calculations as a singular method. The upshot of this is both more accurate assessment of anthropogenic stresses on the soil-plant system and a deeper engagement with farmers as environmental actors and knowledge producers in their own right.

Similarly, Mara Goldman's chapter puts multiple ways of knowing into conversation to ultimately produce a novel understanding of wildlife movement and population sizes in Tanzania's Tarangire-Manyara Ecosystem. Tacking back and forth between scientific inquiry (conducting wildlife walking transects and analyzing data collected by other researchers), ethnography (consulting with local Maasai), and reflection on the politics of knowledge production, Goldman's research identifies specific limitations and strengths of different forms of data without dismissing any out of hand. Combining disparate methods ultimately yields data that challenge existing assumptions about wildlife habitat. Rather than a stark divide between nature and society, Goldman finds evidence of regular use of village land by wildlife. Such a finding is especially significant as it calls into question scientific justification for conservation policies that seek to separate humans from wildlife in space.

In other words, bringing critique into the fold and reflecting on the complexities involved in measuring wildlife movements actually expands the intellectual and practical value of Goldman's research.

Attention to social and political-economic relations can also help to formulate novel research questions. While neoliberal and market-based environmental policies have been examined in great detail by critical nature-society researchers, they represent a relatively untapped area for physical science. At the same time, political debate continues over the efficacy of market-based strategies, and there is a pressing need for evidence-based research on their varied material impacts. The chapter by Lave et al. shows how entirely new areas of research can be launched when, for example, fluvial geomorphology meets critical political economy. For these authors, geomorphic data on stream channel form is used for empirical testing of theoretical assumptions about stream mitigation banking (SMB) as a neoliberal environmental policy. Importantly, their study uses a critical theoretical approach not to deconstruct science but to generate questions and inform hypotheses which are then tested using physical surveys and geomorphic data. At the same time, they do not selectively use geomorphic data to merely confirm theoretical assumptions regarding neoliberal environments.

Crucially, these chapters demonstrate that a "critical" approach to Physical Geography need not be (and, we believe, ought not to be) a means of undermining science, limiting its explanatory power, or reducing the field of available research questions and methods. The chapters discussed above use social theory and attention to social power relations to widen both existing fields of inquiry and ranges of solutions to environmental problems. By offering radically different framings, questions, and methods through which to integrate natural and social science, a CPG approach can allow researchers to document unseen and underlying processes, challenge institutional failures, and ultimately contribute to social and environmental justice in our field sites, sciences, and worlds. However, this also raises challenging questions about the place of values in science and the complexity of enacting a normative science that seeks to produce knowledge in the service of social and environmental transformation. It is to these questions that we turn now.

Can science be normative, and what are critical physical geographers trying to change?

It is through recognizing that *all* science is embedded in social relations and structures of power (King and Tadaki, this volume) and that our research has material effects on society and environment, whether we approve of them or not (Law, this volume), that we are inspired to perform research under conditions that run counter to current norms of science. As such, many of the chapters of this Handbook envision and enact modes of inquiry that differ radically from the modernist ideal of science as the production of objective, value-neutral knowledge about the world by experts through technical analytic frameworks.

Much CPG work might indeed be termed "normative": it is explicitly value-laden and self-conscious of its own commitments, norms, and preferences about the world, and may be activist in arguing for what the world should become. While this may seem utterly heretical to some within the scientific community, examples of normative concepts already abound in the environmental sciences: ecosystem health, ecological integrity, and ecosystem services, for example, are all value-laden concepts, yet their normativity is generally concealed by claims of objectivity (Lackey 2004; Landis 2007). Similarly, the ways in which research questions are formulated and posed as needing-to-be answered are not exempt from values or politics, even as if they might appear to emerge from a political vacuum. Even the very claim that excluding values from research leads to some kind of authority for scientific knowledge over other kinds of knowledge is itself a normative position.

One response to accusations of values in science has been to seek the removal of all values from science; in the discipline of ecology, for example, many have lamented that the categories of native and non-native are normative, and that species should be judged by criteria that are not imbued with human preferences (Davis et al. 2011). In this line of reasoning, the task of the expert scientist is to produce policy-relevant but policy-neutral knowledge (e.g. the IPCC) and to leave the value judgments and prescriptions to other parties. By avoiding political engagement, it is assumed that scientists enhance public trust in the knowledge they produce (Lackey 2004). CPG differs because rather than seeking to foster trust by removing values from science, it offers a disciplinary home for integrative work that explicitly and reflexively grapples with values and transformative ambitions. Effectively, CPG research argues that if we cannot avoid values we should engage them.

This is not to say that all CPG work shares the same values or ambitions; indeed there are numerous different goals at play in this Handbook based on several different overlapping transformative ambitions. First, we note a broad commitment, informed by Critical Human Geography's radical and activist traditions, to fundamentally challenge existing social and environmental injustices through scientific inquiry. For example, we see this commitment implicit in Greta Marchesi's historical account of twentieth-century soil research programs in Colombia, in which she demonstrates how the research and management priorities of the Federation of Colombian Coffee Growers

(FEDECAFE) supported peasant smallholders, in contrast to US Green Revolution science, which encouraged plantation-scale industrialization of Colombian coffee farms.

A second shared transformative ambition is to contest environmental management failures and expand the range of possible solutions to environmental problems. In the chapter by Dufour et al., for example, the authors find that maps of ecosystem services often exhibit tremendous inaccuracies and inconsistencies, yet are widely assumed to be accurate, objective, and neutral. Here, a CPG approach highlights the fraught nature of the increasingly dominant ecosystem services framework, and explores the political implications of map production in ecosystem management and research. In so doing this chapter shows how blind faith in tools like maps can circumscribe the range of available responses to environmental problems and, in the process, can contribute to management failures and socially unjust outcomes.

Finally, we recognize a common project of transforming the scientific process from the inside, for example, by grappling with uncertainty, engaging diverse publics, rethinking pedagogy and training, and acknowledging the politics of knowledge production. Justine Law, for example, urges us to reconsider scientific norms that impose an artificial separation between the researcher and the researched. With regard to invasion science, Christian Kull exhorts researchers to continually reflect on words and labels, to consider how science is used and who its applications benefit and harm, and to question the voice of expertise. Similarly, the chapter by Knitter et al. calls for a more integrative and reflexive landscape archaeology that is able "to learn from the strategies, techniques, and behavioral patterns of former people and their landscape, about current issues in our own eco-social systems" (p. 196).

Often these three transformative ambitions intersect; Barron's (this volume) work on conservation values, for example, is committed both to effective biodiversity conservation and to ethical and socially just environmental decision-making. Similarly, Biermann and Grissino-Mayer (this volume) aim both to grapple with uncertainty in the field of dendroclimatology and to consider how the field can better inform climate change action and adaptation. A final example comes from the University of Massachusetts RiverSmart project (Gillett et al., this volume), which "combines social and river science, institutional and policy research, and community outreach, to research and address river floods in New England" (p. XX). The RiverSmart team has worked simultaneously toward multiple transformative aims—toward fostering a culture of interdisciplinarity in graduate education, toward increasing awareness and understanding of flood risk among the public, and toward more effective management of river flooding.

In a sense, CPG's transformative ambitions might be understood as a new formulation of engaged research for the Anthropocene. Certainly many researchers want their work to have real world impacts or implications, yet often the options for achieving these impacts feel constrained. Some critical human geographers, for example, hesitate to perform applied research, fearing that they may be indirectly furthering injustices or violence. Conversely, physical geographers likely have more opportunities for applied research on environmental problems, but might find that this research involves far messier social and political dynamics than expected. Perhaps, however, the social, political, and practical relevance of both physical and critical human geographies can be expanded by integrating the unique approaches and forms of expertise that each field brings to the table.

Yet, and as noted above, geographers are certainly not the only ones pursuing politically and socially engaged science, and this transformative impulse appears to be increasing in a variety of scientific fields. The high planetary stakes implied by the Anthropocene and global climate change have in particular pushed a number of prominent climatologists (e.g. James Hansen and Jason Box, to name a few) into advocacy and activism. Indeed the Anthropocene has been an arena in which scientists, social scientists, and the public are already discussing values and charting courses for our planetary future (e.g. Asafu-Adjaye et al. 2015, Haraway 2015, Moore 2015). Society relies upon science to help us parse the possible routes our worlds may take and come to grips with exactly how we got here, recognizing that "here" is experienced and understood differently across different social groups. Yet it is increasingly clear that coming to grips with the Anthropocene necessitates new approaches to knowledge production. Such new approaches need to attend to the limits of scientific knowledge and to consider seriously "when to look beyond science for ethical solutions," which "new facts to seek and when to resist asking science for clarification," and how to "reframe problems so their ethical dimensions are brought to light" (Jasanoff 2007: 33). As a discipline that already includes both biophysical and social researchers, Geography is well-placed to contribute to these discussions.

Conclusions

There remain many challenges and potential pitfalls for those who wish to practice CPG or participate in such normative discussions, but these challenges are not unsurmountable. First, there are not always existing structures through which scientists and social scientists can readily communicate the ethical dilemmas raised by particular situations, the uncertainties inherent to their findings, or the complex interactions through which an environmental phenomenon becomes a problem to be solved. Many researchers perceive (often rightfully so) that policymakers and the public demand firm answers in the form of easily digestible sound bites. If researchers do acknowledge the ethical and scientific assumptions, uncertainties, or complexities inherent to their findings, they may be simply ignored, or these uncertainties may be exaggerated to diminish the power of the findings, or even the findings may "bite back" when, in the future, necessary assumptions are no longer found to hold.

Second, the integrative nature of CPG raises additional practical concerns. Instead of incorporating ideas and concepts from a single intellectual community, a project may be drawing on two or three entirely separate fields, each with their own epistemologies, histories, debates, and so on, for example, fluvial geomorphology and political economy (Lave et al., this volume), or vegetation ecology and postcolonial theory (Duvall, this volume). With this expansion, there is an equally large reduction in the number of other researchers who have been trained in these combinations and can confidently gauge the merit of such work. Finding capable reviewers may prove exceedingly difficult for some CPG scholarship. For students who endeavor to do this type of integrative work, it is challenging to find advisors and mentors who are able and willing to guide such projects (Kelley et al., this volume). Even after research is successfully published, there is a risk that it will reach only a limited community of other CPG researchers. The challenge, then, for those who undertake CPG research is to strive to be outward- rather than inward-oriented. Reaching out across the human-physical divide should not be seen as buffering a project from critique but rather expanding the field to which that research is responsible. In other words, research that combines geomorphology and political economy needs to be responsible to both research communities rather than to CPG as a field.

In addition, CPG cannot be blind to the nature of the academy as it has become. As argued elsewhere (e.g. Pain 2014; Mountz et al. 2015; Lane 2017), Geography, as with all disciplines, finds itself within a neoliberal academy. There is ever more intervention in day-to-day academic practices, interventions that can be highly judgmental about what research is valued and how (e.g. what constitutes "important" research questions, how research should be disseminated to the public, etc.). Geography does not escape such judgments, reflected in the often-perceived weakness of research that does not easily fit predominant wisdom as to what is novel or cutting edge in human geography or physical geography. Faced with such constraints, how should we advise and support someone who wishes to become a CPG researcher? If doing CPG implies taking risks, precisely because CPG research cuts across disciplinary traditions, who should take those risks? How do we create the safe and secure working environments that sustain such risk-taking? How do we train those who might see a CPG approach as necessary? In this volume, the chapter by Kelley et al. begins to flesh out some responses to these challenges with regard to graduate training, but these questions remain important points of conversation for all who are interested in the integrative and transformative mission of CPG.

A final challenge relates to the particular transformative aims of some CPG work. By looking at the histories of mission-oriented disciplines like conservation biology and restoration ecology, we can see that it is by no means simple to "[balance] scientific legitimacy with political efficacy" (Galusky 2000: 226). Indeed, there is a perceived risk that engaging with values and politics will contribute to the erosion of public trust in science. This risk is particularly acute when scientists act as stealth advocates, using what appears to be objective, value-neutral science to covertly advocate for particular value-laden aims (Pielke 2007). Invoking scientific authority, stealth advocacy works to limit the choices available to decision-makers and to close off debate about the relative value of each possible choice, attempting to narrow decisions down to purely technical issues. But this approach runs counter to CPG's self-conscious recognition that what we study and how we study it are shaped by politics and power relations. Such an approach also implies that value-neutral scientific inquiry can provide clear direction on value-laden decisions about what the world ought to be and how to bring it into being. For CPG researchers, the challenge will be to avoid pre-empting debate or concealing choices and values as irrefutable scientific fact, even when it may prove tempting and politically expedient to do so. We would perhaps do well to heed the advice of Jasanoff (2008): to "[disclose] the limits of [our] information and the extent of [our] uncertainty in a spirit of professional humility" (p. 240). In so doing we enable scientific inquiry to earn its authority and privilege through relentless reflexivity and openness rather than claims of value-neutral objectivity.

References

Asafu-Adjaye, J., L. Blomqvist, S. Brand, B. Brook, R. DeFries, E. Ellis, C. Foreman, et al. 2015. *An ecomodernist manifesto*. [Online]. http://www.ecomodernism.org/manifesto-english/. Accessed 2 Mar 2017.

Blue, B., and G. Brierley. 2016. 'But what do you measure?' Prospects for a constructive Critical Physical Geography. *Area* 48 (2): 190–197.

Castree, N. 2015. Geographers and the discourse of an earth transformed: Influencing the intellectual weather or changing the intellectual climate? *Geographical Research* 53 (3): 244–254.

Crowley, K. 2016. Our thinking about crossover scholarship is wrong. *Inside Higher Ed.* [Online]. https://www.insidehighered.com/views/2016/04/05/value-cross-over-scholarship-academics-essay. Accessed 2 Mar 2017.

Davis, M., M. Chew, R. Hobbs, A. Lugo, J. Ewel, G. Vermeij, J. Brown, et al. 2011. Don't judge species on their origins. *Nature* 474: 153–154.

Dyer, J.M. 2010. Land-use legacies in a central appalachian forest: Differential response of trees and herbs to historic agricultural practices. *Applied Vegetation Science* 13: 195–206.

Fischer-Kowalski, M., and H. Haberl. 2002. Sustainable development: Socio-economic metabolism and colonization of nature. *International Social Science Journal* 50: 573–587.

Galusky, W. 2000. The promise of conservation biology: The professional and political challenges of an explicitly normative science. *Organization and Environment* 13 (2): 226–232.

Haraway, D. 2015. Anthropocene, capitalocene, plantationocene, chthulucene: Making kin. *Environmental Humanities* 6 (1): 159–165.

Jasanoff, S. 2007. Technologies of humility. *Nature* 450: 33.

———. 2008. Speaking honestly to power. *American Scientist* 6 (3): 240.

Johnston, R.J. 1986. Fixations and the quest for unity in geography. *Transactions of the Institute of British Geographers* 11: 449–453.

Klepeis, P., and B.L. Turner. 2001. Integrated land history and global change science: The example of the southern yucatan peninsular region project. *Land Use Policy* 18 (1): 27–39.

Kristof, N. 2014. Smart minds, slim impact. *The New York Times*, 15 February. SR11.

Lackey, R. 2004. Normative science. *Fisheries* 29: 38–39.

Lahsen, M. 2005. Seductive simulations? uncertainty distribution around climate models. *Social Studies of Science* 35 (6): 895–922.

Lambin, E.F., B.L. Turner, H.J. Geist, S.B. Agbola, A. Angelsen, J.W. Bruce, O.T. Coomes, et al. 2001. The causes of land-use and land-cover change: Moving beyond the myths. *Global Environmental Change* 11 (4): 261–269.

Landis, W. 2007. The Exxon Valdez oil spill revisited and the dangers of normative science. *Integrated Environmental Assessment and Management* 3 (3): 439–441.

Lane, S.N. 2017. Slow science, the geographical expedition, and Critical Physical Geography. *The Canadian Geographer/Le Géographe Canadien* 61 (1): 84–101.

Lave, R., M. Wilson, E. Barron, C. Biermann, M. Carey, C. Duvall, L. Johnson, et al. 2014. Intervention: Critical Physical Geography. *The Canadian Geographer/Le Géographe Canadien* 58 (1): 1–10.

Moore, A. 2015. The Anthropocene: A critical exploration. *Environment and Society* 6 (1): 1–3.

Mountz, A., A. Bonds, B. Mansfield, J.M. Loyd, J. Hyndman, M. Walton-Roberts, R. Basu, et al. 2015. For slow scholarship: A feminist politics of resistance through collective action in the neoliberal university. *ACME: An International E-Journal for Critical Geographies* 14 (4): 1235–1259.

Pain, R. 2014. Impact: Striking a blow or working together? *ACME: An International E-Journal for Critical Geographies* 13 (1): 19–23.

Pielke, R. 2007. *The honest broker: Making sense of science in policy and politics*. Cambridge: Cambridge University Press.

Proctor, J.D. 1998. The social construction of nature: Relativist accusations, pragmatist and critical realist responses. *Annals of the Association of American Geographers* 88 (3): 352–376.

Soulé, M. 1995. *Reinventing nature?: Responses to postmodern deconstruction*. Chicago: Island Press.

Sundberg, M. 2009. The everyday world of simulation modeling: The development of parameterizations in meteorology. *Science, Technology and Human Values* 34: 162–181.

Index[1]

[1] Note: page number followed by 'n' refers to note.

© The Author(s) 2018
R. Lave et al. (eds.), *The Palgrave Handbook of Critical Physical Geography*,
https://doi.org/10.1007/978-3-319-71461-5

Printed by Printforce, the Netherlands